May 22, 1979.

American Polo

Paul J Bunnell

Electrical Transients in Power Systems

ELECTRICAL TRANSIENTS IN POWER SYSTEMS

ALLAN GREENWOOD
Consulting Engineer
Power Transmission Division
General Electric Company

WILEY-INTERSCIENCE
a Division of John Wiley & Sons, Inc.
New York · London · Sydney · Toronto

Library of Congress Catalogue Card Number: 70-127664

ISBN 0 471 32650 X

Printed in the United States of America

10 9 8 7 6 5 4

To Stephen, who is just beginning

Preface

This book is a distillation of my experience in teaching electrical transients to successive classes of college students and practicing engineers. It also reflects fourteen years of considerable involvement with practical transient problems on electric utility and industrial power systems. Its purpose is to teach students and engineers the fundamentals of this vital subject and to equip them to recognize and solve transient problems in power networks and components. Practicality has been a paramount concern in its preparation.

Many of the basic notions concerning the transient behavior of electric circuits were well explored by Steinmetz and other early pioneers. What is new is the emergence and re-emergence of perennial problems in different guises with new applications and new equipment. Like successive generations of cigarettes and candy bars, these problems are much the same in different wrappers. I have attempted to set out the fundamental ideas at the beginning of the book and made a consistent effort to show thereafter how one peels away the superficial differences in practical transient studies, to a point where basic principles can be applied.

Where formal mathematical analysis is called for, I have chosen to use the Laplace transform method. This is explained but not justified, in Chapter 2. However, there are many places in the book where solutions to problems are reached by a relatively simple process of deduction, which stresses physical insight. In such instances mathematical rigor has been subordinated to physical understanding; mathematics is often used to facilitate this understanding rather than as a substitute for it. It is my experience that the majority of students and engineers, especially those who do not have a mathematical turn of mind, proceed best by first considering the particular and then progressing to the more general.

The material tends to increase in complexity as the book progresses; single-phase circuits are studied before three-phase circuits, and lumpy circuits before distributed circuits. This has one added advantage when the book is used for course text purposes. Certain chapters can be used as a basis for an undergraduate course, which could stand by itself, or lead naturally to a graduate course based on the material of other chapters.

The presentation is broader in scope than most other texts on this subject, for it combines the experimental with the analytical and supplements both with many examples from actual investigations. Though basic knowledge of transients may not have advanced in recent years at the same rate as formerly, there has been a tremendous proliferation in the techniques used to study transients. The use of computers is a good example. Chapter 14 is devoted to this topic. Recent advances in instrumentation for measuring transients has been spectacular; oscilloscopes with storage tubes, sampling tubes, and traveling wave tubes are good examples. These devices and their capabilities are described in Chapter 16.

Two other areas of knowledge add to the breadth of this book. In Chapter 8 I have attempted to draw from diverse places in the literature and put together as a consistent whole a collection of facts regarding certain electromagnetic phenomena that play a significant part in many transient electric disturbances. These relate to electric and magnetic coupling between circuits, more especially to the transient penetration of current and flux into conductors. These have an important bearing on such matters as pickup, shielding, attenuation or damping, and losses. The second area concerns the circuit characteristics of power system components. One may be very adept at manipulating equations, but this will be of little value unless the results can be reduced to practical terms. I have therefore included a compilation of typical characteristics of system elements, such as the capacitance and inductance of transformers, reactors, buswork, and the natural frequencies and time constants of such apparatus.

I wish to acknowledge the considerable contributions that many of my colleagues and associates in the General Electric Company have made indirectly to this book through countless discussions over the years on the subject of transients. I would like to make special mention of Dr. T. H. Lee, W. F. Skeats, and E. J. Tuohy, for most stimulating exchanges on many topics.

ALLAN GREENWOOD

Media, Pennsylvania
March 1970

Contents

Electrical Transients in Power Systems

1 Fundamental Notions about Electrical Transients

1.1 Introduction

An electrical transient is the outward manifestation of a sudden change in circuit conditions, as when a switch opens or closes or a fault occurs on a system. The transient period is usually very short. The fraction of their operating time that most circuits spend in the transient condition is insignificant compared with the time spent in the steady state. Yet these transient periods are extremely important, for it is at such times that the circuit components are subjected to the greatest stresses from excessive currents or voltages. In extreme cases damage results. This may disable a machine, shut down a plant, or black out a city, depending upon the circuit involved. For this reason a clear appreciation of events taking place during transient periods is essential for a full understanding of the behavior of electric circuits.

It is unfortunate that many electrical engineers have only the haziest conception of what is happening in the circuit at such times. Indeed, some appear to view the subject as bordering on the occult. Yet transients can be understood: they can be calculated and sometimes prevented, or at least controlled, so as to be innocuous to the circuit or power system on which they appear. In this chapter we consider some basic ideas about electrical transients which will lay the ground work for their study in greater depth.

1.2 Circuit Parameters

Examination of any electric circuit shows that it is made up of three kinds of parameter:

Resistance, R
Inductance, L
Capacitance, C

All components, whether in a utility system, industrial circuit, or elsewhere, possess each of these attributes to a greater or lesser degree. Under steady-state conditions one will frequently predominate, for example, inductance in a reactor. In the transient state, however, conditions may be very different. On occasion the distributed capacitance of the reactor winding will momentarily be its most important feature.

The resistance, inductance, and capacitance of a circuit are distributed quantities; that is, each small part of the circuit possesses its share. But it is frequently found that they can be treated as "lumped" constants, concentrated in particular branches, without seriously impairing the accuracy of calculations. We shall so treat them in much of this book. In circumstances where this technique is not suitable, as in dealing with long transmission lines, a different approach will be used.

The parameters L and C are characterized by their ability to store energy, L in the magnetic field and C in the electric field of the circuit. These stored energies are functions of the instantaneous current I and voltage V, and are, respectively,

$$\tfrac{1}{2}LI^2 \qquad \text{and} \qquad \tfrac{1}{2}CV^2$$

In contrast, the parameter R is a dissipater of energy, the rate of dissipation being RI^2 at any instant.

Under steady-state conditions the energy stored in the various inductances and capacitances of a direct current circuit are constant, whereas in an alternating current circuit energy is being transferred cyclically between the Ls and Cs of the circuit as the current and voltage rise and fall at the frequency of the supply. This latter process is attended by certain losses, depending upon the resistance present. The losses will be supplied by the various sources in the system.

When any sudden change occurs in a circuit, there is generally a redistribution of energy to meet the new conditions, and in a way, it is this that we are studying when we inquire into the nature of transients. It is very important to realize that this redistribution of energy cannot take place instantaneously for two reasons:

1. To change the magnetic energy requires a change of current. But change of current in an inductor is opposed by an emf of magnitude $L\,dI/dt$. An instantaneous change of current would therefore require an infinite voltage to bring it about. Since this is unrealizable in practice, currents in inductive circuits do not change abruptly and consequently there can be no abrupt change in the magnetic energy stored. Another way of stating this is that the magnetic flux linkage of a circuit cannot suddenly change.

2. To change the electric energy requires a change in voltage. The voltage across a capacitor is given by $V = Q/C$, where Q is the charge, and its rate

of change is

$$\frac{dV}{dt} = \frac{1}{C}\frac{dQ}{dt} = \frac{I}{C}$$

For an instantaneous change of voltage an infinite current must flow. This too is unrealizable; consequently the voltage across a capacitor cannot change abruptly nor can the energy stored in its associated electric field.

The redistribution of energy following a circuit change takes a finite time, and the process during this interval, as at any other time, is governed by the principle of energy conservation, that is, the rate of supply of energy is equal to the rate of storage of energy plus the rate of energy dissipation.

These three simple facts—current through an inductor cannot suddenly change; voltage across a capacitor cannot suddenly change; energy conservation must be preserved at all times—are fundamental to understanding electrical transients. To fully appreciate the implications of these facts is to touch the essence of the subject.

1.3 Mathematical Statement of the Problem and Its Physical Interpretation

The statement of any circuit transient problem properly starts with the setting down of the differential equation or equations describing the behavior of the system when excited by the particular stimulus being studied. This is usually done quite readily with the aid of Kirchhoff's laws. Consider the very simple problem depicted in Fig. 1.1. As a consequence of closing a switch, a capacitor is charged through a resistor. To find the current, we might express the circuit equation using Kirchhoff's first law as follows:

$$V = IR + \frac{1}{C}\int I\,dt \tag{1.3.1}$$

To find voltage across the capacitor, the differential equation might be written

$$V = RC\frac{dV_1}{dt} + V_1 \tag{1.3.2}$$

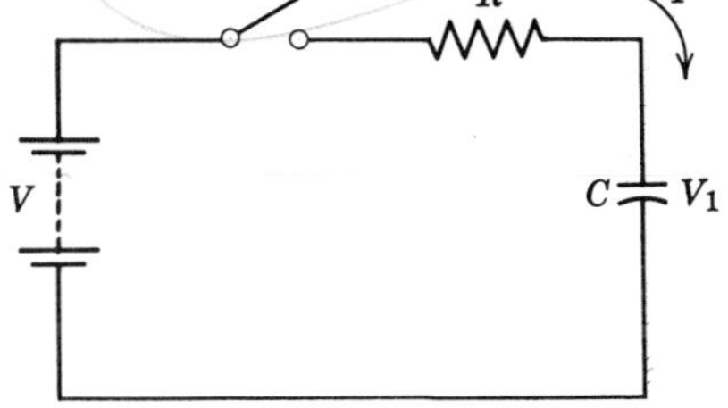

R
I
V
C
V_1

Fig. 1.1. The *RC* circuit.

inasmuch as

$$I = \frac{dQ}{dt} = C\frac{dV_1}{dt}$$

Solving Eq. 1.3.2 by separating the variables,

$$\frac{dV_1}{V - V_1} = \frac{dt}{RC}$$

$$\ln(V - V_1) = -\frac{t}{RC} + \text{constant}$$

or

$$V_1 = V - A\epsilon^{-t/RC} \tag{1.3.3}$$

where A is a constant to be evaluated from the initial conditions in the circuit. If C is precharged to $V_1(0)$ before the switch is closed, setting $t = 0$ yields

$$V_1 = V - [V - V_1(0)]\epsilon^{-t/RC} \tag{1.3.4}$$

This solution is shown graphically in Fig. 1.2, which illustrates a point made in the last section. When the capacitor is connected to the battery it does not instantaneously assume the potential of the battery but proceeds to that value through a transient, which in this instance has an exponential form. This is a simple problem, but it has all the important attributes of far more complicated problems. For this reason we will look at it in more detail.

There are two recognizable parts to the solution given in Eq. 1.3.4. The first term, V, represents the final steady state when the capacitor is charged to the battery voltage. The second term is the true transient which links the initial conditions to this final steady state in a smooth, continuous manner consistent with the physical restrictions of the circuit. The form of this transient term depends essentially upon the circuit itself. The magnitude depends upon the manner in which the stored energy is disposed at time

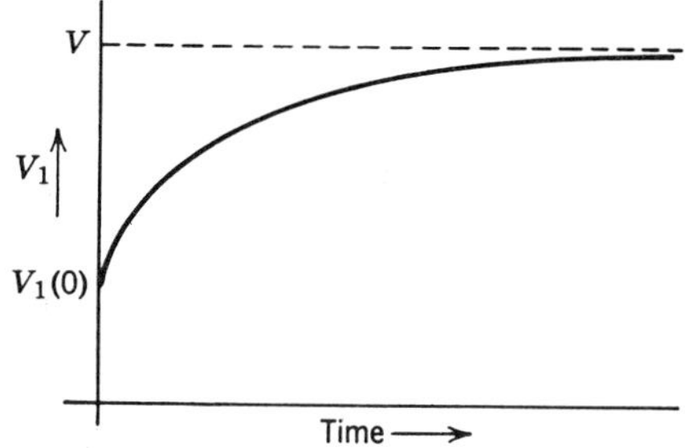

Fig. 1.2. Capacitor voltage in the circuit of Fig. 1.1 after the switch is closed.

zero. This exponential will manifest itself regardless of the stimulus or drive creating the disturbance. Indeed, such a circuit with no stimulus at all, left to dissipate its stored energy, would do so in this same characteristic manner: the capacitor voltage would decline exponentially if the battery was short circuited. For the circuit of Fig. 1.1, this term would be derived from the simpler equation

$$RC\frac{dV_1}{dt} + V_1 = 0 \tag{1.3.5}$$

which yields for a solution

$$V_1 = V_1(0)\epsilon^{-t/RC}$$

Mathematical texts dealing with differential equations refer to the solution obtained when the drive is set equal to zero as the *complementary solution.* Its physical significance is now clear; it describes the transient bridge between initial and final steady-state conditions. As stated earlier, it reflects the character of the circuit. In this instance the term $\epsilon^{-t/RC}$ may be described as the "trademark" of the RC circuit. The so-called *particular solution*, on the other hand, reflects the drive or stimulus creating the disturbance. When applying analytical methods to the solution of circuit problems, it is important to consider the physical interpretation of the solution reached. This will be attempted throughout the book.

1.4 Circuit Characteristics or Trademarks

It was pointed out in Section 1.3 that the single characteristic of the RC circuit which distinguishes it from other circuits is its exponential response, $\epsilon^{-t/RC}$ to any disturbance. Since $-t/RC$ must be dimensionless, RC has the dimensions of time; hence it is referred to as the *time constant.* As we have pointed out, a circuit takes a finite time to adjust from one condition to another following any disturbance. At the instant of closing or opening a switch, for example, we have certain initial conditions. Ultimately we reach a new steady state. The time constant is a measure of how rapidly this change takes place. After one time constant, $1/\epsilon$ of the change remains to be accomplished, or $(1 - 1/\epsilon)$ has already taken place. This is an appropriate time to look into the characteristics of other combinations of circuit elements. These elementary circuits are shown in Fig. 1.3.

Close examination of these circuits reveals some startling facts. The only kind of response that is evoked when an electric circuit comprising lumped elements is disturbed takes the form of exponential functions or combinations thereof with real or imaginary exponents. These will sometimes combine to give sine or cosine functions. This is the case in the LC circuit. Now sines and cosines are periodic functions, which suggests the idea of a frequency.

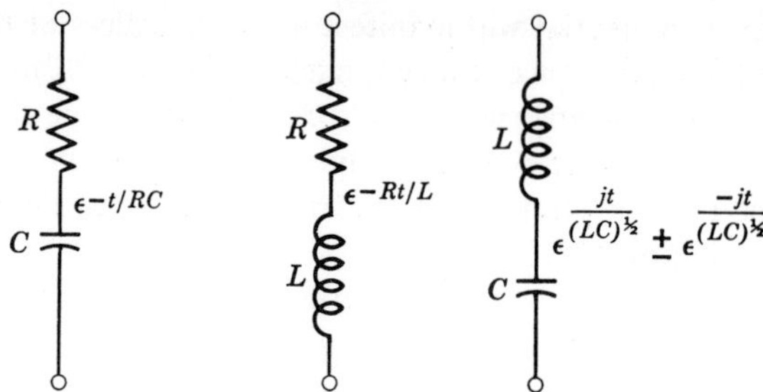

Fig. 1.3. Trademarks of some simple circuits.

This so-called natural frequency is the trademark of the *LC* circuit. Thus we find that when such circuits are excited, no matter how, they oscillate at their natural frequencies. The *LC* circuit does not have a time constant because when it is stimulated it does not achieve a final steady condition but instead continues to oscillate about such a position. The period of the oscillation, which will be shown to be $2\pi(LC)^{1/2}$, replaces the time constant. The *RL* circuit is similar to the *RC* circuit except that its time constant is L/R rather than *RC*.

Through experience in handling transient problems and familiarity with solutions, the amount of formal calculation required is diminished. It becomes possible to construct solutions in what might at first appear to be an intuitive manner. In fact, it is a consequence of consciously or unconsciously recognizing the trademarks thus far discussed and applying the several other fundamental concepts outlined in the first four sections of this chapter.

The only combinations of components not shown in Fig. 1.3 are the series and parallel *RLC* circuits. But here again such circuits react to a drive in the same manner as the simpler circuits, albeit the exponents may be more abstruse. These two circuits are given special treatment in Chapter 4. More extensive circuits are made up of combinations of the simple circuits, so that in the transient state they continue to demonstrate the same forms as their component parts. Their responses may be more complicated but they are no more complex.

1.5 The Principle of Superposition

Superposition is a very important principle in many branches of physical science and a very powerful tool for solving problems. It states that in any linear system if a stimulus S_1 produces a response R_1, and a stimulus S_2 produces a response R_2, then S_1 and S_2 applied simultaneously will evoke a response $R_1 + R_2$. The principle is not restricted to two stimuli but is true for any finite number. A linear system is one in which the response is proportional to the stimulus. A simple example is Hooke's Law, which states that the extension of a spring is proportional to the force applied to it. Thus the principle of superposition tells us that if a weight W_1 hung on a spring extends the spring δ_1, and a weight W_2 causes an extension δ_2, then the extension will be $(\delta_1 + \delta_2)$ if the weights W_1 and W_2 are attached to the spring simultaneously.

The application of superposition in steady-state circuit theory is based on the linear relationship between emf and current. Thus, in a network comprising numerous branches, with say n sources disposed around the network, the currents can be calculated in any particular branch by determining the sum of the currents that each source emf would drive individually. The procedure is to short circuit every source but one, leaving only the internal impedance of the remaining $(n - 1)$ sources. It is then possible to obtain the current the one remaining source gives rise to in the branch of interest. The procedure is repeated for the other sources in turn. With all n sources operating simultaneously, the current in the branch in question is the sum of the individual currents just calculated, paying due regard to their sign. Note that the principle of superposition is just as valid for the transient state as it is for the steady state, so that transients can be added to transients. We shall take advantage of this on many occasions.

There are two particular applications of the principle of superposition that are of fundamental importance. Earlier in this chapter it was stated that most transients are the result of switching operations. The term "switching operation" is used in its broadest sense. It includes the accidental application and removal of faults as well as the closing and opening operations of switches or circuit breakers. Such operations are very conveniently studied by the principle of superposition.

Consider the opening of a switch in an alternating current circuit (Fig. 1.4*a*) and the subsequent interruption of the current. Usually the current is not interrupted by simply parting the switch contacts. It continues to flow through an arc that forms between the contacts; actual interruption is effected when the current comes to zero, as it does regularly twice each cycle. The current might appear as shown in Fig. 1.4*b*. A current of this form would also be realized if, at a current zero, a current which we will designate I_2 were superimposed on top of the existing current, which we might designate I_1 (Fig. 1.4*c*). Up to instant A, I_1 is flowing in the circuit. After this instant the *net current* flowing in the circuit is zero. Physically, we can think of this process as one in which interruption is simulated by injecting into the circuit at the contacts of the switch a current equal in magnitude but opposite in sign to the existing current. When I_1 alone is flowing there is a certain distribution of voltage about the circuit as a consequence of the emf E. If we remove that emf and inject current I_2 into the circuit, another distribution

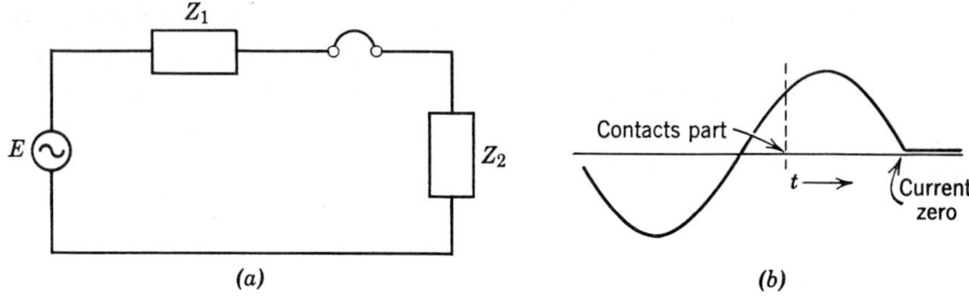

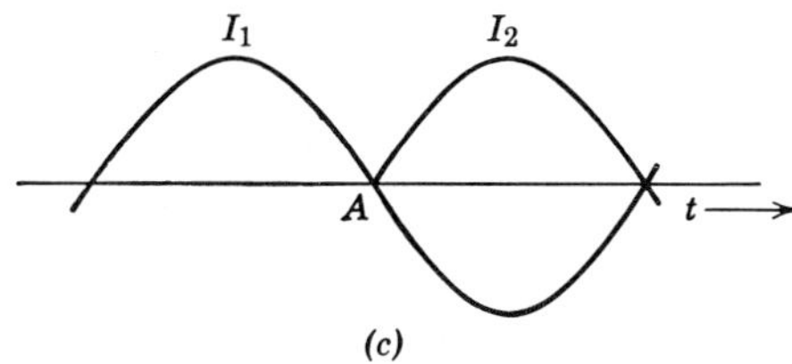

Fig. 1.4. The principle of superposition applied to the opening of a switch. (*a*) The circuit. (*b*) The current. (*c*) Superposition of an injected current.

of voltage would be evident. The Principle of Superposition states that when both of these stimuli, the emf E and the injected current I_2, are applied simultaneously, the total response will be the sum of the individual responses. Since the combination of these stimuli effectively simulates current interruption, this combined response will give the circuit's response to the interruption and will include all the transient effects thereby evoked.

The closing of a switch can be treated in a similar manner. Before closing there will be a certain voltage across the switch; it could, for example, be varying at power frequency. When the switch closes, this voltage disappears. It is as if a voltage exactly equal and opposite to that formerly existing across the switch contacts was suddenly applied at these points. By superposition currents and voltages about the circuit after closing the switch can be obtained by adding to the currents and voltages existing with the contacts open those stimulated by applying at the switch, the voltage appearing at the contacts before the switch was closed with its sign reversed.

Recall the restriction that was placed on the application of superposition when the principle was introduced at the beginning of this section: the method can be applied only in linear circuits. There are some components in utility and industrial power systems that are nonlinear, for example, any saturable device such as an iron-cored reactor or an unloaded transformer. Here the current is not directly proportional to the voltage, though this condition may be approximated over limited ranges. The application of superposition must be restricted to these ranges. Nonlinear resistors are used from time to time, especially as protective devices. Again, the principle of superposition should not be applied where these are located. Finally, any type of rectifier is an extremely nonlinear device since it presents almost zero impedance to the flow of current in one direction, but an almost infinite impedance to current flow in the other direction. Superposition cannot be applied indiscriminately, although it will be shown that, with care, it can be used over certain intervals, even in circuits containing rectifiers.

2 The Laplace Transform Method of Solving Differential Equations

2.1 The Concept of a Transform

Having obtained some physical notion of what an electrical transient is, we now proceed to show how transients can be studied in a quantitative manner. Kelvin once remarked:

> I often say that when you can measure what you are speaking about, and express it in numbers, you know something about it; but when you cannot measure it, when you cannot express it in numbers, your knowledge is of a meager and unsatisfactory kind; it may be the beginning of knowledge, but you have scarcely in your thoughts, advanced to the stage of Science, whatever the matter may be.

Today we would probably state this verity in a different way: "To understand something, you must be able to hang a number on it." In this chapter we lay the groundwork for "hanging numbers" on electrical transients. The initial approach is rather formal. We use the concept of a transform.

The name transform is really a contraction of a more descriptive title, a *functional transformation*. It implies the performing of some operation on a function to change it into a new function, frequently in a different variable. The new function is referred to as the transform of the old. Such a transformation is carried out for a purpose, in our case to simplify the solution of differential equations.

There are many transform operations in everyday use in engineering which are not formally given the name, but which nevertheless are functional transformations. Whenever we use a phasor notation to represent a sinusoidally time-varying quantity we are making a functional transformation. This might also be said of the process of taking the logarithm of a number. The number is the function, its logarithm is its transform. This transformation is made to replace the processes of multiplication and division by the simpler manipulations of addition and subtraction. When such an operation has been performed, the product obtained is the transform of the solution, in this case its logarithm. To obtain the solution proper one must go through a reverse process, or inverse transformation, that is, take the antilogarithm. In most instances this last step will not be carried out formally in our analyses. Instead, we will refer to a table of functions and their transforms and extract the applicable function for our particular transform. This is akin to entering a table of logarithms to find the antilogarithm of a specific logarithm.

The simple transient in the *RC* circuit, reviewed in Chapter 1, is a useful illustrative example, although it does not represent the type of problem regularly encountered in power systems. Practical circuits are far more complicated, so that, even after simplification for the purpose of analysis, they often retain many circuit elements in series-parallel combination. Consequently, it will require several differential or integro-differential equations (one for each mesh) to describe the behavior of the circuit and each may be more complicated than Eq. 1.3.1. These equations must be solved simultaneously to evaluate the variables of interest. To do this efficiently, some systematic technique must be employed. We use the Laplace transform method for this purpose.

The Laplace transformation, when applied to terms of an ordinary differential equation, converts the equation into an *algebraic equation*. In so doing the variable t disappears and a new variable, s, is introduced. The Laplace transformation has the added virtue of drawing attention to the initial conditions by providing just enough terms for these conditions to be satisfied. When operated upon in this manner the equations of the problem lose their transient aspect and appear more like equations of a steady-state problem in the new variable s.

The procedure is as follows. After setting down the differential equations describing a problem, the terms are transformed one by one to obtain an algebraic equation for each of the initial differential equations. These are then solved simultaneously for the variable of interest, to give what is called the *operational solution*. The time function corresponding to this operational solution is then found from a table of transforms, or on rare occasions by applying the inverse Laplace transformation (1), which is a means for inverting transforms from first principles.

2.2 The Laplace Transform

The Laplace transform of a function $F(t)$ is defined mathematically as follows:

$$\mathcal{L}F(t) = \int_0^\infty F(t)\epsilon^{-st}\,dt \tag{2.2.1}$$

or, more precisely,

$$\lim_{\substack{T\to\infty \\ a\to 0}} \int_a^T F(t)\epsilon^{-st}\,dt \tag{2.2.2}$$

Another symbol used for the Laplace transform of $F(t)$ is $f(s)$. For currents and voltages it is usual to write $\mathcal{L}I(t) = i(s)$ and $\mathcal{L}V(t) = v(s)$, reserving an uppercase letter for the function itself and a lowercase letter for its transform.

We proceed with the minimum of justification for the way in which we apply and manipulate the transform, since our purpose is to use the transform as a tool rather than to study it for itself. However, there are certain questions that arise when one first applies this method. For example, are there any restrictions on $F(t)$, or does every function have a transform? The mathematical answer to this question is that the Laplace transform can be obtained for any function of exponential order. This means any function that does not increase with t more quickly than e^{-st} diminishes. This is another way of saying that the transform has meaning only if it is possible to perform the integrating operation described by Eq. 2.2.2. Thus we find that t^2 is of exponential order, but e^{t^2} is not, since, regardless of the value of s (as long as it is finite), as t increases, $e^{t^2} \cdot e^{-st}$ eventually increases indefinitely. In practical problems of circuit analysis we are investigating the behavior of a real physical system, and to any real physical stimulus there will be a real physical response, thus in our area of interest the integral will always converge.

Another question that arises is whether the Laplace transform follows the distributive law. That is, is the transform of a sum the sum of the transforms of the parts? The answer is yes, and it can be stated thus:

$$\mathcal{L}[F_1(t) + F_2(t)] = \mathcal{L}F_1(t) + \mathcal{L}F_2(t) \tag{2.2.3}$$

It will be observed that the operation of taking the transform (Eq. 2.2.2) brings about a change in variable. We start with a function of t, $F(t)$, and finish with a function of s, $f(s)$. The character of s itself is relatively unrestricted. In general it can be said that s can be real or complex. It is often written

$$s = \sigma + j\omega \tag{2.2.4}$$

Further discussion of this is left until we have developed a number of transforms. To do this, we start with some of the more common stimuli encountered in circuit problems.

1. The constant V:

$$\begin{aligned}
\mathcal{L}V &= \int_0^\infty V\epsilon^{-st}\,dt \\
&= V\int_0^\infty \epsilon^{-st}\,dt \qquad \text{(since } V \text{ does not vary with } t\text{)} \\
&= V\left|\frac{\epsilon^{-st}}{-s}\right|_0^\infty \\
&= \frac{V}{s}
\end{aligned} \tag{2.2.5}$$

2. The ramp (typically a current ramp), a function which increases uniformly with time, $I(t) = I't$:

$$\mathcal{L}I't = \int_0^\infty I't\epsilon^{-st}\,dt$$

$$= I'\int_0^\infty t\epsilon^{-st}\,dt$$

Integrate by parts:

$$= I'\left\{\frac{-t\epsilon^{-st}}{s}\Bigg|_0^\infty - \int_0^\infty \frac{\epsilon^{-st}}{s}\,dt\right\}$$

$$= I'\left\{\frac{\epsilon^{-st}}{s}\left[\frac{1}{s} - t\right]\right\}_0^\infty$$

$$= \frac{I'}{s^2} \qquad (2.2.6)$$

3. The exponential $\epsilon^{\alpha t}$ (the great prevalence of exponential functions in electric circuit theory has already been stressed):

$$\mathcal{L}\epsilon^{\alpha t} = \int_0^\infty \epsilon^{\alpha t}\epsilon^{-st}\,dt$$

$$= \int_0^\infty \epsilon^{(\alpha - s)t}\,dt$$

$$= \left|\frac{\epsilon^{(\alpha-s)t}}{\alpha - s}\right|_0^\infty$$

$$= \frac{-1}{\alpha - s} \qquad (\text{for } s > \alpha)$$

$$= \frac{1}{s - \alpha} \qquad (2.2.7)$$

Note that if $s < \alpha$, $\int_0^\infty \epsilon^{\alpha t}\,\epsilon^{-st}\,dt$ does not converge.

No restriction has been placed on α; if it is negative, then from Eq. 2.2.7,

$$\mathcal{L}\epsilon^{-\alpha t} = \frac{1}{s + \alpha} \qquad (2.2.8)$$

This opens the way to construct the following transforms.

4.
$$\sin \omega t = \frac{\epsilon^{j\omega t} - \epsilon^{-j\omega t}}{2j}$$

Substituting $j\omega$ for α in Eqs. 2.2.7 and 2.2.8 gives

$$\mathcal{L} \sin \omega t = \frac{1}{2j}\left(\frac{1}{s - j\omega} - \frac{1}{s + j\omega}\right)$$

$$= \frac{1}{2j}\left(\frac{2j\omega}{s^2 + \omega^2}\right)$$

$$= \frac{\omega}{s^2 + \omega^2} \tag{2.2.9}$$

5. $$\cos \omega t = \frac{\epsilon^{j\omega t} + \epsilon^{-j\omega t}}{2}$$

therefore by the same substitution

$$\mathcal{L} \cos \omega t = \frac{1}{2}\left(\frac{1}{s - j\omega} + \frac{1}{s + j\omega}\right)$$

$$= \frac{s}{s^2 + \omega^2} \tag{2.2.10}$$

These are some of the more common stimuli. But to solve differential equations we must also be able to take the transforms of derivatives of functions.

6. $d/dt\, F(t)$; this is obtained indirectly by the following device. By definition,

$$\mathcal{L}F(t) = \int_0^\infty F(t)\epsilon^{-st}\, dt$$

Integrate by parts letting $u = F(t)$ and $dv = e^{-st}\, dt$:

$$\mathcal{L}F(t) = -F(t)\frac{\epsilon^{-st}}{s}\bigg|_0^\infty - \int_0^\infty \frac{-\epsilon^{-st}}{s} F'(t)\, dt$$

or (2.2.11)

$$\mathcal{L}F(t) = \frac{F(0)}{s} + \frac{1}{s}\int_0^\infty F'(t)\epsilon^{-st}\, dt$$

The first term on the right-hand side of Eq. 2.2.11 is $1/s$ times the value of $F(t)$ at $t = 0$, obtained from the lower limit. The second term on the right is, by definition, what we are looking for, $\mathcal{L}F'(t)$. Rearranging Eq. 2.2.11,

$$\mathcal{L}F'(t) = s\mathcal{L}F(t) - F(0) \tag{2.2.12}$$

The transform of the second derivative of $F(t)$ is obtained in like manner:

$$\mathcal{L}F'(t) = \int_0^\infty F'(t)\epsilon^{-st}\, dt$$

Again, integrating by parts in the same sequence as before,

$$\mathcal{L}F'(t) = \frac{F'(0)}{s} + \frac{1}{s}\int_0^\infty F''(t)\epsilon^{-st}\,dt$$

or

$$\mathcal{L}F''(t) = s\mathcal{L}F'(t) - F'(0)$$

Substituting for $\mathcal{L}F''(t)$ from Eq. 2.2.12,

$$\mathcal{L}F''(t) = s^2\mathcal{L}F(t) - sF(0) - F'(0) \tag{2.2.13}$$

It is apparent that this procedure can be carried out indefinitely for consecutive derivatives. By deduction we can write

$$\mathcal{L}F^{(n)}(t) = s^n\mathcal{L}F(t) - s^{n-1}F(0) - s^{n-2}F'(0) \cdots F^{n-1}(0) \tag{2.2.14}$$

The various terms $F(0)$, $F'(0)$, $F''(0)$, and so on, are the values of $F(t)$, $F'(t)$ and $F''(t)$ at $t = 0$. In a circuit problem they would probably represent the initial values of currents and voltages and their rates of change at the initial instant when the transient begins. It was pointed out in Section 1.3.2 that the solution to a differential equation has two parts, the particular integral, which depends on the drive or stimulus, and the complementary solution, which is independent of this drive, being characteristic of the circuit itself. The complementary solution contains as many constants of integration as the order of the equation; a second-order equation will have two, a third-order three, etc. The solution is general until these are specified, when we obtain the solution to a particular problem. It is reasonable to suppose that when a circuit is disturbed by some stimulus, its behavior will depend upon the way in which it is disturbed. It is also to be expected that its response will reflect the condition of the circuit at the time of the disturbance ($t = 0$), whether, for example, this capacitor is charged, or that branch is carrying a current. Another way of putting this is that the initial distribution of energy among all the Ls and Cs in the circuit must be specified. There will be just enough constants of integration to allow this to be done. A beauty of the Laplace transform method of solving differential equations is that in the process of carrying out the transformation the precise number of initial conditions required by the problem will appear in the form $F(0)$, $F'(0)$, $F''(0)$, etc. This very clearly ties together the operational equations and their solutions with the physical state of the circuit.

7. The Laplace transform of an integral is found as follows. By definition,

$$\mathcal{L}\left[\int_{-\infty}^{t} F(t)\,dt\right] = \int_0^\infty \left[\int_{-\infty}^{t} F(\tau)\,d\tau\right]\epsilon^{-st}\,dt$$

(the variable is changed here to avoid confusion). The integration on the right is performed by parts, letting

$$u = \int_{-\infty}^{t} F(\tau)\, d\tau, \qquad dv = \epsilon^{-st}\, dt$$

$$\mathcal{L}\left[\int_{-\infty}^{t} F(t)\, dt\right] = -\int_{-\infty}^{t} F(\tau)\, d\tau \frac{\epsilon^{-st}}{s}\Bigg|_{0}^{\infty} + \frac{1}{s}\int_{0}^{\infty} F(t)\epsilon^{-st}\, dt \tag{2.2.15}$$

Consider the first term on the right in Eq. 2.2.15. As $t \rightarrow \infty$ this term approaches zero because of the negative exponential. Inserting the lower limit, $t = 0$, makes $e^{-st} = 1$. Thus the first term becomes

$$\frac{1}{s}\int_{-\infty}^{0} F(\tau)\, d\tau$$

The second term in Eq. 2.2.15 is simply $1/s$ times the Laplace transform of $F(t)$. The equation can therefore be rewritten

$$\mathcal{L}\left[\int_{-\infty}^{t} F(t) dt\right] = \frac{1}{s}\mathcal{L}F(t) + \frac{1}{s}\int_{-\infty}^{0} F(\tau) d\tau \tag{2.2.16}$$

Once again this method acknowledges the initial state of the circuit, for that is what the second term represents. An example will make this clear.

Suppose that $F(t)$ is a current $I(t)$; then $\int_{t_1}^{t_2} I(t)\, dt$ represents a charge which has flowed, or perhaps accumulated on a capacitor, during the interval specified. By Eq. 2.2.16,

$$\mathcal{L}\left[\int_{-\infty}^{t} I(t)\, dt\right] = \frac{1}{s}\mathcal{L}I(t) + \frac{1}{s}\int_{-\infty}^{0} I(t)\, dt = \mathcal{L}Q(t)$$

or

$$\mathcal{L}\left[\int_{-\infty}^{t} I(t)\, dt\right] = \frac{i(s)}{s} + \frac{Q(0)}{s} = q(s) \tag{2.2.17}$$

where $Q(0)$ is the initial charge on the capacitor, that is, the integral of all the current that had flowed in or out of the capacitor up to $t = 0$.

Enough transforms have been derived to enable us to apply them to a few simple problems. A useful table of transforms is provided in Appendix 1. Memorizing the more common transforms helps in gaining facility in their use.

2.3 Some Simple Applications of the Laplace Transform in Circuit Problems

The examples chosen to illustrate the Laplace transform method are simple and could surely be solved with less sophistication. However, their very simplicity assures that the method is not obscured by a mass of algebra. A good place to start is with the circuits in Fig. 1.3. We will study the response of each to the application of a battery voltage V. The first is the RC circuit which was solved more conventionally in Section 1.3.

If the current is I and the capacitor voltage V_C,

$$V_C + IR = V \tag{2.3.1}$$

and

$$I = \frac{dQ_c}{dt} = \frac{C\,dV_C}{dt} \tag{2.3.2}$$

In Section 1.3 a solution was found for V_C. We now would like to find the current. From Eq. 2.3.1,

$$\frac{dV_C}{dt} = -R\frac{dI}{dt} \tag{2.3.3}$$

Substituting Eq. 2.3.3 in Eq. 2.3.2,

$$I = -RC\frac{dI}{dt}$$

$$\text{or} \tag{2.3.4}$$

$$\frac{dI}{dt} + \frac{I}{RC} = 0$$

Transforming the equation,

$$si(s) - I(0) + \frac{i(s)}{RC} = 0 \tag{2.3.5}$$

Note how the problem has been reduced to solving an algebraic equation. Observe also that one is obliged to take cognizance of the initial value of current, $I(0)$. This will depend on the value of $V_C(0)$, which may or may not be zero. When the switch is first closed it is clear from Eq. 2.3.1 that

$$I(0) = \frac{V - V_C(0)}{R}$$

Substituting this in Eq. 2.3.5,

$$i(s)\left(s + \frac{1}{RC}\right) = \frac{V - V_C(0)}{R}$$

$$\text{or} \tag{2.3.6}$$

$$i(s) = \frac{V - V_C(0)}{R}\frac{1}{s + 1/RC}$$

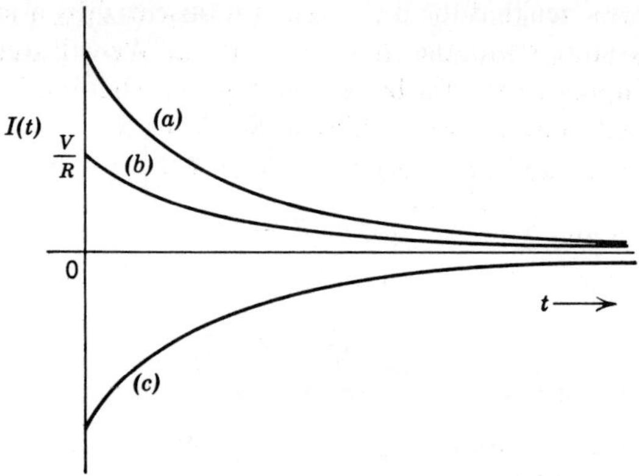

Fig. 2.1. The current in an RC circuit when the capacitor is connected to a battery through a resistor. Initial conditions: (a) $V_c(0) = -V$; (b) $V_c(0) = 0$; (c) $V_c(0) = +3V$.

This is the so-called operational solution for the current. The transform is of the form $1/(s + \alpha)$, so the solution can be written with the aid of Eq. 2.2.8:

$$I(t) = \left[\frac{V - V_C(0)}{R}\right]\epsilon^{-t/RC} \tag{2.3.7}$$

The solutions for several different values of $V_c(0)$ are given in Fig. 2.1. In each instance the current is asymptotic to zero. This follows from the fact that regardless of the initial charge on the capacitor, it will ultimately be charged to the battery voltage.

Consider now the current in the RL circuit (Fig. 1.3b) when this circuit is excited by a similar battery voltage. The differential equation is

$$RI + L\frac{dI}{dt} = V \tag{2.3.8}$$

This transforms as follows:

$$\begin{aligned} Ri(s) + Lsi(s) - LI(0) &= \frac{V}{s} \\ i(s)\left(s + \frac{R}{L}\right) &= \frac{V}{Ls} + I(0) \end{aligned} \tag{2.3.9}$$

If the circuit is simply connected to a battery, $I(0)$ must be zero, for the inductance precludes any discontinuity in the current. In these circumstances,

$$i(s) = \frac{V}{L}\frac{1}{s(s + R/L)} \tag{2.3.10}$$

This transform, which might be written $1/s(s+\alpha)$ is one we have not thus far encountered. However, it can be written as the sum of two familiar transforms:

$$\frac{1}{s(s+\alpha)} = \frac{1}{\alpha}\left[\frac{1}{s} - \frac{1}{s+\alpha}\right] \tag{2.3.11}$$

From Eqs. 2.2.5 and 2.2.8,

$$\mathcal{L}^{-1}\frac{1}{s(s+\alpha)} = \frac{1}{\alpha}\,[1 - \epsilon^{-\alpha t}] \tag{2.3.12}$$

The symbol $\mathcal{L}^{-1}$ represents the inverse transformation, that is, if

$$f(s) = \mathcal{L}F(t), \qquad F(t) = \mathcal{L}^{-1}f(s)$$

Applying Eq. 2.3.12 in Eq. 2.3.10 gives for the current

$$I(t) = \frac{V}{R}\,[1 - \epsilon^{-Rt/L}] \tag{2.3.13}$$

In most problems $I(0)$ in Eq. 2.3.9 will be zero for the reason stated, but the Laplace method allows for the possibility of a finite current at time zero and shows how it will affect the solution. It is possible in practical circumstances for such a condition to exist where one switching operation follows another and the initial transient has not died out. Another example occurs when the current in the highly inductive field circuit of a machine is to be interrupted. It is usual to open the field switch isolating the field winding from the supply and simultaneously close the field winding on a resistor in order to dissipate the energy stored in the magnetic circuit of the machine. On this occasion $I(0)$ would be finite but V would be zero. If $I(0)$ is finite, it leads to an extra term in Eq. 2.3.9,

$$\frac{I(0)}{s + R/L} = \mathcal{L}I(0)\epsilon^{-Rt/L}$$

The physical interpretation of Eq. 2.3.13 warrants some attention. The current settles down in due course to a steady value of V/R, but it cannot adjust to this value from zero, or from any other value $I(0)$, at the instant the switch is closed, because of the inductance of the circuit. The transition takes place in the characteristic manner of such circuits, through an exponential with a time constant L/R. The solution has the circuit's trademark clearly stamped upon it. This is the only way this circuit can respond to any sudden stimulus. The effect of varying the time constant is illustrated in Fig. 2.2. With a longer time constant it takes correspondingly longer to reach any given fraction of the asymptotic value V/R.

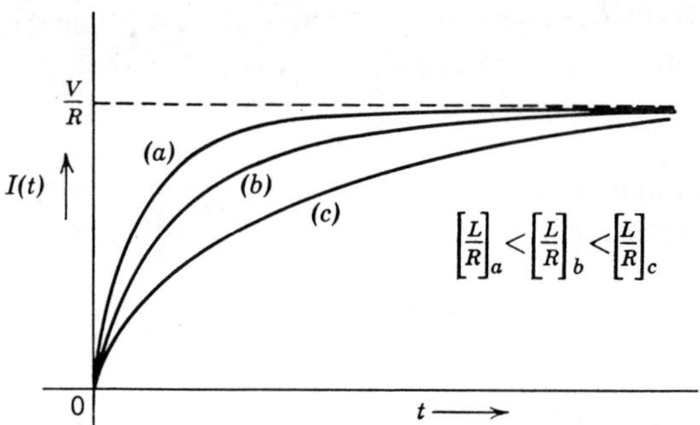

Fig. 2.2. The current in an *RL* circuit when the inductor is connected to a battery through a resistor.

We now turn our attention to the *LC* circuit, which introduces a new dimension. In the previous two circuits there was only one seat for stored energy, the inductor or the capacitor. Such circuits give rise to single energy transients identifiable by their single exponential response. In a circuit with both inductance and capacitance, double energy transients appear and the response involves two exponential terms. Depending upon the circuit, the exponents may be real, imaginary, or conjugate complex. In those cases where they are imaginary or complex, they combine to give a sine or cosine function which manifests itself physically as an oscillation in the circuit. The natural frequency of the circuit is excited by the switching operation. Energy oscillates between inductance and capacitance.

Because of the two seats for energy, the *LC* circuit gives rise to a second-order differential equation when it is stimulated. On closing the switch in the circuit shown in Fig. 2.3, Kirchoff's law gives for the circuit equation

$$L\frac{dI}{dt} + V_C = V \tag{2.3.14}$$

If the quantity of interest is the voltage, I is replaced by $C\,dV_C/dt$:

$$LC\frac{d^2V_C}{dt^2} + V_C = V \tag{2.3.15}$$

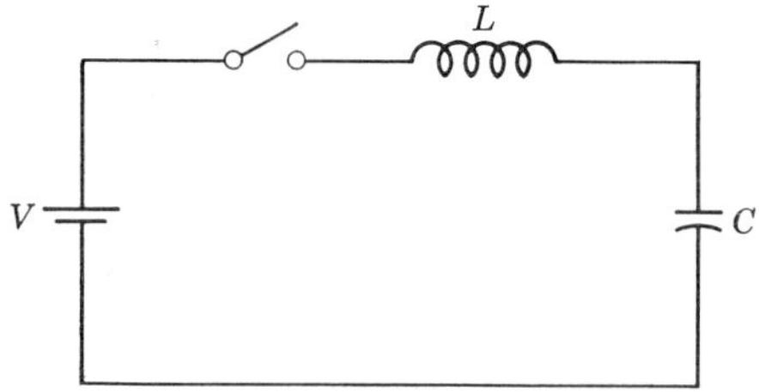

Fig. 2.3. The oscillatory *LC* circuit.

To find the current I, V_C is replaced by $(1/C)\int I\,dt = Q_c/C$:

$$L\frac{dI}{dt} + \frac{1}{C}\int I\,dt = V \tag{2.3.16}$$

Then Eq. 2.3.16 must be transformed to obtain an operational solution. This introduces the transform of an integral given in Eq. 2.2.17:

$$Lsi(s) - LI(0) + \frac{i(s)}{sC} + \frac{Q_c(0)}{sC} = \frac{V}{s} \tag{2.3.17}$$

The initial voltage on the capacitor is $Q_c(0)/C$. This equation could have been derived by differentiating Eq. 2.3.16

$$L\frac{d^2I}{dt^2} + \frac{I}{C} = 0 \tag{2.3.18}$$

and then transforming

$$s^2Li(s) - sLI(0) - LI'(0) + \frac{i(s)}{C} = 0 \tag{2.3.19}$$

But from Eq. 2.3.14

$$I'(t) = \frac{dI}{dt} = \frac{V - V_C}{L}$$

therefore

$$LI'(0) = V - V_C(0) \tag{2.3.20}$$

Substituting Eq. 2.3.20 in Eq. 2.3.19 gives

$$s^2Li(s) - sLI(0) + V_C(0) + \frac{i(s)}{C} = V \tag{2.3.21}$$

which is essentially the same as Eq. 2.3.17. It should be noted that since this is a second-order equation two initial conditions, $I(0)$ and $V_C(0)$ [or $I'(0)$], must be specified to obtain a complete solution. Rearranging Eq. 2.3.21,

$$i(s)\left[s^2 + \frac{1}{LC}\right] = \frac{V - V_C(0)}{L} + sI(0)$$

or (2.3.22)

$$i(s) = \frac{V - V_C(0)}{L}\,\frac{1}{(s^2 + 1/LC)} + I(0)\,\frac{s}{(s^2 + 1/LC)}$$

There could be no initial current in the circuit of Fig. 2.3 [$I(0) = 0$], but $V_C(0)$ might have any value. Suppose that C is discharged initially, and let $1/LC = \omega_0^2$; then

$$i(s) = V\left(\frac{C}{L}\right)^{1/2}\frac{\omega_0}{s^2 + \omega_0^2} \tag{2.3.23}$$

This is the operational solution. The solution proper can be written at once from Eq. 2.2.9:

$$I(t) = V\left(\frac{C}{L}\right)^{1/2} \sin \omega_0 t \tag{2.3.24}$$

This states that the current oscillates sinusoidally at the *natural frequency* ω_0 of the circuit, which is a function of the L and C of the circuit only. Another important point is that the ratio of the voltage to the current is given by $(L/C)^{1/2}$, which apparently has the dimensions of impedance. This is called the *surge impedance* of the circuit and is written

$$Z_0 = \left(\frac{L}{C}\right)^{1/2} \tag{2.3.25}$$

It is a very important characteristic of any LC circuit.

To calculate the voltage of the capacitor, we proceed from Eq. 2.3.15, which may be rewritten:

$$\frac{d^2V_C}{dt^2} + \omega_0^{\,2} V_C = \omega_0^{\,2} V \tag{2.3.26}$$

Transformed, this gives

$$(s^2 + \omega_0^{\,2})v_c(s) = \omega_0^{\,2} V + sV_C(0) + V_C'(0)$$

Again, $V_c'(0) = 0$ since $I(0) = 0$, and $I(0) = CV_c'(0)$. Allowing a finite value for $V_c(0)$,

$$v_c(s) = \frac{V\omega_0^{\,2}}{s(s^2 + \omega_0^{\,2})} + \frac{sV_C(0)}{s^2 + \omega_0^{\,2}} \tag{2.3.27}$$

The second of the transforms on the right-hand side of Eq. 2.3.27 is familiar. Its inverse transform is $\cos \omega_0 t$. The first transform has not been encountered before. However, it readily reduces to something we can handle:

$$\frac{\omega_0^{\,2}}{s(s^2 + \omega_0^{\,2})} = \frac{1}{s} - \frac{s}{s^2 + \omega_0^{\,2}}$$

thus

$$\mathcal{L}^{-1} \frac{\omega_0^{\,2}}{s(s^2 + \omega_0^{\,2})} = 1 - \cos \omega_0 t \tag{2.3.28}$$

From Eqs. 2.3.27 and 2.3.28, the solution for V_C can be written

$$\begin{aligned} V_C(t) &= V(1 - \cos \omega_0 t) + V_C(0) \cos \omega_0 t \\ &= V - [V - V_C(0)] \cos \omega_0 t \end{aligned} \tag{2.3.29}$$

This is plotted for several values of $V_C(0)$ in Fig. 2.4.

In a practical circuit, there would be some resistance, which gradually damps out the oscillation so that the capacitor finally settles down to the

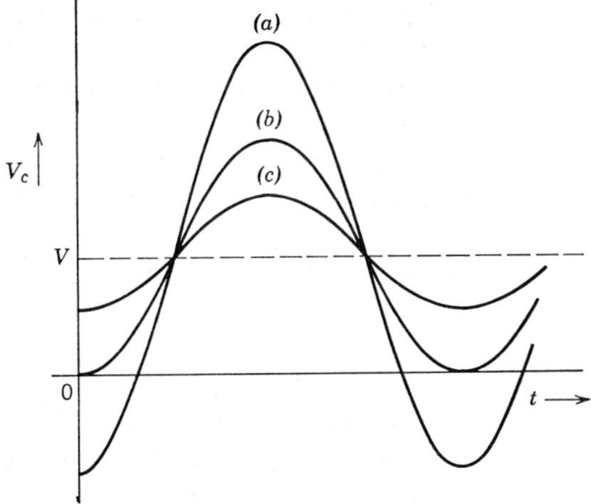

Fig. 2.4. The capacitor voltage in the circuit of Fig. 2.3 after the switch is closed: (*a*) $V_c(0) = -V$; (*b*) $V_c(0) = 0$; (*c*) $V_c(0) = +V/2$.

battery voltage. It is interesting to note how, in Fig. 2.4, the capacitor overshoots this value. The further below the battery voltage the capacitor voltage starts, the further above will the capacitor swing. In fact, if there is no damping, the capacitor voltage will swing just as far above the battery voltage as it started below. In Fig. 2.4, curve *a* therefore reaches a peak of 3 V.

This process is physically illustrated in Fig. 2.5 as follows. When the capacitor has been charged by the current to the supply voltage, the current is well established in the circuit inductance, and therefore it cannot suddenly drop to zero. It continues to flow, but after this instant V_C becomes greater

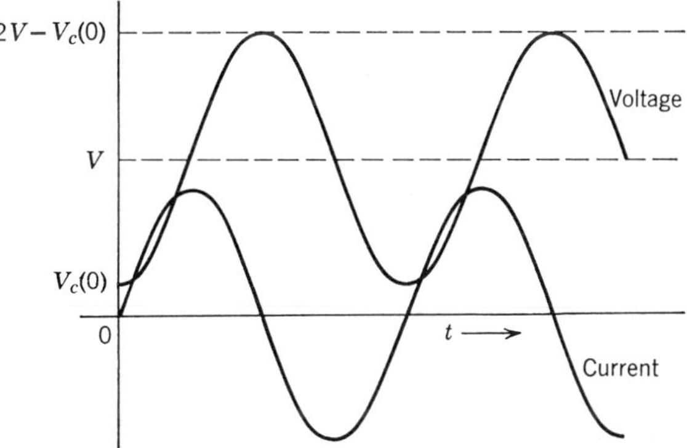

Fig. 2.5. Voltage and current relationship in the *LC* circuit.

than V, so the net voltage in the circuit is such as to reduce the current, which declines, coming to zero a quarter of a cycle later. At this point, since $I = C\,dV/dt$, the voltage has reached a peak, that is, dV/dt changes sign, and the capacitor commences to discharge.

2.4 Building Other Transforms

It was shown in Eq. 2.3.11 how the transform $1/s(s + a)$ could be evaluated by observing that it was the difference between two other simple transforms:

$$\frac{1}{s(s+\alpha)} = \frac{1}{\alpha}\left[\frac{1}{s} - \frac{1}{s+\alpha}\right]$$

This method of developing new transforms from simpler, known transforms, by taking partial fractions, is general in its application and very useful. The rules for forming partial fractions are given in reference 2. We will utilize these to find the inverse of

$$\frac{-s^3 + \omega_1 s^2 - \omega_1^{\,2} s + \omega_1\omega_2^{\,2}}{s^4 + (\omega_1^{\,2} + \omega_2^{\,2})s^2 + \omega_1^{\,2}\omega_2^{\,2}}$$

Before proceeding note that the order of s in the denominator is greater than that in the numerator. Only transforms of this type arise in practical power system problems.

The denominator of the transform may be factored:

$$\begin{aligned}\frac{-s^3 + \omega_1 s^2 - \omega_1^{\,2} s + \omega_1\omega_2^{\,2}}{s^4 + (\omega_1^{\,2} + \omega_2^{\,2})s^2 + \omega_1^{\,2}\omega_2^{\,2}} &= \frac{-s^3 + \omega_1 s^2 - \omega_1^{\,2} s + \omega_1\omega_2^{\,2}}{(s^2 + \omega_1^{\,2})(s^2 + \omega_2^{\,2})} \\ &= \frac{As + B}{(s^2 + \omega_1^{\,2})} + \frac{Cs + D}{(s^2 + \omega_2^{\,2})} \qquad (2.4.1)\end{aligned}$$

where A, B, C, and D are constants to be evaluated. The following identity can now be written by equating numerators:

$$\begin{aligned}-s^3 + \omega_1 s^2 - \omega_1^{\,2} s + \omega_1\omega_2^{\,2} &\equiv (As + B)(s^2 + \omega_2^{\,2}) + (Cs + D)(s^2 + \omega_1^{\,2}) \\ &\equiv (A + C)s^3 + (B + D)s^2 \\ &\quad + (A\omega_2^{\,2} + C\omega_1^{\,2})s + B\omega_2^{\,2} + D\omega_1^{\,2}\end{aligned}$$

Equating like coefficients:

$$\left.\begin{aligned} A + C &= -1 \\ B + D &= \omega_1 \\ A\omega_2^{\,2} + C\omega_1^{\,2} &= -\omega_1^{\,2} \\ B\omega_2^{\,2} + D\omega_1^{\,2} &= \omega_1\omega_2^{\,2} \end{aligned}\right\} \qquad (2.4.2)$$

Solving Eqs. 2.4.2 simultaneously:

$$A = D = 0, \qquad B = \omega_1, \qquad C = -1$$

Thus Eq. 2.4.1 can be rewritten:

$$\frac{-s^3 + \omega_1 s^2 - \omega_1^2 s + \omega_1\omega_2^2}{s^4 + (\omega_1^2 + \omega_2^2)s^2 + \omega_1^2\omega_2^2} = \frac{\omega_1}{(s^2 + \omega_1^2)} - \frac{s}{(s^2 + \omega_2^2)}$$

which can be evaluated from Eqs. 2.2.9 and 2.2.10:

$$\mathcal{L}^{-1} \frac{-s^3 + \omega_1 s^2 - \omega_1^2 s + \omega_1\omega_2^2}{s^4 + (\omega_1^2 + \omega_2^2)s^2 + \omega_1^2\omega_2^2} = \sin \omega_1 t - \cos \omega_2 t \tag{2.4.3}$$

2.5 Operational Impedance

There is another approach to finding solutions to transient problems that we shall find useful from time to time. Suppose we wish to calculate a particular transient current. It is reasonable to suppose that we would ultimately find it by dividing a voltage by an impedance:

$$I = \frac{V}{Z}$$

Of course, V and Z may be complicated functions. If an operational method is employed to solve this problem, a prior step would be to obtain the operational solution, which might be written

$$i(s) = \frac{v(s)}{z(s)} \tag{2.5.1}$$

In this expression $v(s)$ is the transform of the voltage V and $z(s)$ is what we will call the *operational impedance*. If the circuit being studied is stimulated by a simple battery voltage V, then $v(s) = V/s$. How then is $z(s)$ determined? A clue to this will be found by reviewing a couple of previous examples.

Consider the application of a voltage V to an RC circuit, the first problem solved in Section 2.3. Let us assume in the first instance that the capacitor is uncharged. Then the operational solution given in Eq. 2.3.6 can be rewritten:

$$i(s) = \frac{V}{s}\frac{1}{R + 1/Cs} \tag{2.5.2}$$

It is apparent from Eq. 2.5.2 that for this problem $z(s) = R + 1/Cs$.

Consider next the application of a constant voltage V to an RL circuit, the second problem in Section 2.3. The operational solution here is given in Eq. 2.3.10, which for our present purpose is best stated

$$i(s) = \frac{V}{s}\frac{1}{R + Ls} \tag{2.5.3}$$

In this expression $z(s) = R + Ls$.

Finally, the third problem in Section 2.3 derives the current in an LC circuit, excited by applying a voltage V. Equation 2.3.23 can be rearranged in the form

$$i(s) = \frac{V}{s}\,\frac{1}{Ls + 1/Cs} \tag{2.5.4}$$

In this instance,

$$z(s) = \frac{1}{Ls + 1/Cs}$$

These examples indicate that $z(s)$ is formed by writing Ls for each inductance and $1/Cs$ for each capacitance in the circuit. Resistors are unchanged, that is, they appear simply as R. There is a similarity between these expressions and the symbolic representation of inductive and capacitive reactances by $j\omega L$ and $1/j\omega C$ in steady-state a-c analysis. In fact, the latter is a special case of the former; the Laplace representation includes the steady state. Thus to solve a transient problem by use of the operational impedance, proceed as if solving for the alternating current in the branch of interest, with an alternating voltage applied. An example will make this clear. It is desired to calculate the current that would flow in the circuit shown in Fig. 2.6*a* when a voltage V is applied at A and B. If this was a steady-state a-c problem and V a steady-state alternating voltage, the representation of Fig. 2.6*b* would be used. The corresponding operational impedance diagram for the transient problem is shown in Fig. 2.6*c*. Ls and $1/Cs$ in parallel have an impedance:

$$\frac{L/C}{Ls + 1/Cs} = \frac{s}{C(s^2 + 1/LC)}$$

thus,

$$z(s) = R + \frac{s}{C(s^2 + 1/LC)}$$

$$i(s) = \frac{V}{s\left[R + \dfrac{s}{C(s^2 + 1/LC)}\right]} \tag{2.5.5}$$

Had the stimulus been some other function, for example, a decaying exponential, for $Ve^{-\alpha t}$ we would have used $V/(s + \alpha)$ instead of the V/s used for the step function in Eq. 2.5.5.

When the subject of investigation is a voltage, the operational impedance is used in the form $v(s) = i(s)z(s)$.

The method of solution just described is best suited for problems when the circuits are initially dead, that is, for circuits that contain no stored energy. This would be the case if initially all the currents were zero and all the capacitors were discharged. Its use is not restricted to such circuits, but one

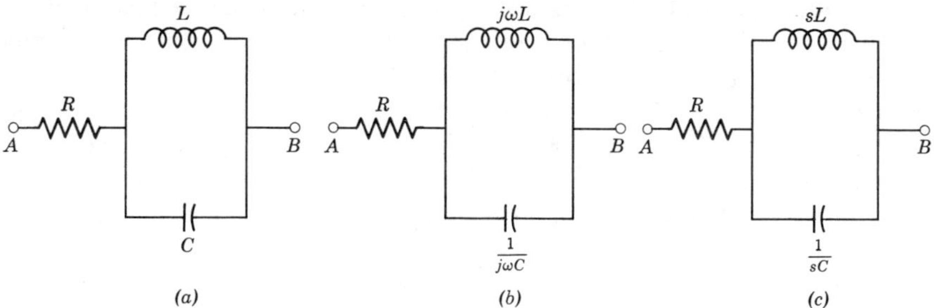

Fig. 2.6. Comparison of a-c symbolic representation (*b*) and the operational impedance (*c*) for an *RLC* circuit (*a*).

must apply superposition with great care for circuits with nonzero initial conditions. For this reason, the author prefers in most instances the method described earlier of setting up the differential equations from Kirchhoff's laws and applying the Laplace transform.

2.6 Duhamel's Integral—Response of a Circuit to an Arbitrary Stimulus

The material in Sections 2.5 and 1.5 (the principle of superposition) can be utilized to determine the transient response of a circuit to a stimulus of arbitrary form. The method is formalized in Duhamel's integral, which we will introduce shortly.

Consider Fig. 2.7*a*, which represents the waveform of a voltage surge $U(t)$. This can be approximated by the stepped waveform shown in Fig. 2.7*b*. The degree of approximation will improve as the number of steps increases. Now superposition tells us that the response to a succession of stimuli can

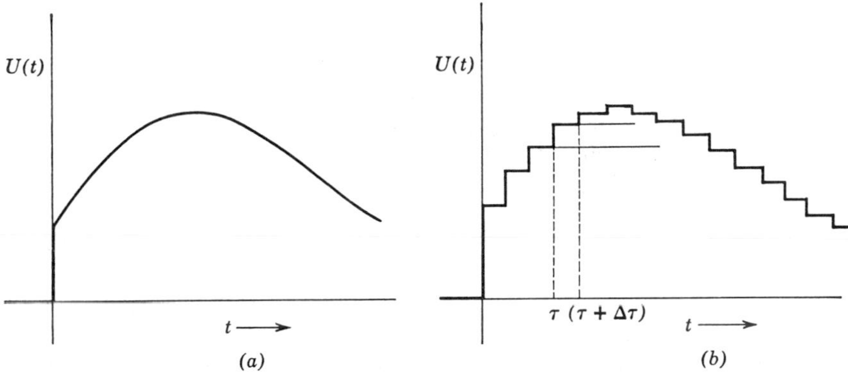

Fig. 2.7. Surge waveform approximated by a succession of steps.

be obtained by adding the responses of the individual stimuli. In this instance, the stimuli are step functions and as Carter (2) puts it, "If the stimulus applied to a circuit consists of a succession of shocks, the response to the stimulus may be obtained by adding together the responses to these shocks." We therefore need the response of the circuit to a step, or more precisely to a unit function, or step of unit height, often written simply **1**. If we designate this response $U_1(t)$, the response to a step of height V, that is, to $V \cdot \mathbf{1}$, will be $V \cdot U_1(t)$.

In the notation of the Laplace transform the unit function is written

$$\mathcal{L}^{-1}\,\mathbf{1} = \frac{1}{s} \tag{2.6.1}$$

thus, when such a step is applied to a circuit whose operational impedance is $Z(s)$, the operational expression for the current will be

$$i_1(s) = \frac{1}{sz(s)} \tag{2.6.2}$$

The inverse transform of this the current, or what we have called $u_1(t)$ in general terms, is

$$u_1(t) = I_1(t) = \mathcal{L}^{-1}\frac{1}{sz(s)} \tag{2.6.3}$$

Turning now to Fig. 2.7*b*, the initial value of $U(t)$, $U(0)$, evokes a response $U(0) \cdot u_1(t)$. To this must be added, at appropriate intervals, the responses to the other steps. Consider the one that starts at time τ; a time $\Delta\tau$ elapses before the next step is applied. It follows, therefore, that the height of this step is $U'(\tau) \cdot \Delta\tau$, where $U'(\tau)$ is the value of dU/dt at the instant τ. *Measured from that instant*, the circuit's response to this shock will be $U'(\tau)u_1(t - \tau)\,\Delta\tau$. Consequently the response of the circuit to the whole succession of steps up to time t is

$$U(0)u_1(t) + \sum_{\tau=0}^{\tau=t} U'(\tau)u_1(t - \tau)\,\Delta\tau \tag{2.6.4}$$

Where $U(t)$ is declining, the steps are negative, but they are treated in exactly the same way. Proceeding to the limit where $\Delta\tau$ becomes indefinitely small, we find that $U(t)$ causes a response $u(t)$, given by

$$u(t) = U(0)u_1(t) + \int_0^t U'(\tau)u(t - \tau)\,d\tau \tag{2.6.5}$$

This is Duhamel's integral. Notice that in evaluating the integral, τ is the variable; t is treated as a constant.

Carter (2) points out that by integrating by parts and by other elementary means, we may prove that Duhamel's integral can be written in the following alternative ways:

$$u(t) = U(0)u_1(t) + \int_0^t U'(\tau)u_1(t-\tau)\,d\tau \tag{2.6.5a}$$

$$u(t) = u_1(0)U(t) + \int_0^t u_1'(\tau)U(t-\tau)\,d\tau \tag{2.6.5b}$$

$$u(t) = U(0)u_1(t) + \int_0^t u_1(\tau)U'(t-\tau)\,d\tau \tag{2.6.5c}$$

$$u(t) = u_1(0)U(t) + \int_0^t U(\tau)u'(t-\tau)\,d\tau \tag{2.6.5d}$$

$$u(t) = \frac{d}{dt}\left\{\int_0^t U(\tau)u_1(t-\tau)\,d\tau\right\} \tag{2.6.5e}$$

$$u(t) = \frac{d}{dt}\left\{\int_0^t u_1(\tau)U(t-\tau)\,d\tau\right\} \tag{2.6.5f}$$

The choice between these different alternatives is often determined by the problem to be solved. An example will make this clear.

For this example we will consider the response of the RL circuit shown in Fig. 2.8 to a stimulus

$$U(t) = V\epsilon^{-\alpha t} \tag{2.6.6}$$

We will solve this first by the regular Laplace transform method developed in the first few sections of this chapter. Subsequently, the Duhamel's integral will be applied to show this alternative approach.

The differential equation describing the circuit behavior is

$$IR + L\frac{dI}{dt} = V\epsilon^{-\alpha t} \tag{2.6.7}$$

Transforming this equation gives

$$Ri(s) + Lsi(s) - LI(0) = \frac{V}{s+\alpha}$$

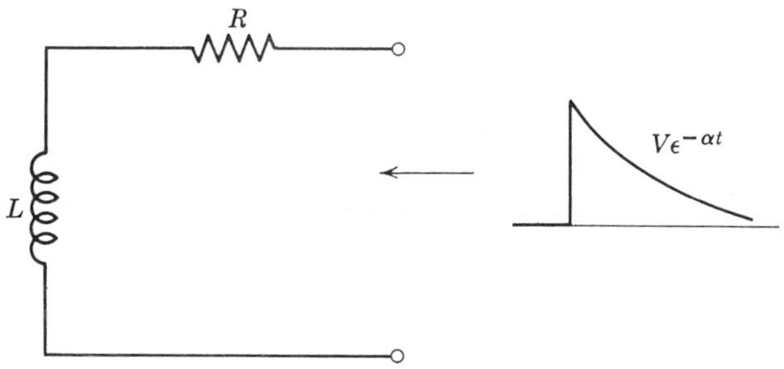

Fig. 2.8. An RL circuit stimulated by an exponential drive.

If the circuit is initially dead [i.e., $I(0) = 0$],

$$i(s) = \frac{V}{(R + Ls)(s + \alpha)} \tag{2.6.8}$$

Introducing $\lambda = R/L$, this may be written

$$i(s) = \frac{V}{L(s + \lambda)(s + \alpha)} \tag{2.6.9}$$

It is apparent that Eq. 2.6.9 can be rewritten

$$i(s) = \frac{V}{L(\alpha - \lambda)} \left\{ \frac{1}{s + \lambda} - \frac{1}{s + \alpha} \right\} \tag{2.6.10}$$

which contains the now familiar transform of Eq. 2.2.8 and leads to the solution

$$I(t) = \frac{V}{L(\alpha - \lambda)} [\epsilon^{-\lambda t} - \epsilon^{-\alpha t}] \tag{2.6.11}$$

To solve the problem by Duhamel's integral, we must first find $u_1(t)$, which we have already done in Section 2.3. It is apparent from Eq. 2.3.13 that

$$u_1(t) = I_1(t) = \frac{1}{R} [1 - \epsilon^{-\lambda t}] \tag{2.6.12}$$

From Eqs. 2.6.6 and 2.6.12, it seems that the form of Duhamel's integral given in Eq. 2.6.5*b* best suits this problem since the fact that $u_1(0) = 0$ simplifies the expression. Now,

$$u_1'(t) = \frac{\lambda \epsilon^{-\lambda t}}{R}$$

Therefore, from Eq. 2.6.5*b*,

$$\begin{aligned} u(t) = I(t) &= \int_0^t V \epsilon^{-\alpha \tau} \frac{\lambda \epsilon^{-\lambda(t-\tau)}}{R} d\tau \\ &= \frac{V}{L} \epsilon^{-\lambda t} \int_0^t \epsilon^{-(\alpha - \lambda)\tau} d\tau \\ &= \frac{V \epsilon^{-\lambda t}}{L(\alpha - \lambda)} \left| -\epsilon^{-(\alpha-\lambda)\tau} \right|_0^t \\ &= \frac{V}{L(\alpha - \lambda)} [\epsilon^{-\lambda t} - \epsilon^{-\alpha t}] \end{aligned}$$

which accords with solution 2.6.11.

We have been thinking in terms of a voltage stimulus, but it should be clearly understood that Duhamel's integral can be applied equally well if

$U(t)$ is a current stimulus. In this case $u_1(t)$ will be a voltage and will be obtained by multiplying the current by the impedance:

$$u_1(s) = \frac{1}{s} \cdot z(s)$$

or

$$u_1(t) = \mathcal{L}^{-1} \frac{z(s)}{s}$$

REFERENCES

1. S. Goldman, *Laplace Transform Theory and Electrical Transients*, Dover Publications, New York (1966).
2. G. W. Carter, *The Simple Calculation of Electrical Transients*, Cambridge University Press, New York (1944).

3 Simple Switching Transients

3.1 Introduction

It was pointed out in Chapter 1 that a transient is initiated whenever there is a sudden change of circuit conditions; this most frequently occurs when a switching operation takes place. All the examples studied thus far have been concerned with switching transients, but they have been confined to circuits with dc drives. In this chapter we consider a-c circuits, restricting ourselves to two problems: (1) the closing of a switch or circuit breaker to energize a load and (2) the opening of a breaker to clear a fault. This provides an opportunity to discuss some of the practical details of switching transients.

A prime concern throughout this text is the emphasis on the physical aspects of what is occurring in the circuit. Mathematics is used as an adjunct to this end, not as a substitute for it. The Laplace transform method developed in Chapter 2 has helped to simplify and systematize the mathematics. However, where there are opportunities to simplify the mathematics further, we shall take them. Ultimately, we will find that we can circumvent the mathematics entirely in many of the problems we are about to discuss, proceeding directly to the solution by deduction.

3.2 The Circuit Closing Transient

The circuit involved in this example is that of Fig. 3.1. It has been reduced to its barest essentials in the interest of initial simplicity. The load is represented by a series combination of resistance and inductance, which has a steady-state power factor given by

$$\cos \varphi = \frac{R}{|Z|} = \frac{R}{(R^2 + \omega^2 L^2)^{1/2}} \tag{3.2.1}$$

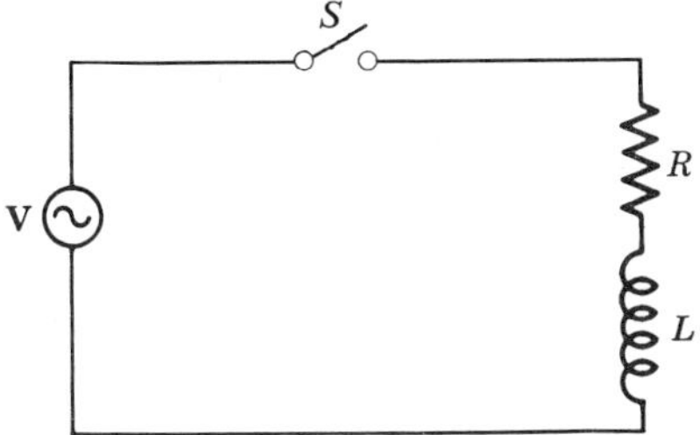

Fig. 3.1. An *RL* circuit with a sine wave drive.

The source is assumed to have negligible impedance compared with the load. The source voltage is **V**, indicating a phasor varying at the supply frequency ω. When the switch S is closed, the equation expressing the current is

$$RI + L\frac{dI}{dt} = \mathbf{V} = V_m \sin(\omega t + \theta) \tag{3.2.2}$$

The inclusion of the arbitrary phase angle θ permits closing of the switch at any instant in the voltage cycle.

Before attempting to solve Eq. 3.2.2, let us consider for a moment what we have already discovered about this circuit. It is clear that in due course the current will attain a steady-state value of V/Z, and that it will lag in phase the voltage by an angle φ defined by Eq. 3.2.1. However, it is equally clear that except perhaps for some special circumstance, the current cannot achieve this value instantaneously, because the circuit inductance demands that the current start at zero. We would suppose, therefore, that there is some transient that leads the current to its steady-state value in a smooth, continuous way and that since this is an *RL* circuit, the exponential $\epsilon^{-Rt/L}$ will play an important part in the solution. These observations say a great deal about the solution of Eq. 3.2.2, although as yet we have made no attempt to solve it. We now proceed to this task.

Equation 3.2.2 can be rewritten

$$RI + L\frac{dI}{dt} = V_m\left\{\sin \omega t \cos \theta + \cos \omega t \sin \theta\right\}$$

Transforming both sides,

$$Ri(s) + Lsi(s) - LI(0) = V_m\left\{\frac{\omega \cos \theta}{s^2 + \omega^2} + \frac{s \sin \theta}{s^2 + \omega^2}\right\} \tag{3.2.3}$$

Remember that $\sin \theta$ and $\cos \theta$ are constants once the value of θ has been assigned. In this circuit $I(0) = 0$, so the operational solution for current is

$$i(s) = \frac{V_m}{L}\frac{1}{(s + R/L)}\left\{\frac{\omega \cos \theta}{s^2 + \omega^2} + \frac{s \sin \theta}{s^2 + \omega^2}\right\} \tag{3.2.4}$$

This can be written more concisely as follows:

$$i(s) = \frac{A}{(s+\alpha)(s^2+\omega^2)} + \frac{Bs}{(s+\alpha)(s^2+\omega^2)} \tag{3.2.5}$$

where

$$A = \frac{V_m}{L}\,\omega\cos\theta, \qquad B = \frac{V_m}{L}\sin\theta, \qquad \text{and} \quad \alpha = \frac{R}{L}$$

These are new transforms, but they can be reduced readily by the method of partial fractions outlined in Section 2.4. Simple manipulation reveals that

$$\frac{1}{(s+\alpha)(s^2+\omega^2)} \equiv \frac{1}{(\alpha^2+\omega^2)}\left\{\frac{1}{s+\alpha} - \frac{s}{s^2+\omega^2} + \frac{\alpha}{s^2+\omega^2}\right\} \tag{3.2.6}$$

Therefore

$$\mathcal{L}^{-1}\frac{1}{(s+\alpha)(s^2+\omega^2)} = \frac{1}{(\alpha^2+\omega^2)}\left\{\epsilon^{-\alpha t} - \cos\omega t + \frac{\alpha}{\omega}\sin\omega t\right\} \tag{3.2.7}$$

The other terms, $Bs/(s+\alpha)(s^2+\omega^2)$ in Eq. 3.2.5 can be evaluated by the same method. Instead let us turn around Eq. 2.2.12 and obtain the inverse transform of this second term by differentiating Eq. 3.2.7:

$$\mathcal{L}^{-1}\frac{s}{(s+\alpha)(s^2+\omega^2)} = \frac{1}{(\alpha^2+\omega^2)} \times \{-\alpha\epsilon^{-\alpha t} + \omega\sin\omega t + \alpha\cos\omega t + 1 - 1 + 0\} \tag{3.2.8}$$

Equation 3.2.4 can now be evaluated with the aid of Eqs. 3.2.7 and 3.2.8:

$$\begin{aligned} I(t) &= \frac{V_m}{L(\alpha^2+\omega^2)}\left\{\omega\cos\theta\left[\epsilon^{-\alpha t} - \cos\omega t + \frac{\alpha}{\omega}\sin\omega t\right]\right. \\ &\qquad\left. + \sin\theta[\alpha\cos\omega t + \omega\sin\omega t - \alpha\epsilon^{-\alpha t}]\right\} \\ &= \frac{V_m}{L(\alpha^2+\omega^2)}\{[\omega\cos\theta - \alpha\sin\theta]\epsilon^{-\alpha t} + [\omega\cos\theta - \alpha\sin\theta]\cos\omega t \\ &\qquad + (\alpha\cos\theta + \omega\sin\theta)\sin\omega t\} \end{aligned} \tag{3.2.9}$$

Now from Eq. 3.2.1, $\tan\varphi = \omega L/R = \omega/\alpha$, so that $\sin\varphi = \omega/(\alpha^2+\omega^2)^{1/2}$ and $\cos\phi = \alpha/(\alpha^2+\omega^2)^{1/2}$. Equation 3.2.9 simplifies further:

$$\begin{aligned} I(t) &= \frac{V_m}{L(\alpha^2+\omega^2)^{1/2}}\{-\sin(\theta-\varphi)\epsilon^{-\alpha t} + \sin(\omega t + \theta - \varphi)\} \\ &= \frac{V_m}{(R^2+\omega^2L^2)^{1/2}}\{\sin(\omega t + \theta - \varphi) - \sin(\theta - \varphi)\epsilon^{-\alpha t}\} \end{aligned} \tag{3.2.10}$$

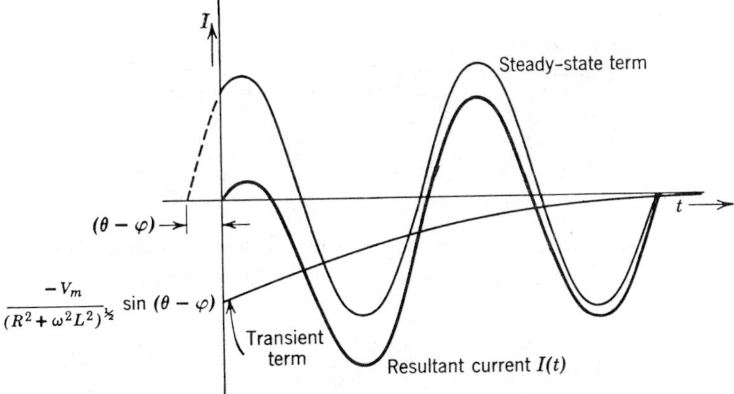

Fig. 3.2. Asymmetrical alternating current from Eq. 3.2.10.

This is precisely the form we predicted before embarking on the analysis. The first term is the steady-state final value. Its amplitude is $V_m/|z|$ and it indeed has a phase angle $-\varphi$ with respect to the voltage. The second term is the transient. It involves, as expected, the exponential $\epsilon^{-Rt/L}$, moreover, at $t = 0$, it is equal and opposite to the steady-state term, thus assuring that the current start from zero. Equation 3.2.10 is depicted graphically in Fig. 3.2.

In the very special case where the switch closes at the instant when $\theta = \varphi$, the transient term will be zero and the current wave will be symmetrical. On the other hand, if the switch closes when $\theta - \varphi = \pm\pi/2$, the transient term attains its maximum amplitude and the first peak of the resulting composite current wave will approach twice the peak amplitude of the steady-state sinusoidal component. This has some significant practical implications for circuit breakers.

An important function of a circuit breaker is to close circuits as well as open them and on occasions a breaker will close in on a short circuit. The closing angle θ is arbitrary, so it is possible to hit the condition where the current is fully offset for which the peak of the first loop approaches a maximum value of twice the peak short circuit current. In a three-phase circuit, the closing angles in the different phases are 120° apart if the poles close in mechanical synchronism, so there is a good chance that this condition, or something close to it, will be met on some phase at every closing operation. The circuit breaker must be designed electrically, mechanically, and thermally to withstand this duty and perform the subsequent operations (such as opening and reclosing) without its effectiveness being impaired. Electromagnetic forces set up by such currents could cause buckling of parts in an inadequately designed circuit breaker.

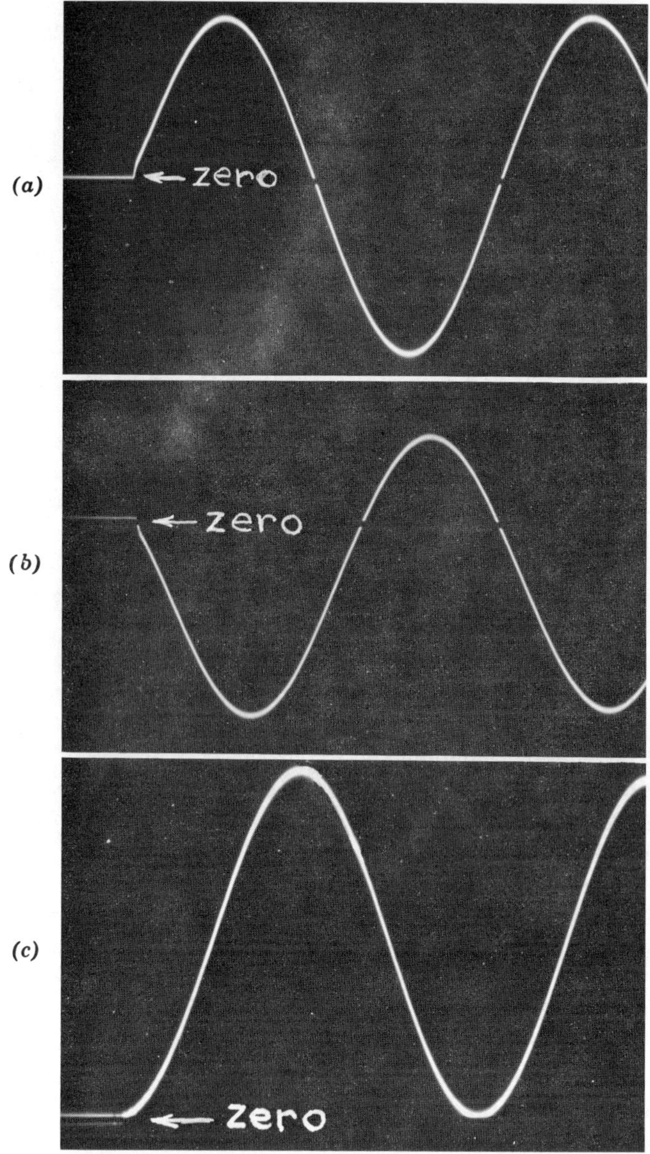

Fig. 3.3. Examples of transient currents showing various degrees of symmetry. (*a*) Symmetrical. (*b*) Partially asymmetrical. (*c*) Fully asymmetrical.

Similar asymmetrical currents can flow through the contacts of a closed breaker if a short circuit occurs on the system it is protecting. These so-called high "momentary currents" will encourage the contacts to weld at the points where they touch, due to the energy dissipated there in contact resistance. In configurations where the contacts are butted together this condition may be made worse by the contacts "popping," that is, being momentarily driven apart, with the establishment of an arc, by the electromagnetic forces of the currents converging on and diverging from the points of contact. Again, any weld so formed must not impair the efficiency of the circuit breaker. Its mechanism must be able to break the weld and subsequently rupture the fault current if called upon to do so.

Circuit breakers frequently have to interrupt asymmetrical currents, for a fault can occur at any point in the cycle. Some may not be able to do this at the current zero following a major loop of current. The arc will reignite and interruption will be effected after the minor loop. However, when we consider recovery voltages it will be apparent that current amplitude is not the only influential factor in determining whether a breaker will successfully interrupt.

The rate of decay of the transient component of an asymmetrical current depends upon the time constant of the circuit. It is often referred to as the d-c decrement of the circuit. It varies with the type of circuit and is usually expressed in terms of the X/R ratio rather than the L/R time constant. Typical X/R ratios for different circuits are given in Chapter 15.

If we know the time from the initiation of the fault to the parting of the circuit breaker contacts and the available fault current on the system, it is possible to determine the actual current the breaker has to interrupt. Clearly in a circuit with a high decrement the duty will be less severe than in one where the time constant is long. It is not usual for the opening of a breaker to be deliberately delayed in order that the current may become more symmetrical. Such considerations as limiting of damage at the fault, thermal overloading of lines and equipment, and system stability make it mandatory to remove the fault as soon as possible. Figure 3.3 gives some oscilloscopic records of closing transients showing the effect of varying the instant in the cycle at which current is initiated.

3.3 The Recovery Transient Initiated by the Removal of a Short Circuit

The simplest circuit that can be chosen to illustrate this phenomenon is that shown in Fig. 3.4. It contains only what is necessary for the purpose. It is assumed that the load being fed through the circuit breaker is suddenly isolated by the occurrence of a fault and that the fault is a dead short circuit.

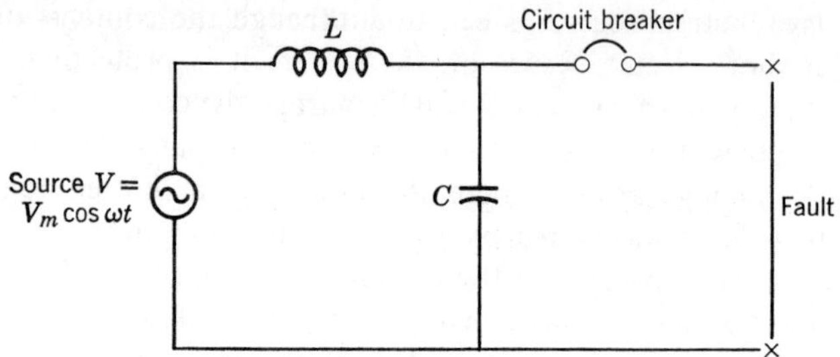

Fig. 3.4. Equivalent circuit for studying the transient recovery voltage when a circuit breaker clears a fault.

The L is all the inductance limiting the current to the point of fault, and C is the natural capacitance of the circuit adjacent to the circuit breaker. Such capacitance always exists. It comprises capacitance to ground through bushings, current transformers, and so forth, and perhaps the capacitance of an adjacent transformer, as well as capacitance across the breaker contacts that will be grounded on one side by the fault. Circuit resistance and other forms of loss will be neglected at this time.

When a fault occurs on a system a substantial fault current usually flows. The parting of the circuit breaker contacts does not in itself interrupt the current, for an arc will be established between the parting contacts, through which the current will continue to flow. Successful interruption depends upon controlling and finally extinguishing the arc. Different types of circuit breaker use a variety of different agencies to bring this about. Fortunately, in alternating current circuits at least, the current passes through zero twice each cycle. It is at one of these instants that interruption is actually effected.

In the circuit of Fig. 3.4 the current is assumed to be symmetrical and will be completely reactive since it is limited entirely by inductance. This means that at the instant of current zero, the circuit voltage will be at a maximum value, but the voltage at the switch contacts, and therefore across the capacitor C, will be the arc voltage. The relative importance of the arc voltage varies. In high-voltage circuits it is usually only a small percentage of the system voltage. In low-voltage circuits it may be much more significant. For the first examination of the ensuing events, it will be assumed that the arc voltage is negligible.

In this analysis, time will be measured from the instant of interruption when the fault current has just come to zero. Since the source voltage is a sinusoidally varying quantity and is at its peak at this moment, it is expressed as $V_m \cos \omega t$. The circuit equation is therefore

$$L\frac{dI}{dt} + V_C = V_m \cos \omega t \tag{3.3.1}$$

There are two unknowns, I and V_C, so another relationship between them is required. This is

$$I = C\frac{dV_C}{dt} \tag{3.3.2}$$

for after the switch has cleared, the only path for current is into the capacitor. The combination of these equations gives

$$\frac{d^2V_C}{dt^2} + \frac{V_C}{LC} = \frac{V_m}{LC}\cos\omega t \tag{3.3.3}$$

Before solving this equation, let us consider, as we did with the closing transient in Section 3.2, what kind of a solution we might expect from the physical conditions in the circuit. Ultimately, after the circuit breaker has completely cleared the fault, we would expect to see across its contacts the voltage of the supply. But at the moment of clearing ($t = 0$) the voltage is the previous arc voltage, which we have chosen to neglect. The voltage cannot change discontinuously because capacitor C must be charged. We would therefore expect that following current zero, a transient would be initiated whereby C is charged from the supply through the inductance L. Our experience with LC circuits in Chapter 2 leads us to expect that this will cause an oscillation at the natural frequency of the circuit.

The analysis is very similar to that used in Section 2.3. Let $1/LC = \omega_0{}^2$, and transform Eq. 3.2.3:

$$s^2 v_C(s) - sV_C(0) - V_C'(0) + \omega_0{}^2 v_C(s) = \omega_0^2\, V_m\frac{s}{s^2 + \omega^2}$$

or

$$v_C(s) = \omega_0^2\, V_m\frac{s}{(s^2 + \omega^2)(s^2 + \omega_0{}^2)} + V_C(0)\frac{s}{(s^2 + \omega_0{}^2)} + \frac{V_C'(0)}{(s^2 + \omega_0{}^2)} \tag{3.3.4}$$

If we neglect arc voltage the second term on the right is zero. The third term is also zero because, from Eq. 3.3.2,

$$V_C'(0) = \frac{I(0)}{C} = 0$$

Therefore only the first term remains. Now

$$\frac{s}{(s^2 + \omega^2)(s^2 + \omega_0{}^2)} = \frac{1}{(\omega_0{}^2 - \omega^2)}\left\{\frac{s}{(s^2 + \omega^2)} - \frac{s}{(s^2 + \omega_0{}^2)}\right\}$$

Therefore

$$v_C(s) = V_m\frac{\omega_0{}^2}{(\omega_0{}^2 - \omega^2)}\left\{\frac{s}{(s^2 + \omega^2)} - \frac{s}{(s^2 + \omega_0{}^2)}\right\}$$

and

$$V_C(t) = \frac{\omega_0{}^2}{(\omega_0{}^2 - \omega^2)}V_m[\cos\omega t - \cos\omega_0 t] \tag{3.3.5}$$

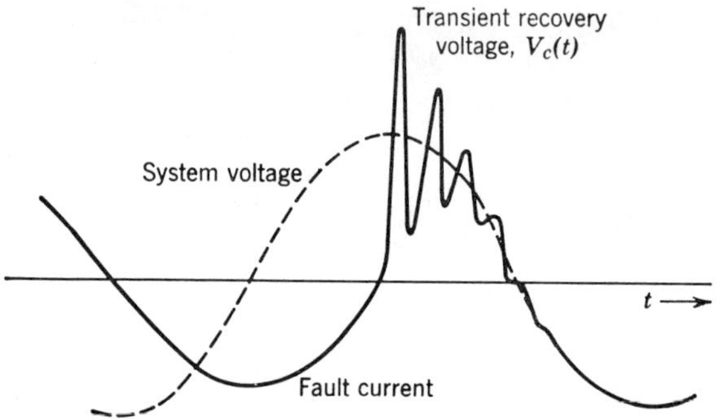

Fig. 3.5. Transient recovery voltage across the circuit breaker in Fig. 3.4 following interruption of the fault current.

Equation 3.3.5 is the voltage appearing across the switch contacts after current zero. It is the classical recovery transient described by Park and Skeats (1). Almost invariably $\omega_0 \gg \omega$, so that $\omega_0^2/(\omega_0^2 - \omega^2) \approx 1$. Thus to a very close approximation,

$$V_C(t) = V_m[\cos \omega t - \cos \omega_0 t] \tag{3.3.6}$$

Indeed, it often happens that over the period of interest—the time for which the natural frequency oscillation persists—there is little change in the power frequency term. In this case Eq. 3.3.6 can be further reduced to give

$$V_C(t) = V_m[1 - \cos \omega_0 t] \tag{3.3.7}$$

Events before and after current zero are depicted in Fig. 3.5. This diagram shows the accuracy of the predictions we made before starting the analysis. It was pointed out then that there would be losses in the system that would damp out any disturbance created by the opening operation. This damping is not included in Eq. 3.3.6 but has been introduced into Fig. 3.5. Observe how the voltage approaches twice the peak system voltage when the two components interfere constructively.

Up to the instant of current zero, the capacitance in Fig. 3.4 has been short circuited by the fault. With the successful operation of the circuit breaker this constraint is removed; the source will therefore attempt to charge the capacitance to its own potential. The situation is quite similar to that depicted as curve (b) in Fig. 2.4 and we see the inertial effect of the circuit inductance causes the overswing of voltage we observed there.

If the natural frequency ω_0 is high (if L or C or both are very small), the voltage across the switch contacts will rise very quickly. If this rate of application of the recovery voltage should exceed the rate of build-up of dielectric strength in the medium between the contacts, the breaker will be unable to hold off the voltage and a so-called reignition will occur. This usually results in the switch carrying fault current for at least another half cycle. It is for this reason that in British literature the recovery transient is often referred to as the restriking voltage. (Perhaps the two different names for the same phenomenon, recovery voltage and restriking voltage, reflect the basic optimism or pessimism of American and British switchgear engineers!)

It is apparent that the rate of rise of the recovery voltage, r.r.r.v., is an important factor in switchgear application. It gives a measure of circuit severity from a switchgear point of view. As already mentioned, circuits that lead to the highest r.r.r.v.s are those with high natural frequencies. An air-cored reactor is a good example. Consider such a choke with an inductance of 1 mH and an effective capacitance of 400 pF, its natural frequency will be

$$f_0 = \frac{1}{2\pi(10^{-3} \times 4 \times 10^{-10})^{1/2}} = \frac{10^6}{3.99} \approx 250 \text{ kHz}$$

This represents a period of 4 μsec. On a 13.8-kV circuit, neglecting damping, the voltage will swing to twice the system peak in half a period. The mean rate of rise of voltage during this time is therefore

$$\frac{2 \times 13.8\sqrt{2}}{\sqrt{3} \times 2} = 11.3 \text{ kV}/\mu\text{sec}$$

This is beyond the clearing capability of most circuit breakers. Another example of a very fast recovery transient is that initiated on the clearing of a "kilometric fault." This will be discussed in Chapter 13.

Two typical recovery transients, obtained oscillographically, are shown in Fig. 3.6. The first is of the kind we have just been discussing. The second

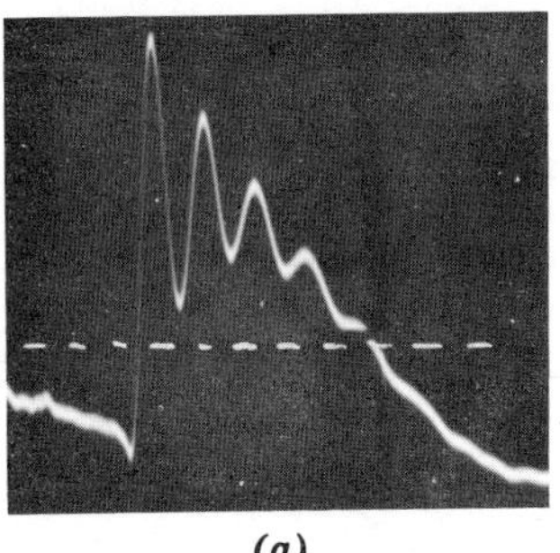

(a)

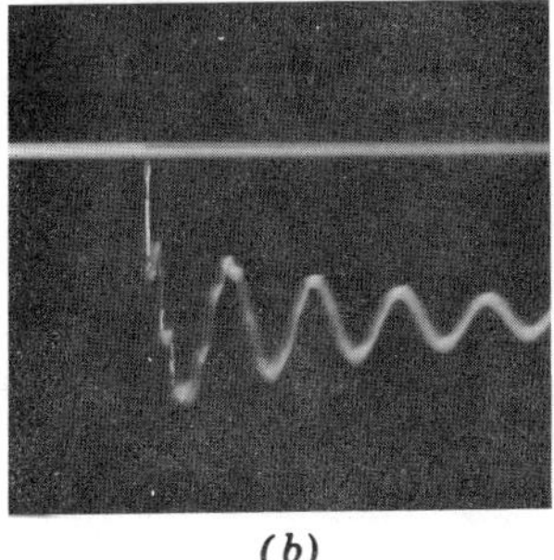

(b)

Fig. 3.6. Circuit breaker transient recovery voltages obtained oscillographically. (*a*) Single frequency, (*b*) Double frequency.

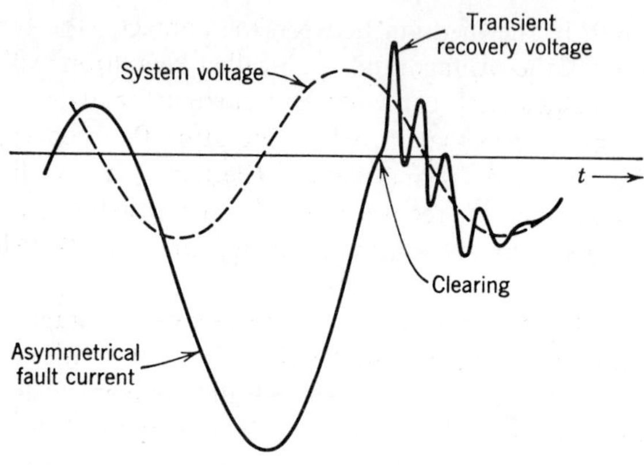

Fig. 3.7. Voltage transient following the interruption of an asymmetrical current.

is, for obvious reasons, called a double frequency transient. This will be dealt with in Section 3.4.

It was shown in Section 3.2 that when a switch is closed at random it is likely that the current will lack symmetry. It can have any degree of asymmetry, depending upon the instant in the voltage cycle at which the switch is closed. Similarly, a fault current can have any degree of asymmetry depending upon the time in the cycle at which the fault occurs. The circuit breaker will again interrupt at current zero and the recovery voltage will oscillate about the instantaneous value of the supply voltage. But with an asymmetrical fault current the voltage will no longer be at its peak. The transient recovery voltage is therefore not as high. This is illustrated in Fig. 3.7.

In the preceding analysis, specifically in Eq. 3.3.4, the arc voltage in the circuit breaker was assumed negligible, although it was pointed out that sometimes this is not the case. For example, consider the arc voltage term in Eq. 3.3.4. Its inverse transform gives a cosine at the natural frequency ω_0,

$$\mathcal{L}^{-1} V_C(0) \frac{s}{s^2 + \omega_0^2} = V_C(0) \cos \omega_0 t$$

Now $V_c(0)$ is the value of the arc voltage at the time of current zero ($t = 0$), since the capacitor always has the arc voltage impressed across it as long as the arc persists. This adds to a similar term in the recovery voltage tending to increase this switching transient. The effect is offset by a second effect of the arc voltage, which is to oppose the current flow and thereby change the phase of the current, bringing it more into phase with the supply voltage. Thus, when the switch clears the supply voltage is not at its peak. These two

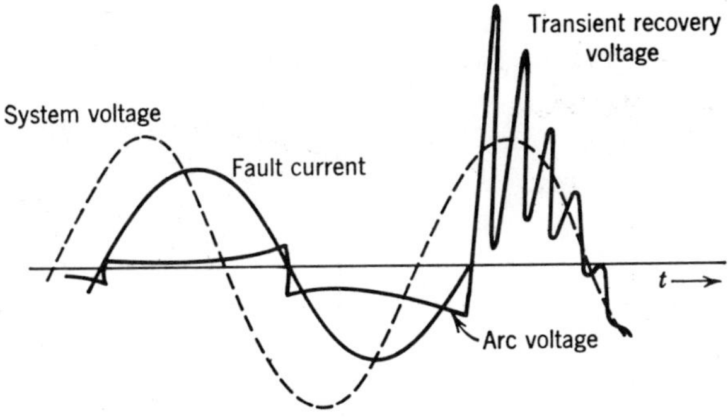

Fig. 3.8. The effect of arc voltage on the recovery transient after switching.

effects are illustrated in Fig. 3.8, which should be compared with Fig. 3.5, where the arc voltage has been ignored.

3.4 Double-Frequency Transients

The simplest form of the double-frequency transient is that initiated by opening the circuit breaker in the circuit shown in Fig. 3.9. In this diagram L_1 and C_1 are the inductance and stray capacitance on the source side of the breaker. Here L_2 and C_2 might represent an inductive load and its stray capacitance, an unloaded transformer, for example. When the switch operates in such a circuit it completely divorces the load from the supply. Thereafter the two halves of the circuit behave independently. It is possible to deduce what happens following such a switching operation without resorting to any mathematical analysis. Before the switch opening the 60 cps voltage will divide in proportion to the inductances, that is, to a close approximation the voltage of the capacitors will be

$$\frac{L_2}{L_1 + L_2} V$$

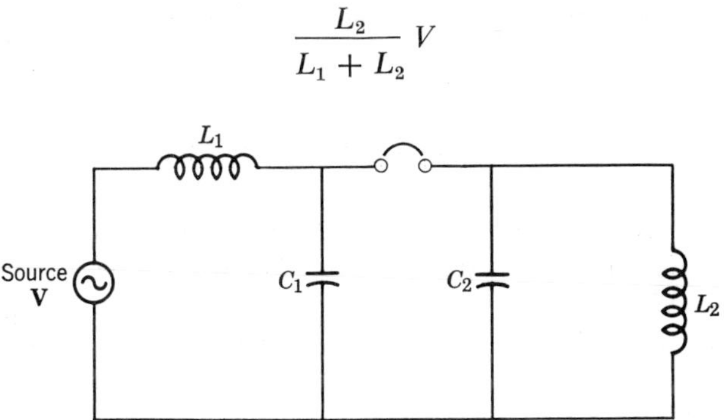

Fig. 3.9. Circuit with two natural frequencies.

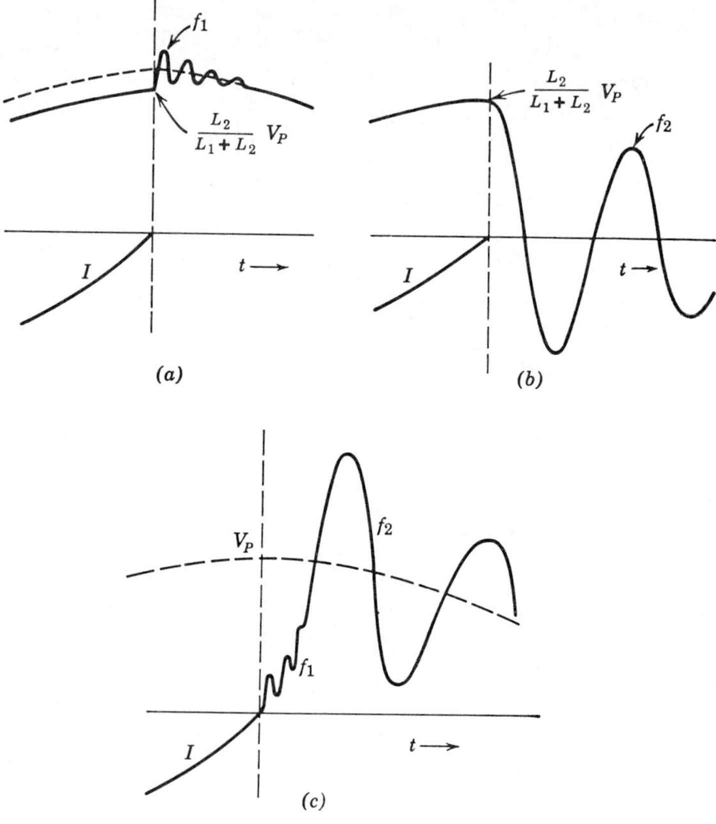

Fig. 3.10. Double frequency recovery transient. (*a*) Source side transient. (*b*) Load side transient. (*c*) Recovery voltage across the switch.

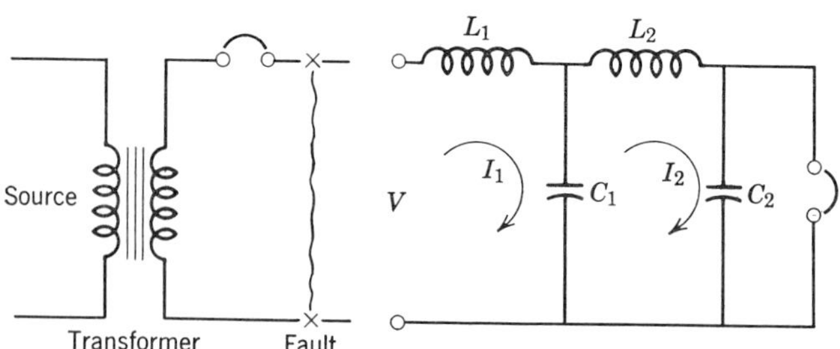

Fig. 3.11. Example of a two-frequency circuit.

It is likely that $L_2 \gg L_1$, otherwise the regulation would be very poor, so that C_1 and C_2 are charged to something a little less than the instantaneous system voltage. When the current passes through zero, this voltage will be at its peak. Following current interruption C_2 will discharge through L_2 with a natural frequency given by

$$f_2 = \frac{1}{2\pi(L_2C_2)^{1/2}} \tag{3.4.1}$$

Meanwhile C_1, now free to take up the source potential, will oscillate about that value until the losses of the system damp out the disturbance. The frequency of this oscillation will be

$$f_1 = \frac{1}{2\pi(L_1C_1)^{1/2}} \tag{3.4.2}$$

Source side and load side transients are depicted in Fig. 3.10*a* and 3.10*b*. The recovery voltage across the circuit breaker contacts will be the difference between these two, as shown in Fig. 3.10*c*.

There are many other two-frequency circuits that occur in practice. One that is encountered quite often is drawn in Fig. 3.11, which also shows one situation in which it might arise. This shows a circuit breaker clearing a short circuit on the secondary side of a transformer. In this case L_1 represents the inductance up to the transformer L_2 the leakage inductance of the transformer, while C_1 and C_2 are the inherent system capacitances on either side of the transformer.

We recognize this as a double-frequency circuit by the two *LC* loops. However, it is by no means clear what the natural frequencies are because the loops are mutually coupled. Still more obscure is the relative amplitude of the two component frequencies. To determine the recovery transient we must analyze the circuit.

Contrary to what one might suppose, the algebra rapidly becomes quite complicated. We assume that during the period of interest the change in the source voltage is negligible. It can then be represented by a constant voltage V. Because the reactance of C_1 at 60 Hz is very much higher than the corresponding reactances of L_1 and L_2, it is safe to assume that the initial voltage distribution is determined by the *L*s or, more specifically, that the initial voltage on C_1 is given by

$$V_{c_1}(0) = \frac{L_2}{L_1 + L_2}V \tag{3.4.3}$$

The following equations can be written:

$$V_{C_1} = V - L_1 \frac{dI_1}{dt} = V_{C_1}(0) + \frac{1}{C_1}\int (I_1 - I_2)\, dt \tag{3.4.4}$$

$$V_{C_2} = \frac{1}{C_2}\int I_2\, dt \tag{3.4.5}$$

$$V_{C_2} = V - L_1 \frac{dI_1}{dt} - L_2 \frac{dI_2}{dt} \tag{3.4.6}$$

When transformed, these equations read as follows:

$$\frac{V}{s} - L_1 s i_1(s) - L_1 I_1(0) = \frac{V_{C_1}(0)}{s} + \frac{1}{C_1 s}\,[i_1(s) - i_2(s)]$$

or

$$-i_1(s)\left[L_1 s + \frac{1}{C_1 s}\right] + \frac{i_2(s)}{C_1 s} = \frac{V_{C_1}(0)}{s} + L_1 I_1(0) - \frac{V}{s} \tag{3.4.7}$$

$$v_{C_2}(s) = \frac{i_2(s)}{sC_2} \tag{3.4.8}$$

$$v_{C_2}(s) = \frac{V}{s} - L_1 s i_1(s) - L_1 I_1(0) - L_2 s i_2(s) + L_2 I_2(0) \tag{3.4.9}$$

Time is measured from the instant when the switch clears at current zero, so that $I_1(0) = I_2(0) = 0$. From Eqs. 3.4.8 and 3.4.9,

$$i_1(s) = \frac{V}{L_1 s^2} - \left[\frac{L_2 C_2 s^2 + 1}{L_1 s}\right] v_{C_2}(s) \tag{3.4.10}$$

Substituting Eqs. 3.4.10 and 3.4.8 in Eq. 3.4.7,

$$\left\{\left[\frac{L_2 C_2 s^2 + 1}{L_1 s}\right] v_{C_2}(s) - \frac{V}{L_1 s^2}\right\}\left[L_1 s + \frac{1}{C_1 s}\right] + \frac{C_2}{C_1} v_{C_2}(s) = \frac{V_{C_1}(0) - V}{s} \tag{3.4.11}$$

Multiplying through Eq. 3.4.11 by s^2/L_2C_2 and rearranging,

$$\left\{s^4 + s^2\left[\frac{1}{L_1 C_1} + \frac{1}{L_2 C_2} + \frac{1}{L_2 C_1}\right] + \frac{1}{L_1 C_1 L_2 C_2}\right\} v_{C_2}(s) = \frac{s V_{C_1}(0)}{L_2 C_2} + \frac{V}{L_1 C_1 L_2 C_2 s}$$

or, from Eq. 3.4.3,

$$\left\{s^4 + s^2\left[\frac{1}{L_1 C_1} + \frac{1}{L_2 C_2} + \frac{1}{L_2 C_1}\right] + \frac{1}{L_1 C_1 L_2 C_2}\right\} v_{C_2}(s) = V\left[\frac{s}{(L_1 + L_2) C_2} + \frac{1}{L_1 C_1 L_2 C_2 s}\right] \tag{3.4.12}$$

The expression on the left-hand side of Eq. 3.4.12 is a quadratic in s^2. The equation can therefore be rewritten:

$$(s^2 + \omega_1^2)(s^2 + \omega_2^2)v_{C_2}(s) = AV\left\{\frac{1}{s} + Bs\right\}$$

or

$$v_{C_2}(s) = AV\left\{\frac{1}{s(s^2 + \omega_1^2)(s^2 + \omega_2^2)} + \frac{Bs}{(s^2 + \omega_1^2)(s^2 + \omega_2^2)}\right\} \quad (3.4.13)$$

in which ω_1, ω_2, A, and B are constants, dependent upon the circuit parameters L_1, C_1, L_2, and C_2. Inverse transforms for the two terms can be obtained by taking partial fractions or by looking up the inverse transform in the table in Appendix 1. From there we find the solution to be

$$\begin{aligned} V_2(t) &= AV\left\{\frac{1}{\omega_1^2\omega_2^2} + \frac{1}{\omega_1^2(\omega_1^2 - \omega_2^2)}\cos\omega_1 t - \frac{1}{\omega_2^2(\omega_1^2 - \omega_2^2)}\cos\omega_2 t \right. \\ &\qquad \left. + \frac{B}{(\omega_1^2 - \omega_2^2)}[\cos\omega_2 t - \cos\omega_1 t]\right\} \\ &= AV\left\{\frac{1}{\omega_1^2\omega_2^2} + \left[\frac{1 - \omega_1^2 B}{\omega_1^2(\omega_1^2 - \omega_2^2)}\right]\cos\omega_1 t - \left[\frac{1 - \omega_2^2 B}{\omega_2^2(\omega_1^2 - \omega_2^2)}\right]\cos\omega_2 t.\right. \end{aligned} \quad (3.4.14)$$

It is clearly possible to write Eq. 3.4.14 in terms of the circuit constants L_1, C_1, L_2, and C_2, but the expressions are very lengthy because of the complicated relationships between these quantities and ω_1 and ω_2. At this point (i.e., Eq. 3.4.14) in a practical problem, one is well advised to substitute numbers before proceeding.

This exercise shows, if nothing else, how quickly the algebra and arithmetic burgeon with a modest increase in the complexity of the problem. For this reason we explore in later chapters methods whereby some of this labor can be avoided.

REFERENCE

1. R. H. Park and W. F. Skeats, "Circuit Breaker Recovery Voltages, Magnitudes and Rates of Rise," *Trans. AIEEE*, Vol. 50 (1931), p. 204.

4 Damping

4.1 Some Observations on the *RLC* Circuit

All *LC* circuits that we have examined have been loss-free in that no dissipative element has been included. But all practical circuits have losses arising primarily from circuit resistance and iron losses in equipment. In addition, the system loads represent very important dissipative elements. Whether loads or losses, the dissipation is accommodated by including resistance in the circuits. In making transient analyses, all losses usually are neglected in the first instance, greatly reducing the complication of the calculations. Moreover, this approach leads to solutions that give severely high overvoltages. Once the general behavior of the circuit has been established, the modification introduced by the system losses can be considered separately. Introducing resistance always has the effect of damping out the natural oscillations of a circuit. How quickly this occurs, or indeed whether the circuit can oscillate at all, will depend upon the extent of the losses, or, put another way, the value of R relative to the values of L and C.

This chapter is devoted exclusively to studying two very important circuits, the parallel and series *RLC* circuits, which are illustrated in Fig. 4.1. These circuits are considered important because the networks involved in many practical transient problems in power systems can be safely reduced to one or the other of these configurations for the purpose of analysis. In many other instances, the network can be reduced to a number of these simple circuits which are so loosely coupled that, on being treated independently, they yield results of acceptable engineering accuracy (1).

The behavior of electric circuits is surprisingly simple. This is especially true of the two circuits now under consideration. The differential equations describing the two circuits in their transient state are essentially similar. For then the quotient of the time constants T_P and T_S is equal to η^2:

$$\frac{T_P}{T_S} = \frac{R^2C}{L} = \eta^2 \tag{4.1.6}$$

These relationships lead to a duality in the analysis of the series and parallel circuits.

While the basic phenomena in these two circuits are simple, the solutions to the equations can be complicated. There is ample scope for error in manipulating the algebra. To reduce the computational labor once the fundamental ideas have been grasped, we use the following approach to short cut several stages of the normal analysis of a problem.

There is a small number of transforms that appear regularly in the operational solutions for problems involving these circuits. We shall invest the effort required to find their inverse transforms and then we shall plot these inverse transforms in a series of dimensionless curves using η, defined in Eq. 4.1.6, as a parameter. Thereafter, when we encounter these transforms in solving practical problems, the solutions can be extracted from these curves with about the same effort as one would expend in using a graph of a trigonometric function.

4.2 The Basic Transforms of the *RLC* Circuits

Before proceeding to the general approach just described, we look at a few specific problems. This will help verify some of the assertions that have been made and introduce the basic transforms referred to in the last section.

Consider the solution for the current in the inductor in the parallel circuit, when a switch is closed in the capacitor branch, allowing C to discharge through R and L. Let this current be I_L. The current from the capacitor must be equal to the sum of the currents in the other two branches. Calling the capacitor voltage V_C,

$$-C\frac{dV_C}{dt} = I_L + \frac{V_C}{R} \tag{4.2.1}$$

We may also write

$$V_C = L\frac{dI_L}{dt} \tag{4.2.2}$$

Eliminating V_C from Eqs. 4.2.1 and 4.2.2,

$$-CL\frac{d^2I_L}{dt^2} = I_L + \frac{L}{R}\frac{dI_L}{dt}$$

or

$$\frac{d^2I_L}{dt^2} + \frac{1}{RC}\frac{dI_L}{dt} + \frac{I_L}{LC} = 0$$

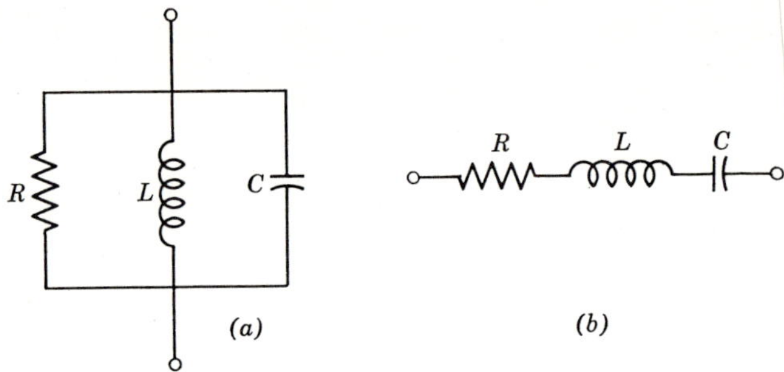

Fig. 4.1. Parallel and series *RLC* circuits.

the parallel circuit, Fig. 4.1*a*, the equation can be written as

$$\frac{d^2\Phi}{dt^2} + \frac{1}{RC}\frac{d\Phi}{dt} + \frac{\Phi}{LC} = F(t)$$

where Φ can be the current in any of the branches, or the voltage acr circuit. The $F(t)$ depends upon the drive.

For the series circuit, Fig. 4.1*b*, the equation is

$$\frac{d^2\psi}{dt^2} + \frac{R}{L}\frac{d\psi}{dt} + \frac{\psi}{LC} = F(t)$$

where ψ is the voltage across any component or the current throu circuit. We note that the only difference between Eqs. 4.1.1 and in the coefficients of their second terms.

These coefficients are themselves interesting. To satisfy Eqs. 4.1 4.1.2 they must have the dimensions of T^{-1}. Accordingly we designa as follows:

$$\left.\begin{aligned} \text{parallel circuit time constant} &= RC = T_P \\ \text{series circuit time constant} &= \frac{L}{R} = T_S \end{aligned}\right\}$$

Observe that the product of these time constants is the square of the period of the undamped circuit:

$$T_P T_S = LC = T^2$$

If we define a parameter η as the ratio of the resistance R to t impedance, $Z_0 = (L/C)^{1/2}$ that is,

$$\eta = \frac{R}{Z_0} = R\left(\frac{C}{L}\right)^{1/2}$$

or

$$\frac{d^2 I_L}{dt^2} + \frac{1}{T_P}\frac{dI_L}{dt} + \frac{I_L}{T^2} = 0 \tag{4.2.3}$$

which accords with the assertion made in Eq. 4.1.1. Here $F(t)$ is zero because there is no external drive.

In the symbols of the Laplace transform, Eq. 4.2.3 may be written

$$\left(s^2 + \frac{s}{T_P} + \frac{1}{T^2}\right) i_L(s) = \left(s + \frac{1}{T_P}\right) I_L(0) + I'_L(0) \tag{4.2.4}$$

But $I_L(0) = 0$, and from Eq. 4.2.2, $I'_L(0) = V_C(0)/L$. With these substitutions in Eq. 4.2.4,

$$i_L(s) = \frac{V_C(0)}{L} \cdot \frac{1}{s^2 + s/T_P + 1/T^2} \tag{4.2.5}$$

Equation 4.2.5 contains the basic transform of the parallel RLC circuit. It, or variations of it, always appear in the operational solution when such a circuit is disturbed. The transform is evaluated in the following manner:

$$\frac{1}{s^2 + s/T_P + 1/T^2} = \frac{1}{(s - s_1)(s - s_2)}$$
$$= \frac{1}{(s_1 - s_2)}\left\{\frac{1}{s - s_1} - \frac{1}{s - s_2}\right\}$$

where s_1 and s_2 are the roots of $s^2 + s/T_P + 1/T^2 = 0$ and are given by

$$\left.\begin{aligned} s_1 &= -\frac{1}{2T_P} + \frac{1}{2}\left[\frac{1}{T_P{}^2} - \frac{4}{T^2}\right]^{1/2} \\ s_2 &= -\frac{1}{2T_P} - \frac{1}{2}\left[\frac{1}{T_P{}^2} - \frac{4}{T^2}\right]^{1/2} \end{aligned}\right\} \tag{4.2.6}$$

Thus

$$i_L(s) = \frac{V_C(0)}{L(1/T_P{}^2 - 4/T^2)^{1/2}}\left[\frac{1}{(s - s_1)} - \frac{1}{(s - s_2)}\right] \tag{4.2.7}$$

We recognize a simple transform here and can write the inverse transform readily:

$$I_L(t) = \frac{V_C(0)}{L(1/T_P{}^2 - 4/T^2)^{1/2}}\left[\epsilon^{+s_1 t} - \epsilon^{+s_2 t}\right] \tag{4.2.8}$$

The form of this solution will depend upon the values of s_1 and s_2. From Eq. 4.2.6, if $1/T_P{}^2 > 4/T^2$, then s_1 and s_2 are real. If, on the other hand, $1/T_P{}^2 < 4/T^2$, then s_1 and s_2 are complex. These conditions can be expressed in terms of the parameter η, defined in Eq. 4.1.5:

$$\eta = \frac{R}{Z_0} \tag{4.2.9}$$

Thus, if $1/T_P^2 > 4/T^2$, then $\eta < 1/2$; conversely, if $1/T_P^2 < 4/T^2$, then $\eta > 1/2$. Consider the case where the roots are complex. Rewriting Eq. 4.2.6 in terms of η,

$$\left.\begin{aligned} s_1 &= -\frac{1}{2T_P}[1 - j(4\eta^2 - 1)^{1/2}] \\ s_2 &= -\frac{1}{2T_P}[1 + j(4\eta^2 - 1)^{1/2}] \end{aligned}\right\} \tag{4.2.10}$$

So that for the condition where $\eta > 1/2$, the solution (4.2.8) can be written

$$I_L(t) = \frac{V_C(0)T_P\epsilon^{-t/2T_P}}{Lj(4\eta^2 - 1)^{1/2}}\left[\exp\frac{j(4\eta^2 - 1)^{1/2}t}{2T_P} - \exp\frac{-j(4\eta^2 - 1)^{1/2}t}{2T_P}\right]$$

or (4.2.11)

$$I_L(t) = \frac{V_C(0)}{L}\frac{2T_P\epsilon^{-t/2T_P}}{(4\eta^2 - 1)^{1/2}}\sin(4\eta^2 - 1)^{1/2}\frac{t}{2T_P}$$

For the inverse of the basic transform with which we started (Eq. 4.2.5), we can now write

$$\mathcal{L}^{-1}\frac{1}{s^2 + s/T_P + 1/T^2} = \frac{2T_P\epsilon^{-t/2T_P}}{(4\eta^2 - 1)^{1/2}}\sin(4\eta^2 - 1)^{1/2}\frac{t}{2T_P} \tag{4.2.12}$$

This looks rather formidable, but it will soon be apparent that it is not as bad as it would appear. There are two features about the expression that are worth noting. First, it involves only the parameter η and the time constant T_P. Second, wherever t appears it is divided by $2T_P$. We shall take advantage of these facts shortly. Meanwhile, consider the case in which s_1 and s_2 are real.

When $\eta < \frac{1}{2}$,

$$\left.\begin{aligned} s_1 &= -\frac{1}{2T_P}[1 - (1 - 4\eta^2)^{1/2}] \\ s_2 &= -\frac{1}{2T_P}[1 + (1 - 4\eta^2)^{1/2}] \end{aligned}\right\} \tag{4.2.13}$$

Therefore, the inverse transform is somewhat different:

$$\begin{aligned} \mathcal{L}^{-1}\frac{1}{s^2 + s/T_P + 1/T^2} &= \frac{2T_P\epsilon^{-t/2T_P}}{(1 - 4\eta^2)^{1/2}}\left[\exp\frac{(1 - 4\eta^2)^{1/2}t}{2T_P} - \exp\frac{(1 - 4\eta^2)^{1/2}t}{2T_P}\right] \\ &= \frac{2T_P\epsilon^{-t/2T_P}}{(1 - 4\eta^2)^{1/2}}\sinh(1 - 4\eta^2)^{1/2}\frac{t}{2T_P} \end{aligned} \tag{4.2.14}$$

The corresponding solution to the problem (Eq. 4.2.8) is obtained by multiplying Eq. 4.2.14 by $V_C(0)/L$.

There remains one other possibility, $\eta = 1/2$. The transform is then of slightly different form, $1/(s + 1/2T_P)^2$, which we have not evaluated. We can readily see what the result will be by considering the limit of Eq. 4.2.14 as $\eta \to 1/2$. Let $(1 - 4\eta^2)^{1/2} = a^{1/2}$, and let $t/2T_P = x$

$$\sinh a^{1/2}x = a^{1/2}x + \frac{a^{3/2}x}{\lfloor 3} + \frac{a^{5/2}x^5}{\lfloor 5} \cdots \frac{a^{(2n-1)/2}x^{2n-1}}{\lfloor 2n-1}$$

$$\frac{\sinh a^{1/2}x}{a^{1/2}} = x + \frac{ax^3}{\lfloor 3} + \frac{a^2x^5}{\lfloor 5} \cdots \frac{a^{n-1}x^{2n-1}}{\lfloor n-1} \tag{4.2.15}$$

as $\eta \to 1/2$, $a^{1/2} \to 0$. Therefore as $\eta \to 1/2$,

$$\frac{\sinh a^{1/2}x}{a^{1/2}} \to x$$

Thus

$$\mathcal{L}^{-1}\frac{1}{(s + 1/2T_P)^2} = 2T_P\epsilon^{-t/2T_P}\frac{t}{2T_P} \tag{4.2.16}$$

$$= t\epsilon^{-t/2T_P}$$

There are therefore three solutions for the current in the inductor of a parallel *RLC* circuit when the capacitor is discharged through the other two branches. The form of the solution depends upon the value of η. These solutions are collected together again here for convenience:

$$I_L(t) = \frac{V_C(0)}{L}\frac{2T_P\epsilon^{-t/2T_P}}{(4\eta^2 - 1)^{1/2}}\sin(4\eta^2 - 1)^{1/2}\frac{t}{2T_P} \tag{4.2.11}$$

$$I_L(t) = \frac{V_C(0)}{L}t\epsilon^{-t/2T_P} \tag{4.2.16}$$

$$I_L(t) = \frac{V_C(0)}{L}\frac{2T_P\epsilon^{-t/2T_P}}{(1 - 4\eta^2)^{1/2}}\sinh(1 - 4\eta^2)^{1/2}\frac{t}{2T_P} \tag{4.2.14}$$

Examples of each solution are shown in Fig. 4.2.

Suppose now that instead of the inductor current I_L we wish to find the capacitor voltage V_C. Then starting with the same Eqs. 4.2.1 and 4.2.2 we eliminate the current.

Differentiating Eq. 4.2.1:

$$\frac{d^2V_C}{dt^2} + \frac{1}{C}\frac{dI_L}{dt} + \frac{1}{RC}\frac{dV_C}{dt} = 0 \tag{4.2.17}$$

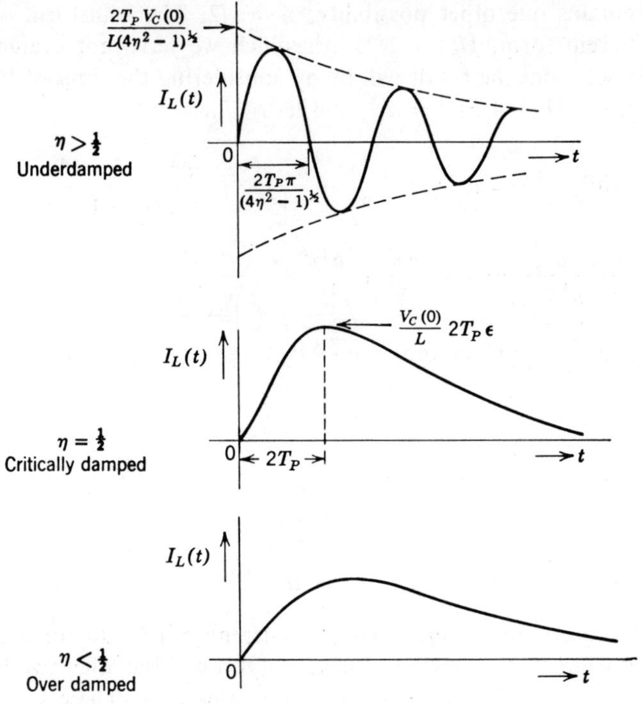

Fig. 4.2. The inductor current in a parallel *RLC* circuit for different degrees of damping.

Substitute for dI/dt from Eq. 4.2.2,

$$\frac{d^2V_C}{dt^2} + \frac{1}{RC}\frac{dV_C}{dt} + \frac{V_C}{LC} = 0$$

or

$$\frac{d^2V_C}{dt^2} + \frac{1}{T_P}\frac{dV_C}{dt} + \frac{V_C}{T^2} = 0 \tag{4.2.18}$$

This is identical in form to the equation for current. Transforming Eq. 4.2.18 gives

$$\left(s^2 + \frac{s}{T_P} + \frac{1}{T^2}\right)v_C(s) = \left(s + \frac{1}{T_P}\right)V_C(0) + V_C'(0) \tag{4.2.19}$$

It is evident from Eq. 4.2.1 that

$$V_C'(t) = -\frac{I_L(t)}{C} - \frac{V_C(t)}{RC}$$

Therefore

$$V_C'(0) = -\frac{I_L(0)}{C} - \frac{V_C(0)}{RC}$$

but $I_L(0) = 0$, so

$$V_C'(0) = -\frac{V_C(0)}{RC} = -\frac{V_C(0)}{T_P} \tag{4.2.20}$$

When this is substituted in Eq. 4.2.19, it cancels with part of the first term on the right and gives for the operational solution

$$v_C(s) = V_C(0)\frac{s}{(s^2 + s/T_P + 1/T^2)} \tag{4.2.21}$$

Equation 4.2.21 contains the second important transform associated with the parallel RLC circuit:

$$\frac{s}{(s^2 + s/T_P + 1/T^2)}$$

This is the transform of Eq. 4.2.1 multiplied by s. It is apparent from Eq. 2.2.12 that multiplying a transform by s has the effect of differentiating inverse transform:

$$\mathfrak{L}^{-1}\frac{s}{s^2 + s/T_P + 1/T^2} = \frac{d}{dt}\mathfrak{L}^{-1}\frac{1}{s^2 + s/T_P + 1/T^2} + F(0) \tag{4.2.22}$$

Equation 4.2.22 allows the transform to be evaluated at once from the first basic transform:

$$\left.\begin{aligned}
&\text{For } \eta > 1/2:\\
&\mathfrak{L}^{-1}\frac{s}{s^2 + s/T_P + 1/T^2}\\
&\quad = \epsilon^{-t/2T_P}\left\{\cos(4\eta^2 - 1)^{1/2}\frac{t}{2T_P} - \frac{\sin(4\eta^2 - 1)^{1/2}}{(4\eta^2 - 1)^{1/2}}\frac{t}{2T_P}\right\}\\
&\text{For } \eta = 1/2:\\
&\mathfrak{L}^{-1}\frac{s}{(s^2 + 1/2T_P)^2}\\
&\quad = \epsilon^{-t/2T_P}\left(1 - \frac{t_2}{2T_P}\right)\\
&\text{For } \eta < 1/2:\\
&\mathfrak{L}^{-1}\frac{s}{s^2 + s/T_P + 1/T^2}\\
&\quad = \epsilon^{-t/2T_P}\left\{\cosh(1 - 4\eta^2)^{1/2}\frac{t}{2T_P} - \frac{\sinh(1 - 4\eta^2)^{1/2}}{(1 - 4\eta^2)^{1/2}}\frac{t}{2T_P}\right\}
\end{aligned}\right\} \tag{4.2.23}$$

There is one other important transform that is frequently encountered with the parallel RLC circuit, and that is

$$\frac{1}{s(s^2 + s/T_P + 1/T^2)}$$

As an example of how this can arise, consider what happens if such a circuit is excited by a ramp of current, $I't$ as shown in Fig. 4.3. The current increases uniformly at a rate of I' amp/sec. We assume the circuit is initially dead, that is, there are no currents flowing in any branch and C is discharged.

After the ramp of current is applied, the currents in the three branches are:

In the resistor: $\dfrac{V}{R}$

In the inductor: $\dfrac{I}{L}\displaystyle\int V\,dt$

In the capacitor: $C\dfrac{dV}{dt}$

Their sum is the injected current $I't$:

$$\frac{V}{R} + C\frac{dV}{dt} + \frac{1}{L}\int V\,dt = I't$$

or, differentiating,

$$\frac{d^2V}{dt^2} + \frac{1}{T_P}\frac{dV}{dt} + \frac{V}{T^2} = \frac{I'}{C} \tag{4.2.24}$$

Taking transforms,

$$\left(s^2 + \frac{s}{T_P} + \frac{1}{T^2}\right)v(s) = \frac{I'}{sC} + \left(s + \frac{1}{T_P}\right)V(0) + V'(0) \tag{4.2.25}$$

Now, $V(0) = 0$, and $V'(0) = 0$ also, since $V'(0) = I_C(0)/C$ and initially there is no current in any branch of the circuit. Equation 4.2.25 therefore reduces to

$$v(s) = \frac{1}{s(s^2 + s/T_P + 1/T^2)}\frac{I'}{C} \tag{4.2.26}$$

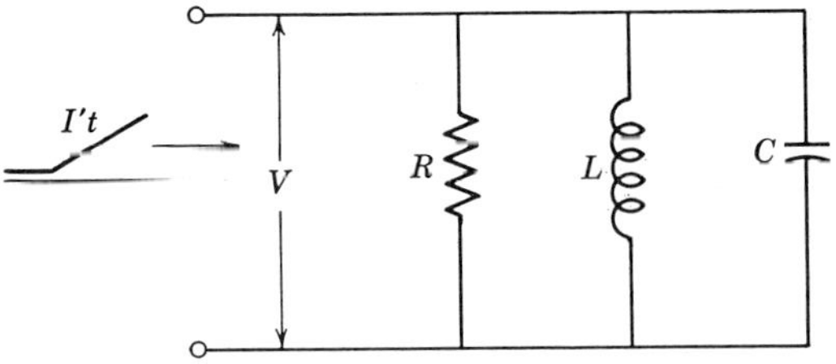

Fig. 4.3. Exciting an RLC circuit with a ramp of current.

This is the third transform we spoke of. This transform can be evaluated with the aid of Eq. 2.2.16, which says that to multiply a transform by $1/s$ is to integrate its inverse transform. So the inverse transform of Eq. 4.2.26 can be obtained by integrating Eqs. 4.2.12, 4.2.13, or 4.2.16, depending on the value of η. Alternatively, we can take partial fractions:

$$\frac{1}{s(s^2 + s/T_P + 1/T^2)} = T^2\left[\frac{1}{s} - \frac{s}{s^2 + s/T_P + 1/T^2} - \frac{1}{T_P(s^2 + s/T_P + 1/T^2)}\right] \quad (4.2.27)$$

The inverse transform of each term on the right has been obtained before. The result is as follows:

For $\eta > 1/2$:

$$\mathcal{L}^{-1}\frac{1}{s(s^2 + s/T_P + 1/T^2)} = T^2\left\{1 - \epsilon^{-t/2T_P}\left[\frac{\sin (4\eta^2 - 1)^{1/2}}{(4\eta^2 - 1)^{1/2}}\frac{t}{2T_P} + \cos (4\eta^2 - 1)^{1/2}\frac{t}{2T_P}\right]\right\}$$

For $\eta = 1/2$:

$$\mathcal{L}^{-1}\frac{1}{s(s + 1/2T_P)^2} = -2T_P\epsilon^{-t/2T_P}(t + 2T_P)$$

For $\eta < 1/2$:

$$\mathcal{L}^{-1}\frac{1}{s(s^2 + s/T_P + 1/T^2)} = T^2\left\{1 - \epsilon^{-t/2T_P}\left[\frac{\sinh (1 - 4\eta^2)^{1/2}}{(1 - 4\eta^2)^{1/2}}\frac{t}{2T_P} + \cosh (1 - 4\eta^2)^{1/2}\frac{t}{2T_P}\right]\right\}$$

(4.2.28)

We how have expression for the inverse transforms of the three basic transforms

$$\frac{1}{s^2 + s/T_P + 1/T^2}, \qquad \frac{s}{s^2 + s/T_P + 1/T^2}, \qquad \frac{1}{s(s^2 + s/T_P + 1/T^2)}$$

of the parallel *RLC* circuit. As predicted in Section 4.1, the expressions are complicated. To simplify we reduce the solutions to sets of dimensionless curves for different values of η.

4.3 The Generalized Damping Curves

The best approach to this subject is to consider a specific example. In Eq. 4.2.11 it was shown that when the capacitor in a parallel RLC circuit is discharged through the other two branches, the inductor current is given by

$$I_L(t) = \frac{V_C(0)}{L}\frac{2T_P\epsilon^{-t/2T_P}}{(4\eta^2 - 1)^{1/2}}\sin(4\eta^2 - 1)^{1/2}\frac{t}{2T_P} \qquad (\eta > 1/2) \quad (4.2.11)$$

We observe that wherever t appears, it appears in conjunction with the time constant T_P. If we let $t'/2\eta = t/2T_P$, so that

$$t' = \frac{t\eta}{T_P} = \frac{t}{T} \qquad (4.3.1)$$

where as before, T is the angular period $1/\omega_0$ of the undamped wave, and time is therefore measured in units of this amount.

Substituting Eq. 4.3.1 in Eq. 4.2.11 and observing that $T_P/L = \eta/Z_0$, we write

$$I_L(t) = \frac{V(0)}{Z_0}\frac{2\eta\epsilon^{-t'/2\eta}}{(4\eta^2 - 1)^{1/2}}\sin(4\eta^2 - 1)^{1/2}\frac{t'}{2\eta} \qquad (4.3.2)$$

Consider the situation where there is no damping. This condition is reached in the limit as $\eta \rightarrow \infty$. From Eq. 4.3.2 it is apparent that in this special case

$$I_L(t) = \frac{V(0)}{Z_0}\sin t' \qquad (4.3.3)$$

The peak amplitude of this sinusoid is $V(0)/Z_0$. Whenever there is damping, that is, wherever $\eta < \infty$, the peak will be less than $V(0)/Z_0$. We can use this peak of the undamped wave as a standard against which to compare any degree of damping, thereby devising a kind of per unit value as follows:

$$\text{per unit} \quad I_L(t) = \frac{I_L(t)}{V(0)/Z_0} = \frac{2\eta\epsilon^{-t'/2\eta}}{(4\eta^2 - 1)^{1/2}}\sin(4\eta^2 - 1)^{1/2}\frac{t'}{2\eta} \qquad (4.3.4)$$

This is for the underdamped current. Should $\eta < 1/2$,

$$\text{per unit} \quad I_L(t) = \frac{2\eta}{(1 - 4\eta^2)^{1/2}}\epsilon^{-t'/2\eta}\sinh(1 - 4\eta^2)^{1/2}\frac{t'}{2\eta} \qquad (4.3.5)$$

The only variables in these expressions are the parameters η and t', both of which are dimensionless. A family of curves can be drawn for this function

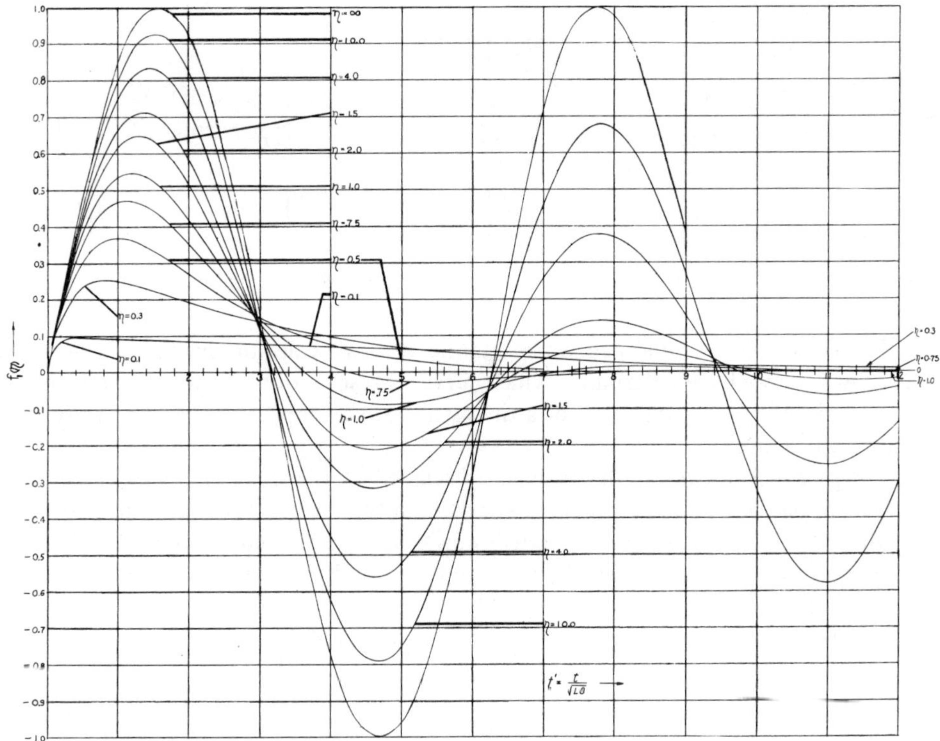

Fig. 4.4. Generalized plot of the inverse transform

$$\mathfrak{L}^{-1} \frac{1}{s + s/T_P + 1/T^2}$$

representing the function

$$f_1(\eta) = \frac{2\eta}{(4\eta^2 - 1)^{1/2}} \epsilon^{-t/2\eta} \sin (4\eta^2 - 1)^{1/2} \frac{t'}{2\eta}$$

where

$$\eta = \frac{R}{Z_0} = R\left(\frac{C}{L}\right)^{1/2} = \frac{1}{\lambda}$$

For series circuits, substitute λ for η.

for different values of η with t' as the independent variable (Fig. 4.4). To find the value of these curves, whenever we come upon the transform

$$\frac{1}{s^2 + s/T_P + 1/T^2}$$

and have to obtain its inverse transform, we ask ourselves what the value of the peak of the undamped wave and the value of η in this problem would be.

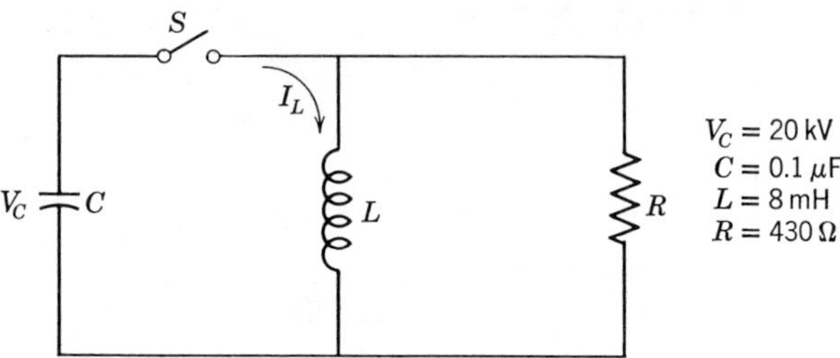

Fig. 4.5. Subsidence transient in a parallel RLC circuit.

The peak of the undamped wave sets the vertical scale for the curves, and knowing η we have determined which curve represents our solution. Equation 4.3.1 simplifies the obtaining of a time scale. An example will help to clarify this procedure.

Figure 4.5 illustrates the problem discussed at the beginning of Section 4.2. In that section an expression was derived for the current in the inductor when the switch S is closed. We will now make use of that analysis and the dimensionless curves of Fig. 4.4 to obtain a numerical solution for the values shown in Fig. 4.5.

The operational solution for I_L was given in Eq. 4.2.5:

$$i_L(s) = \frac{V_C(0)}{L} \cdot \frac{1}{s^2 + s/T_P + 1/T^2}$$

The structure of the transform tells us that our curves in Fig. 4.4 are appropriate. The first step is to calculate the surge impedance of the circuit:

$$Z_0 = \left(\frac{L}{C}\right)^{1/2} = \left[\frac{8 \times 10^{-3}}{10^{-7}}\right]^{1/2} = 284\ \Omega$$

Next, obtain η:

$$\frac{R}{Z_0} = \frac{430}{284} = 1.51$$

This means that the curve identified as $\eta = 1.5$ in Fig. 4.4 closely represents the form of the current I_L.

Without any damping the current peak would be

$$\frac{V_C(0)}{Z_0} = \frac{20{,}000}{284} = 70.5\ \text{A}$$

Because of the damping introduced by R, the curve for $\eta = 1.5$ peaks at only 0.65 of this value:

$$I_L\ (\text{peak}) = 70.5 \times 0.65 = 45.8\ \text{A}$$

To obtain a time scale we must find the angular period T:

$$T = (LC)^{1/2} = (8 \times 10^{-3} \times 10^{-7})^{1/2} = 28.4\ \mu\text{sec}$$

This means that each unit of t' in Fig. 4.4, *for this specific problem*, represents 28.4 μsec. This means that I_L reaches its first peak after approximately 38 μsec, and the frequency of the damped oscillation is approximately 5350 Hz.

The numerical answers to this problem were gleaned directly from the generalized curves, without recourse to evaluating or even reaching the expression for I_L derived in Eq. 4.2.11. In a very real sense the curves of Fig. 4.4 are a graphical representation of the inverse transform

$$\mathcal{L}^{-1} \frac{1}{s^2 + s/T_P + 1/T^2}$$

There is no need to evaluate the transform for each problem, since this has been done. We now go directly from the operational solution of a problem to the solution proper in graphical form.

In precisely the same manner, and with exactly the same substitutions (letting $t/2T_P = t'/2\eta$) that were applied in developing the curves of Fig. 4.4, the inverse transforms

$$\mathcal{L}^{-1} \frac{s}{s^2 + s/T_P + 1/T^2} \quad \text{and} \quad \mathcal{L}^{-1} \frac{1}{s(s^2 + s/T_P + 1/T^2)}$$

can also be represented by sets of dimensionless curves for different values of η. This has been done in Figs. 4.6 and 4.7. We shall make extensive use of all these curves in later chapters of this book. In the meantime we direct attention to the *RLC* circuit shown in Fig. 4.1*b*.

4.4 The Series *RLC* Circuit

At the beginning of this chapter it was asserted that the basic equations describing the behavior of parallel and series *RLC* circuits were identical in form, as given in Eqs. 4.1.1 and 4.1.2. An example will now be taken to demonstrate this point. Suppose a battery of voltage V is connected to a series *RLC* circuit in which the capacitor is initially discharged. What current will flow?

Let the unknown current be I. Adding the voltage around the circuit leads to the following equation:

$$IR + L\frac{dI}{dt} + \frac{1}{C}\int I\,dt = V \tag{4.4.1}$$

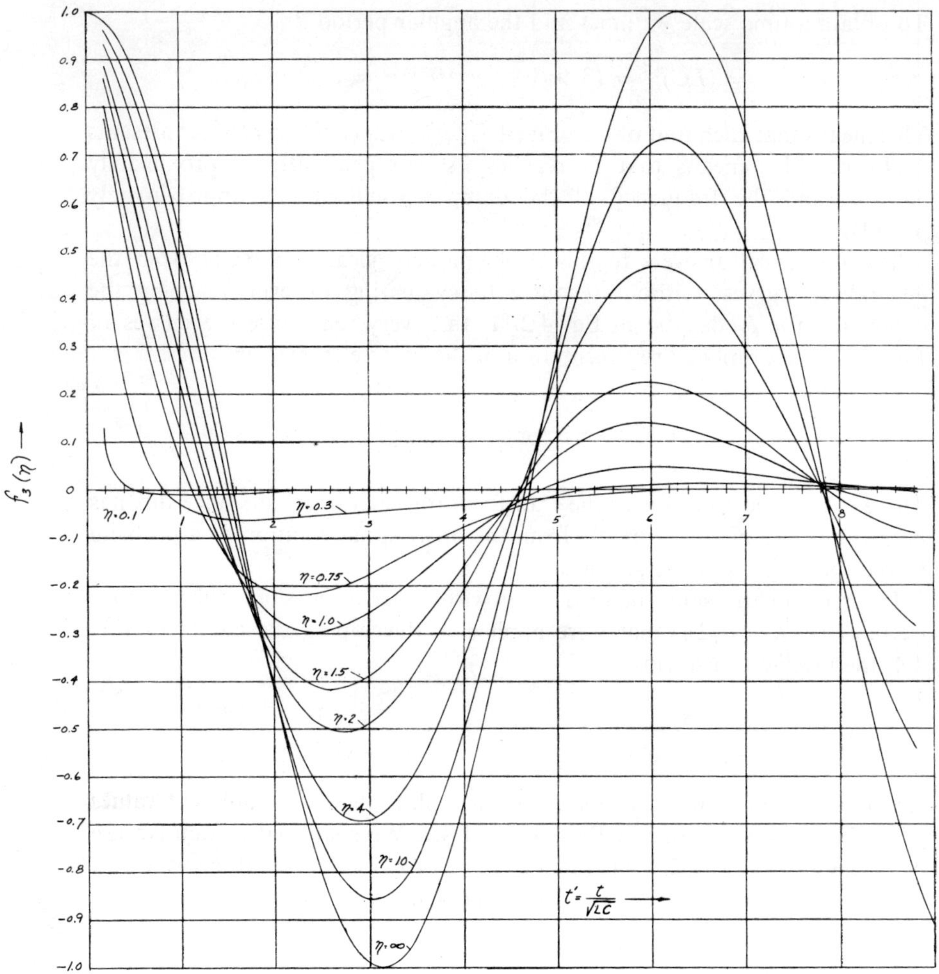

Fig. 4.6. Generalized plot of the inverse transform

$$\mathfrak{L}^{-1} \frac{s}{s^2 + s/T_P + 1/T^2}$$

representing the function

$$f_3(\eta) = \epsilon^{-t'/2\eta} \left\{ \cos (4\eta^2 - 1)^{1/2} \frac{t'}{2\eta} \frac{- \sin (4\eta^2 - 1)^{1/2} (t'/2\eta)}{(4\eta^2 - 1)^{1/2}} \right\}$$

where

$$\eta = \frac{R}{Z_0} = R\left(\frac{C}{L}\right)^{1/2} = \frac{1}{\lambda}$$

For series circuits, substitute λ for η.

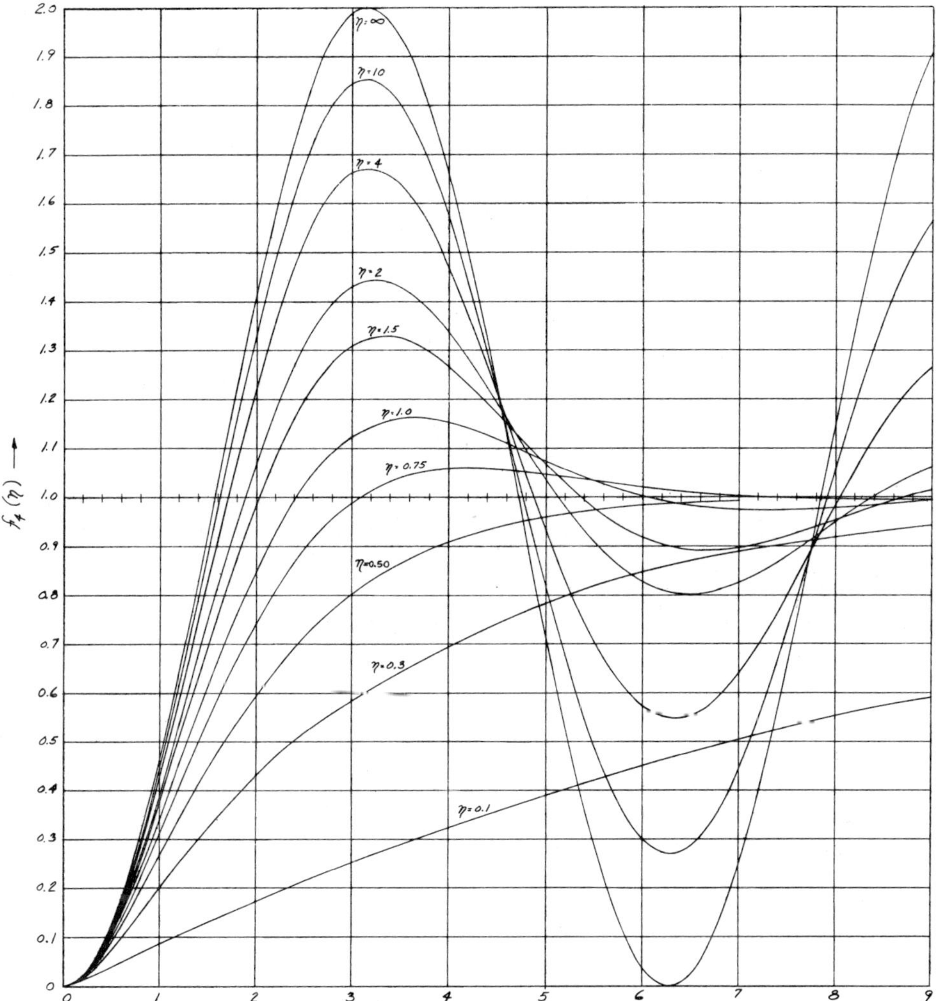

Fig. 4.7. Generalizing plot of the inverse transform

$$\mathfrak{L}^{-1}\frac{1}{s(s^2 + s/T_P + 1'/T^2)}$$

representing the function

$$f_4(\eta) = \left\{1 - \epsilon^{-t'/2\eta}\left[\frac{\sin (4\eta^2 - 1)^{1/2}(t'/2\eta)}{(4\eta^2 - 1)^{1/2}} + \cos (4\eta^2 - 1)^{1/2}\frac{t'}{2\eta}\right]\right\}$$

where

$$\eta = \frac{R}{Z_0} = R\left(\frac{C}{L}\right)^{1/2} = \frac{1}{\lambda}$$

for series circuits, substitute λ for η.

Differentiating Eq. 4.4.1 and rearranging,

$$\frac{d^2I}{dt^2} + \frac{R}{L}\frac{dI}{dt} + \frac{I}{LC} = 0 \tag{4.4.2}$$

This is indeed of the same form as Eq. 4.1.2. Introducing the series time constant $L/R = T_s$, Eq. 4.4.2 can be rewritten as

$$\frac{d^2I}{dt^2} + \frac{1}{T_S}\frac{dI}{dt} + \frac{I}{T^2} = 0 \tag{4.4.3}$$

Transforming this equation,

$$s^2i(s) - sI(0) - I'(0) + \frac{si(s)}{T_S} - \frac{I(0)}{T_S} + \frac{i(s)}{T^2} = 0$$

or

$$\left(s^2 + \frac{s}{T_S} + \frac{1}{T^2}\right)i(s) = \left(s + \frac{1}{T_S}\right)I(0) + I'(0) \tag{4.4.4}$$

Inasmuch as the circuit is inductive, the current must start from zero when the battery is connected, therefore $I(0) = 0$. Also, from physical reasoning, it is clear that $I'(0) = V/L$, since at the moment of closing the switch, there is no voltage on the capacitor, and $IR = 0$, therefore the only term in Eq. 4.4.1 is the second. Inserting this information into Eq. 4.4.4,

$$i(s) = \frac{V}{L}\cdot\frac{1}{s^2 + s/T_S + 1/T^2} \tag{4.4.5}$$

This is the same basic transform that we obtained from the parallel *RLC* circuit. The form is precisely the same, the difference in detail is the appearance of T_S rather than T_P in the coefficient of s. The inverse transform can be derived by the same steps that were used with the parallel circuit, and perhaps not surprisingly, the results are similar. For the oscillatory condition,

$$I = \frac{V}{L}\frac{2T_S}{(4\lambda^2 - 1)^{1/2}}\,\epsilon^{-t/2T_s}\sin\,(4\lambda^2 - 1)^{1/2}\frac{t}{2T_S} \tag{4.4.6}$$

This is the same as I_L in the parallel circuit with the substitutions of T_S for T_P and λ for η. The parameter λ is the reciprocal of η, that is,

$$\lambda = \frac{1}{\eta} = \frac{Z_0}{R} \tag{4.4.7}$$

Similar expression for the current can be developed for the critically damped and overdamped conditions:

$$I = \frac{V}{L}\,t\epsilon^{-t/2T_S} \qquad (\lambda = 1/2) \tag{4.4.8}$$

$$I = \frac{V}{L}\frac{2T_S}{(1 - 4\lambda^2)^{1/2}}\,\epsilon^{-t/2T_S}\sin\,(1 - 4\lambda^2)^{1/2}\frac{t}{2T_S} \qquad (\lambda < 1/2) \tag{4.4.9}$$

We note that for critical damping $\lambda = 1/2$ or $R = 2Z_0$. If $R < 2Z_0$ the circuit is oscillatory, if $R > 2Z_0$ the circuit is overdamped. Contrast this with the parallel circuit. Here for critical damping $\eta = 1/2$ or $R = Z_0/2$. The circuit is oscillatory if $R > Z_0/2$; it is overdamped if $R < Z_0/2$. This interesting duality extends much further. It is clear from the example just studied that

For $\lambda > 1/2$:

$$\mathcal{L}^{-1}\frac{1}{s^2 + s/T_S + 1/T^2} = \frac{2T_S\epsilon^{-t/2T_s}}{(4\lambda^2 - 1)^{1/2}} \sin (4\lambda^2 - 1)^{1/2}\frac{t}{2T_S}$$

For $\lambda = 1/2$:

$$\mathcal{L}^{-1}\frac{1}{(s + 1/2T_S)^2} = t\epsilon^{-t/2T_s} \qquad (4.4.10)$$

For $\lambda < 1/2$:

$$\mathcal{L}^{-1}\frac{1}{s^2 + s/T_S + 1/T^2} = \frac{2T_S\epsilon^{-t/2T_s}}{(1 - 4\lambda^2)^{1/2}} \sinh (1 - 4\lambda^2)^{1/2}\frac{t}{2T_S}$$

With the substitution of λ for η and T_S for T_P, these are identical to the corresponding inverse transforms for the parallel circuit set down in Eqs. 4.2.12, 4.2.14 and 4.2.16. *Consequently, making the same substitutions, the generalized curves of Fig. 4.4 developed for the parallel circuit can be used equally well for the series circuit.* By an extension of the same reasoning this is also true of the curves of Figs. 4.6 and 4.7. For the series case

$$\frac{t}{2T_S} = \frac{t'}{2\lambda}, \qquad \text{or} \quad t' = \frac{t\lambda}{T_S} = \frac{Z_0 t}{L} = \frac{t}{T}$$

That is, as before, time is measured in units equal to the period of the undamped wave.

We will now illustrate how the generalized damping curves are used to solve a practical power system problem.

4.5 Resistance Switching

In certain types of circuit breakers the contacts are shunted by resistors. Such resistors serve one of two functions (2). In a multibreak circuit breaker they may be used to help to distribute the transient recovery voltage more uniformly across the several breaks. Alternatively, their purpose is to reduce the severity of the transient recovery voltage at the time of interruption by introducing damping into the oscillation. A resistor of comparatively high ohmic value would suffice for the first of these duties. The only requirement would be that its resistance be low compared with the reactance of the capacitance shunting the breaks at the frequency of the recovery transient.

To reduce the transient recovery voltage requires a considerably lower value of resistor. Figure 4.8 shows a typical resistance switching circuit reduced to its barest essentials. In this figure, L represents the system inductance, C the stray capacitance shunting the breaker, and R the resistor used to modify the recovery transient. When the fault current proper has been switched, a residual current will remain flowing through R. This must be interrupted subsequently by opening the auxiliary interrupter S. We now show how to determine the value of R to achieve a desired modification of the transient recovery voltage in any particular situation.

It was explained in Section 1.5 and illustrated in Fig. 1.4 how the principle of superposition can be used to calculate transient recovery voltages. This is done by determining the response of the circuit to a current equal in magnitude but of opposite sign to the fault current, when it is injected into the circuit at the switch contacts. Viewed from the switch contacts, the circuit elements R, L, and C appear in parallel. Let us suppose that the fault current being interrupted is symmetrical. It will be given by $V/\omega L$. The problem thus reduces to that represented symbolically in Fig. 4.9. The transient period of interest is usually short compared with the time for a half cycle of the injected current wave. This is because the natural frequency of the circuit is usually much higher than the power frequency. For this reason, the injected current can be treated as a ramp, of slope V/L amp/sec or

$$I = \frac{V}{L}t \tag{4.5.1}$$

where V is the instantaneous system voltage at the moment of interruption. This is precisely the same circuit and the same stimulus as we examined in Section 4.2 and illustrated in Fig. 4.3. It was shown there that the operational solution for the switch voltage was given by

$$v(s) = \frac{1}{s[s^2 + s/T_P + 1/T^2]}\frac{I'}{C} \tag{4.2.26}$$

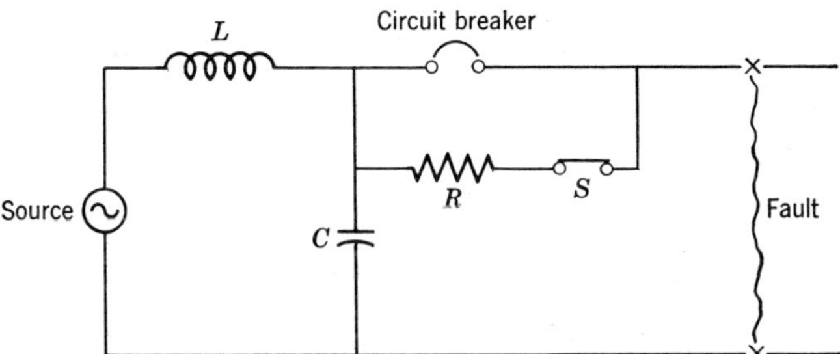

Fig. 4.8. Circuit breaker, with shunt resistor, clearing a fault.

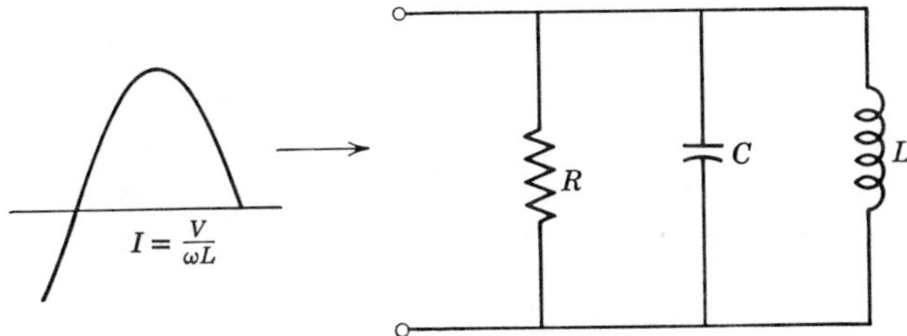

Fig. 4.9. Equivalent circuit for the resistance switching problem.

in which $I' = V/L$. The inverse transforms for this expression were given in Eq. 4.2.28 for different values of η, but, more important, their form was shown in the generalized curves of Fig. 4.7. The (1 − cosine) wave, for which $\eta = \infty$, is the unmodified, undamped recovery transient. The other curves show the effect of introducing various values of shunting resistor, which the different values of η imply. A specific example will make this clear. Consider a 345-kV, 25,000-MVA, 3-phase air blast circuit breaker. It is capable of interrupting approximately 40,000 A, which is usually accomplished by having six breaks or interrupters in series. Without damping of any kind, interrupting a symmetrical current, it was shown in Section 3.3 that the transient recovery voltage would reach a peak of twice the system voltage. In this example this would be $2 \times 345\sqrt{2}/\sqrt{3} = 562$ kV, for a system with grounded neutral. Suppose that it is desired to reduce the voltage to 70% of this value by means of shunt resistance. Figure 4.7 indicates that this will require a value of approximately 1.8 for η. Before this can be translated into a specific ohmic value for the resistor, we must know more about the circuit. A 345-kV system with an available fault current of 40,000 A has a 60 Hz reactance of $345{,}000/40{,}000\sqrt{3} = 5\ \Omega$, which represents an inductance of 13.2 mH. This then is the L in Fig. 4.9. A typical value for the capacitance C of a 345-kV bus to which the breaker is connected might be 25,000 pF. This gives, for the surge impedance of the bus,

$$Z_0 = \left(\frac{L}{C}\right)^{1/2} = \left[\frac{1.32 \times 10^{-2}}{2.5 \times 10^{-8}}\right]^{1/2} = 725\ \Omega$$

Now, if $\eta = 1.8$,

$$R = \eta Z_0 = 1.8 \times 725 = 1300\ \Omega$$

Thus, to achieve the required reduction of transient recovery voltage, each of the six interrupters in the circuit breaker should be shunted with 217 Ω. Figure 4.10 shows an actual resistor designed for this kind of duty. This has a lower ohmic value so that it would effect an even greater reduction in the transient voltage peak in the situation just examined. Note how the

Fig. 4.10. Low-inductance ribbon-wound shunt resistor for an air blast circuit breaker.

resistor is fashioned to reduce its inductance. Any inductance will decrease its effectiveness. When the resistor current is subsequently interrupted a second transient will be initiated. To study this it is necessary to introduce the capacitance C', shunting the resistor break, and, if it is significant, the inductance of the local loop to the fault. This is shown as an equivalent circuit in Fig. 4.11. This appears to be a more complicated circuit than we have analyzed so far, but it can be solved along similar lines.

4.6 Load Switching

The most frequent functions performed by some switching devices arc to switch on and switch off loads, which in many instances can be represented by a parallel *RL* circuit. Low power factor loads will be predominantly inductive, high power factor loads predominantly resistive. When such a load is switched off, the effective capacitance of the load becomes important in determining the form of the transient generated. This is illustrated in

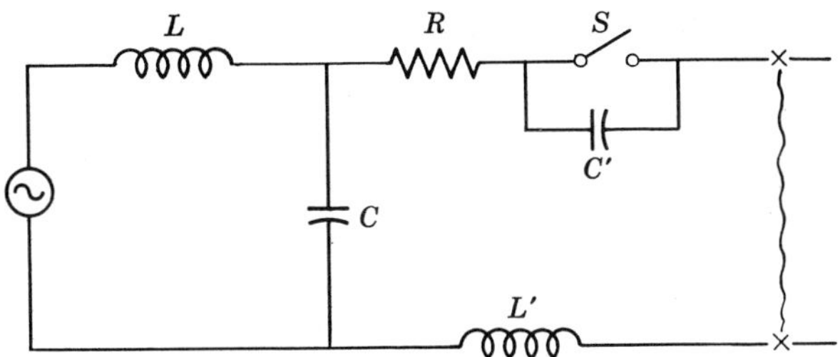

Fig. 4.11. Resistance switching—equivalent circuit for interrupting the resistor current.

Fig. 4.12. It is obvious that this resembles the resistance switching just described (cf. Fig. 4.9).

The load depicted in Fig. 4.12 has a relatively high power factor. When the current extinguishes, the instantaneous voltage, and therefore the voltage across the load, is V_0. Now C will be charged to this voltage and will subsequently discharge through L and R. In Fig. 4.12*b* this is shown as a damped oscillatory discharge and is in fact a damped cosine wave, the form of which can be extracted from Fig. 4.6 for the appropriate value of η.

The effect of power factor is interesting to observe. As the power factor improves, the current comes more and more into phase with the voltage, so that V_0 diminishes. At unity power factor voltage is zero when current is zero, so there is no transient at all. Thus power factor is the major controlling

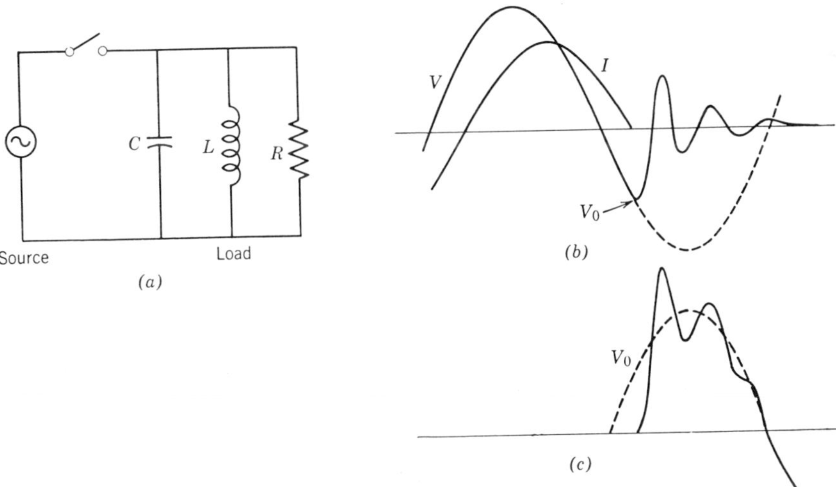

Fig. 4.12. Transient evoked when switching a load. (*a*) Simple equivalent circuit. (*b*) Transient voltage across the load. (*c*) Transient voltage across the switch.

influence in the magnitude of this switching transient. Damping further reduces the potential peak voltage, but C as well as R and L is important here, because all three determine the value of η.

4.7 Other Forms of Damping

In this chapter we have treated damping as if it could be represented by a linear resistor connected either in parallel or in series with the L and C branches of a circuit. This is an adequate representation for many practical situations, for although the energy may be dissipating in a piece of neighboring steel work, as a consequence of eddy currents induced therein by the magnetic field of the circuit, it is coupled into the circuit somewhat like a resistive load.

There are instances where such a treatment is insufficient. These occur especially with fast transients, where the time for flux to penetrate adjacent structures, or for that matter the circuit conductors themselves, is not insignificant on the time scale of the transients. Problems of this kind are more difficult to handle. They are given special treatment as a part of Chapter 8. A practical example of this is the penetration of traveling waves into the ground. This is further discussed in Chapter 9.

Another form of damping, one also associated with transmission lines and one that cannot be handled by the techniques of this Chapter, is corona. Energy is extracted from the circuit in this case by currents flowing out from the conductor in small discharges into the surrounding air. This too is discussed in Chapter 9.

REFERENCES

1. A. N. Greenwood and T. H. Lee, "Generalized Damping Curves and Their Use in Solving Power Switching Transients," *Trans, IEEE*, Vol. 82, Part III (1963), p. 527.
2. H. E. Cox and T. W. Wilcox, "The Performance of High Voltage Oil Circuit Breakers Incorporating Resistance Switching," *JIEE*, Vol. 94, Part II (1947) p. 351.

5 Abnormal Switching Transients

5.1 Normal and Abnormal Switching Transients

We have seen how, when a switch opens in a single-phase circuit, it is possible for the recovery voltage to reach a value twice as high as the normal peak voltage of the system. Similarly, we have seen that when a switch closes, if the point of closing occurs at a voltage zero, the peak current can reach a value twice that of the eventual steady-state current. These are considered normal current and voltage transients. In fact, as has been pointed out, currents and voltages of this theoretical magnitude are not achieved in practical circuits because of circuit damping that is always present. However, there are other circumstances in which voltages and current far in excess of these values can arise. Such transients are, by our definition, referred to as abnormal current and voltage transients. There are several possible ways in which these abnormal disturbances can come about, but they all have one thing in common: they all involve the trapping of energy somewhere in the circuit and its subsequent release. This could be due to charge on a capacitor or line and/or current in an inductor. It follows that if a circuit is completely quiescent when a transient is initiated, the transient will be a normal transient. Of course it could be that this transient stores energy in the system so that subsequently when a second transient is initiated, it would be abnormal.

We have seen in earlier chapters how the initial condition of a circuit enters into the analysis of a transient through such terms as $I_1(0)$, $V_2(0)$, and $I_2'(0)$. When any of these terms is finite, the possibility exists that an abnormal transient will develop. Abnormal transients have considerable practical significance.

5.2 Current Suppression

When a relatively small current is interrupted by a circuit breaker, the overzealous action of its arc suppression devices may cause the current to be

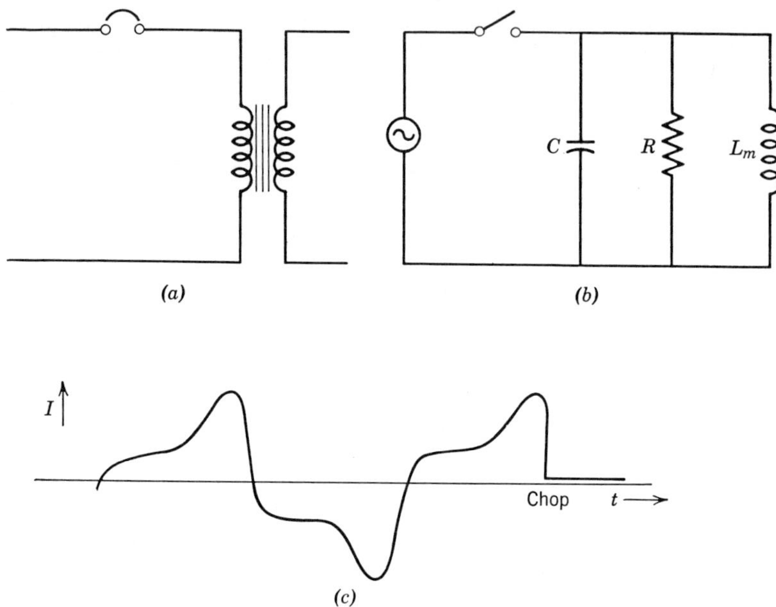

Fig. 5.1. Current chopping. (*a*) The circuit breaker and an unloaded transformer. (*b*) The effective equivalent circuit. (*c*) The current.

brought to zero abruptly and prematurely ahead of the normal zero. This is referred to as "current chopping" and is a form of current suppression. It can give rise to an abnormal overvoltage by virtue of the magnetic energy associated with the current being trapped in the circuit. This phenomenon often is observed when the no-load or magnetizing current of a transformer is being switched. How the overvoltage is generated can be understood by referring to Fig. 5.1.

Let us suppose that at the time the chop occurs the instantaneous current is I_0. This is flowing in the transformer winding and is associated with a certain amount of magnetic energy, most of which resides in the transformer core:

$$\text{energy} = \tfrac{1}{2} L_m I_0^2$$

This may be considerable, for although I is only about 1% of the normal full load current, the magnetizing inductance L_m is quite high. The current cannot cease suddenly in such an inductive circuit, yet it has no complete path through the switch. It is therefore diverted into the system capacitance on the transformer side of the switch, which is designated C in Fig. 5.1. This consists primarily of the transformer winding capacitance, about which much more will be said in Chapter 11, together with any capacitance that may be in the connections between the switch and the transformer. When the current is diverted into this capacitance, the energy of the magnetic field of the transformer is transferred to the electric field of the capacitance. If this capacitance is known, it is possible to calculate the voltage to which C will be charged:

$$\tfrac{1}{2}CV^2 = \tfrac{1}{2}L_m I_0^2$$

or

$$V = I_0 \left(\frac{L_m}{C}\right)^{1/2} \qquad (5.2.1)$$

This states that the peak voltage reached across the capacitor, and therefore across the winding, is given by the product of the instantaneous current chopped and the surge impedance of the transformer. A striking fact about Eq. 5.2.1 is that the transient voltage is independent of the system voltage. To put a figure on this problem, consider a 1000-kVA, 13.8-kV transformer, of the kind found in substations of industrial plants. The magnetizing current is typically 1.5 A; thus

$$L_m = \frac{V}{\omega I_m} = \frac{13{,}800}{\sqrt{3} \times 377 \times 1.5} = 14\ H$$

The effective capacitance will vary depending on the type of winding and the insulation, whether oil, air, or askerel, but would be in the range 1000–7000 pF. Suppose we choose 5000 pF. Then

$$Z_0 = \left(\frac{L_m}{C}\right)^{1/2} = \left(\frac{14}{5 \times 10^{-9}}\right)^{1/2} = 54{,}000\ \Omega$$

If the circuit breaker chops the peak current, which because of harmonic distortion might be 2.5A, the theoretical transient voltage peak would reach 135kV. This is indeed an abnormal overvoltage for a 13.8-kV system.

In practice the voltage would not reach nearly as high a value as this, partly because of losses causing damping, but, more important, because only a fraction of the energy trapped in the core at the time of the chop is released. This can be understood from Fig. 5.2. While the transformer is energized, the core is being taken through a hysteresis cycle at the frequency of the supply. Energy is being put into the core during the intervals QX and ZP in Fig. 5.2 and returned to the source during intervals XZ and PQ. Energy proportional to the area of the hysteresis loop is expended in hysteresis losses each cycle. When the current is at its peak, point X, energy proportional to the triangular area OXY is stored in the iron. As the current declines to zero at the point Z, energy proportional to the shaded area XYZ is retrieved; the remainder is spent. If the current is chopped at the peak and is forced to divert

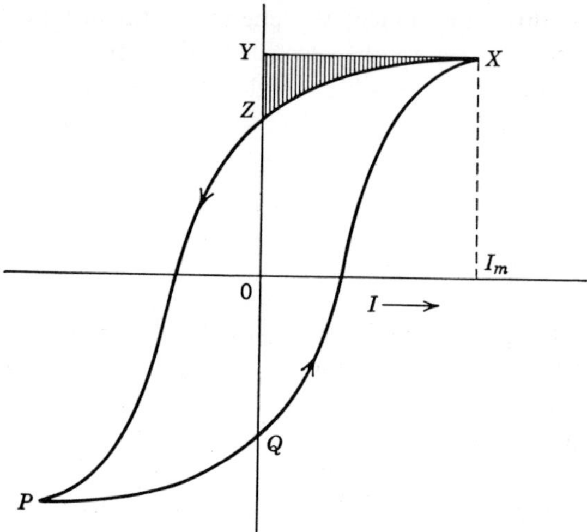

Fig. 5.2. Energy released by a transformer core when the magnetizing current is chopped.

into the capacitance, it follows approximately the same route X to Z and it is the energy represented by ZYX that charges the capacitor. As mentioned earlier, losses other than hysteresis reduce this still further. In modern grain-oriented steels, the shaded area represents 40% or less of the triangular area OXY. Thus at worst the chopping voltage will not exceed $I_0(0.4L_m/C)^{1/2}$ which is 64% of what we calculated before. It has been pointed out (1) that the potential danger from current chopping tends to diminish as the voltage class of the transformer increases.

Air-cored reactors, or reactors with significant air gaps do not behave in this way. Practically all their energy is recoverable. Reactors of this kind are sometimes used as a shunt to ground on long transmission lines to compensate for the line's capacitance. In such applications the equipment is usually protected by a lightning arrester, which can operate to limit any overvoltage that may occur due to current chopping when the reactor is switched out.

It is worthwhile to examine the current chopping transient more formally because it is a good exercise in the use of the generalized damping curves. Consider the circuit shown in Fig. 5.1*b*. Suppose the currents in the CRL branches are I_C, I_R, and I_L, respectively, and let the value of the current chopped be I_0. Immediately after the chop occurs there is no path through the switch, so that thereafter the sum of the currents in the three branches must be zero:

$$I_C + I_R + I_L = 0 \tag{5.2.2}$$

Now, each of these currents can be expressed in terms of the voltage across the transformer winding, V. Thus

$$\left.\begin{aligned} I_C &= C\frac{dV}{dt} \\ I_R &= \frac{V}{R} \\ I_L &= \frac{1}{L_m}\int V\,dt \end{aligned}\right\} \tag{5.2.3}$$

Substituting Eq. 5.2.3 in Eq. 5.2.2, differentiating, and rearranging gives the familiar form

$$\frac{d^2V}{dt^2} + \frac{1}{RC}\frac{dV}{dt} + \frac{V}{L_mC} = 0 \tag{5.2.4}$$

Transforming Eq. 5.2.4,

$$v(s)\left[s^2 + \frac{s}{RC} + \frac{1}{L_mC}\right] = \left(s + \frac{1}{RC}\right)V(0) + V'(0) \tag{5.2.5}$$

where, following our former notation, $V(0)$ is the value of voltage when the switch chops and $V'(0)$ is its rate of change at that time. A little consideration reveals that

$$V'(0) = \frac{I_C(0)}{C} = \frac{I_0}{C} \tag{5.2.6}$$

since at the moment the switch chops the chopped current must be diverted into the capacitor. Combining Eqs. 5.2.6 and 5.2.5 leads to

$$v(s) = \frac{sV(0)}{s^2 + s/RC + 1/L_mC} + \frac{V(0)}{RC}\frac{1}{(s^2 + s/RC + 1/L_mC)} + \frac{I_0}{C(s^2 + s/RC + 1/L_mC)} \tag{5.2.7}$$

We have two of the transforms here that we studied in Chapter 4, where it was asserted that they would be encountered quite frequently. The inverse transforms are given in Eqs. 4.2.12 and 4.2.13, and their form is shown in the generalized damping curves of Fig. 4.4 and 4.6.

The first two terms of Eq. 5.2.7 taken together give a damped cosine wave:

$$V(0)\epsilon^{-t/2RC}\cos\,(4\eta^2 - 1)^{1/2}\frac{t}{2RC}$$

where $\eta = R/Z_0$. This is the normal transient that would occur if the transformer was disconnected from the supply with no current chopping. It represents the subsidence transient of the transformer's capacitance discharging through its magnetizing inductance. The second term is a direct consequence of the current chopping and is the one potentially capable of creating an abnormal overvoltage. Note how it arose as a consequence of $V'(0)$ or $I(0)$ having a finite value in Eq. 5.2.5. This term is

$$\frac{2I_0R}{(4\eta^2-1)^{1/2}}\epsilon^{-t/2RC}\sin(4\eta^2-1)^{1/2}\frac{t}{2RC}$$

As $\eta(=R/Z_0)$ tends to infinity (the undamped case), this expression reduces to $I_0Z_0 \sin t/(L_mC)^{1/2}$, which has the amplitude predicted by Eq. 5.2.1. The transformer oscillates at its natural frequency and the amplitude of this disturbance for any degree of damping can be seen at once from Fig. 4.4 by entering the appropriate value of η. A practical example of the current chopping transient is shown in Fig. 5.3.

We have supposed that the current was chopped instantaneously or, put another way, that the chop was vertical. This is often the case in a circuit breaker. However, current-limiting fuses are known to advance the current zero when they blow, but this is more in the nature of a suppression than a chop. The current declines over a readily measurable time. It can be shown, as we would expect, that the voltage developed by such a nonvertical chop is less than that for the instantaneous chop. How much less depends upon the rate of decline or, more specifically, on the time to accomplish the chop, compared with the natural period T of the oscillation excited. A curve illustrating this point is given in Fig. 5.4. It will be seen that for $0 < t_C < T/4$, there is little diminution but as t_C exceeds $T/4$, the transient voltage peak declines rapidly.

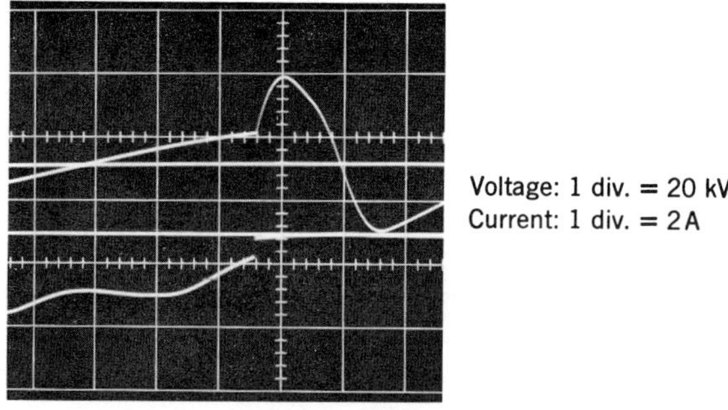

Fig. 5.3. Transient voltage evoked by chopping the magnetizing current of a 13.8-kV transformer.

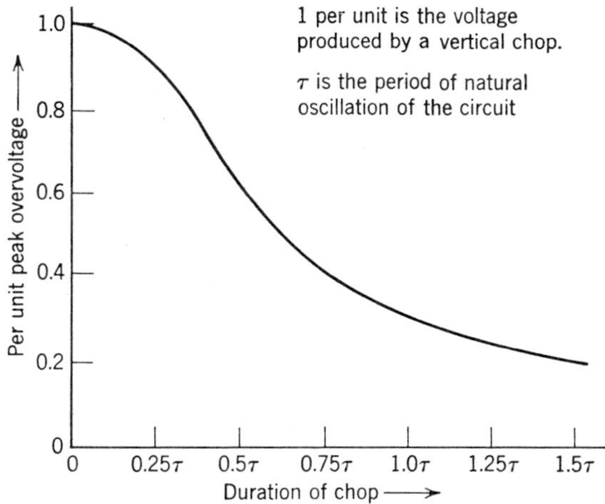

Fig. 5.4. Diminution of overvoltage when current suppression takes a finite time.

Quite often a circuit breaker will chop the current, if the current is small, soon after its contacts part. With only a small contact separation the dielectric is unable to support the transient voltage developed and the switch reignites. Sometimes the switch will make successive attempts to clear the circuit in this manner, reaching progressively higher voltages as the contact gap increases, until isolation is finally established. This phenomenon has been described by Young (2). An example is shown in Fig. 5.5. The switch in question was a vacuum interrupter. The maximum voltage attained when this occurs is not as great as when the recovery is clean. This is because the stored energy of the transformer core is not allowed to accumulate to any degree in the capacitance before a reignition occurs, which pitches some of the energy into the source and dissipates the remainder in the reignition. Should a

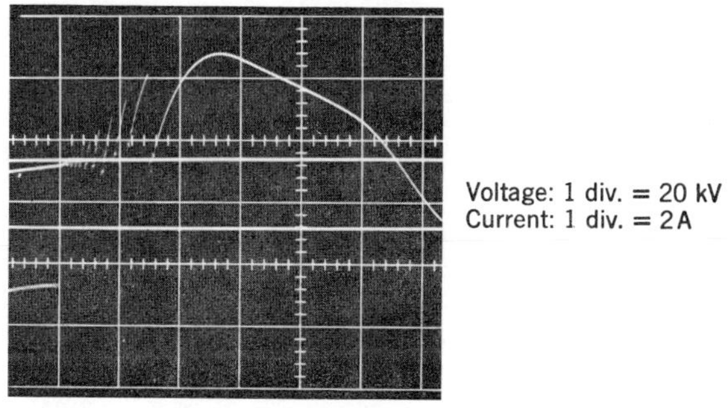

Fig. 5.5. Current-chopping transient limited by reignitions.

magnetizing inrush current (Section 5.5) be interrupted, this self-protecting feature of the breaker may not be operative, since it is likely to pause until near current zero before chopping. By this time the contact separation may no longer be small. This kind of switching should therefore be avoided if possible, with breakers prone to chop.

Although current chopping is a potential hazard, practical considerations of power circuits often remove the danger. For example, it is not unusual in industrial power systems to find a significant length of cable between the circuit breaker and the transformer it is switching. This can greatly reduce the surge impedance of the circuit and so reduce the overvoltage from a given current chop. A 15-kV class of power cable typically has a capacitance of the order of 100 pF/ft. Thus a transformer with an effective winding capacitance of 3000 pF, connected by 100 ft of cable, will have a surge impedance of only half that of the transformer alone.

Motors can also be subjected to overvoltages due to current chopping, but generally their surge impedance is lower than a transformer's of corresponding size. This is because the inductance is lower and the capacitance is higher than for the transformer. The voltage for a given current chop is therefore smaller, but this is offset in many cases by the insulation strength of the motor being inferior to that of the transformer.

Power circuit breakers are not the only devices that suppress current. The simple household mercury switch does so, and in inductive circuits, perhaps where a lamp ballast is involved, the surge generated can be considerable. In the contactor and ignition coil of an automobile ignition system we actually make use of high voltage to fire the spark plugs.

Mercury arc rectifiers are known to suppress current on occasions. It is believed that this is a consequence of "arc starvation," which results in the current path through the mercury plasma suddenly assuming a high impedance (3). The effect is somewhat like the hammer-knock in a hydraulic system when a valve is suddenly closed. In the electric circuit the different components in the path of the current are set into oscillation at their natural frequencies and with an amplitude proportional to their respective surge impedances. These disturbances can be observed oscilloscopically across the components involved, or, by connecting an oscilloscope across the offending mercury arc tube, the total response of the system can be recorded. This, of course, is the sum of the individual responses. Such a response, artificially induced on the rectifier supply to an aluminum smelter, is described in reference (4).

We are inclined to think of semiconductor devices as being the victims of voltage surges rather than the originators of them. This is not always true. Some silicon diodes can create considerable overvoltages by current chopping, sufficient, for example, to destroy themselves. How this comes about is as

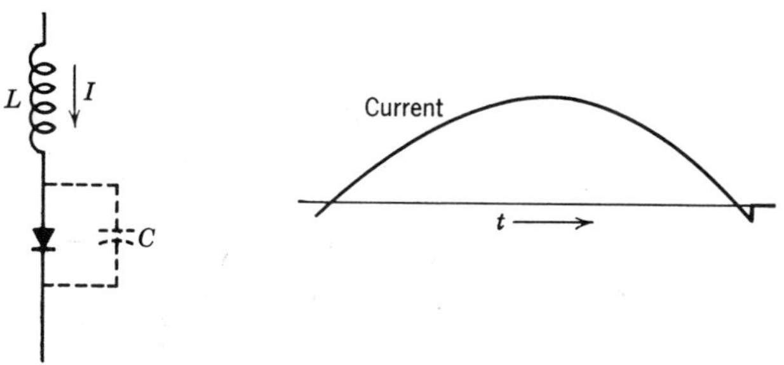

Fig. 5.6. Current suppression in a silicon diode.

follows (refer to Fig. 5.6). At the end of a half cycle of conduction in a diode, carriers remaining in the junction region continue to allow current to flow, that is, the current momentarily reverses. But the flow of this reverse current sweeps the carriers from the junction and returns the device to its blocking state. This change of impedance can be very rapid, so that this so-called "clean-up current" collapses very quickly. If the circuit is inductive the stored magnetic energy is transferred to the relatively small capacitance of the diode junction, across which it develops a considerable voltage (5). What is seen is a high-frequency oscillation involving L and C. To prevent the diode destroying itself a small "snubber" capacitor can be placed across the device to provide an additional sink for the energy.

5.3 Capacitance Switching

The switching of capacitance, as when a long open circuited line or cable is dropped, or when a capacitor bank is disconnected, can present some potentially hazardous conditions and has traditionally been a source of considerable chagrin to the switchgear engineer (6, 7). Figure 5.7 depicts events occurring before and after such a switching operation, which in this case was performed successfully. Because of the relative phase of current and voltage (current leads the voltage by approximately 90°) the capacitor is fully charged to maximum voltage when the switch interrupts. The capacitance, now isolated from the source, retains its charge as shown in Fig. 5.7*b*. As a consequence of this trapping of the charge, it can be seen from Fig. 5.7*c* that half a cycle after current zero the voltage across the switch reaches a peak value of 2 V, which is potentially dangerous. Figure 5.7 tends to oversimplify conditions to some extent in that when a capacitor is connected to a system, the leading current that it draws, flowing through the inductance of the system, causes the capacitor voltage to be somewhat higher than the open-circuit system voltage, a negative regulation sometimes referred to as the "Ferranti Rise." When the capacitor is disconnected, the potential of the

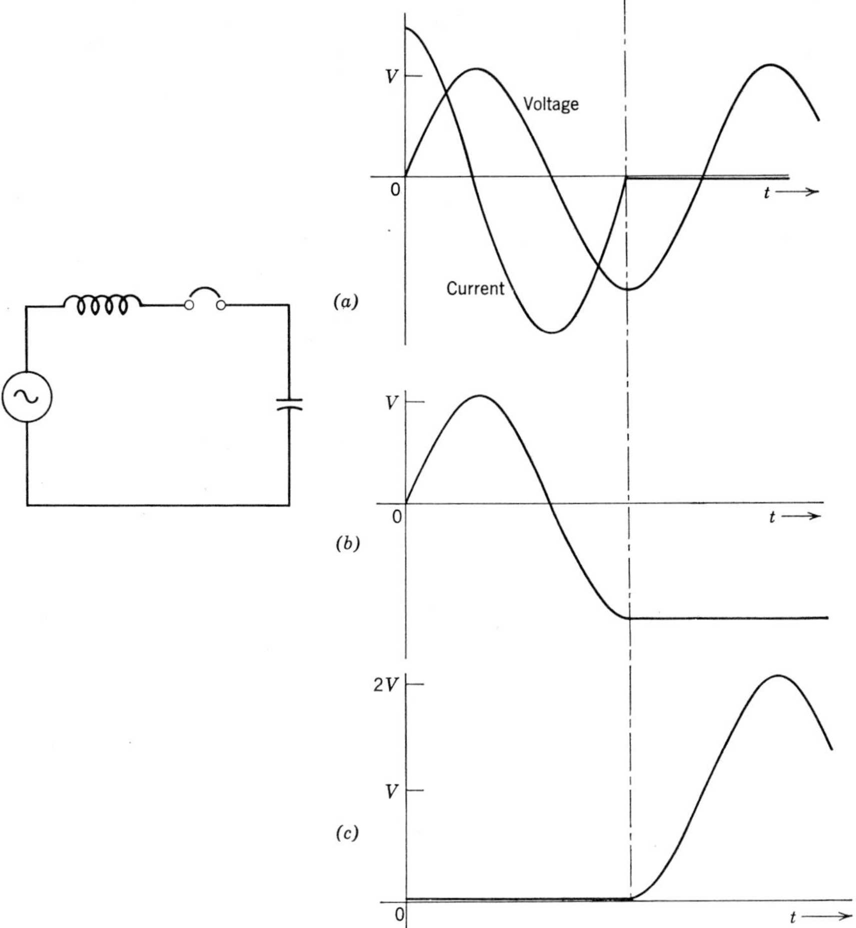

Fig. 5.7. Capacitance switching. (*a*) System voltage and current. (*b*) Capacitor voltage. (*c*) Voltage across the switch.

source side of the circuit breaker will return to this lower value, but will do so by way of an oscillation involving the source inductance and the stray capacitance adjacent to the breaker on the source side. A more accurate representation of the disconnecting event is shown in Fig. 5.8. Here ΔV is the aforementioned negative regulation. In order to simplify the ensuing analysis we will choose to ignore this, recognizing, however, that it does exist and can be important on relatively weak systems.

Many circuit breakers, when called upon to interrupt a load or fault current, do not do so at the first current zero, but instead wait until sufficient gap has been established between their contacts for their various arc-extinguishing agencies to have a better chance of operating successfully. The current involved in capacitance switching is frequently small, so that more often than not the circuit breaker is capable of interrupting it at the first current zero. If this should occur soon after the contacts have parted, the voltage of 2 V will appear across the contacts while their separation is small, so there is an increased likelihood of the device reigniting. Let us suppose that a reignition takes place precisely when the voltage reaches its peak, which is tantamount to reclosing the switch at that instant. This is an *LC* circuit, so we would expect it to respond to this sudden disturbance by going into an oscillation at its natural frequency, which is

$$f_0 = \frac{\omega_0}{2\pi} = \frac{1}{2\pi(LC)^{1/2}}$$

where L is the inductance of the supply and C the capacitance of the bank. The circuit is similar to the one shown in Fig. 3.4, so the basic equation of the circuit is the same. Differences in the solutions to the two problems are due strictly to the different initial conditions. We note specifically that this circuit is not quiescent, in that at the time of restrike the capacitor has a charge. The equation for the transient current following the restrike is

$$V_m \cos \omega t - V_C = L \frac{dI}{dt} \tag{5.3.1}$$

but

$$V_C = V_C(0) + \frac{1}{C} \int I \, dt \tag{5.3.2}$$

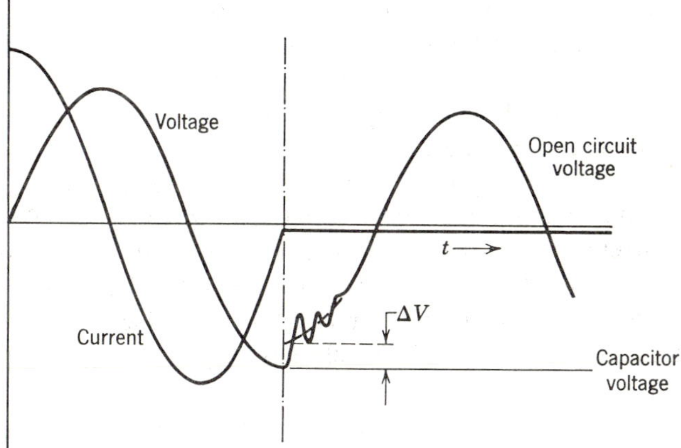

Fig. 5.8. Capacitance switching showing the effect of source regulation.

Substituting Eq. 5.3.2 in Eq. 5.3.1 and assuming that during the period of interest the supply voltage remains essentially constant at its peak value, we may write

$$L\frac{dI}{dt} + \frac{1}{C}\int I\,dt = V_m - V_C(0) \tag{5.3.3}$$

This transforms to

$$Lsi(s) - LI(0) + \frac{i(s)}{sC} = \frac{V_m - V_C(0)}{L} \tag{5.3.4}$$

But $I(0) = 0$, so that Eq. 5.3.4 can be rearranged to read

$$\begin{aligned}(s^2 + \omega_0^{\,2})i(s) &= \frac{V_m - V_C(0)}{L}\\ i(s) &= \frac{V_m - V_C(0)}{L(s^2 + \omega_0^{\,2})}\end{aligned} \tag{5.3.5}$$

The inverse transform of Eq. 5.3.5 is

$$\begin{aligned}I(t) &= \frac{V_m - V_C(0)}{L\omega_0}\sin\omega_0 t\\ &= [V_m - V_C(0)]\left(\frac{C}{L}\right)^{1/2}\sin\omega_0 t\end{aligned} \tag{5.3.6}$$

The quantity $[V_m - V_C(0)]$ represents the voltage across the circuit breaker when it restrikes. In the problem being considered this is $2V_m$. The quantity $(L/C)^{1/2}$ is, of course, the surge impedance of the circuit. Thus the transient current is sinusoidal; it oscillates as we expected and is equal to the original voltage across the switch divided by the surge impedance.

It is illuminating to insert a few figures to see the magnitude of the current we are talking about. Suppose $C = 60\ \mu F$ and the system voltage is 15 kV. This corresponds to a 3-phase capacitor bank approximately 5000 kVA. The capacitive current to be switched when such a bank is disconnected is 200 A. Let us suppose that the source inductance is 1 mH. If a reignition occurs when the voltage across the breaker is twice peak line-to-neutral voltage, the current according to Eq. 5.3.6 would have a peak of

$$\begin{aligned}2V_P\left(\frac{C}{L}\right)^{1/2} &= \frac{2 \times 15{,}000\sqrt{2}}{\sqrt{3}}\left(\frac{6 \times 10^{-5}}{10^{-3}}\right)^{1/2}\\ &= 6000\ \text{A}\end{aligned}$$

which we note is many times the normal 60 Hz current. The frequency of the restrike current would be

$$f_0 = \frac{1}{2\pi(LC)^{1/2}} = \frac{1}{2\pi(6 \times 10^{-5} \times 10^{-3})^{1/2}}$$
$$= 650 \text{ Hz}$$

A practical circuit would have some damping, so that these calculated figures would be modified to some extent. We could determine this modification from the damping curves of Chapter 4 given the value for the circuit. As this current rushes into the capacitor we can calculate the capacitor voltage from Eqs. 5.3.2 and 5.3.6:

$$V_C = V_C(0) + \frac{1}{C}\int_0^t [V_m - V_C(0)]\left(\frac{C}{L}\right)^{1/2} \sin \omega_0 t \, dt \qquad (5.3.7)$$

If the voltage trapped on the capacitor, $V_C(0)$, is $-V_m$ as in Fig. 5.7*b* and the restrike occurs when the system voltage is at $+V_m$, Eq. 5.3.7 can be rewritten as

$$V_C = -V_m + \frac{2V_m}{(LC)^{1/2}}\int_0^t \sin \omega_0 t \, dt$$
$$= -V_m + 2V_m(1 - \cos \omega_0 t) \qquad (5.3.8)$$

This has a maximum value of $+3V_m$. The initial clearing, the trapping of the charge on the capacitor, and the subsequent restrike are shown in Fig. 5.9, which shows that this is not necessarily the end of the problem.

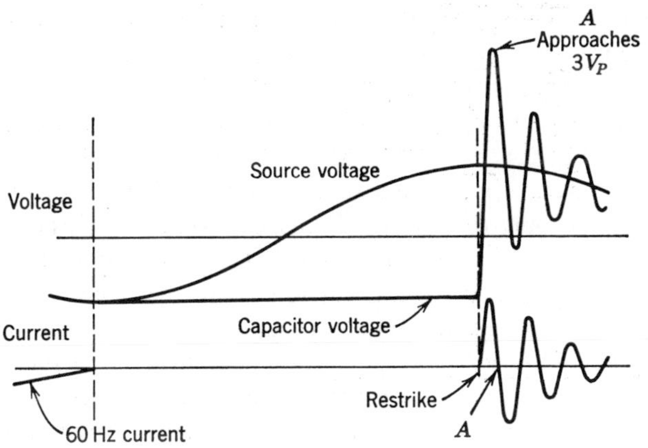

Fig. 5.9. Capacitance switching with a restrike at peak voltage.

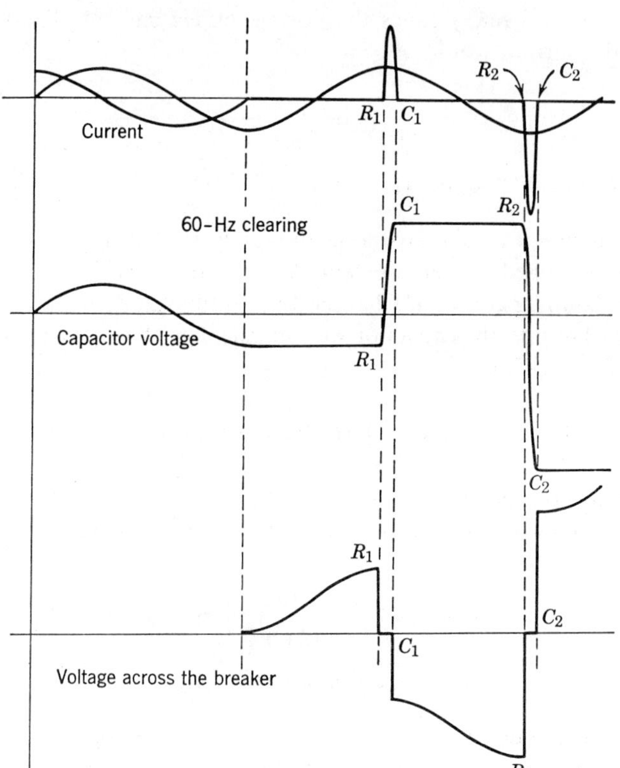

Fig. 5.10. Capacitance switching with multiple restrikes.

At the time that the transient voltage reaches its peak, identified by A in Fig. 5.9, the transient current in the lower trace passes through zero. Some breakers are capable of interrupting at such a current zero. If this happens, the high voltage is left trapped on the capacitor. The source voltage, on the other hand, would continue on its way, so that after another half cycle there would be approximately $4V_P$ across the interrupter. If this causes a second breakdown, a second oscillatory discharge would be initiated. However, since there is now twice the voltage across the switch, the current would be twice as high, and the voltage excursion would be from $+3V_P$ to $-5V_P$ (the voltage excursion, neglecting damping, is always twice the voltage across the switch). It is technically possible for the voltage to escalate still further by the same mechanism until an external flashover occurs or the capacitor fails. This happens from time to time. In the sequence drawn in Fig. 5.10 the Rs represent sequential restrikes and the Cs subsequent clearings. The sequence is idealized and to some extent oversimplified. For example, in practice restrikes will not always occur precisely at the voltage peak, so that the voltage, if it escalates, does so more slowly. Again, the circuit is more complicated. Some capacitance will exist on the source side of the breaker, which will introduce higher frequency disturbances, as was pointed out in Fig. 5.9. When the switch recovers after point A, the potential at the switch is quite high. But the source would have it be at its potential. The source side of the switch, therefore, goes through a high-frequency transient involving an oscillation of the aforementioned capacitance and the inductance of the source. In fact, *at this time*, it is possible for a voltage of 4 per unit to be developed across the switch, a point which is often overlooked. A reignition may occur at this time rather than half a cycle later, which will probably result in the switch conducting current half another cycle.

Problems of this kind do not arise often when capacitors or cables are being disconnected. On the contrary, most circuit breakers are virtually restrike-free. But on the occasions when switching disturbances do occur they can lead to some very substantial overvoltages by the sequences described.

Some mention has been made of the very considerable currents that flow when restrikes take place. This is true when energizing a bank, if the switch happens to close near the peak of the voltage cycle. These so-called inrush currents are particularly marked when a second bank is switched in parallel with the first, for then current flows from one capacitor into other. The surge impedance of this local circuit is quite low, since L is small when the banks are close together, and C is the capacitance of the series combination of the two banks, which is probably high.

Current suppression, discussed in Section 5.2, and the problems that can arise during capacitance switching operations are examples of overvoltages caused by the release of stored energy in the system. In the first case this was originally trapped in the magnetic field; in the second, it was in the electric field of capacitance. Section 5.4 treats similar problems. The subject of 3-phase capacitance switching is treated in Chapter 6.

5.4 Other Restriking Phenomena

The sequential restriking and clearing of circuit breakers can create dangerous overvoltages without benefit of a capacitor bank. The principle is the same, but the circumstances are different. The escalation of voltage is usually much more rapid. Consider the example shown in Fig. 5.11. The load in this diagram might be an arc furnace. When the electrodes become accidentally shorted by the scrap, a furnace can be represented by such a tank circuit with relatively little damping. If the circuit breaker would open under these conditions, the power factor being very low, the voltage would be near its peak at the moment of circuit interruption. The local capacitance,

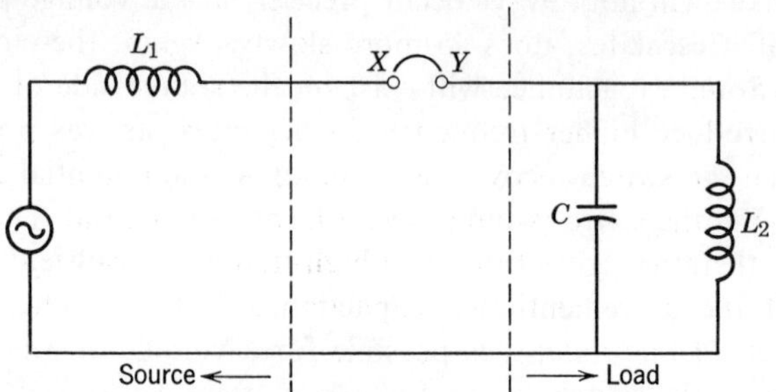

Fig. 5.11. Circuit to illustrate multiple restriking transients.

represented by C, which might be a length of cable, or simply the inherent capacitance of the furnace transformer, discharges through L_2, oscillating about its final ground potential at a frequency

$$f_2 = \frac{\omega_2}{2\pi} = \frac{1}{2\pi(L_2C)^{1/2}}$$

This is usually in the range of hundreds of cycles to tens of kilocycles. The oscillation is shown at A in Fig. 5.11. Suppose that at B the switch can no longer support the voltage across it and breaks down. This is most likely to happen if the initial clearing takes place when the contacts have just parted. Then the capacitance will be reconnected to the supply and will swing in potential about the instantaneous supply voltage at a frequency

$$f_1 = \frac{\omega_1}{2\pi} = \frac{1}{2\pi[L_1L_2C/(L_1 + L_2)]^{1/2}}$$

This is shown at B in Fig. 5.12. Two things can happen. Either this oscillation will persist to a point where the 60-Hz current from the source is well

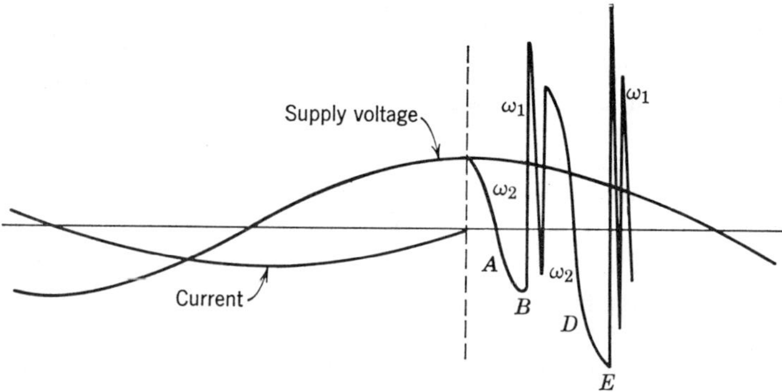

Fig. 5.12. Recurrent restriking phenomenon involving the repetitive restriking and clearing of a switch supplying an oscillatory load.

established and the switch carries another half cycle of the current, or the switch may clear on one of the current zeros associated with the frequency ω_1.

The second is potentially dangerous, and is shown in Fig. 5.12. We see that C can be left charged to quite a high voltage. Once more it discharges through L as shown at D. Suppose now that a second breakdown occurs at E. With separation of the switch contacts increasing this breakdown is likely to take place at a higher voltage, and the subsequent swing at frequency ω_1 will lead to a correspondingly higher peak. If repeated breakdowns and clearings like this occur as the contacts of the circuit breaker separate, the voltage can build up to a high value.

It will be apparent that what has just been described is very similar to capacitor switching. The only difference is that charge is not trapped on C, for now there is a discharge path through L. This allows a voltage reversal much more quickly than was the case with capacitance switching where we must wait for the 60-Hz voltage to reverse. Any resistance in parallel with L and C would damp the oscillations at ω_1 and ω_2. This has the effect of reducing the swings and lessening the likelihood of restrikes.

There is yet another form of sequential clearing and restriking which can create vicious overvoltages. In principle it is very similar to the phenomenon just described, but it does not involve a circuit breaker. This is the so-called *arcing ground.* It occurs principally on ungrounded power systems and its prevalence there has been an important factor acting against the practice of operating systems in this manner.

The operation of a power system completely isolated from ground appears to have an important advantage in that when a single line-to-ground fault occurs (and these are by far the most prevalent), there is no return path for a fault current, so, in a manner of speaking, the systems cannot be said to be truly faulted. On a system with a grounded neutral, such a fault could develop a heavy fault current, possibly cause considerable damage, and require a circuit-breaker operation and attendant outage. But all systems are grounded in some way, most particularly through capacitance and therein lies the seed of the present problem.

Consider Fig. 5.13*a*, which shows a 3-phase system with isolated neutral. The system capacitance referred to ground is also shown. In a balanced system, these capacitances will be equal and will act to maintain the system neutral near ground potential. A symmetrical set of 3-phase capacitance currents will flow in the lines, as indicated by the adjacent phasor diagram. A ground fault of any phase will disturb this symmetry. The phase involved will be brought to ground potential, causing a shift in potential of the neutral and the other two phases as shown in Fig. 5.13*b*. The capacitance of phases B and C are now excited by the line-to-line voltages, and the currents flowing in them are therefore $\sqrt{3}$ times as great and have a changed phase relationship

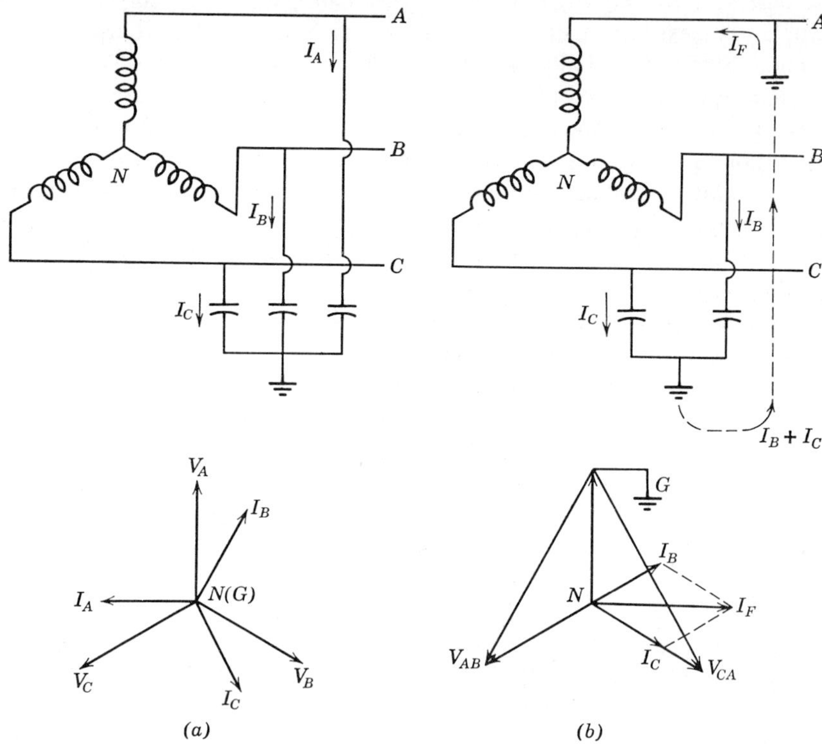

Fig. 5.13. (*a*) Three-phase ungrounded system showing capacitance currents. (*b*) The arcing ground showing fault current and phasor relationships.

as indicated by the second phasor diagram. These currents return by the fault so there really is a fault current, albeit quite small. However, it is often sufficient to support an arc, and we have an arcing ground established. It could, for example, be a flashover of a support insulator. Trouble arises when this acts as its own circuit breaker, for if the arc is momentarily extinguished at a current zero, the system will attempt to relax to its former balanced condition. However, there are two things that we observe about this. First, at the moment of extinction, capacitance C_A is discharged, while C_B and C_C have considerable voltages across them. The charges associated with these voltages will remain trapped, so that the voltages will appear superimposed on the power frequency voltage variations that will be impressed subsequently by the supply. Second, the relaxation process following clearing is likely to be oscillatory, since the circuit involves inductances and capacitances, that is, overshoots will occur. As a consequence of these factors, the arc may reignite and cause further transient disturbances. The precise details of events will vary, but the circuit contains the elements for cumulative escalation of voltage by sequential breakdown and clearing at the point of fault.

To counter this disagreeable state of affairs the system neutral is sometimes connected to ground through a reactor of quite high inductance, variously known as a Petersen coil, arc suppression coil, or ground fault neutralizer. On the incidence of the ground fault, this will provide an alternative path for current, and inasmuch as this current will be lagging on the voltage in the faulted phase by close to 90°, it will be opposite in phase to the capacitance current feeding the fault (see Fig. 5.13*b*). By an appropriate choice of reactor, it is possible to virtually eliminate the capacitance current and suppress the arc.

5.5 Transformer Magnetizing Inrush Current

Nonlinear properties of circuit elements can be a potential source of abnormalities. Transformer magnetizing inrush current is an example. Under normal excitation a transformer draws a magnetizing current of between 0.5 and 2% of its rated current. Because of saturation effects in the iron, this is not sinusoidal. The amount of distortion depends on the flux density to which the core is worked. The core is taken around a hysteresis loop like that shown in Fig. 5.2, each cycle, the rate of change, of flux at every instant producing just sufficient emf to counterbalance the instantaneous voltage of supply. The hysteresis loop is redrawn in Fig. 5.14 and a magnetization curve has been added. This is the locus of the apices of hysteresis loops taken for different steady-state applied voltages. It is evident that as the

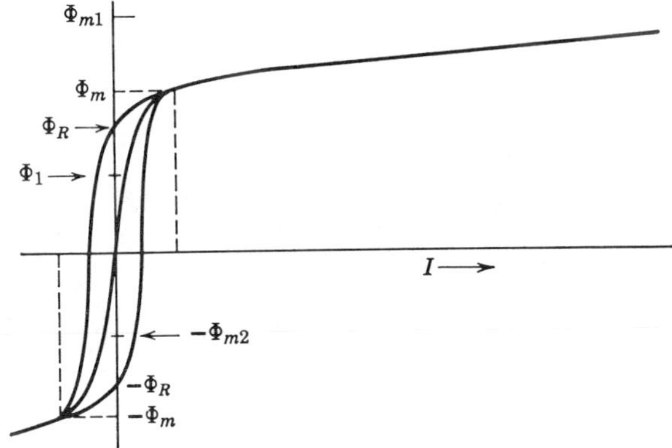

Fig. 5.14. Hystersis loop and magnetization curve for predicting transient inrush currents.

voltage is increased and more and more flux is demanded from the core, the peak current increases sharply, because the core becomes saturated. For the loop shown in Fig. 5.14, the excursions of the flux are between $+\Phi_m$ and $-\Phi_m$, while the current excursions are between $+I_P$ and $-I_P$. What we have been describing here is a steady condition. What we wish to examine now is the transient condition that occurs when voltage is first applied to the transformer winding.

To do this we have to go back to the last time the transformer was energized and, more particularly, to the time when it was switched off. Figure 5.14 shows that the instant the current passes through zero there is considerable remanent flux $\pm\Phi_R$ in the core, which is removed only by a reversal of the current. Thus we may expect that after a transformer has been disconnected there is a significant flux left in its core. It will usually be less than Φ_R because a transient current will flow in the winding after current ceases in the disconnecting device, as a consequence of the transformer discharging its own capacitance.

Let us suppose that this value of flux is $+\Phi_1$. Let us further suppose that when next the transformer is energized, the polarity of the voltage is such that it calls for the flux to increase positively. If, in addition, the applied voltage wave is just passing through zero, the flux will have to increase through an increment equal to $+\Phi_m$ before the voltage peak is reached. Since the flux started from an initial value of $+\Phi_1$ it will have to reach Φ_{m1}, before reversing. It is very clear from Fig. 5.14 that because of magnetic saturation an enormous current would be drawn from the supply. On the following negative half cycle, the peak flux will be only $-\Phi_{m1}$, so that the

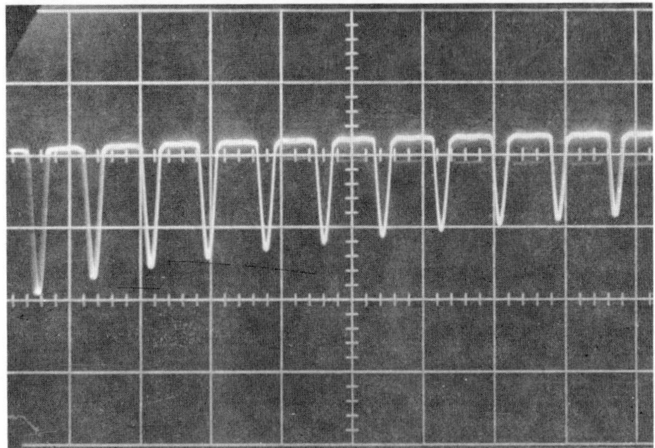

Fig. 5.15. Transient inrush current to 1000 kVA, 13.8-kV transformer. 1 div. = 80 A.

peak current will be less than normal. We have considered here the very worst condition for energizing the transformer, but statistically this situation or a condition close to it has equal probability with all others of occurring. When this happens, it is not unusual for the peak inrush current to be several times the *rated load current* of the transformer. Figure 5.15 is an example of this. The apparatus involved was a 1000-kVA, 13.8-kV, 3-phase transformer, which has a normal peak magnetizing current of less than 2 A and a rated load current of 42 A. It will be seen from the oscillogram that the inrush current is over 150 A.

The current does not persist at such a gross level or with such a measure of asymmetry, because what we have here is a series *RL* circuit. It was shown in Section 3.2 that the asymmetrical current in an *RL* circuit contains a d-c component that declines exponentially with a time constant of L/R, making the total current more and more symmetrical as it does so. This is true in principle for the transformer inrush current, but because L is nonlinear and R reflects the core losses as well as the winding resistance the time constant cannot be so readily defined. In practice it may take several seconds for the asymmetrical inrush current to approach its steady-state final value.

5.6 Ferroresonance

In the phenomenon of series resonance, a very high voltage can appear across the elements of a series *LC* circuit when it is excited at or near its natural frequency. From Fig. 5.16*a* it is evident that the voltages V_L and V_C add to give the applied voltage V. But because the voltage across the inductor leads the current in phase by 90°, and the capacitor voltage lags the current by the same amount, the phasor diagram appears as in Fig. 5.16*b*. It is seen that both V_L and V_C can far exceed V. Voltage conditions of this kind can be sustained and are therefore more properly called dynamic overvoltages, rather than transients. However, they are sometimes brought about by transient fault conditions.

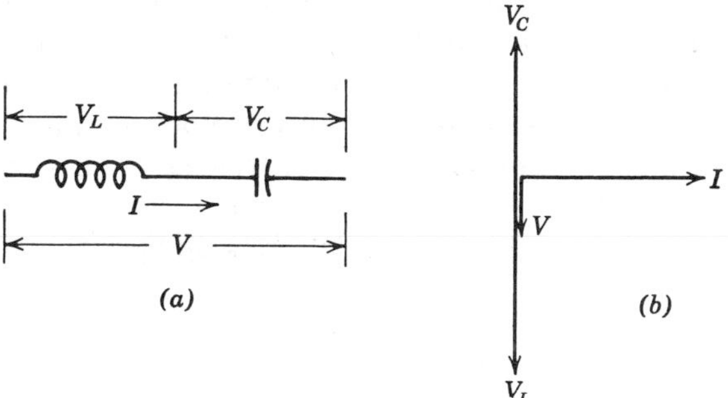

Fig. 5.16. Simple series resonance.

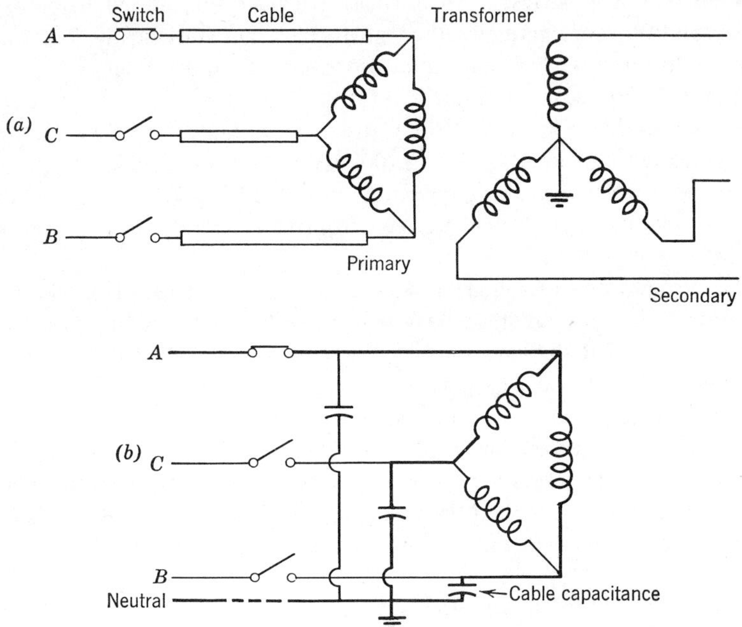

Fig. 5.17. Circumstance in which ferroresonance can occur.

Such resonant conditions are to be avoided in power circuits for obvious reasons, but they occur inadvertently on occasions. The phenomenon is referred to as *ferroresonance*, since the inductance involved is usually iron cored, and more often than not a transformer. The nonlinear character of an iron-clad inductance also introduces some peculiar effects.

A typical example will be described to illustrate how the phenomenon may arise. Figure 5.17*a* shows a switch used to energize and de-energize the primary of a transformer. The two are interconnected by a length of cable, a common practice in distribution systems, where the switching device may be mounted at the top of a pole and the transformer on a nearby pad at ground level. The switch is shown with only one pole closed. This condition will prevail momentarily on closing as the first pole completes its circuit. Some circuits of this kind use a fuse in series with the switch to interrupt fault currents. The condition in Fig. 5.17*a* could be sustained, therefore, if two of the fuses had blown.

It might appear at first that with only one pole of the switch closed the transformer is not energized. In a way this is true. There nevertheless remains a path for current through two of the phase windings and the cable capacitance, as indicated by the bolder lines in Fig. 5.17*b*. It is this current that can produce resonance and impress excessive voltages across the transformer and the cables on the unenergized phases. It can, for instance, cause lightning arresters connected at the B and C bushings of the transformer to operate. If the condition is sustained, repeated operation can destroy the arresters.

For further details on this subject, the reader is referred to Clarke (8), Clarke et al. (9), or, more recently, Hopkinson (10). In Section 14.7 ferroresonance is used as an example to illustrate the use of the digital computer for solving circuit transient problems involving nonlinear elements. The results are illustrated in Fig. 14.21.

REFERENCES

1. C. G. Damstra, "Current Chopping and Overvoltages in Relation to System Parameters," CIGRE Report No. 201 (1964).
2. A. F. B. Young, "Some Researches in Current Chopping in High Voltage Circuit-Breakers," *Proc. IEE, London*, Vol. 100, No. 76 (1953), p. 337.
3. H. C. Stiener and R. W. Strecker, "Voltage Transients due to Arc Extinction," *Trans. AIEE*, Vol. 79, Part I (1960), p. 139.
4. A. N. Greenwood, W. C. Kotheimer and C. A. Langlois, "An Investigation of the Transient Behavior of an Aluminum Pot-Line Installation," *Trans. AIEE*, Vol. 80, Part II (1961), p. 1.
5. A. N. Greenwood, Discussion of Paper "Commutation and Destructive Oscillation in Diode Circuits" by I. Somos, *Trans. AIEE*, Vol. 80, Part I (1961), p. 162.
6. R. C. Van Sickle and J. Zaborsky, "Capacitance Switching Phenomena," *Trans. AIEE*, Vol. 70, Part I (1951), p. 151.
7. I. B. Johnson et al., "Some Fundamentals of Capacitance Switching," *Trans. AIEE*, Vol. 74, Part III (1955), p. 727.
8. E. Clarke, *Circuit Analysis of AC Power Systems*, Vol. 2, John Wiley & Sons, New York (1943).
9. E. Clarke, H. A. Peterson and P. L. Light, "Abnormal Voltage Conditions in Three-Phase Systems Produced by Single-Phase Switching," *Trans. AIEE*, Vol. 60 (1941), p. 325.
10. R. H. Hopkinson, "Ferroresonant Overvoltage Control Based on T.N.A. Tests of Three-Phase Wye-Delta Transformer Banks" *Trans. IEEE*, Vol. 87 (1968), p. 352.

6 Transients in Three-Phase Circuits

6.1 Introduction

Thus far we have discussed transients in single-phase circuits only. Yet there is no inherent limitation in the methods, which is fortunate since practically all electric power is generated, transmitted, distributed, and utilized in three-phase systems. Rather we have restricted ourselves in this way because while establishing principles and gaining experience in techniques, we did not wish to be cluttered by the complications of polyphase circuits. These are generally more complicated, though little more complex than the single-phase circuits. The complication arises from the proliferation of components or branches introduced by the other phases and also because of the need sometimes to consider mutual coupling between phases. We have already seen how calculations rapidly become very protracted in single-phase circuits as the number of branches increases, and how this forces us to make approximations. This is more necessary with three-phase calculations, so we will devote some space to discussing circuit reduction for three-phase transient calculations.

There are two basic methods for calculating transients in three-phase circuits: one is to simply extend the single-phase approach; the other is to use symmetrical components. The first approach recognizes that the three-phase circuit is really one entity and that a disturbance created at one point affects all points to some degree. The oscillatory frequencies excited are not the power frequency and do not bear some fixed phase relationship in the three phases as do the power currents and voltages. The theory of symmetrical components has been used for many years in calculating steady-state fault currents and voltages when unsymmetrical faults, such as phase-to-phase short circuits, occur in a three-phase system. The method virtually removes the unsymmetrical condition and with little extra complication allows the

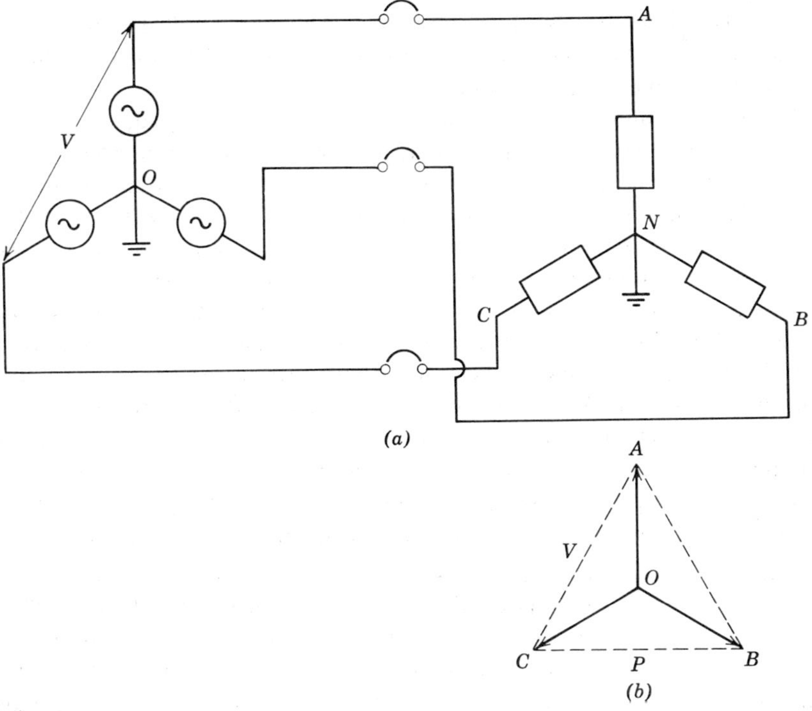

Fig. 6.1. (*a*) Simple three-phase circuit with grounded neutrals. (*b*) Phasor diagram.

computation to proceed much as it would for a symmetrical three-phase short circuit. The procedure for transient calculations is somewhat similar.

6.2 Importance of the Type of Neutral Connection

Three-phase systems can be solidly grounded at their neutrals, they can be completely isolated from ground, or they can be grounded through a neutral impedance of some kind. The transient voltages created by switching operations or other disturbances will often, though not always depend upon which of these conditions prevails. This is not surprising because we are dealing with different circuits under these different conditions. In a system where the neutral is solidly grounded, the three phases are virtually independent and behave like three independent single-phase circuits if the ground impedance itself is negligible. Thus, if a circuit breaker opens to clear a fault or shed a load, the transient recovery voltage across the breaker or load can be determined by the single-phase methods of Sections 3.3 and 3.4 without modification. In the very simple circuit shown in Fig. 6.1*a* we note that the active rms voltage in each phase is $V/\sqrt{3}$, and if we neglect damping, we would expect the breaker recovery voltage to reach $2\sqrt{2}$ times this value on each phase, as they clear in sequence.

That the situation is different when the neutral is ungrounded, or grounded through an impedance is made obvious by a simple example. Suppose we are switching off a Y-connected load (Fig. 6.1b), which has no connection at the neutral. Because of the balanced nature of the load the neutral N will be held close to the potential of 0, the neutral of the supply, during normal operation, as if in fact the two points were interconnected. However, once a switching operation is initiated, the situation quickly changes. Because the currents in the three phases pass through zero in sequence, one phase, say A, will always interrupt first. When this occurs, the constraint on the load neutral, just described, will no longer exist. Instead the neutral of the load will tend to move to the mid-potential P, of the two phases, to which the load remains connected. If the other poles of the circuit breaker would remain closed, the voltage AP, eventually appearing across the phase which has interrupted, would be $\sqrt{3}\,V/2$, so that during the transient period it can instantaneously approach $2\sqrt{2}$ times this value. Under normal circumstances, of course, the other two phases would interrupt shortly after the first. Their duty would be relatively less severe, inasmuch as the two remaining phases are now in series, with two breaks to interrupt against a voltage V. With a grounded neutral system, therefore, the three poles of the breaker have identical jobs to do. But with an ungrounded system, when interrupting a balanced load or three-phase fault, the duty of the first phase to clear is the most severe and is more severe than the grounded case.

In the next few sections we will explore some practical three-phase switching operations.

6.3 Switching a Three-Phase Reactor with an Isolated Neutral

As a first example of switching with an isolated neutral, consider the dropping of a three-phase reactor load. We will show how the operations can be reduced to a single-phase circuit problem. Each reactor will possess inductance, and of equal importance in any transient calculation, it will have a certain capacitance. We choose to represent the reactors therefore by π circuits each having an inductance L, and a capacitance C to ground at each end. When connected together the reactors will appear as in Fig. 6.2. This raises an interesting point. We described the neutral as being isolated. In fact it is grounded through the capacitance $3C$. There is really no such thing as a truly isolated neutral. What we are studying here is a special case of the impedance-grounded neutral where the neutral impedance happens to be a capacitor.

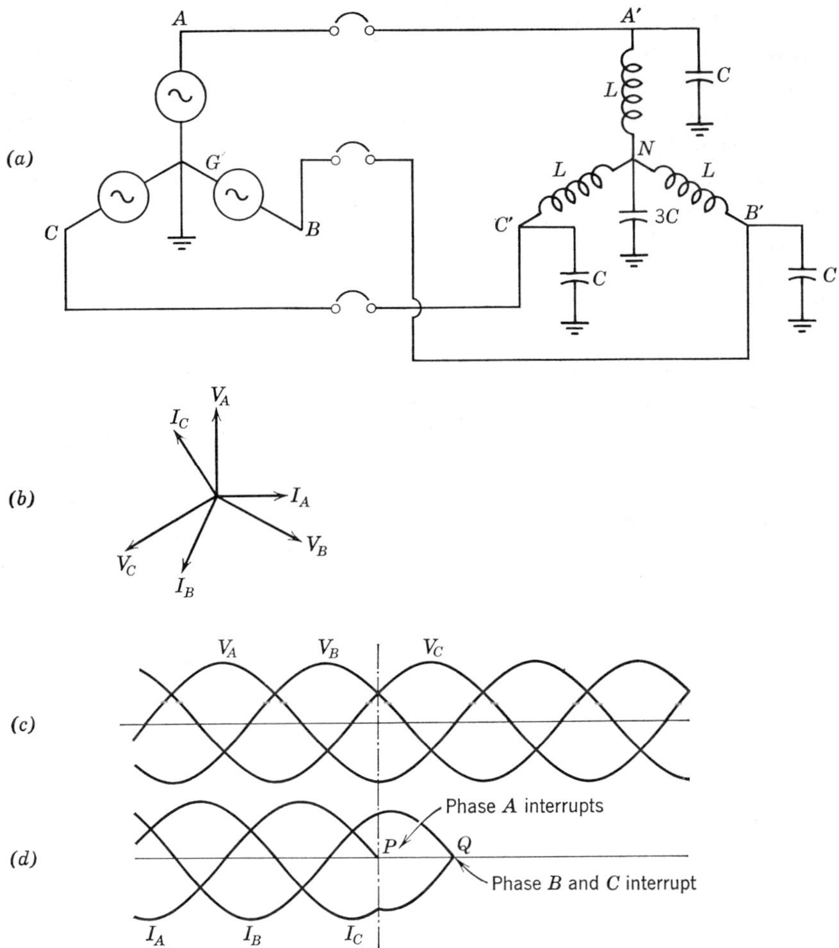

Fig. 6.2. (*a*) Equivalent circuit for switching a three-phase reaction. (*b*) Phasor relationships. (*c*) Three-phase voltages. (*d*) Three-phase currents.

Let us suppose that phase *A* interrupts first. The phasor diagram Fig. 6.2*b* shows the phase relationships of the three-phase currents and voltages at this instant. The instantaneous values of these various quantities are obtained by projecting the phasors onto the vertical axis in the usual manner. Such variations of the currents and voltages are shown in Fig. 6.2*c* and 6.2*d*. The voltage of phase *A* at the moment of clearing is at a negative maximum, whereas those of phase *B* and *C* are halfway to a positive peak with *C* increasing and *B* decreasing. Momentarily there is no voltage between *B* and *C*, that is, the phase-to-phase voltage *BC* is passing through zero. The currents in these phases are equal and opposite having an instantaneous value of $(\sqrt{3}/2)I$ peak. After phase A has interrupted, the currents I_B and I_C must continue to be equal and opposite, since they are now the same current and the only current flowing. As will be seen from Fig. 6.2*d* they suffer a change of slope when I_A is interrupted and come to zero 90° later.

Let us concern ourselves for the moment with the transient voltage appearing across the switch in phase A, following interruption. We assume that the impedance of the source is so much lower than that of the load that it can be safely considered an infinite bus. If we look into the circuit at AA' through the contacts of the switch in phase A, we observe that the circuit possesses a certain symmetry if the circuit is redrawn as in Fig. 6.3*a*. This is a common situation in three-phase problems of this kind, and one which we can take advantage of. If we wish to solve this problem by the current injection method, that is, determine the response of the circuit to the injection at AA' of a current equal and opposite to the current in phase A, we recognize that there are two identical paths for such a current, through phase B or through phase C. Presumably, the injected current would divide equally as suggested by the arrows, and the response of the two halves of the circuit

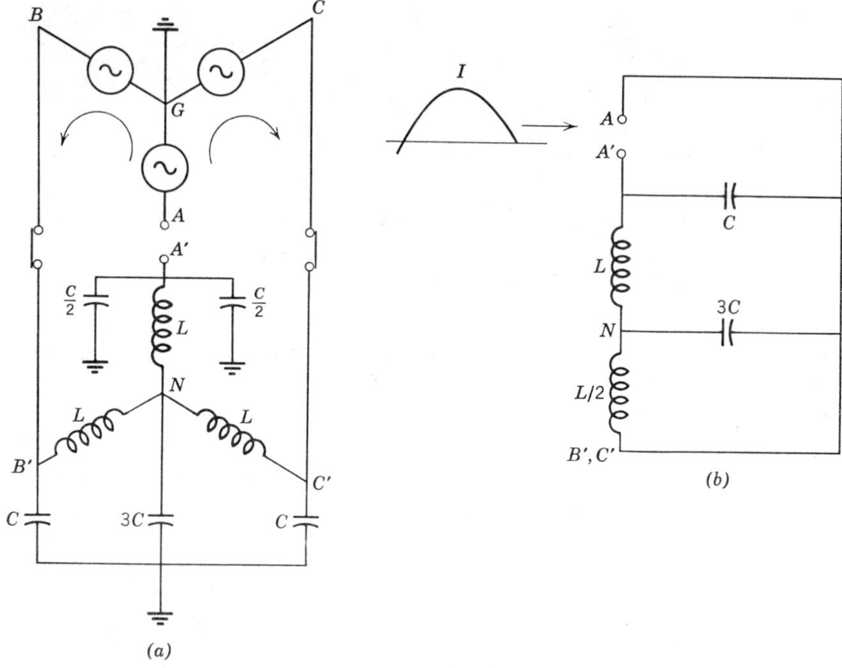

Fig. 6.3. Rearrangements of the circuit in Fig. 6.2.

would be identical. For example, points B' and C' would suffer the same transient variations in potential. This being so, such points can be joined together from a transient point of view. This is akin to folding the circuit along its line of symmetry. The result is Fig. 6.3*b*. We observe that the capacitors at B' and C' become shorted out in the process, because in the current injection method of solving opening transients, superposition demands that we remove all other sources, or replace them by their internal impedances, which in this example we have chosen to neglect. The circuit of Fig. 6.3*b* is one we have examined before in Section 3.4 (see Fig. 3.11), where a solution for the recovery transient was obtained (Eq. 3.4.14). An alternate method of reaching a final solution is given in Section 14.2. It is sufficient here that we show how the three-phase problem can be reduced to a familiar single-phase equivalent.

6.4 Three-Phase Capacitance Switching

This is an important subject, for an increasing number of capacitor banks are being located on three-phase buses at almost all voltage levels to improve the power factor or, in the case of transmission systems, to improve regulation and the power handling capability of lines. These banks are switched quite frequently as load conditions vary and if the switch used for this purpose should restrike or reignite while performing such a switching operation, severe transient overvoltages may occur. The subject has been well treated in a paper by Johnson et al. (1). It is of additional interest to us here in that it provides a good example of how to study relatively simple transients in three-phase circuits without recourse to heavy mathematical tools.

The fundamentals of the capacitance switching problem were worked through in Section 5.3 for a single-phase circuit. If the neutral of three-phase capacitor bank and the source neutral are solidly interconnected, that analysis would apply to the individual phases of the three-phase circuit without modification. Our purpose here is to study the case where the capacitor bank neutral is isolated from the source. It is quite common for capacitor banks to have an ungrounded neutral.

In Fig. 6.4*a* each phase of the capacitor bank is represented by capacitors C_A, while C_N is the effective capacitance to ground of the bank neutral. Under steady-state conditions when the bank is energized, the symmetry of the circuit will cause N to be at ground potential, so that C_N is discharged. The phasor relationships for the currents and voltages are given in Fig. 6.4*b*. Suppose that when the switch is opened, phase A interrupts first. The current I_A will reach zero when the voltage V_A is at its peak. This is the instant illustrated in Fig. 6.4*b*. Since there is nowhere for the charge on C_A to go, it will remain trapped and this capacitor will retain the peak line to neutral

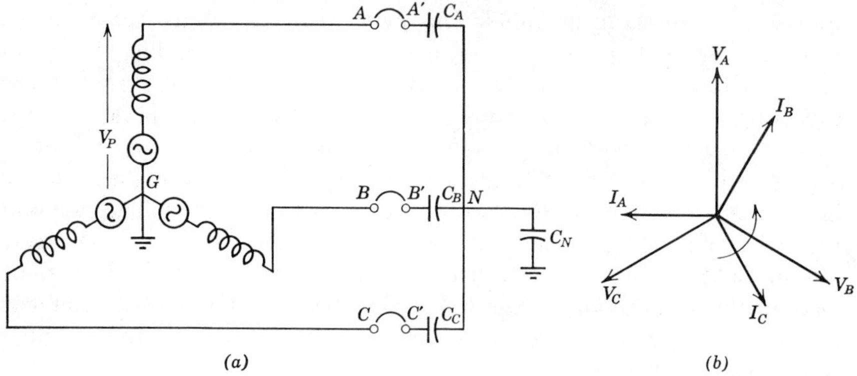

Fig. 6.4. (*a*) Simple equivalent circuit for three-phase capacitor bank switching. (*b*) Phasor relationships when I_A passes through zero.

voltage V_P. The instantaneous values of the currents I_B and I_C will be $(\sqrt{3}/2)V_P\omega C$, and the currents will be opposite as well as equal. We can make the following further observations about the circuit at that time and thereafter:

1. The voltages V_B and V_C are instantaneously equal at $-V_P/2$.
2. Phases B and C form a single circuit in which the current $I = I_B = -I_C$ is in such a direction initially as to charge C_B and discharge C_C.
3. The effect of (2) is to cause the potential of the neutral, N, to rise with respect to ground. Current therefore flows into C_N.
4. The current I will start at its maximum value at the instant phase A interrupts, for then $V_{BC} = 0$. Then I will diminish, coming to zero one quarter cycle later when V_{BC} is at its peak of $\sqrt{3}/V_P$.
5. During this quarter cycle the charge added to C_B and removed from C_C will be

$$Q = \int_0^t I\,dt = \frac{\sqrt{3}}{2} V_P\omega C \int_0^{\pi/2\omega} \cos \omega t\,dt$$

$$= \frac{\sqrt{3}}{2} V_P C$$

The change in potential of these capacitors will be $\sqrt{3}\,V_P/2$.

Voltage conditions existing on the bank at this time are shown in Fig. 6.5*b*, which is an interpretation of the phasor diagram Fig. 6.5*a*. If poles B and C of the switch interrupt at this time, these voltages will remain frozen or trapped on the capacitors. It is worth pointing out that we have in effect assumed that $C_N \ll C$, so that the current taken by C_N is negligible. This is likely to be the case with capacitor banks, but there are other instances where this is not true. These will be considered later.

It is of interest to consider the maximum voltage that appears across the poles of the circuit breaker thereafter, for this may lead to restrikes or reignitions in the breaker, which in turn can lead to transient disturbances of much greater amplitude. With the voltages shown in Fig. 6.5*b* trapped on the capacitors, the potential of one side of each pole of the switch is fixed. The other sides continue to be driven by the supply, so the maximum voltage appears across the switch when the source side reaches its maximum value in the opposite polarity to the trapped charge on the capacitor bank. A little consideration will show that these maximum voltages will be the following:

Phase A:

$2.5V_P$ Occurring 90° after phases B and C clear (and every 360° thereafter)

Phase B:

$\left(1+\frac{\sqrt{3}}{2}\right)V_P$ Occurring 210° after phases B and C clear (and every 360° thereafter)

Phase C:

$\left(1+\frac{\sqrt{3}}{2}\right)V_P$ Occurring 150° after phases B and C clear (and every 360° thereafter)

Suppose that phases B and C do not interrupt 90° after phase A, as we have assumed. It can readily be shown that this leads to higher voltages. Consider the voltage $V_{AA'}$ across pole A in Fig 6.4*a* (our convention is that $V_{AA'}$ is the potential of A' with respect to A):

$$V_{AA'} = V_{A'C'} - V_{AC} \tag{6.4.1}$$

but, since we are now considering that C has not interrupted, C and C' will be at about the same potential. The capacitor C_A has V_P trapped on it at the time

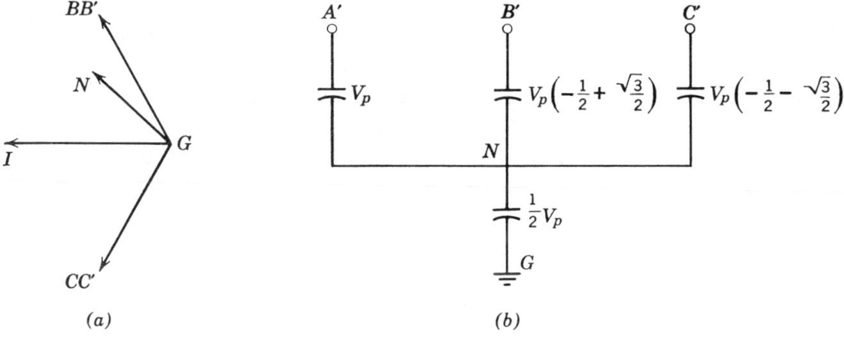

Fig. 6.5. (*a*) Phasor diagram when current in phases B and C reaches zero. (*b*) Voltage trapped on the capacitor if phases B and C interrupt.

of clearing, that is, $V_{A'N} = V_P$. We measure time from this instant. Now,

$$V_{A'C'} = V_{A'N} + V_{NC'} \tag{6.4.2}$$

Initially,

$$V_{NC'} = 0.5V_P$$

and thereafter it has applied to it half the phase-to-phase voltage

$$V_{BC} = \sqrt{3}\, V_p \sin \omega t,$$

therefore

$$V_{NC'} = V_P\left(0.5 + \frac{\sqrt{3}}{2} \sin \omega t\right) \tag{6.4.3}$$

From Eqs. 6.4.2 and 6.4.3,

$$V_{A'C'} = V_P\left(1.5 + \frac{\sqrt{3}}{2} \sin \omega t\right) \tag{6.4.4}$$

At the source,

$$V_{AC} = \sqrt{3}\, V_P \sin\left(\omega t + \frac{\pi}{3}\right)$$

$$= \sqrt{3}\, V_P\left(\tfrac{1}{2}\sin \omega t + \frac{\sqrt{3}}{2} \cos \omega t\right) \tag{6.4.5}$$

Combining Eqs. 6.4.1, 6.4.3, and 6.4.5:

$$V_{AA'} = 1.5V_p(1 - \cos \omega t) \tag{6.4.6}$$

This shows that 180° after pole A interrupts, it has a voltage of $3V_P$ impressed across its contacts, if phases B and C have not interrupted. This contrasts with the $2.5V_P$ we determined for the condition when these phases do clear.

We now examine what happens if the circuit breaker is unable to support the voltage to which it is momentarily subjected, when one or more interrupters restrike and re-establish current. Specifically, we consider the three-phase capacitor bank with isolated neutral, where phase A has cleared first, phases B and C have interrupted 90° later, and the charge pattern of Fig. 6.5*b* is trapped on the capacitors. Let us suppose that phase A restrikes when the voltage across it has just attained its peak value of $2.5V_P$. Figure 6.6 shows condition on this phase. Phases B and C have been omitted for the sake of clarity. The capacitor C_N, as before, represents the neutral capacitance, that is, the stray capacitance to ground of the capacitor bank structure. The source inductance including the loop to the bank is denoted by L.

The restrike completes this circuit, initial conditions being:

$$\text{source voltage} = V_S = -V_P$$

$$\text{voltage on } C_A = V_{A'N} = +V_P$$

$$\text{voltage on } C_N = V_{NG} = 0.5V_P$$

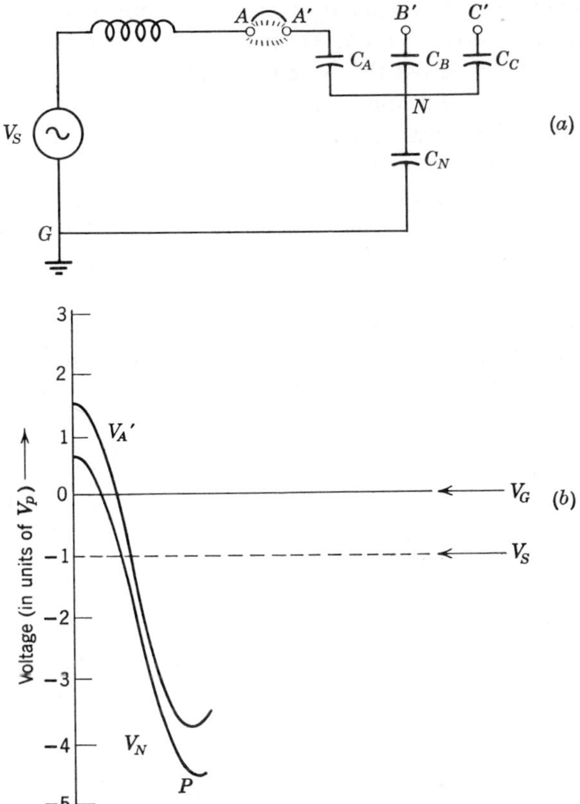

Fig. 6.6. (*a*) Equivalent circuit for a restrike when a three-phase capacitor bank with isolated neutral. (*b*) Voltage excursion following a restrike during three-phase capacitance switching.

Events are closely analogous to those described for single-phase capacitance switching in Section 5.3. A transient oscillatory current will flow between the source and the bank, L will be the inductance involved and the series combination of C_A and C_N, or $C_A C_N/(C_A + C_N) = C_1$, say, is the capacitance. Since $C_A \gg C_N, C_1 \approx C_N$. The frequency of the current will be

$$f_0 = \frac{\omega_0}{2\pi} = \frac{1}{2\pi(LC_1)^{1/2}} \tag{6.4.7}$$

When a switch is closed in such a circuit, the magnitude of current is given by the quotient of the voltage across the switch and the surge impedance of the circuit, in this case

$$I = \frac{2.5V_P}{(L/C_1)^{1/2}} \sin \omega_0 t \tag{6.4.8}$$

The current charges the capacitance, causing its potential to swing about the potential of the source, the total excursion, neglecting damping, being twice the voltage initially across the switch. Inasmuch as $C_A \gg C_N$, most of this change in potential will appear across C_N. Figure 6.6*b*, which shows this transient, has been constructed on the assumption that C_N takes all of the voltage. We see that the neutral of the capacitor bank swings to $-4.5V_P$ at point P. This instant corresponds to a current zero in the high-frequency current. Some types of circuit breaker might interrupt the current at this time, in which case this high potential would remain trapped on C_N. What more often happens is that one or both of the other poles of the circuit breaker restrike, for in its swing the neutral N has dragged around with it the potentials of points B' and C' to which it is tightly coupled through the capacitors C_B and C_C (Fig. 6.6). Thus, when V_{NG} reaches $-4.5V_P$, it is apparent from Fig. 6.5*b* that

$$V_{B'G} = \left[-4.5 + \left(\frac{\sqrt{3}}{2} - \frac{1}{2}\right)\right]V_P = -4.134V_P$$

and

$$V_{C'G} = \left[-4.5 + \left(-\frac{1}{2} - \frac{\sqrt{3}}{2}\right)\right]V_P = -5.866V_P$$

Considering the instantaneous values of the source voltages on these phases at this time, $V_{BG} = 0.5V_P = V_{CG}$, it is clear that the voltage momentarily across the breaks in phases B and C will be

$$V_{B'B} = -4.634V_P; \qquad V_{CC'} = -6.366V_P$$

As stated above, one of these phases is likely to restrike, thereby connecting the line-to-line voltage across two phases of the bank.

The analysis of this situation is again much like the single-phase case. It would be as if poles A and C were closed simultaneously. The capacitance involved would be the series combination of C_A and C_C or $\frac{1}{2}C_A$. The inductance would be $2L$. The change of potential on this combination would be twice the sum of the voltages originally across the breaks A and C when the first restrike occurred, $2(2.5 + 1.366)V_P$, if the initial restrike on pole A occurs with peak voltage across it. This will be shared equally by the capacitors C_A and C_C. The analysis would proceed much as for the single-phase case described in Section 5.3. The point to remember is that the charges trapped on the capacitors at the first clearing will be retained through the subsequent transients and will affect the voltage distributions during these transients. The principle of superposition is directly applicable.

Earlier in this section it was pointed out that we had tacitly assumed that the neutral capacitance C_N was very small compared with phase capacitors

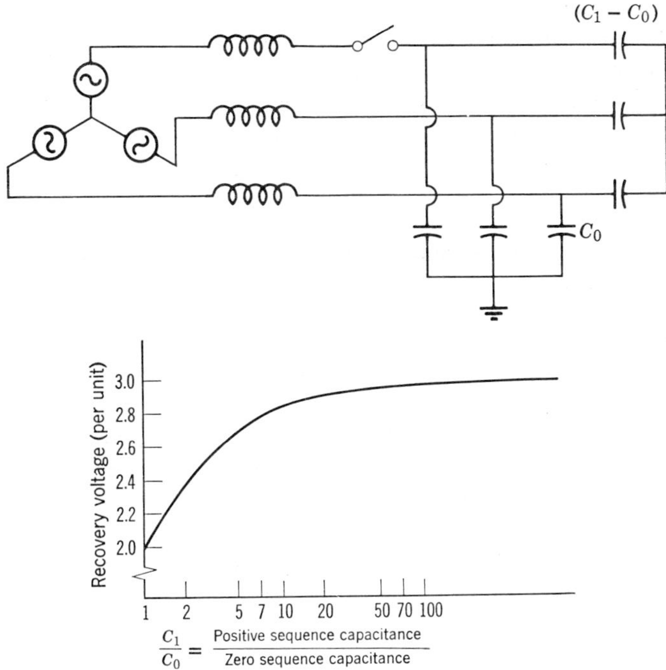

Fig. 6.7. Crest of recovery voltage across first pole to interrupt without restriking.

constituting the bank itself. This was implied by the fact that until the restrike was considered, no allowance was made for the current required to charge and discharge C_N. This condition does not always prevail. A notable exception arises with long open-circuited transmission lines. Here there is considerable capacitance to ground as well as between the lines. The total capacitance per phase, line-to-line and line-to-ground, is the positive phase sequence capacitance, which will be designated C_1. That part involving line and ground only is the zero sequence capacitance C_0. The ratio C_1/C_0 is never less than 1. It is infinite when the neutral is isolated and approaches unity asymptotically as C_0 is increased. The consequence of this, as it affects the voltage across the switch after clearing, is shown in Fig. 6.7, which is adapted from the paper of Johnson et al. (1). When $C_1/C_0 = 1$, corresponding to the solidly grounded case, the maximum voltage across the open switch is 2 V per unit. This also applies, of course, to the single-phase case. The voltage increases with increasing value of C_1/C_0, approaching 3 V for the completely isolated neutral case. This is important in that it has a strong influence on whether a switch or circuit breaker will restrike and, if it does, on what voltage will be generated by the restrike. This subject is discussed further in Section 13.3 as it pertains to transmission line switching.

We have been able to analyze effectively the important subject of three-phase capacitance switching by application of relatively simple principles developed for single-phase circuits. We now consider the alternative method of symmetrical components.

6.5 The Symmetrical-Component Method for Solving Three-Phase Switching Transients

The theory of symmetrical components, developed by C. L. Fortescue (2), has proved most valuable for studying asymmetrical conditions in polyphase networks, most notably for calculating fault currents when asymmetrical faults occur. Some familiarity with this method is assumed in this text. The unfamiliar reader will find Wagner and Evans *Symmetrical Components* (3) a helpful introduction to the subject.

The asymmetry introduced by the fault is removed by replacing the unsymmetrical or unbalanced three-phase circuit by three balanced circuits, the so-called sequence networks, in which balanced currents flow and balanced voltages are generated. By correctly interconnecting these networks it is possible to synthesize the asymmetrical components in the sequence networks. Fortescue (4) pointed out that it should be possible to extend the method to include transient conditions, and Evans and Monteith (5) and later Hammarlund (6) described the method. The natural frequency oscillations that constitute the transient are resolved into positive, negative, and zero sequence components in the same way as the currents and voltages in power frequency problems, and for the same reason: the sequence components do not interact in a symmetrical system.

There is more than one way in which the method can be applied. The first example, which is due to Hammarlund (6), illustrates an approach well suited to three-phase faults. It recognizes that the voltages between the contacts of a circuit breaker are very asymmetrical when the first pole has interrupted and the other two are still closed or arcing. In Fig. 6.8, the breaker is located somewhere in a system in the manner shown. The breaker voltages are designated V_A, V_B, and V_C in phases A, B, and C, respectively. They have symmetrical components V_0, V_1, and V_2. The corresponding breaker currents are I_0, I_1, and I_2.

Opening pole A of the breaker is simulated by inserting an impedance Z between the contacts in that phase. Making Z infinite is equivalent to interrupting the current in that opening pole, but generally

$$V_A = I_A Z$$

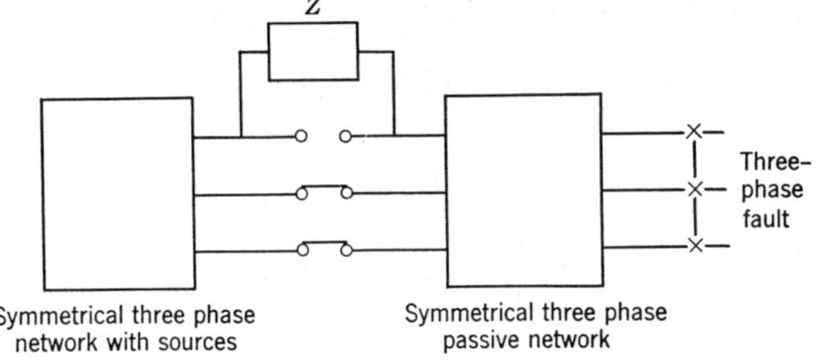

Fig. 6.8. Clearing the first phase of a three-phase fault in a balanced network.

If we assume that the other poles are still closed or have negligible voltage,

$$V_B = V_C = 0$$

Recalling the definition of the symmetrical components,

$$\left.\begin{aligned} V_0 &= \tfrac{1}{3}[V_A + V_B + V_C] \\ V_1 &= \tfrac{1}{3}[V_A + aV_B + a^2V_C] \\ V_2 &= \tfrac{1}{3}[V_A + a^2V_B + aV_C] \end{aligned}\right\} \tag{6.5.1}$$

where a is the operator $\epsilon^{j2\pi/3}$, it is apparent that

$$V_0 = V_1 = V_2 = \frac{V_A}{3} \tag{6.5.2}$$

moreover,

$$I_0 + I_1 + I_2 = \frac{V_A}{Z} \tag{6.5.3}$$

In terms of the sequence impedance networks we now can write

$$\left.\begin{aligned} V_0 &= I_0Z_0 \\ V_1 &= I_1Z_1 + E_1 \\ V_2 &= I_2Z_2 \end{aligned}\right\} \tag{6.5.4}$$

The sequence impedances designated by the Zs are those seen looking into the balanced system through the contacts of any phase of the breaker. The sequence networks must include all the capacitances of the system, since these are so important under transient conditions.

It is convenient to rewrite Eqs. 6.5.2 and 6.5.4 as

$$3V_0 = 3V_1 = 3V_2 = V_A \tag{6.5.2a}$$

and

$$\left.\begin{aligned} 3V_0 &= I_0 3Z_0 \\ 3V_1 &= I_1 3Z_1 + 3E_1 \\ 3V_2 &= I_2 3Z_2 \end{aligned}\right\} \tag{6.5.4a}$$

Equations 6.5.3, 6.5.2a and 6.5.4a are satisfied by the circuit arrangement of Fig. 6.9. The sudden change of Z from zero to infinity will simulate interruption of the current in that phase, and the voltage V_A appearing across the open circuit will be the transient recovery voltage of the first pole to clear. It will be the sum of the responses of the separate networks, but they are interconnected to combine correctly to give the true transient voltage. A model switch in a model of the network, shown in Fig. 6.9, would produce this recovery voltage at its terminals upon being opened. This is exploited in the Transient Network Analyzer, which is discussed in Chapter 14.

From an analytical point of view the recovery transient could be calculated by any of the methods we have studied so far. For example, we could short-circuit the generator E_1 and inject minus the fault current at the switch contacts. Hammarlund (6) gives the sequence network interconnections for the determining of the transients when the second and third phases clear.

This approach does not seem well adapted to asymmetrical faults such as single or double line-to-ground short circuits. Viewed from the contacts we are no longer looking into a balanced system as we are with a three-phase fault. There are now two degrees of asymmetry, the fault and the switch, since the phases are not interrupted in unison. Such cases are tackled in

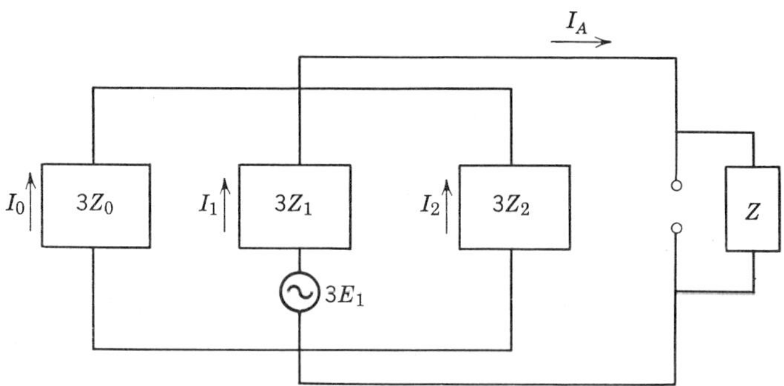

Fig. 6.9. Interconnection of the sequence networks for the first phase to clear of a three-phase fault.

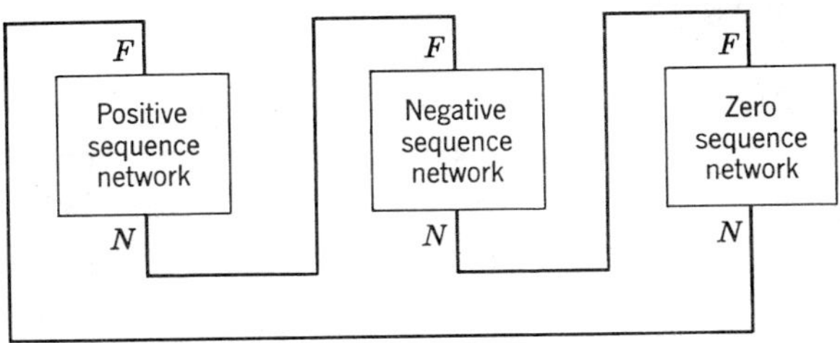

Fig. 6.10. Interconnection of phase sequence networks for a single line-to-ground fault.

a manner more directly akin to the short-circuit calculations using symmetrical components.

Consider a single phase-to-ground fault. The theory of symmetrical components shows that to determine the fault current the sequence networks must be interconnected according to Fig. 6.10. The sequence networks in this instance are as seen from the fault, rather than the breaker. The current flowing in the connection completing the loop is one-third of the fault current, since for this condition it can be shown that

$$I_1 = I_2 = I_0 = \frac{I_F}{3}$$

However, if a switch were located in this branch in Fig. 6.10, there would develop across it, on being opened, the recovery voltage of a circuit breaker located at the fault. This is so since it would accomplish the interruption of I_1 in the positive sequence network, I_2 in the negative sequence network, and I_0 in the zero sequence network, the networks being so connected that the sequence voltages add.

If the circuit breaker clearing the fault is not located at the fault, but elsewhere in the network, switches must be opened at this point, simultaneously, in all three networks. Or, from an analytical standpoint, the sequence components of current must be injected at the breaker terminals in each phase sequence. The recovery voltage of the breaker will be the sum of these three responses.

We shall use this method to solve the very practical problem of determining the recovery voltage transient evoked by clearing a single phase-to-ground fault on a cable system typical of the type found in large metropolitan distribution systems. Figure 6.11*a* shows the circuit. A high-voltage line feeds a primary distribution bus through a step-down transformer. The bus, in turn, feeds a series of cables leading to a secondary distribution system. Each cable circuit is protected by a circuit breaker. The bus also has connected a three-phase capacitor bank for power factor improvement, which has its neutral isolated from ground except through capacitance.

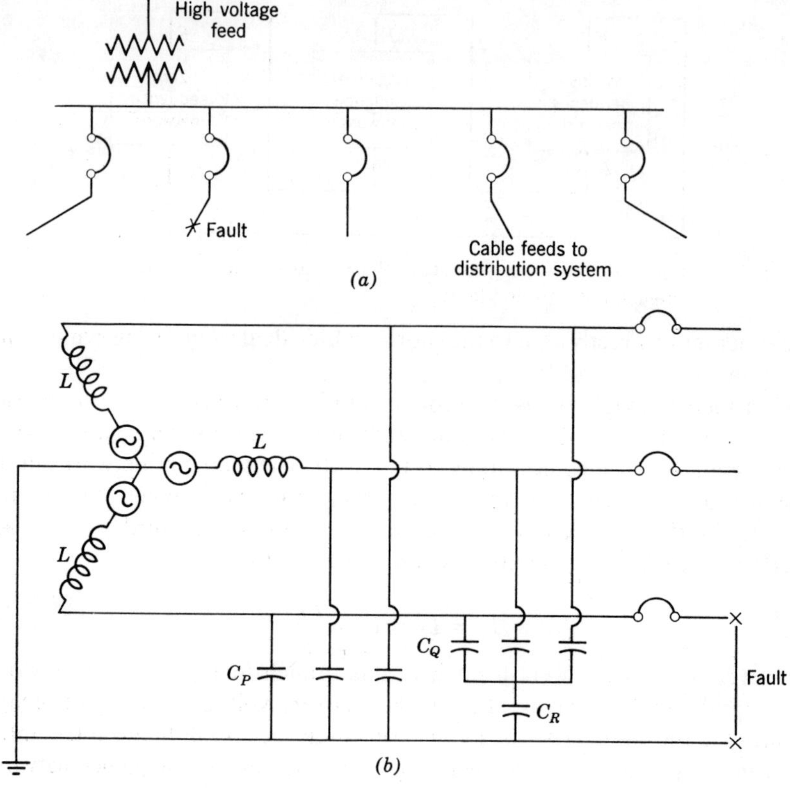

Fig. 6.11. Single phase fault on a cable system. (*a*) One-line diagram. (*b*) Equivalent circuit.

Figure 6.11*b* shows the equivalent circuit of the system. The source has an inductance L and the cable a total capacitance C_P per phase, which is shown connected line-to-ground, since the cable sheath would be grounded. The capacitor bank has a capacitance of C_Q. Its neutral has a capacitance C_R to ground.

To determine the fault current, the correct connection of the impedance networks is that shown in Fig. 6.10. When this is translated into terms of the specific problem we are currently examining, it appears as in Fig. 6.12*a*. Connecting points X and Y would cause a current equal to one-third of the fault current to flow in each sequence network. By short-circuiting the emf E in the positive sequence network and injecting minus this current at X and Y, we could generate across the three impedance networks the recovery voltage that appears at the circuit breaker terminals when the single-phase fault current is interrupted.

Except for the presence of the emf E, which does not figure in the current injection method, the positive and negative sequence impedance networks are identical. They can be conveniently combined as shown in Fig. 6.12*b*. This figure simplifies the zero sequence impedance network also, by combining the three capacitances. In most instances C_R is very much smaller than C_Q and C_P. Where this is so, the capacitance in the zero sequence network, C_0, reduces to C_P. Figure 6.12*b* is a particularly simple one to handle. It is clearly a two-frequency circuit, for there is no coupling between the two tank circuits. The natural frequencies are

$$\omega_1 = \frac{1}{[L(C_P + C_Q)]^{1/2}}, \qquad \omega_2 = \frac{1}{(LC_0)^{1/2}}$$

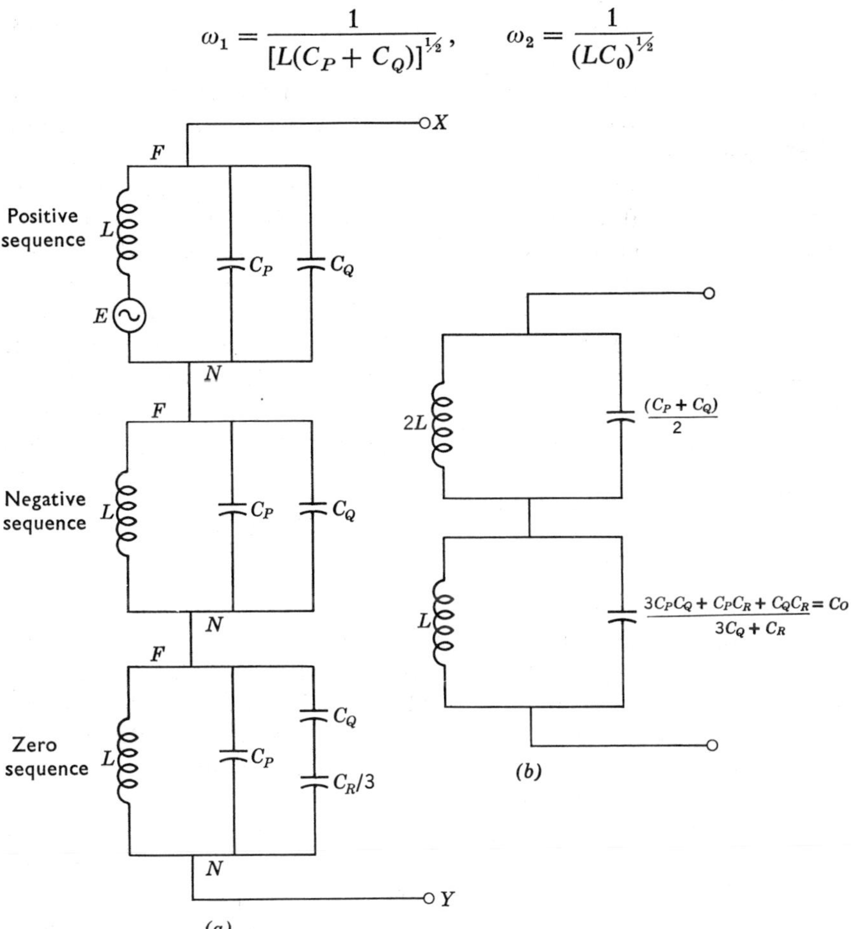

Fig. 6.12. (*a*) Circuit for determining the recovery voltage transient in Fig. 6.11*b* when the single-phase fault is isolated. (*b*) Reduction of circuit of 6.12*a*.

When a ramp of current, $I't$, is injected into a parallel LC circuit, it is readily shown that the response is given by

$$V = I'L(1 - \cos \omega_0 t)$$

But if we are simulating an interruption, I' is the slope of the fault current at current zero. For a symmetrical current this is V_P/L. The response is therefore

$$V = V_P(1 - \cos \omega_0 t)$$

In Fig. 6.12*b* we are injecting one-third of the fault current so the full response becomes

$$V = \frac{2V_P}{3}(1 - \cos \omega_1 t) + \frac{V_P}{3}(1 - \cos \omega_2 t) \tag{6.5.5}$$

Thus, what appeared in Fig. 6.11*b* as a relatively formidable problem has been quickly resolved and a solution found by the method of symmetrical components.

The supply transformers in power systems of this kind usually have their neutral grounded through a resistance. This is especially true in industrial power systems. The purpose of the resistance is to reduce ground fault current. Typically, in a 13.8-kV distribution system, R would be in the range $4 < R < 20\ \Omega$, which would limit the fault current to the range $2000 > I_F > 400$ A. Since this affects zero sequence currents only, it appears only in the zero sequence impedance network. It appears as $3R$ in series with L. In a typical system at this voltage level, L would be of the order of 1 mH. Depending upon the amount of cable present, C might vary between a fraction of a microfarad to several tens of microfarads. Suppose it is 5 μF, then the surge impedance will be

$$Z_0 = \left(\frac{L}{C_0}\right)^{1/2} = \left(\frac{1000}{5}\right)^{1/2} = 14.1\ \Omega$$

From our knowledge of damping in resonant circuits, discussed in Chapter 4, it is evident that for $R \geqslant 28\ \Omega$ the recovery oscillation in the zero sequence network will be completely suppressed. It should also be pointed out that with a resistor of the kind described, in the neutral, the fault current is largely controlled by the resistor. This will affect the current injected into the circuit to simulate the interruption.

Symmetrical components have been used by others for transient calculations. For example, Lewis (7) has used them to calculate transient fault currents. We have chosen to use symmetrical components because this particular manner of resolving asymmetrical quantities is the one most widely known. It will be apparent to those conversant with other component systems, such as the $\alpha,\beta,0$ system (8), that they could be adapted to this

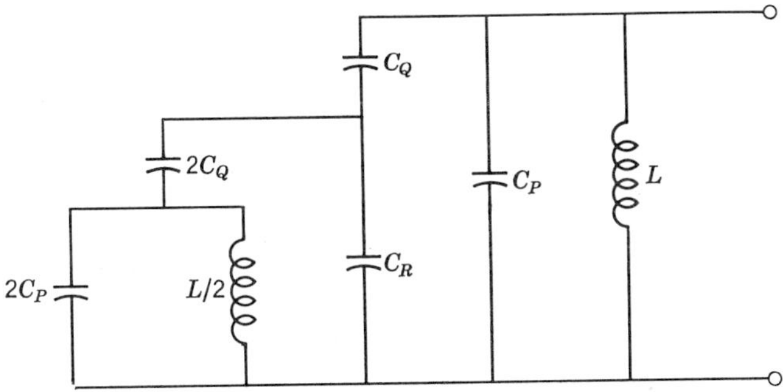

Fig. 6.13. Equivalent circuit for solving the recovery transient in Fig. 6.11*b* using the "classical" method.

approach. As far as the author is aware solving transient problems with components of this kind has not been reported in the literature.

Any loads on the feeders of Fig. 6.11*a* that can be represented by resistance will appear as resistors across both circuits if the load neutral is grounded, and they will contribute to the damping of both circuits. If the load neutral is isolated, it will shunt the positive and negative sequence networks only.

It is instructive to examine this same problem by the classical approach, that is, the extension of the single-phase method. For this we must find the equivalent circuit seen looking into the breaker terminals on the faulted phase in Fig. 6.11*b*. This is drawn in Fig. 6.13. Again, injecting minus the fault current into this circuit will produce the recovery voltage at the circuit breaker terminals. Whether this will evoke the same response as injecting $I_{F/3}$ into Fig. 6.12*b* is left as an exercise for the industrious reader.

6.6 Dealing with $Y\Delta$ Transformers

From time to time transient problems must be solved in circuits containing $Y\Delta$ or ΔY transformers. The method of setting about such a situation is as follows. The transformer is treated as an ideal 1:1, three-phase mutual inductor. The leakage inductance, divided equally between primary and secondary windings, is placed in series. The voltage, the inductance values, and all other impedance values are referred to that side of the transformer of immediate interest. If we are concerned with computing transients in the primary circuit, then voltage and impedance values are referred to that side, if in the secondary, then they are referred to the secondary side. Such a transformer representation is shown in Fig. 6.14.

Referring to this figure, the important features to notice are that windings on the same limb have voltages identical in phase and magnitude, according to our rules, whereas the currents are equal but in phase opposition, in order

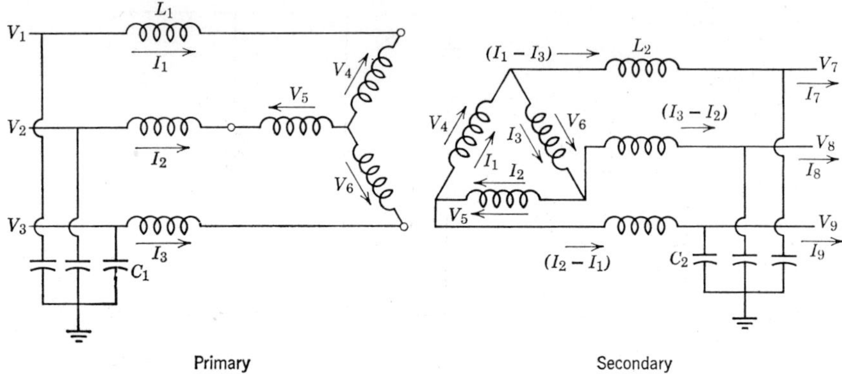

Fig. 6.14. Equivalent circuit for performing a transient analysis on a $Y\Delta$ transformer.

to balance the ampere turns. We can now write nodal and mesh equations such as

$$\left.\begin{aligned} V_1 &= L_1 \frac{dI_1}{dt} + V_4 \\ V_2 &= L_1 \frac{dI_1}{dt} + V_5 \\ V_3 &= L_1 \frac{dI_2}{dt} + V_6 \end{aligned}\right\} \tag{6.6.1}$$

and for currents

$$\left.\begin{aligned} \text{in line } 1 &= I_1 + C_1 \frac{dV_1}{dt} \\ \text{in line } 2 &= I_2 + C_1 \frac{dV_2}{dt} \\ \text{in line } 3 &= I_3 + C_1 \frac{dV_3}{dt} \end{aligned}\right\} \tag{6.6.2}$$

For the secondary circuit, we obtain such equations as

$$\left.\begin{aligned} V_4 - L_2 \frac{d}{dt}(I_1 - I_3) + L_2 \frac{d}{dt}(I_2 - I_1) &= V_7 - V_9 \\ V_5 - L_2 \frac{d}{dt}(I_2 - I_1) + L_2 \frac{d}{dt}(I_3 - I_1) &= V_9 - V_8 \\ V_6 - L_2 \frac{d}{dt}(I_3 - I_2) + L_2 \frac{d}{dt}(I_2 - I_1) &= V_8 - V_7 \end{aligned}\right\} \tag{6.6.3}$$

and for the output currents

$$\left.\begin{aligned} I_7 &= I_4 - I_3 - C_2\frac{dV_7}{dt} \\ I_8 &= I_3 - I_2 - C_2\frac{dV_8}{dt} \\ I_9 &= I_2 - I_1 - C_2\frac{dV_9}{dt} \end{aligned}\right\} \tag{6.6.4}$$

It is also apparent that

$$V_4 + V_5 + V_6 = 0 \tag{6.6.5}$$

and

$$I_1 + I_2 + I_3 = 0 \tag{6.6.6}$$

Before we can put down all of the equations in a practical problem we must know the nature of the source which provides V_1, V_2, and V_3 and the load, if any, to which V_7, V_8, and V_9 are applied. Source and load will involve further sets of equations. To give a simple example, suppose that the source can be considered an infinite bus and that a line-to-line fault occurs on the secondary terminals 7 and 8, the transformer being an open circuit at the time. If we wish to determine the primary currents when this occurs, we can use the preceding equations and the appropriate initial conditions:

$$\left.\begin{aligned} V_7 &= V_8 \\ (I_1 - I_2) + C_2\frac{dV_7}{dt} &= -\left[(I_3 - I_2) + C_2\frac{dV_8}{dt}\right] \\ (I_2 - I_1) &= C_2\frac{dV_9}{dt} \end{aligned}\right\} \tag{6.6.7}$$

The source would have zero impedance and we would have to know at what instant in the cycle the fault occurred.

Subsequently, we might wish to calculate the transient voltages appearing at the transformer primary terminals if a circuit breaker on the primary side would open to interrupt the fault current. We already know the fault current flowing in the breaker; we could calculate the voltage by injecting minus this current into the circuit at the breaker contacts. For this purpose we would replace each phase of the source by a short circuit. Equations 6.6.1 to 6.6.7, appropriately modified to meet the circuit conditions, would again form the basis for the analysis. For example, we would have to set $V_1 = V_2 = V_3 = 0$. A different solution of the problem would be obtained, depending upon which pole of the circuit breaker cleared first.

6.7 Circuit Reduction

Most of the transients examined thus far have been in relatively simple circuits with a modest number of components. Practical power systems are rarely like this. To make a complete transient study of a typical industrial or utility system, without the aid of a large computer, would be an intractable task, for the algebra proliferates at a prodigious rate as the number of branches increases. Fortunately, this is rarely necessary, because an acceptable solution can usually be found from a greatly simplified circuit. How to derive such a circuit will be the subject of this section.

Circuit reduction is partly a matter of common sense and partly a matter of experience. It is accomplished in two ways. First, we include in the circuit only those components that are really effective and then only those attributes of these components that are truly influential in a transient situation (this is the part that calls for experience). Second, we rearrange and collapse the equivalent circuit to reduce the number of elements; this is fairly straightforward.

There are a few simple guidelines that are useful in circuit reduction. First, it is often evident on proceeding through the circuit from the point of disturbance toward the energy source, or sources, that points are reached at which the system becomes very "stiff." In an industrial circuit, for example, this could be the supply side of a substation transformer. For a disturbance some distance out on a branch circuit, it might be the main supply bus. It is usually sufficient to cut off the circuit beyond such points and replace it by an infinite bus. This sets a boundary to the problem.

Within the boundary there are sometimes situations where two or more capacitances are so closely coupled they act as a single capacitance except for very short times. This occurs where the impedance between them is small compared to the impedances to the next adjacent significant capacitances. In such cases, little error will be introduced by lumping the capacitors together. This eliminates one or more loops from the circuit.

Another useful generalization is that the further away from the point of disturbance, the less effect the disturbance has. This is analogous to dropping a stone into a pond. It is consequently permissible to take more liberties at remote points when approximating the circuit.

Occasionally apparently complicated circuits have surprisingly simple transient responses. Figure 17.16*b* is evidence of this. It shows a transient recorded by the author on a 13.8-kV bus at a utility substation. The system was similar to that shown in Fig. 6.11*a*, except that there were many more cables involved, which fed a metropolitan network. Also, there were three incoming feeds to the bus. This particular transient was initiated by closing the circuit breaker to energize the capacitor bank. The bank is initially discharged and therefore causes the bus voltage to collapse momentarily. The voltage to ground of all three phases is shown. First we see the undisturbed 60-Hz wave and then we observe the transient created by two poles of the breaker closing to complete a circuit, phase-to-phase, through the ungrounded capacitor bank. This is followed by a second disturbance when the third pole of the circuit break closes. It will be seen that despite its extensiveness, this system behaves like a two-frequency circuit. The higher frequency is caused by a sharing of charge between the many cables and the capacitor bank. This oscillation takes place more or less independently of the supply. Subsequently, the sources take the combined cable/bank capacitance and charge it, again in an oscillatory manner. This is the low frequency in the oscillogram. This example should encourage the reader to prune quite severely when setting up equivalent circuits for transient studies.

By way of an introduction to circuit reduction techniques, consider the simple circuit of Fig. 6.15*a*. It shows a generator joined to a transformer through a length of cable and a circuit breaker. The transformer and generator windings are connected according to the symbols shown. Our purpose is to study the transient created by opening the circuit breaker to isolate the transformer from the generator. We start by setting down an equivalent circuit for each of the three principal components, generator, cable, and transformer (Fig. 6.15*b*). Although the inductance seems to be the outstanding attribute of an unloaded transformer, our experience with transients has taught that the transformer capacitance is also important. Indeed, we will show later that there are situations wherein the capacitance is the only attribute of consequence. Except where we are concerned with what is happening within a winding, it is usually adequate to represent a winding by a π circuit, half the winding capacitance being symmetrically disposed at both ends of the winding inductance. We show capacitance to ground only, for the cable, indicating that it is of the shielded type. If the phase conductors do not have individual ground sheaths, the conductors will possess capacitance from conductor to conductor as well as from conductor to ground.

When the generator, cable, and transformer are connected together the capacitances at the points where they join can be combined. This has been done in Fig. 6.15*c*. To determine the transient stimulated by opening the center pole of the circuit breaker, we could proceed as we have done on several occasions, by injecting at the contacts minus the current flowing through that pole. However, this is still quite a formidable network to attack. It is better to make further reductions along the lines applied to the circuit in Fig. 6.3.

Standing, as it were, at the circuit breaker, and looking into the circuit, it is clear that the top and bottom halves of the circuit are identical, and that without any loss of accuracy the circuit can be folded on itself along its

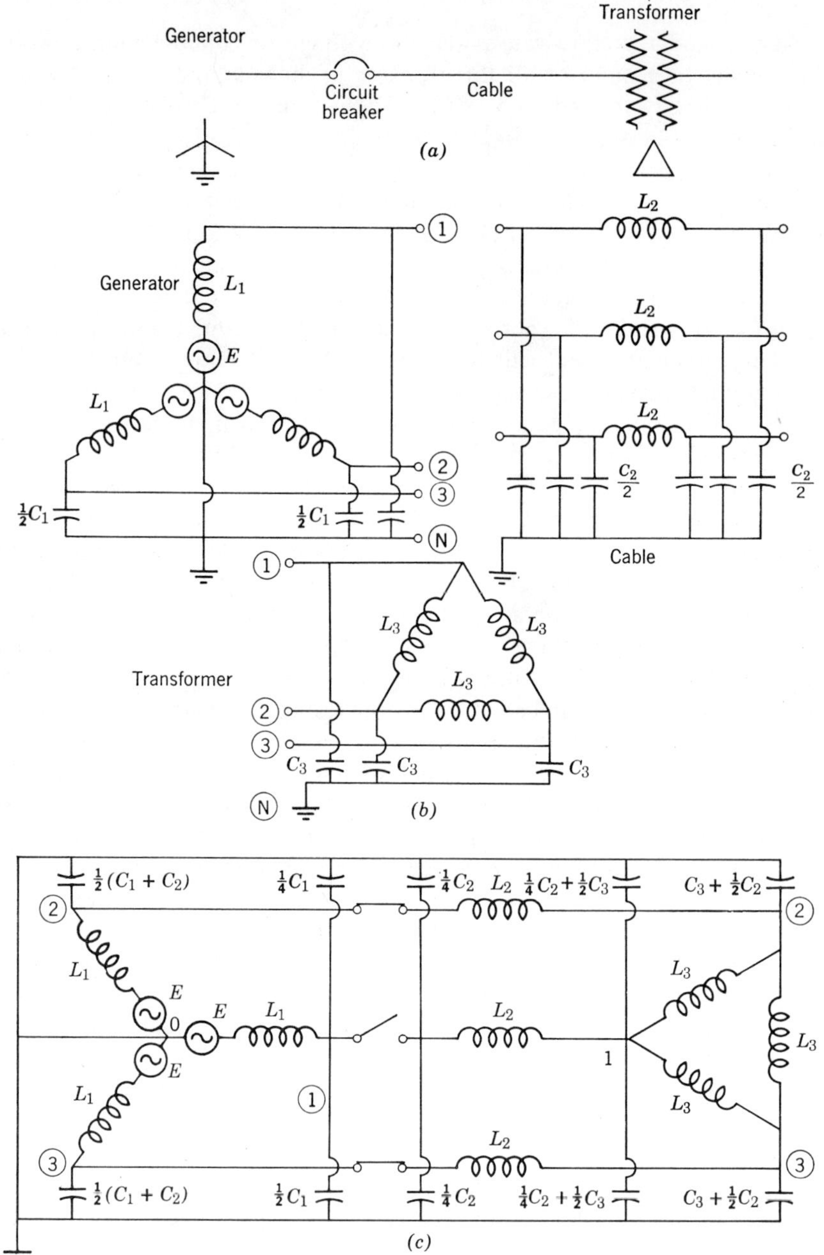

Fig. 6.15. Equivalent circuit for a generator/cable/transformer network: successive reductions. (*a*) One-line diagram. (*b*) Equivalent circuits of individual components. (*c*) Combined equivalent circuit.

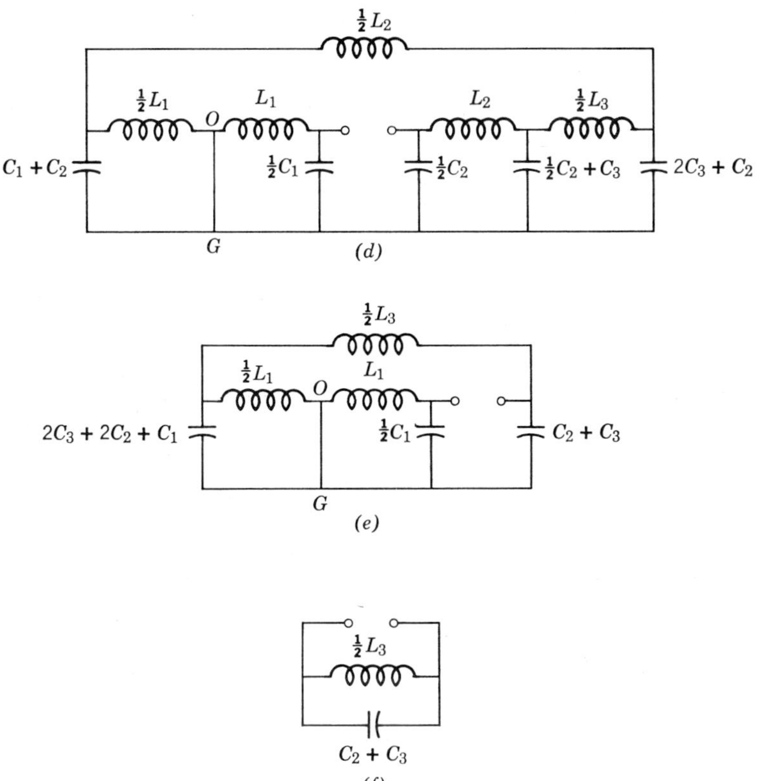

Fig. 6.15. (*continued*) (*d*) Consequence of "folding" the network. (*e*) L_2 negligible. (*f*) L_1 and L_2 negligible.

horizontal center line (Fig. 6.15*d*). This provides an opportunity to parallel certain elements. In the process of making this simplification the winding of the delta between lines 2 and 3 has disappeared altogether. This is a reasonable consequence of the symmetry, for if a current is injected into the circuit at the switch it will find the paths 7–3 and 1–2 through the delta equally attractive, and will divide equally. From a transient point of view no voltage will exist between 3 and 2, so it is legitimate to ignore the inductance L_3 between these points when making this transient calculation.

The circuit of Fig. 6.15*d* cannot be further reduced without introducing some approximations. But these may be justified by the circumstances. For example, if the cable is short, its inductance L_2 will be negligible compared with L_1 and L_3. If L_2 is omitted, the circuit simplifies significantly (Fig. 6.15*e*). Again, if the transformer is unloaded, L_3 would be its magnetizing inductance, which may be much greater than the generator inductance. If

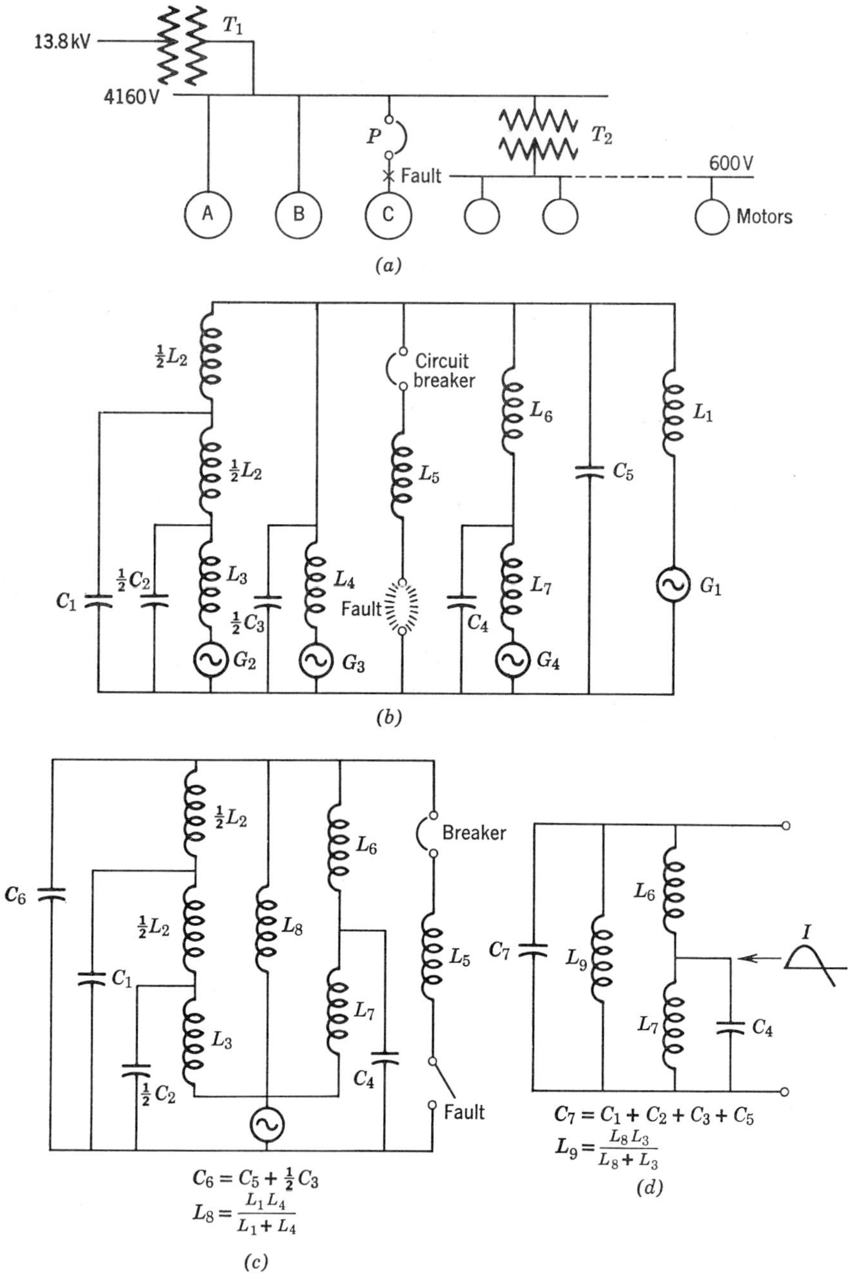

Fig. 6.16. (*a*) Industrial power circuit with motors. (*b*) Equivalent circuit for three-phase fault at *X*. (*c*) Reduced equivalent circuit. (*d*) Equivalent circuit for circuit breaker recovery transient.

this is so, we would be justified in ignoring L_1. What then remains of the circuit is the single tank circuit shown in Fig. 6.15*f*.

Note that the choice of component representation and decisions on approximations depend very much on the circuit conditions being studied. To emphasize this point we will anticipate some of the problems of Chapters 7 and 11 without going into details, using the circuit we have just examined. Suppose that the transformer in Fig. 6.15*a* is connected on its secondary side to equipment which is very sensitive to voltage surges, which might be a semiconductor rectifier. We wish to enquire into the dangers, if any, to this equipment when the circuit breaker is closed to energize the transformer. The circuit representations of Figs. 6.15*b* to 6.15*f* are inadequate for this problem. A representation that acknowledges the distributed properties of the cable and the capacitive coupling between the windings of the transformer is now required. Solutions to problems of this kind must await attention until later chapters. Suffice it to say here that when the circuit breaker closes, a traveling wave of voltage of magnitude proportional to the system voltage on closing spreads along the cable at a speed approaching the speed of light. On reaching the high impedance of the transformer, the voltage wave may almost double in amplitude, and, depending upon the capacitance between the transformer windings, a considerable fraction of the surge may be coupled into the secondary circuit. Considerable damage could result, but we would be unaware of this danger if we based our calculation on the circuits of Fig. 6.15 and assumed only magnetic coupling between the transformer windings.

As another exercise in circuit reduction we will consider part of a typical industrial power system (Fig. 6.16*a*). It operates at three voltage levels. There is a 4160-V bus fed through a transformer T_1 from a 13.8-kV supply. Connected to the bus are three motors *A*, *B*, and *C*, and a transformer T_2, which feeds other motors at 600 V. Suppose a three-phase fault occurs at *X*. The equivalent circuit we use will depend upon the question being posed and the accuracy required in the computation. We will consider two different problems: (1) the transient current that flows into the fault; and (2) transient voltage disturbance on the bus when a circuit breaker at *P* opens and clears the fault. In both cases the 13.8-kV supply on the primary side of T_1 can be assumed "stiff" and can therefore be represented by a generator of zero impedance.

We next consider the significant capacitances of the system, for they will discharge into the fault and contribute most of the initial transient current. These capacitances will influence the rate of rise of recovery voltage on the 4160-V bus following interruption. If the motors are fed by cable, rather than bus work, the cable capacitance will predominate over all others unless the cables are very short. Typical values of capacitance and inductance for system components are given in Chapter 16. From these data it will be seen that 50 ft of cable has more capacitance than most transformers to which the cable is likely to be connected. In this example we assume that motor A has a significant length of cable but motor B does not. In the first instance the motor capacitances should be considered. They can be neglected later if they are unimportant.

Looking into the circuit from the point of fault we observe an equivalent circuit for each phase, which closely approximates Fig. 6.16*b*. Some simplification has already been introduced. The circuit elements have the following significance:

L_1 = inductance of transformer T_1
L_2 = inductance of cable to motor A
L_3 = locked rotor inductance of motor A
L_4 = locked rotor inductance of motor B
L_5 = inductance of loop between bus and fault
L_6 = inductance of transformer T_2
L_7 = combined locked rotor inductances of 600-V motors

C_1 = capacitance of cable to motor A
C_2 = capacitance of motor A
C_3 = capacitance of motor B
C_4 = combined capacitance of 600-V cables and motors
C_5 = bus capacitance and capacitance of transformers T_1 and T_2

All the inductances and capacitances are referenced to the 4160-V level. Here G_1 represents the primary source, while G_2, G_3, and G_4 are the motors A, B and the combined 600-V motors, respectively. Confronted with a collapse of voltage at their buses, due to the fault, the motors will momentarily behave like generators. For a short time they can be represented by a source, equal to their back emf, in series with their locked rotor impedance. Their characteristics will change with time as the fluxes in their rotors die down and they lose speed. The 600-V motors are so remote from the fault that their impedances can be effectively paralleled, as can their cable capacitances. The several generators, which the motors become, have approximately the same emf, which is close to that of the primary supply. This permits all of those sources to be paralleled. This has been done in Fig. 6.16*c*. Certain capacitances and inductances now appear directly in parallel, so they have been combined.

Figure 6.16*c* is a basic circuit. The next step of reduction will depend upon the problem to be solved and the degree of approximation being sought. To find the transient fault current, the author favors the method of short-circuiting the generation and applying at the point of fault minus the voltage appearing at that point when the fault occurs. The breaker would be closed. To determine the recovery voltage when the breaker is subsequently opened, minus the fault current would be injected at the breaker contacts with the fault applied.

The inductance L_5, although small, is an essential element in calculating the transient fault current, for if it were not there, the first component we would see, looking in at the fault, would be the capacitance C_6. To apply a step of voltage to simulate the fault would be to require an infinite current to flow into this capacitance. If, in fact, L_5 is very small and C_5 is considerable, these two components dominate the picture initially, giving rise to a current $V(C_6/L_5)^{1/2} \sin t/(L_5C_6)^{1/2}$. Subsequently, currents associated with C_1 and C_4 will be evident. All these oscillatory currents will ride on top of the current from G, which will be of ramp form, with a slope of V/L', where L' is the series/parallel combination of the inductances between the source of the fault.

In Fig. 6.16*c* the cable has been represented by a T circuit rather than a π circuit. The reason, again, is that with the π arrangement a capacitance of $\frac{1}{2}C_1$ electrically close to the fault would lead to an unreasonably large high-frequency current. In many instances L_5 and C_6 will be so small and of so little consequence they can be neglected. This results in a helpful simplification of the equivalent circuit.

Just as it is impossible to apply a step of voltage to a capacitor, it is impossible to inject a current into an inductor. Thus in computing the recovery voltage transient in this example, we must either omit L_5 or add some capacitance across the circuit breaker before injecting the negative fault current. Very often L_5 will be insignificant compared with the other inductances of the system, so that it can be left out. Not infrequently, the cable inductance is also small and can be similarly rejected. This reduces the circuit to that shown in Fig. 6.16*d*. This can be handled analytically without too much labor to give a solution of acceptable accuracy.

In this example the concern has been with a three-phase fault in what has been assumed to be a grounded system. For other types of faults or with different grounding arrangements, the formulation of the problem and the process of circuit reduction would be more complicated, but not unduly so, if the techniques of this section are combined with the methods of earlier sections in this chapter. For example, if symmetrical components are used, the circuit reductions described above could be applied to the positive and negative sequence networks. Depending upon winding and grounding configurations, the zero sequence network could evolve as a very simple circuit. For example, if the motors have delta or ungrounded wye windings, they present an infinite impedance to zero sequence currents. Here only the motor capacitance would be effective. Similar considerations as far as the transformers are concerned may confine the zero sequence network to the 4160-V bus.

Our objective in this chapter has been to develop methods for solving transients in three-phase circuits and to devise techniques for reducing the complexity of such circuits in order to facilitate the application of these methods. However, we should not overlook some of the results that have emerged and some of the insights gained in the process. Hopefully, with experience, the reader will develop a "nose" for a solution and be able to tell by looking at a circuit which components are and which components are not contributing significantly to the overall response of the system.

REFERENCES

1. I. B. Johnson, A. J. Schultz, N. R. Schultz and R. B. Shores, "Some Fundamentals on Capacitance Switching," *Trans. AIEE*, Vol. 74, Part III (1955), p. 727.
2. C. L. Fortescue, "Method of Symmetrical Coordinates Applied to the Solution of Polyphase Networks," *Trans. AIEE*, Vol. 37, Part II (1918), p. 1329
3. C. F. Wagner, and R. D. Evans, *Symmetrical Components*, McGraw-Hill Book Co., New York and London (1933).
4. C. L. Fortescue, Discussion of E. W. Boehne's paper, "The Determination of Circuit Recovery Rates," *Trans. AIEE*, Vol. 54 (1935), p. 530; appearing in Vol. 55 (1936), p. 192.
5. R. D. Evans and A. C. Monteith, "System Recovery Voltage Determination by Analytical and A-C Calculating Board Methods," *Trans. AIEE*, Vol. 56, (1937), p. 695
6. P. Hammarlund, "Transient Recovery Voltages," *Acta Polytechnica*, El. Engrg Series, Vol. 1, No. 1 (1947), p. 18.
7. W. P. Lewis, "Solution of Network Transients Using Symmetrical Component Techniques," *Proc. IEE*, Vol. 113 (1966), p. 2012.
8. Edith Clarke, *Circuit Analysis of A.C. Power Systems*, Vol. I, John Wiley & Sons, New York, (1950).

7 Transients in Direct Current Circuits and in Conversion Equipment

7.1 Introduction

Transients occur in direct current circuits for much the same reasons that they occur in alternating current circuits. They are brought about by sudden changes in circuit conditions, most notably by switching operations. Where lightning can intrude, as on HVDC transmission lines, and in certain transportation applications, this too creates disturbances. Why then reserve special treatment for direct current circuits? We do so for several reasons. First, the conventional way of interrupting direct currents results in voltage transients of somewhat different form from those generated on alternating current circuits. Second, converting equipment, rectifiers and inverters, function by performing a continuous sequence of switching operations which evokes a continuous sequence of so-called commutation transients. To understand how such equipments operate, it is important to be acquainted with these transients. Finally, the almost universal use for rectification and invertion of solid-state components, which are notoriously vulnerable to transients, accentuates the need for study, so that these components can be applied safely and economically. The way in which this is done abounds with instances for the application of our transient knowledge.

7.2 Interruption of Direct Currents

We are considering here the disturbances created when a direct current is interrupted by a circuit breaker or a fuse. We have seen that in a-c circuits this is effected at or near a power frequency current zero, which occurs twice each cycle. In the d-c circuit the conventional interrupter must create its own current zero. The circuit breaker accomplishes this by developing an arc voltage, opposed to, and significantly in excess of the system voltage. This drives the current to zero. There is some similarity between this process and that described in Section 5.2; however, there is an important difference. Most of the magnetic energy stored in the circuit when the current has to be interrupted is dissipated in the arc, rather than transferred to the system capacitance. This affects the form of the voltage transient generated.

The simplest circuit for a d-c switching operation is that shown in Fig. 7.1, wherein the switch is clearing a short circuit. If the form of the arc voltage V_a is specified, it is possible to determine the current; conversely, if the current is specified, the arc voltage can be determined. Figure 7.2*a* shows the current and voltage relationship in a typical circuit. It is seen that the current declines as a fairly linear ramp. This we would expect from a constant arc voltage, for the net voltage in the circuit is then $V_S - V_A$, and the equality

$$L\frac{dI}{dt} = V_S - V_A \tag{7.2.1}$$

must always apply. Since $(V_S - V_A) < 0$, the slope is negative.

In Fig. 7.2*a* the arc voltage is relatively constant during the linear portion of the current trace. The ragged appearance of the arc voltage trace arises from complex fluctuating processes within the arc itself. These voltage disturbances are not reflected in the current trace because of the inductance of the circuit. They are accommodated by fluctuating currents in the stray capacitance of the system components.

It is clear from Eq. 7.2.1 that rapid interruption of the current, that is, a high di/dt, can be accomplished only by the generation of a high arc voltage. Where the circuit inductance is high, such voltage transients can be of long duration as well as high in magnitude, for, as pointed out earlier, the energy stored in the circuit inductance must be dissipated in the arc as $\int V_A \cdot I\,dt$. Figure 7.2*b* shows the effect of a higher circuit inductance. Compared with Fig. 7.2*a*, the magnitude of the arc voltage is about the same, being a characteristic of the interrupter, but the duration of the interruption time is increased in proportion to the increase in inductance.

This example shows very clearly the difference between the voltage transients developed when alternating and direct currents are interrupted.

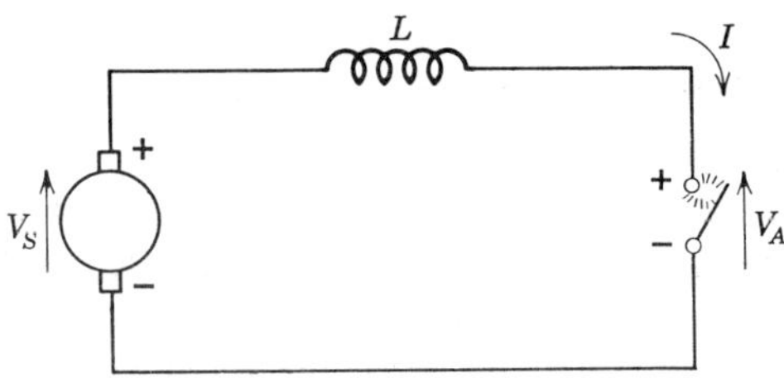

Fig. 7.1. Interruption of a direct-current short circuit.

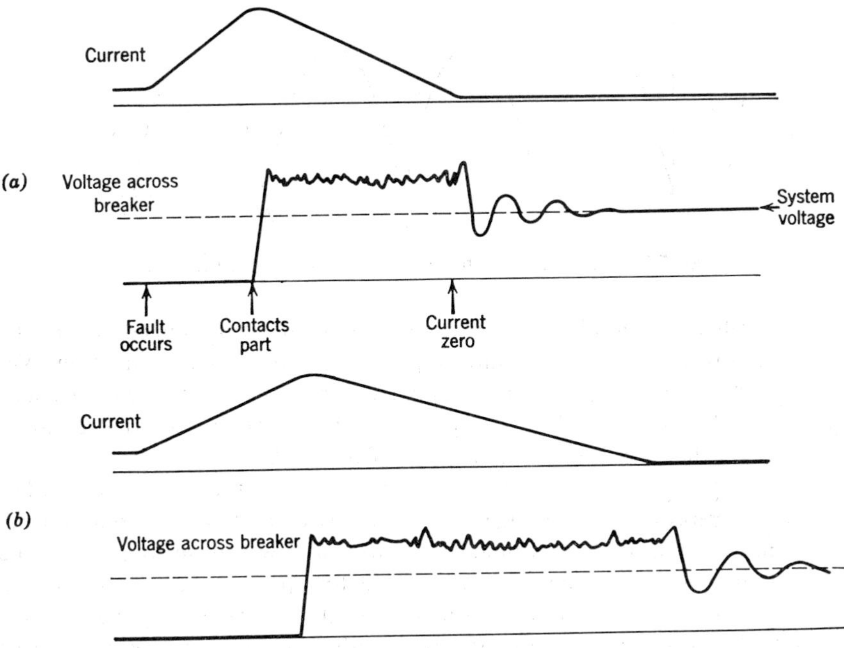

Fig. 7.2. Current and voltage relationships during the interruption of a direct current, showing the influence of circuit inductance. In case (*b*) the circuit inductance is higher than case (*a*).

In the d-c circuit we are obliged to interrupt when the stored magnetic energy is high. In the a-c circuit interruption takes place at current zero, when the magnetic energy is at a minimum. In the a-c circuit the capacitance is a very potent factor, for it largely defines the form of the recovery voltage. In the d-c circuit it provides a path for high-frequency currents generated by rapid fluctuations in the arc voltage and thereby performs a filtering action. It also becomes involved in the transitional process when the arc voltage subsides to the normal supply voltage, once the system stored energy is spent. But it does not control the general level of the overvoltage created during arcing.

7.3 Delayed and Periodic Functions

Before we can study converting equipment we must be able to describe by Laplace transforms the kinds of waveforms that such equipment generates. In general, these waveforms are periodic. A simple, common example is the full-wave rectified output of a sine wave generator, which is shown in Fig. 7.3. Consider how the response of a network to a driving force of this kind could be obtained. An old standby is superposition. However, it has been stated firmly that the principle of superposition applies to linear systems only, and

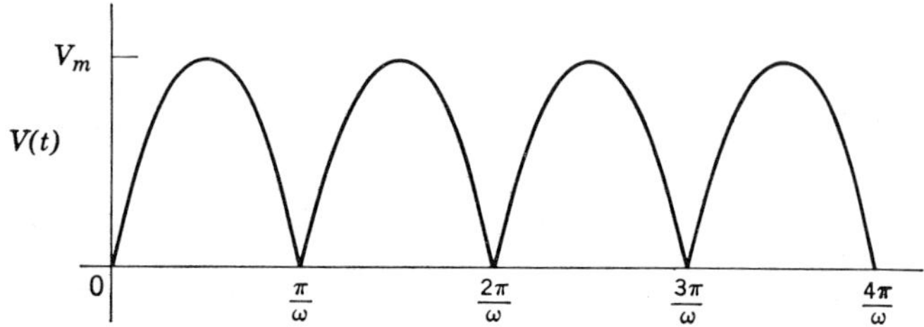

Fig. 7.3. Full wave rectified output of a sine wave generator.

most rectifiers are nonlinear. The ideal rectifier has zero impedance in the forward direction and an infinite impedance in the reverse direction. Yet superposition can be applied if it is applied with caution, taking into account the physical limitations of the components. This point will be explored more fully later, when we have a better grasp of the problem.

Observing Fig. 7.3, we note that for $0 < t < \pi/\omega$ the $V(t)$ is sinusoidal, so we can represent the voltage by the sine wave $V_1(t) = V_m \sin \omega t$ during that time. Thereafter $V_1(t)$ would go negative and would no longer describe $V(t)$. This situation is remedied by adding to $V_1(t)$, or superposing on it, a second sine wave, $V_2(t) = 2V_p \sin \omega t$. Together $V_1(t)$ and $V_2(t)$ accurately describe $V(t)$ for $0 < t < 2\pi/\omega$. If this procedure of adding waves at intervals of π/ω is continued,

$$V_n(t) = V_{n-1}(t) = \cdots = V_3(t) = V_2(t) = 2V_m \sin \omega t$$

we do in fact synthesize the waveform $V(t)$. We may therefore say that the response of a network to $V(t)$ would be the same as the response to the series of stimuli described above. Actually, we need only find the response of a sinusoidal stimulus and then add it to itself at the appropriate half cycle time intervals. A little thought will convince the reader that the half-wave rectifier waveform is even easier to synthesize. All the added waves $V_1(t) \cdots V_n(t)$ are identical; each is $V_m \sin \omega t$.

It would appear that we must add an infinite number of responses, but in fact it usually does not work out that way. Almost invariably we find a pattern emerging, consecutive terms fall into a series, which it is possible to sum.

An alternative, and very similar approach, is to identify the repeating element in the stimulus. In the cases of the full- and half-wave rectified sine waves, these are given in Fig. 7.4. The response to this element is required as it is applied at $t = 0$, $t = \tau$, $t = 2\tau$, etc., where τ is the period of repetition. To do this formally requires the transform of the function, and its transform when delayed $\tau, 2\tau, 3\tau, \ldots$. This leads us to an important theorem sometimes referred to as Heavisides' Translation or Shifting Theorem, by which we can

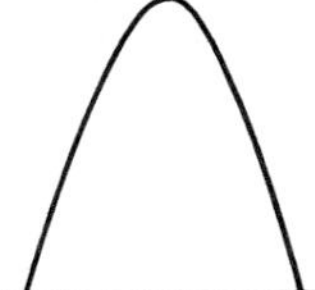

Fig. 7.4. Repeating elements in the full and half wave rectifier waveforms.

obtain the transforms of delayed functions. This theorem may be stated symbolically as follows:

$$\left.\begin{aligned} &\text{If} \qquad \mathcal{L}F(t) = f(s) \\ &\text{then} \qquad \mathcal{L}F(t-\tau) = \epsilon^{-s\tau}f(s) \qquad (\text{for} \quad t \geqslant \tau) \end{aligned}\right\} \tag{7.3.1}$$

$F(t - \tau)$ is $F(t)$ delayed by a time τ; that is, if $F(t)$ has a certain value at a given instant, $F(t - \tau)$ will have the same value τ sec later. This is illustrated in Fig. 7.5.

By definition

$$\mathcal{L}F(t-\tau) = \int_{\tau}^{\infty} \epsilon^{-st}[F(t-\tau)]\, dt$$

(the lower limit is τ because the function is limited to times $t \geqslant \tau$). Let the variable be changed by the substitution $\beta = t - \tau$, so that $t = \beta + \tau$, and $dt = d\beta$. Then

$$\begin{aligned} \mathcal{L}F(\beta) &= \int_0^{\infty} \epsilon^{-s(\beta+\tau)} F(\beta)\, d\beta \\ &= \epsilon^{-s\tau} \int_0^{\infty} \epsilon^{-s\beta} F(\beta)\, d\beta \\ &= \epsilon^{-s\tau} f(s) \end{aligned}$$

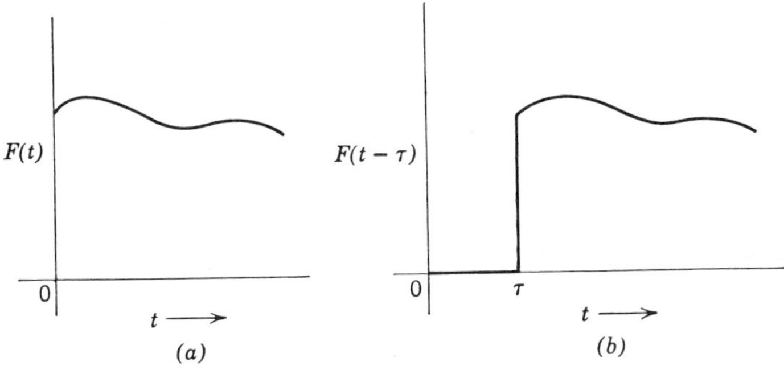

Fig. 7.5. A delayed function. (*a*) The function. (*b*) The function delayed by a time τ.

To cite an example, the theorem states that since the transform of a step function V is V/s, a step function delayed by a time τ has a transform $\epsilon^{-s\tau}(V/s)$. We shall find this very useful when handling traveling waves on transmission lines.

Repeated or periodic functions are treated as a succession of delayed functions. Consider the full-wave rectified sine wave voltage made up by repeating the first element in Fig. 7.4. The transform of a half-cycle sine wave is given by

$$\int_0^{\pi/\omega} \epsilon^{-st} \sin \omega t \, dt = \left| \frac{\epsilon^{-st}}{s^2 + \omega^2} [-s \sin \omega t - \omega \cos \omega t] \right|_0^{\pi/\omega} = \frac{\omega(1 + \epsilon^{-s\pi/\omega})}{s^2 + \omega^2} \tag{7.3.2}$$

The transform of the entire waveform will be formed by adding to Eq. 7.3.2 transforms of delayed functions of similar form:

$$\frac{\omega(1 + \epsilon^{-s\pi/\omega})}{s^2 + \omega^2} [1 + \epsilon^{-s\pi/\omega} + \epsilon^{-2s\pi/\omega} + \cdots] \tag{7.3.3}$$

The infinite series in the bracket can be summed quite readily; it gives

$$\frac{1}{(1 - \epsilon^{-s\pi/\omega})}$$

Substituting this in Eq. 7.3.3 leads to the expression

$$\frac{\omega(1 + \epsilon^{-s\pi/\omega})}{(s^2 + \omega^2)(1 - \epsilon^{-s\pi/\omega})} \tag{7.3.4}$$

for the transform of the complete rectified waveform.

This is a particular example of a form applicable to periodic functions in general. If a function $F(t)$ has a period a, then its transform is given by

$$f(s) = \frac{\int_0^a \epsilon^{-st} F(t) \, dt}{(1 - \epsilon^{-as})} \tag{7.3.5}$$

The proof of Eq. 7.3.5 is as follows. Since $F(t)$ is a periodic function with a period a,

$$F(t + a) = F(t) \qquad (t > 0)$$

We may therefore write for its transform

$$f(s) = \sum_{n=0}^{n=\infty} \int_{an}^{a(n+1)} \epsilon^{-st} F(t) \, dt \tag{7.3.6}$$

Let $\beta = t - na$, then $F(\beta + na) = F(\beta)$ and from Eq. 7.3.6

$$f(s) = \sum_{n=0}^{n=\infty} \epsilon^{-nas} \int_0^a \epsilon^{-s\beta} F(\beta)\, d\beta$$

$$= \int_0^a \epsilon^{-s\beta} F(\beta)\, d\beta \sum_{n=0}^{n=\infty} \epsilon^{-nas}$$

$$= \frac{\int_0^a \epsilon^{-s\beta} F(\beta)\, d\beta}{(1 - \epsilon^{-as})} \tag{7.3.7}$$

As an example of the application of Eq. 7.3.7 consider an RL circuit driven by the output voltage waveform of a six-phase rectified sine wave. The relevant waveform and circuit are shown in Fig. 7.6. If we wish to determine the current, the differential equation is

$$V(t) = RI + L\frac{dI}{dt} \tag{7.3.8}$$

Taking transforms,

$$v(s) = i(s)[R + Ls] - LI(0) \tag{7.3.9}$$

If it is assumed that $I(0) = 0$, Eq. 7.3.9 can be rewritten as

$$i(s) = \frac{v(s)}{R + Ls} \quad \text{or} \quad i(s) = \frac{1}{L}\frac{v(s)}{s + \alpha} \tag{7.3.10}$$

where $\alpha = R/L$. We must now devise an expression for $v(s)$. According to Eq. 7.3.5,

$$v(s) = \int_0^{\pi/3\omega} \frac{\epsilon^{-st} \sin(\omega t + \pi/3)\, dt}{(1 - \epsilon^{-s\pi/3\omega})}$$

$$= \left| \frac{\epsilon^{-st}[-s \sin(\omega t + \pi/3) - \omega \cos(\omega t + \pi/3)]}{(s^2 + \omega^2)(1 - \epsilon^{-s\pi/3\omega})} \right|_0^{\pi/3\omega}$$

$$= \frac{(\sqrt{3}/2)s[1 - \epsilon^{-s\pi/3\omega}] + \omega/2[1 + \epsilon^{-s\pi/3\omega}]}{(s^2 + \omega^2)(1 - \epsilon^{-s\pi/3\omega})}$$

$$v(s) = \frac{\sqrt{3}s}{2(s^2 + \omega^2)} + \frac{\omega(1 + \epsilon^{-s\pi/3\omega})}{2(s^2 + \omega^2)(1 - \epsilon^{-s\pi/3\omega})} \tag{7.3.11}$$

Substituting Eq. 7.3.11 in Eq. 7.3.10, the expression for current becomes

$$i(s) = \frac{\sqrt{3}}{2L}\left[\frac{s}{(s + \alpha)(s^2 + \omega^2)}\right] + \left\{\frac{1 + \epsilon^{-s\pi/3\omega}}{1 - \epsilon^{-s\pi/3\omega}}\right\} \frac{1}{2L}\left[\frac{\omega}{(s + \alpha)(s^2 + \omega^2)}\right] \tag{7.3.12}$$

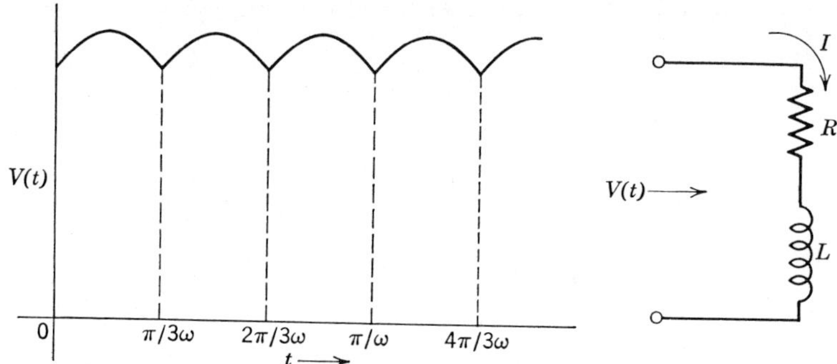

Fig. 7.6. An *RL* circuit driven by a six-phase rectified sine wave.

Equation 7.3.12 may look rather formidable, but in fact there is nothing in it that we have not encountered before. The term in the first square bracket resolves into three simple transforms,

$$\frac{s}{(s+\alpha)(s^2+\omega^2)} = \frac{\alpha}{(\alpha^2+\omega^2)(s+\alpha)} - \frac{\alpha s}{(\alpha^2+\omega^2)(s^2+\omega^2)} + \frac{\omega}{(\alpha^2+\omega^2)(s^2+\omega^2)}$$

each term of which is a recognizable transform,

$$\mathcal{L}^{-1}\frac{s}{(s^2+\omega^2)(s+\alpha)} = \frac{\alpha\epsilon^{-\alpha t}}{(\alpha^2+\omega^2)} - \frac{\alpha}{(\alpha^2+\omega^2)}\cos\omega t + \frac{\sin\omega t}{(\alpha^2+\omega^2)} \quad (7.3.13)$$

The term in the second square bracket of Eq. 7.3.12 can be similarly resolved:

$$\frac{\omega}{(s+\alpha)(s^2+\omega^2)} = \frac{\omega}{(\alpha^2+\omega^2)}\,\frac{1}{(s+\alpha)} - \frac{\omega}{(\alpha^2+\omega^2)}\,\frac{s}{(s^2+\omega^2)} + \frac{\alpha}{(\alpha^2+\omega^2)}\,\frac{\omega}{(s^2+\omega^2)}$$

for which the inverse transform is

$$\mathcal{L}^{-1}\frac{\omega}{(s+\alpha)(s^2+\omega^2)} = \frac{\omega}{(\alpha^2+\omega^2)}\epsilon^{-\alpha t} - \frac{\omega}{(\alpha^2+\omega^2)}\cos\omega t + \frac{\alpha}{(\alpha^2+\omega^2)}\sin\omega t \quad (7.3.14)$$

We now direct attention to the terms in the curly bracket of Eq. 7.3.12. The $(1+\epsilon^{-s\pi/3\omega})$ in the numerator indicates that in addition to Eq. 7.3.14 there is an identical term delayed by a time $\pi/3\omega$ or one-sixth of a cycle. Finally, the presence of $(1-\epsilon^{-s\pi/3\omega})$ in the denominator tells us that the entire second term is periodic and repeats with the same period, $\pi/3\omega$, as the stimulus.

It appears at first sight that a good bookkeeping system is needed to keep track of the growing multiplicity of terms. Actually, the situation is not as complicated as it seems. Equation 7.3.14 comprises, for the most part, sine and cosine terms, which, when added to themselves after an interval of π/ω, annihilate themselves. This follows from the well known property of sine and cosine waves:

$$\sin \omega\left(t + \frac{\pi}{\omega}\right) = -\sin \omega t$$

$$\cos \omega\left(t + \frac{\pi}{\omega}\right) = -\cos \omega t$$

In the example we have been considering the source has a frequency $\omega/2\pi$ and the cycle is divided into six equal intervals of $\pi/3\omega$. A sine or cosine wave starting at the beginning of the fourth interval will cancel an identical wave which started at the first. Waves starting at the beginning of the fifth interval will cancel those that started with the second, and so on. Thus, as far as the components in Eq. 7.3.14 are concerned, we need compute only for a cycle, knowing that subsequent cycles will be identical.

At any instant the exponential terms can be expressed as a summation:

$$\sum \epsilon^{\alpha t}[1 + \epsilon^{-\alpha\pi/3\omega} + \epsilon^{-2\alpha\pi/3\omega} \cdots + \epsilon^{-n\alpha\pi/3\omega}]$$

one term being added every $\pi/3\omega$ sec. As t becomes large compared with the time constant $1/\alpha$, this summation tends to

$$\frac{\epsilon^{-\alpha t}}{(1 - \epsilon^{-\alpha\pi/3\omega})}$$

Expressions 7.3.13 and 7.3.14 contain the complete circuit response including the transient which occurs when the voltage is first applied.

7.4 Characteristics of the Thyristor

It has long been recognized that alternating and direct currents each have their place in the overall scheme of power generation, transmission, distribution, and utilization. For technical and economic reasons generation, transmission and distribution are almost exclusive domains for alternating current, though we are now seeing a limited shift to high-voltage direct current as a mode of transmission for these same reasons. In some areas of utilization direct current has always been favored. But two other factors have become increasingly important in recent years: the ability to control the level of a d-c voltage and the ability to control the frequency of an a-c supply. Such flexibility offers tremendous opportunities for utilization of d-c.

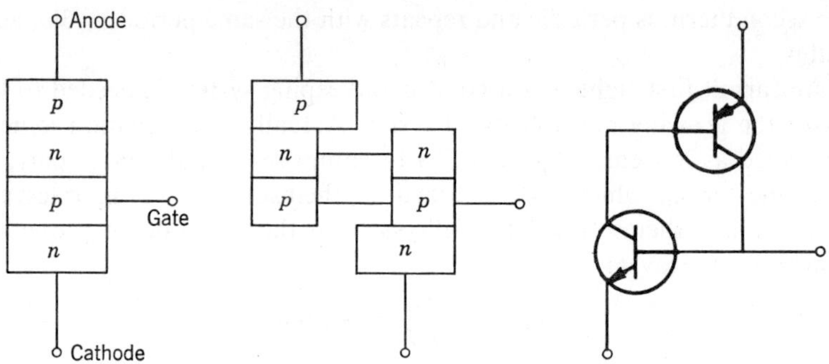

Fig. 7.7. Convenient representation of a thyristor.

For many years this function of control was carried out by gas-filled devices, notably the mercury arc rectifier, the ignitron, and the gas-filled relay or thyratron. These devices are still widely used for control functions. However, the field is being rapidly taken over by the solid-state equivalent, the thyristor, or semiconductor controlled rectifier, which is essentially a high-speed electronic switch. Its ability to be switched from the blocking to the conducting state quickly (turned on), with the expenditure of a relatively small amount of energy, and be subsequently returned to the blocking state (turned off), make it so suitable for the control functions mentioned. It is this aspect that is of interest to us because of the transients it gives rise to, especially since these transients can cause distress to other solid-state devices and to the thyristors themselves. To understand the transient problems it is not necessary to delve deeply in semiconductor theory, but it is desirable to be acquainted with some characteristics of thyristors to see how they interact with the rest of the circuit. This is the goal of this section. Some background in semiconductors will be assumed.

A thyristor has a *p-n-p-n* structure. It is best visualized as consisting of two transistors, a *p-n-p* and an *n-p-n* interconnected to form a regenerative feedback pair (Fig. 7.7). The normal condition of the device is blocking. That is, if voltage is applied between anode and cathode to provide forward or reverse bias, the thyristor will present a high impedance. However, with a forward bias applied, the device will quickly become conducting if a small current is injected at the gate lead, which closes the switch. Once closed, the continued application of gate current is unnecessary. It will remain conducting until the current is brought to zero or a very low value by some outside agency and reverse bias is applied. This is how the "switch" is turned off or opened. The analogy with a conventional circuit breaker is a useful one. In the blocking state, or open position, there is a definite voltage that the junction can support. If this is exceeded, the device will break down; the term "breakover" is used if the offending voltage has a forward bias. This can be destructive especially if the bias is reverse. When a circuit breaker is opened and the current in the arc comes to zero, the plasma between the contacts must be cooled and/or removed to re-establish a blocking condition. Similarly in the thyristor, the plasma must be removed from the junctions by diffusion and recombination before the device can block and accept forward voltage. This requires a finite time, the so-called "turn-off time."

Thyristors have a finite "turn-on time" too. Conduction starts physically near the gate lead and spreads laterally as the plasma spreads laterally across the wafer. If the current builds up too quickly while the conducting area is small, thermal damage will result. The rate of build up of current, dI/dt, must be kept within the device's capability (1).

There is one other important characteristic of the thyristor: it can be turned on unintentionally by a high forward dV/dt without recourse to a gate current. What happens is that the capacitance current provided by changing voltage and the capacitance of the thyristor, $I = C\,dV/dt$, performs the function of the gate current.

Among the characteristics of any thyristor we can therefore expect to find specified the peak forward and reverse voltages it can hold off, the turn-off time, the turn-on time, and the peak forward dV/dt it can accept (2).

With this background we can now proceed to look at a thyristor from two points of view, as a switch opening and closing and thereby causing transients, and as potential victim of transients, not least those which it generates itself. This can be done very usefully by studying the transients in some typical solid-state equipment.

7.5 Commutation Transients—The Current-Limiting Static Circuit Breaker

Commutation transients are those electrical disturbances that are initiated by switching current from one circuit to another. Such transients are therefore occurring continually in any piece of converting equipment in the normal course of its operation whether it be a rotary converter or a static rectifier or inverter. However, as an introduction to the subject, we shall first concern ourselves with the current-limiting static circuit breaker rather than any kind of converter, not because it has greater technical significance, but because it illustrates in very simple capsule form the essence of the commutation transient problem. It is also an excellent vehicle for studying the thyristor in a transient environment.

The simplest conceptualization of this static circuit breaker is shown in Fig. 7.8. The normal current carrying element is thyristor T_1. By applying a signal to its gate, it is switched from the blocking to the conducting state and current I flows from the source on the left to the load on the right. To "open"

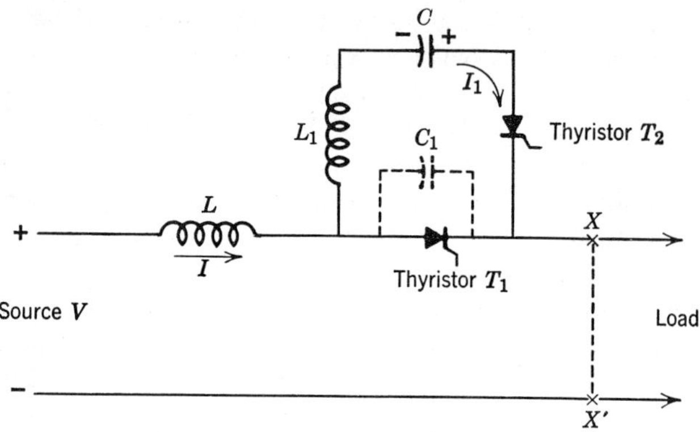

Fig. 7.8. Basic circuit of the current-limiting static circuit breaker.

the circuit breaker, it is necessary to turn off the thyristor. This is done by gating a second thyristor, T_2, which causes the precharged capacitor C to be discharged around the local loop comprising the two thyristors and the loop inductance L_1. The direction of this current I_1 is such as to oppose the existing current in T_1 and drive it to zero. There is now no direct current path from the source to the load. Alternatively, the process can be thought of as one

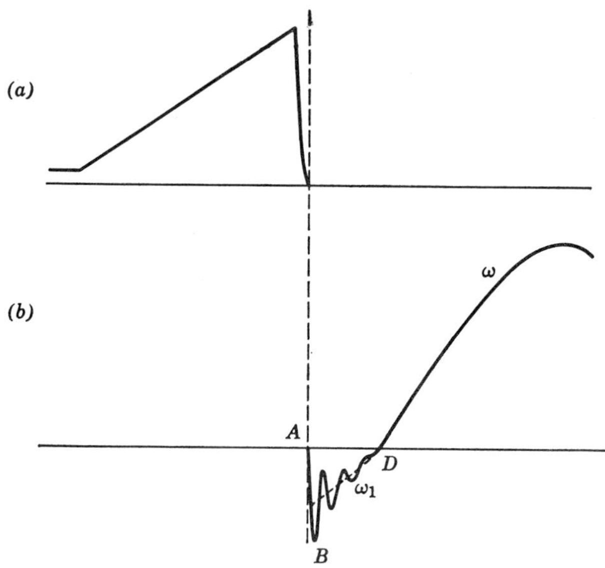

Fig. 7.9. (*a*) Current in T_1. (*b*) Voltage across T_1. (*c*) Voltage and current transients occurring during the operation of current-limiting static circuit breaker.

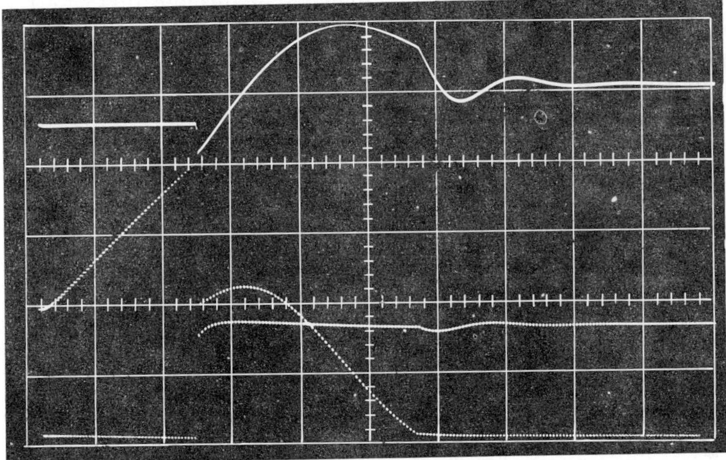

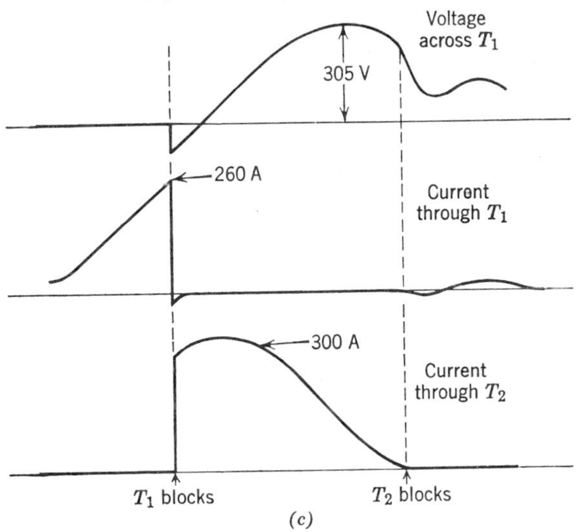

(c)

Fig. 7.9. (continued)

of transferring, or commutating, the current from its path through T_1 into the capacitor.

The current I_1 would be sinusoidal in form and of a peak value $V_C(0) \times (C/L_1)^{1/2}$, where $V_C(0)$ is the precharge voltage on the capacitor. This must exceed the instantaneous value of the current I if interruption is to be effected.

Suppose in the course of supplying a load, a short circuit occurs at XX'. The current I would start to increase at a rate V/L, A/sec, where V and L are the source voltage and source inductance, respectively. This would be sensed and thyristor 2 would be turned on; the current through thyristor 1 would take the form shown in Fig. 7.9*a*. It is clear from this figure why the device is described as current-limiting. Current actually continues to flow in the fault, for it has not been truly interrupted but rather commutated into the capacitor C. However, C will rapidly charge and establish a voltage opposing the current, thereby driving it to zero. What we observe is an oscillation between the source inductance L (augmented by L_1) and the capacitor C, which ultimately is damped out. This is one of the commutation transients involved in this operation. Another, and a more important one in some ways, is the transient that occurs in the local loop comprising the two thyristors when commutation is initiated. As long as thyristor 1 is conducting there is only a small forward voltage across it. But when its current has been driven to zero, it is suddenly exposed to the residual voltage on C, which may be quite significant. The thyristor has a relatively small (and incidentally variable) capacitance, which has been designated C_1 in Fig. 7.8. This is charged from C through L_1. Since $C \gg C_1$, this transient manifests itself as an oscillation at the natural frequency ω_1 of L_1 and C_1, which, once again, will be damped by circuit losses. The voltage that this gives rise to is shown in Fig. 7.9*b*. This figure also shows how this high-frequency oscillation is superimposed on the low frequency (ω) disturbance associated with the fault current flowing into C, described above. Figure 7.9*c* is an experimental oscillogram taken during the operation of such a circuit breaker on a 125-V, d-c circuit. The accompanying legend describes the events. The high-frequency oscillation between L_1 and C_1 was too fast to record with the time scale used. This figure shows yet another transient initiated by the blocking of thyristor T_2 when in due course its current attempts to reverse. What parameters this depends upon is left as an exercise for the reader.

Thus far we have been treating the thyristors simply as ideal switches and have then considered how the circuit responds to their closing and opening. But how do the demands of the circuit tie in with the limitations of the thyristors? When T_2 is first turned on to perform the commutating process, its current will start to rise at a rate $V_C(0)/L$, where $V_C(0)$ is the precharge voltage on the capacitor. In Section 7.4 it was pointed out that there is an upper limit to the initial dI/dt for thyristors. The L_1 must be great enough to assure that this limit is not reached.

Referring to Fig. 7.9*b*, T_1 is subjected to considerable inverse voltage immediately after blocking at point A. The peak of this voltage is indicated by B. This should be within the capabilities of the device. Again, inverse voltage persists across T_1 for a period AD; thereafter it becomes positive. This time, AD, must be longer than the turn-off time of T_1, referred to in Section 7.4. If it is not, the junction will not be clear of carriers and the device will turn on once more.

In Section 7.4 we also referred to the fact that thyristors can be turned

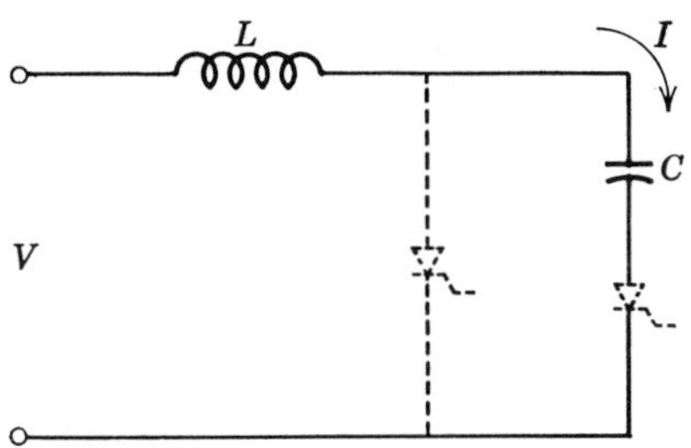

Fig. 7.10. Simple equivalent circuit for the current-limiting static circuit breaker after the fault has been commutated.

on by the capacitance current that flows when forward voltage is applied at an excessive rate. This could occur at point D in Fig. 7.9*b*, if dV/dt there exceeds the permitted limit. Finally, the peak forward voltage attained, apparent from Fig. 7.9*c*, must also be within T_1's capabilities.

To determine whether the thyristors can survive in the transient environment they are helping to create, it is necessary to specify the transients quantitatively. We will conclude discussion of the current-limiting static breaker by looking into this matter. Most of the questions can be answered by considering the very simple circuit of Fig. 7.10 which adequately describes the static breaker clearing a short circuit, from the time the current has been commutated to the capacitor. We have chosen to neglect the small inductance L_1 of the commutating loop in comparison with the source inductance L. The voltage across T_1 is then given by the capacitor voltage V_C. If this would start at zero, as shown in Section 2.3, it would have a form

$$V_C(t) = V_S(1 - \cos \omega t)$$

where $\omega = 1/(LC)^{1/2}$. We also showed there that if the capacitor had an initial voltage $V_C(0)$, the expression would be

$$V_C(t) = (V_S - V_C(0))(1 - \cos \omega t) \tag{7.5.1}$$

Inasmuch as $V_C(0)$ is negative in the situation we are now discussing the peak of Eq. 7.5.1 can exceed $2V_S$. There is another important initial condition, namely, the current is finite; in fact, it is the instantaneous value of the fault current. It is not difficult to show that this introduces a term

$$I(0)\left(\frac{L}{C}\right)^{1/2} \sin \omega t$$

This is identical with the term we obtain in current chopping (Section 5.2), which is not surprising, because this breaker is really chopping the current. The complete expression for the voltage across T_1 following commutation is given by

$$V_C(t) = [V_S - V_C(0)](1 - \cos \omega t) + I(0)\left(\frac{L}{C}\right)^{1/2} \sin \omega t \tag{7.5.2}$$

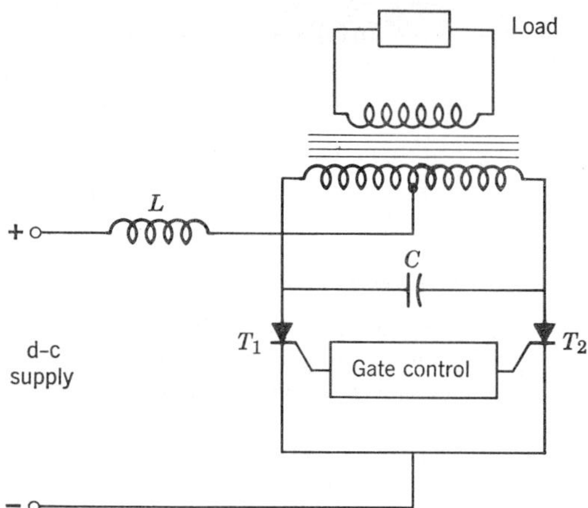

Fig. 7.11. Basic circuit of impulse-commutated inverter.

Given the values of L and C and the initial conditions $V_C(0)$ and $I(0)$, it is possible to evaluate the time AD and the value of dV/dt at D in Fig. 7.9*b* and also the peak forward voltage in Fig. 7.9*c*.

7.6 Commutation Transients in Conversion Equipment

The last section provides a good basis for looking into commutation transients in static convertion equipment. We will use the impulse commutated inverter as an example. This inverter is one of several types used to convert direct to alternating currents. It is quite old in concept. Wagner (3), (4) gives analyses of its operation when it was used with gas tubes. The basic circuit is shown in Fig. 7.11.

The mode of operation is as follows. Assume that thyristor T_1 is conducting, which means that the center point on the transformer winding is positive with respect to the left-hand end. The right-hand end of the winding will be more positive, so that the capacitor C is charged to approximately twice the d-c supply voltage. Suppose now that gate current is fed to thyristor T_2, thereby turning it on. A low-impedance path is created across the capacitor inasmuch as both thyristors are now conducting. A burst of current will flow around this path, the so-called commutating current, which opposes the current already flowing through T_1. The net current in this device is quickly reduced to zero and it is changed to the blocking state. Meanwhile, current has been established in T_2. The current has been commutated from one branch to the other. In due course this will cause the voltage on C to reverse so that when gate current is applied to T_1 a moment later, the transfer can be repeated in the opposite direction. Thus, if the thyristors are gated alternately in the manner described, an alternating voltage will be applied to the transformer primary and to the load on the secondary side. The similarity between these operations and the current-limiting breaker is obvious. The difference is that in the invertor they are occurring repeatedly at the frequency of the inverter. The current-limiting circuit breaker might be described as a "one-shot inverter." The situation in the inverter is complicated by the presence of the load, but a number of features in the two devices have direct correspondence. For example, when commutation is initiated at each half cycle, the inductance of the loop comprising T_1, T_2, and C in Fig. 7.11 corresponds to L_1 in Fig. 7.8. It must be sufficient to hold the dI/dt to a safe value. As a matter of practicality extra inductance is often added in the form of a ferrite core looped around the conductor, which enhances the circuit's natural inductance when the current is low but has little effect when the current is high and the ferrite is saturated. Further treatment of such saturable devices will be found in Chapter 8.

Our discussion so far on the impulse-commutated inverter has been confined to that part of the circuit in Fig. 7.11 that embraces only the capacitor C and the two thyristors. We have been able to look at this in isolation because it is fairly well decoupled from the remainder of the circuit, that is, events can take place quickly within this loop, more or less independently of what is happening outside. For example, the main current I will be relatively unchanged during the fast transients already described, because of the considerable inductance L through which it is flowing. The reader should be alert to the possibility of tackling problems in this way, for it allows one to break down and simplify what otherwise may seem to be intractable situations. We now turn to consider what happens when the current commutation from thyristor 1 to thyristor 2 is a *fait accompli*, and the momentary disturbances created thereby have died down. Conditions at that time can be quite well represented by the equivalent circuit of Fig. 7.12. We are assuming that the load is resistive and can be represented by the resistor R. This will

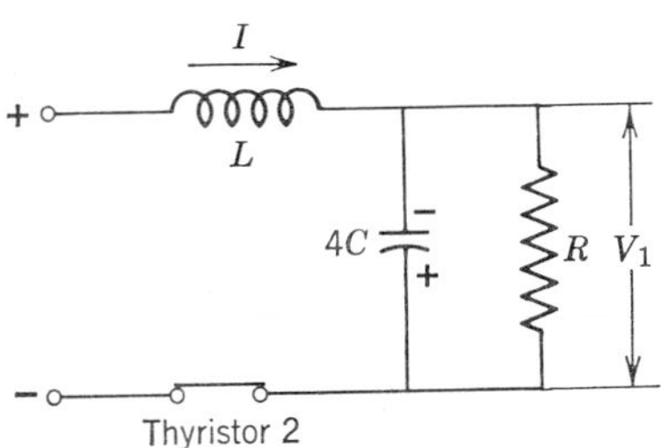

Fig. 7.12. Equivalent circuit for auto-impule inverter with resistive load.

be the value of the load resistance referred to the primary side of the transformer. Likewise, $4C$ is the value of the commutating capacitor now seen through a 1:2 autotransformer. We note that C retains the polarity of the previous half cycle. This is a holdover from our first piece of analysis, which must be used as an initial condition for the next. Let us ignore, for the moment, the resistor R. We are left with a simple oscillatory circuit, with an angular frequency $\omega_0 = (L \cdot 4C)^{-1/2}$ and a surge impedance $Z_0 = (L/4C)^{1/2}$. If the circuit were dead and then connected to the voltage V, the current would be

$$I(t) = \frac{V}{Z_0} \sin \omega_0 t$$

which would charge C and overswing its voltage to $2V$. In Chapter 2 it was shown that if the capacitor has an initial voltage $V_1(0)$, the voltage reached in the transient would be $2V - V_1(0)$. If in addition there is an initial current flowing, there will be a term IZ_0 added to C's voltage, which results from the magnetic energy associated with that current flowing in L, being deposited in C, when the current comes to zero. All of this follows precisely the comments on the static breaker. When we consider R, its effect will be to damp and thereby reduce the currents and voltages. To what degree will depend upon the damping parameter $\eta = R/Z_0$ that was introduced in Chapter 4. To investigate events using the formal approach, we would establish the differential equation describing the behavior of the circuit, which for voltage V_1 is

$$\frac{d^2V_1}{dt^2} + \frac{1}{4RC}\frac{dV_1}{dt} + \frac{V_1}{4LC} = \frac{V}{4LC} \tag{7.6.1}$$

Transforming the equation,

$$v_1(s) = V\frac{\beta}{s(s^2 + \alpha s + \beta)} + V_1(0)\frac{s + \alpha}{(s^2 + \alpha s + \beta)} + V_1'(0)\frac{1}{(s^2 + \alpha s + \beta)} \tag{7.6.2}$$

where $\alpha = 1/4RC$ and $\beta = I/4LC$. Then we would determine the inverse transforms. All the terms in Eq. 7.5.3 we recognize from Chapter 4, and we can extract their solutions from the generalized curves in that chapter, using the appropriate value of η. We would of course arrive at the same terms as we conceived earlier by a more heuristic approach.

What are some of the practical implications of these results for the inverter? It is apparent that the load voltage depends very much upon the load itself. If R is high (light load), it will exert little damping influence and the voltage can become very high. The operating frequency in this type of inverter is dictated by the rate at which gate signals are applied to the thyristors. After one thyristor has been gated, it is clear that sufficient time must be allowed before gating the other, for the voltage on the capacitor to reverse. In fact, the capacitor must be charged sufficiently in the reverse direction to be able to turn-off thyristor 2 when thyristor 1 is gated again. But, on the other hand, if the gate pulse rate is too slow and R is such that the circuit of Fig. 7.12 is oscillatory, the current I would attempt to reverse and turn-off the conducting thyristor without assistance from its mate, that is, the current on the d-c side would become discontinuous.

It would be particularly undesirable for the operating frequency to be at or near the natural frequency of L and C, for this would result in a resonance being struck, which would cause a pumping action on the voltage with each successive cycle. The voltage would rise until it was limited by the saturation of the transformer core.

In most practical inverters, wide excursions of load voltage are prevented by the use of bypass or feedback diodes (5). These are connected so that, in the equivalent circuit of Fig. 7.12, they would appear backbiased across the reactor L. Thus whenever the load voltage attempts to rise above the supply voltage the diode would conduct and act as a clamp. The entrained energy instead of passing into C is returned to the source.

Also of some importance in the behavior of the inverter is the application of linear circuit theory, particularly the principle of superposition, in circuits with obviously nonlinear components, such as diodes and thyristors. We observe that diodes and thyristors are really switches that are open or closed depending upon the direction of current flow (in the case of the thyristor there is the additional requirement of the gate). While current is flowing in the forward direction, the diode is linear. If we forget the small forward volt drop, it is simply a closed switch. When current attempts to reverse, the switch opens. Then it is again linear. Hence it has an infinite impedance. Thus it has two states, in either of which it can be considered linear. Only when it switches from one to the other must we stop and make changes. Taking the turn-off transient as an example, the analysis proceeds until the current comes to zero in the thyristor being commutated; we note conditions at that instant. A new problem is then started in a new circuit. The closed switch is replaced by an open switch, and the initial conditions for the new problem are the final conditions from the old one. In this particular example, we observe that when thyristor T_1 blocks, there is still a considerable voltage on the capacitor which would attempt to reverse the current. The blocking thyristor will be exposed to this voltage, it will appear as an inverse voltage across it, exactly as it occurs in the static breaker, and what happened there happens here, the capacitance of the thyristor and any added capacitance that may be in parallel with it are charged from C, probably in an oscillatory manner. The amount of damping will depend upon the dissipation in the

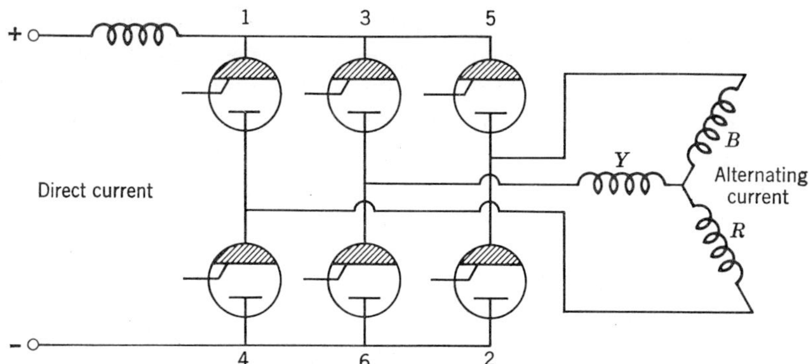

Fig. 7.13. Rectifer/inverter of the line commutated type.

local loop. What has been described here is an example of a general way of treating transients in circuits that have switching devices of this kind.

The preceding analysis was concerned with the inverter in steady-state condition in that certain values were assumed for the charge on C and the current in L at the time of commutation. Another interesting problem to analyze is how one starts such an inverter and proceeds thereafter to move to the steady state. When the very first pulse is applied to a thyristor gate, there is no current in L and no voltage across C. The initial conditions are different, but the approach is the same. It may take several cycles before a steady state is reached. Each cycle will be different.

Again, we have considered only a resistive load. Other types of load have their own peculiar problems, but the same technique applies. Time is broken up into intervals during which the thyristors can be considered as either closed or open switches. Solutions are sought for each interval which will match at the interval boundaries.

The line or system commutated converter is another important type because it is used in high-voltage direct-current (HVDC) transmission. The apparatus we are talking about is used as a rectifier or as an inverter. A simple diagram for such an equipment is sketched in Fig. 7.13 for grid controlled mercury valves. For a detailed discussion of HVDC converting equipment the reader is referred to Adamson and Hingorani (6); our concern here is solely with the electrical transients that occur in the operation of such converters.

Consider first operation in the rectifying mode with power flow from the a-c side on the right to the d-c side on the left, and let us suppose initially that the numbered devices are simple mercury arc rectifiers without grid control. The voltage pattern is that shown in Fig. 7.14*a*, where the numbers indicate the rectifiers carrying the current and the letters the phases on the a-c side supplying that current. Thus, at instant XX', current is being fed to the load by rectifier 1 on the red phase and returned by rectifier 2 on the blue.

An instant later at A, the yellow phase becomes more positive, the current therefore transfers to rectifier 3. Similarly at point B the red phase becomes more negative than the blue and the returning current transfers from rectifier 2 to rectifier 4. The heavier envelope of the waveforms thus becomes the d-c output voltage. Actually, as we are well aware by now, current cannot transfer instantaneously from one path to another if the circuits involved possesses any inductance. In Fig. 7.14a the transfer of current begins at A or B. Thereafter, two phases of the a-c supply are short-circuited by the rectifier. There difference voltage, as it increases, acts to reduce the current in one and increase it in the other until all the current has been commutated. In Fig. 7.14a this takes a time, or angle, designated γ. During this interval the output

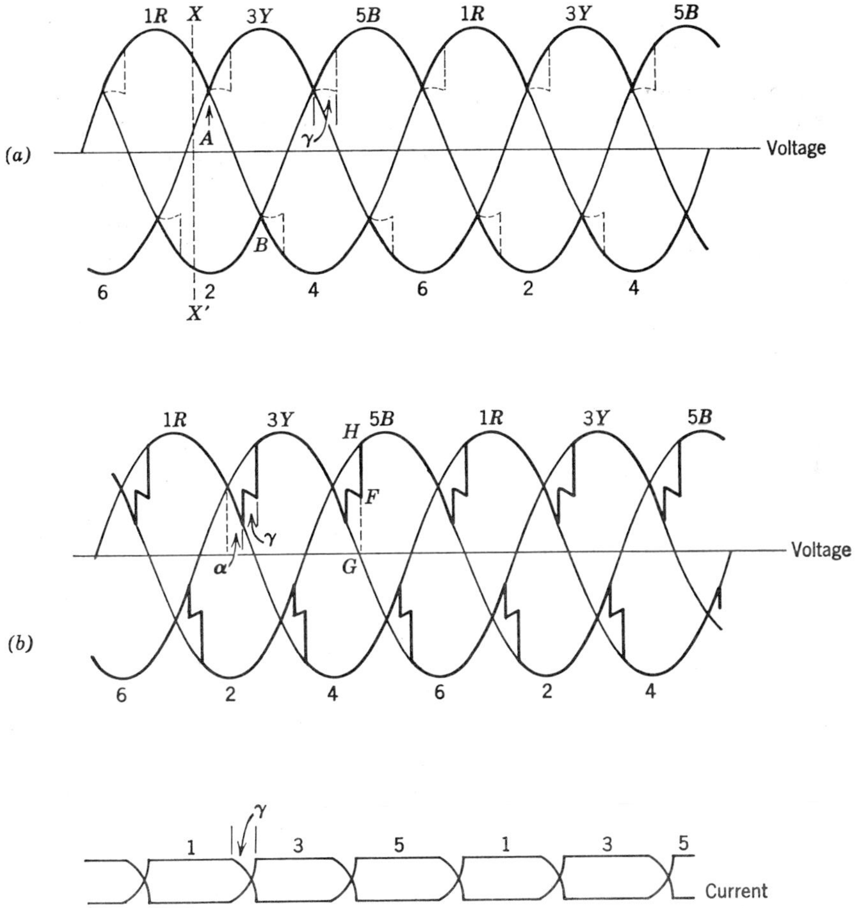

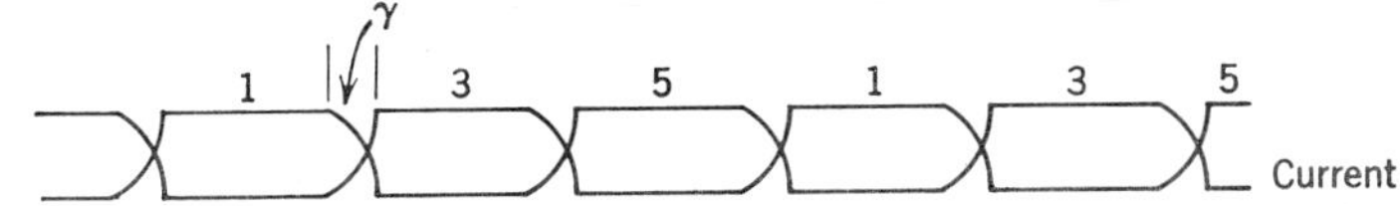

Fig. 7.14. Voltage and current waveforms for three-phase rectifier. (*a*) Idealized voltage form. (*b*) Voltage and current waveforms with phase retard.

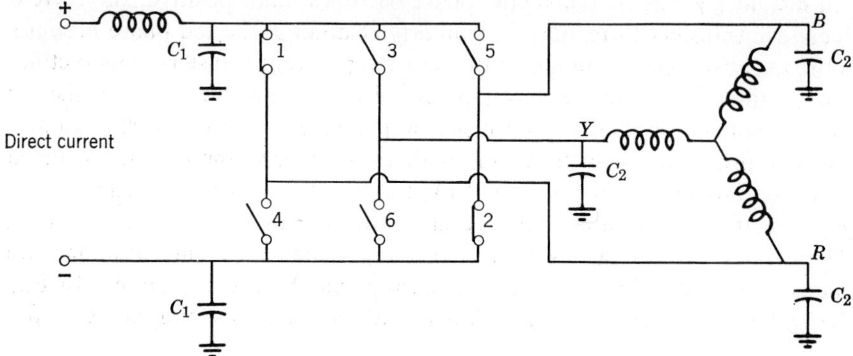

Fig. 7.15. Simple representation of a three-phase bridge rectifier for a transient study.

voltage runs midway between the phase voltage as indicated by the dotted curves in the figure. Gamma (γ) will depend upon the inductance of the a-c supply since the rate at which the current can change is given by the difference voltage divided by the inductance of the two phases involved.

Consider next the situation where the valves have grid control, which allows conduction to be delayed in any valve until its grid is fired. This means that in Fig. 7.14*a* the current would not start to transfer from the red phase to the yellow phase at point A unless the grid of valve 3 was being pulsed at that time. If the pulse is delayed by an angle α there is already a significant voltage across it at the time of firing. The same considerations pertain as described above for the rectifier without grid control. Current transfer is effected more quickly because the driving voltage is greater. Events are depicted in Fig. 7.14*b*, which also shows the currents in the phases.

What we have described so far has been essentially the steady state. Important transients occur superimposed on these power frequency phenomena whenever a grid is pulsed and a new path for current is introduced, or whenever a valve ceases to conduct and an existing path is blocked. To illustrate this point Fig. 7.13 has been modified as shown in Fig. 7.15. The valves have been replaced by switches, which is how they operate as far as the circuit is concerned, and certain capacitors have been added. These represent the natural capacitance of the network or windings to adjacent ground, and any capacitors that may have been added. At the instant shown, valves 1 and 2 are conducting. Following the normal sequence, valve 3, or switch 3, on the yellow phase would be the next to close. This connects C_2 on that phase to C_1 on the positive bus. Since these are charged to different voltages, an equalizing transient current will flow between them. The circuit involved comprises the two capacitors, connected in series, the inductance of the connections between them, through the bus and back by the ground return, and any resistance that these paths may have. Neglecting this resistance for the moment, a simple extension of the analysis of the LC circuit given in Section 2.3 shows that the current will be

$$(V_Y - V_R)\left[\frac{C}{L}\right]^{1/2} \sin \omega_0 t$$

where $C = C_1C_2/(C_1 + C_2)$, L is the inductance of the local loop between the capacitors, and $\omega_0 = 1/(LC)^{1/2}$. The voltage $V_Y - V_R$ is the instantaneous difference between the voltages on the yellow and red phases. If this is high and/or the surge impedance $(L\ C)^{1/2}$ is low, this transient current could be very high, much in excess of the load current. Moreover, it is oscillatory and will therefore attempt to reverse. We have already described the damage that can be done to thyristors by too high an initial buildup of current. In mercury valves a quite different, but equally undesirable phenomenon can occur, called arc starvation (7). In practical circuits extra inductance must be added if need be to preclude these possibilities, and the circuit must either be sufficiently damped to prevent oscillation or some bypass, perhaps in the form of capacitance, must be provided around the valve as a path for reverse current. This will be illustrated by a high-voltage thyristor valve. Since the voltage capability of such devices is limited, a large number must be operated in series. In Fig. 7.16 only those parts of the circuit involved in the transient are shown. Valves 1 and 2 are conducting and the thyristors of string 3 are gated to transfer current from 1 to 3. Now C_2 discharges into C_1 through the reactors L_3 (solid arrows) which have been added in order to limit the value of the transient current. When this current reverses (dotted arrows) it flows in the circuits R_3C_3. The resistance helps to damp the oscillation. It also provides means for limiting the current from C_3 when the thyristor are first gated. The gate pulse to the thyristors must be long enough to assure that they are maintained conducting until the transient current has died down.

This transient current is associated with a transient voltage, which will appear across those strings that are not conducting. The $R_3L_3C_3$ combinations on these strings provide the means for equitably distributing these voltages across the series devices, so that none receives more than it can support.

In Fig. 7.14*b* C_1 and C_2 on the commutating phases have been constrained to maintain approximately the same potential, after the initial transient between them, by virtue of the circulating current flowing between them during the period of commutation. When commutation is completed and phase 1 ceases to conduct, this constraint is removed. At that moment their common potential corresponds to point F. Thereafter, C_2 on the red phase will take up the potential of the instantaneous open circuit voltage of that phase (point G), whereas C_1 and C_2 on the yellow phase will assume the instantaneous potential of the yellow phase (point H). These changes will be brought about through a transient, which will probably be oscillatory and involve an overshoot. In the case of the outgoing phase, the transient frequency will be determined by C_2 and the transformer leakage inductance; in the case of the incoming phase, the sum of C_1 and C_2 with the leakage inductance. The more the firing of the incoming phase is retarded, the greater will these transients be.

Our treatment here has been oversimplified in many ways. The system

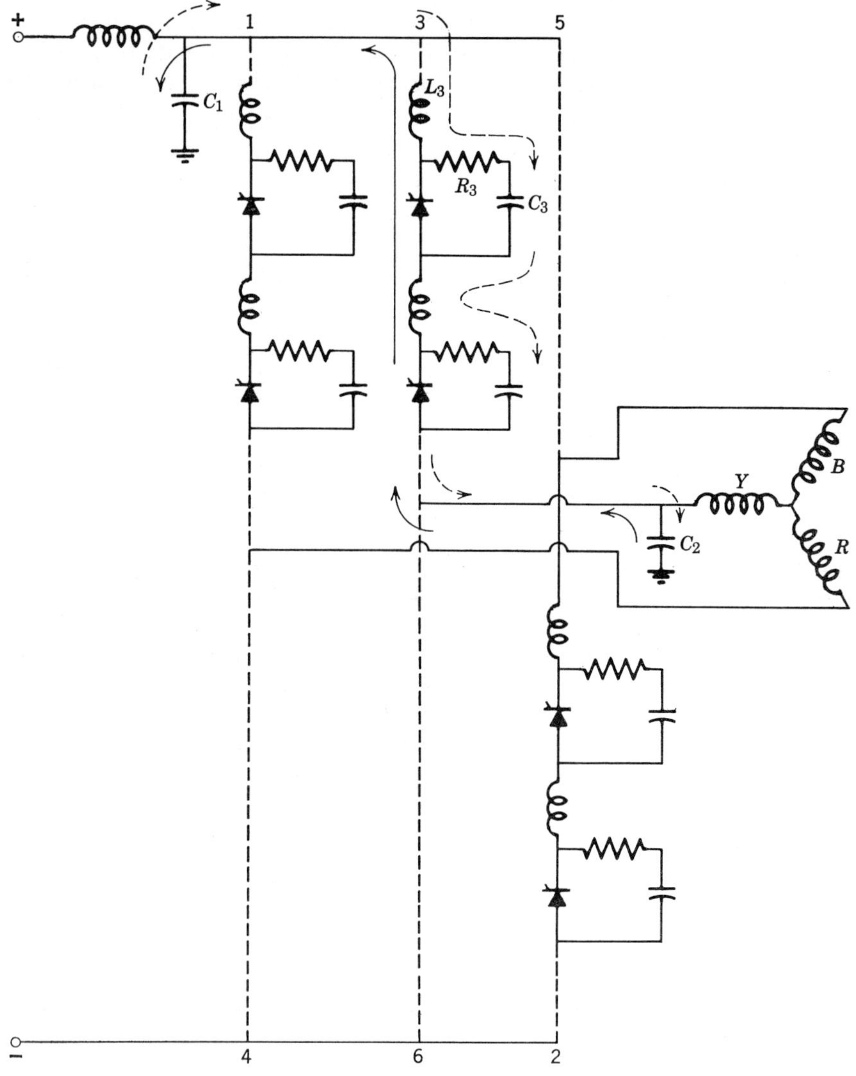

Fig. 7.16. Transient currents flowing when commutating from valve 1 to valve 3.

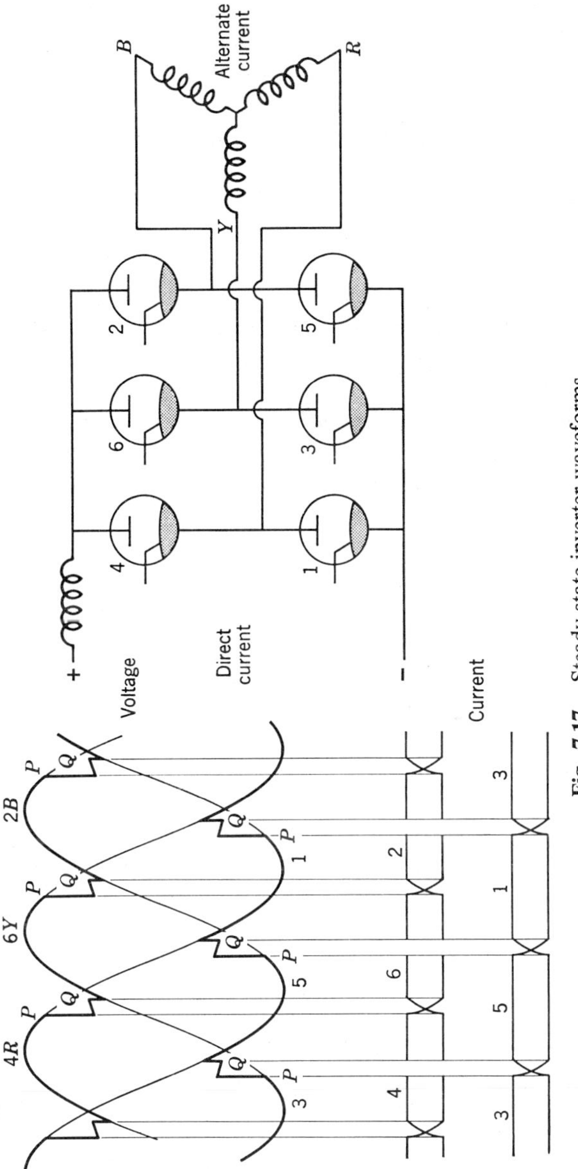

Fig. 7.17. Steady-state inverter waveforms.

capacitance, for example, is only approximated by the formulation we have given. Better approximations could require it be concentrated in more locations, which would produce multiple natural frequencies. It is evident that the transient created by any valve when switching in or switching out is experienced by other valves not conducting at the time, but connected to the valve concerned. The components L_3, C_3, and R_3 across such valves must, of course, influence the form of the transient.

We have so far limited consideration to operation in the rectifying mode, not a serious restriction. The commutation transients evoked in the inverting mode are very similar and can be calculated in the same way. This is not surprising, since, whether a rectifier or an inverter, the equipment is really just a programmed switcher. In the inverting mode we are dealing with the same switches in the same circuit, it is only the program that is different. Refer to Figs. 7.13 and 7.14*b* again and imagine the bus voltage on the d-c side reversed. If the pulsing of the valves is so far delayed that they again have forward voltage across them, power flow will be from the d-c side to the a-c side and the apparatus will be inverting. Figure 7.17 shows the equipment functioning in this manner. Commutation transients will occur at points P and Q, that is, at the beginning and end of the commutation period, as they did in the rectifier, and the same components will be involved. However, at points P the initial discharge will be of C_1 into C_2, rather than the other way around, because C_1 is at the higher potential. Whereas in the rectifier the phase voltages were diverging at the time of commutation, in the inverter they are converging. As a consequence, in the inverter the transients have greater amplitude at the beginning of commutation P than at the end Q. This is opposite to the rectifier.

It is apparent that converting equipments are great generators of transient disturbances. Unless filtering is provided, these disturbances spread into the a-c and d-c systems, where they can cause radio and television interference and perhaps upset neighboring control or communication circuits to which the power circuits may be coupled. The subject of pickup is dealt with in Chapter 8.

REFERENCES

1. S. Ikeda and T. Araki, "The *di/dt* Capability of Thyristors," *Proc. IEEE.*, Vol. 55, No. 8 (1967), p. 1301.
2. I. Somos and D. E. Piccone, "Behavior of Thyristors Under Transient Conditions," *Proc. IEEE*, Vol. 55, No. 8 (1967), p. 1306.
3. C. F. Wagner, "Parallel Inverter with Resistive Load," *Trans. AIEE*, Vol. 54 (1935), p. 1227.
4. C. F. Wagner, "Parallel Inverter with Inductive Load," *Trans. AIEE*, Vol. 55 (1936), p. 970.

5. B. D. Bedford and R. G. Hoft, *Principles of Inverter Circuits*, John Wiley & Sons, New York (1964).
6. C. Adamson and N. G. Hingorani, *High Voltage Direct Current Power Transmission*, Garraway Ltd., London (1960).
7. H. C. Steiner and R. W. Strecker, "Voltage Transients due to Arc Extinction," *Trans. AIEE*, Vol. 79, Part I (1960), p. 139.

8 Electromagnetic Phenomena of Importance under Transient Conditions

8.1 Introduction

The study of electrical transients *is* a study of electromagnetic phenomena. It might seem odd, therefore, to have a chapter with this title. If a distinction can be made, it is perhaps that in this chapter we give more emphasis to the field aspects rather than the circuit aspects involved. The aim is to consolidate some of the basic principles of electrostatic and electromagnetic induction and then consider some practical instances where these principles have considerable influence on the transient behavior of power systems and their associated control circuits.

To cite some examples, every experienced engineer who has investigated switching transients knows that damping is invariably greater than one can account for from the standpoint of steady-state losses. It will be shown that this is attributable, in part, to phenomena taking place within the conductors themselves and in adjacent structures during the transient period. A perennial problem in the areas of relaying, control, and measurement is that of pickup, or noise, as it is sometimes called, whereby unwanted signals are coupled from power circuits into control or measuring circuits, to cause damage to or malfunctions of these circuits. The nature of these couplings will be considered in this chapter, as will the question of shielding, the reduction or elimination of such pickup.

The fundamentals we are about to discuss have many other applications. Some will be used as illustrations in this chapter; others will be introduced at appropriate places elsewhere in the book. Our treatment here cannot be exhaustive. Rather the aim is to present some of the fundamental concepts and then some of the practical consequences, including numerical results, extracted from original papers in this field. For a broader background in the basics of electromagnetic fields the reader should consult a good text on the subject such as Carter (1), Ramo and Whinnery (2), or Stratton (3) (the choice will depend upon his predeliction for mathematics). We start here with a brief review of electrostatic and electromagnetic induction.

8.2 A Review of Electrostatic Induction with Some Transient Applications

Wherever charge separation occurs, that is, where positive and negative charges are separated, an electric field exists between and around the separated charges. This is illustrated in the very simple two-conductor configurations shown in Fig. 8.1*a*. A potential difference, proportional to the charge, will exist between the conductors; the constant of proportionality is the capacitance between 1 and 2:

$$Q = CV \qquad \text{or} \qquad V = \frac{Q}{C} \tag{8.2.1}$$

If another conductor is introduced into the vicinity of the first two (Fig. 8.1*b*) it will distort the field, since its surface forces an equipotential where none existed before. It will also be polarized as shown. Conductor 3 will have no net charge, but there will be charge separation on its surface. It will adopt some potential intermediate between V_1 and V_2, the precise value depending upon the disposition of 3 with respect to 1 and 2. By the same token the physical arrangement will establish the capacitances between conductors 1, 2, and 3. In a typical situation, conductor 1 may be a high-voltage bus, conductor 2 may be ground (literally ground, or a grounded cubicle for example), while conductor(s) 3 may be some independent control or

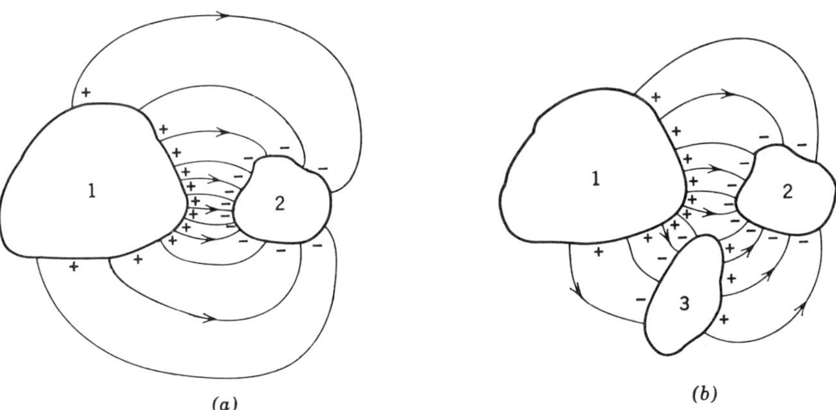

Fig. 8.1. (*a*) Electric field around two charged conductors. (*b*) Modified by the presence of a third conductor.

monitoring circuit. Suppose, now, that a sudden change of potential occurs between conductors 1 and 2. This might be occasioned by some switching operation on the power conductor. This event will call for a redistribution of charge on conductor 3, which will be brought about by current flow in that conductor. This current may bring about unintentional events in the circuit of which conductor 3 is a part. Typically this can result in a false relay operation, perhaps tripping a circuit breaker unnecessarily. Problems arising from this phenomenon can be particularly troublesome in those industrial applications where large blocks of power are controlled by circuits of much lesser power. A thyristor-type drive for a mill motor (see Chapter 7) is a good example. The power required to turn on and turn off these devices is almost infinitesimal compared with that switched by the equipment. If false signals are coupled into the control circuits of such systems, the consequences can be quite devastating. In a reversing drive, for example, the forward power and the reverse power are usually derived through independent groups of thyristors, which are never conducting simultaneously. However, pickup in the control circuit could bring this about. This would result in a short circuit being applied across the a-c bus.

In a quite different area, charges are electrostatically induced on the towers and conductors of transmission lines by the very considerable electric fields set up by thunderclouds. If the cloud is negatively charged, the upper part of the transmission line structure will become positively charged and the lower part negatively charged. The negative charge may leak away in due course, leaving a net positive charge on the line, but a charge still bound by the inducing charge in the cloud. If now there should be a sudden change in the inducing charge, if, for example, there is a lightning flash to a neighboring cloud, a new situation arises, calling for a different charge distribution on the line. As explained in Chapter 10, this will result in traveling waves of current and voltage spreading up and down the line, perhaps causing damage to terminal equipment. Note that in this case, as in the others discussed, the cause of the disturbance was not in the circuit affected—the lightning did not strike the line—but in an adjacent circuit, a circuit established by the lightning between the clouds, and perhaps completed through ground by displacement currents.

Fortunately, in many instances, it is not difficult to minimize electrostatically induced transients. The time to do this is when the equipment is being designed or the system layed out. Referring again to Fig. 8.1*b*, it is apparent that we have a simple potential divider, comprising the series combination of C_{13} and C_{32}. The voltage will divide inversely as the capacitance, thus, a fraction

$$\frac{C_{32}}{C_{32} + C_{13}}$$

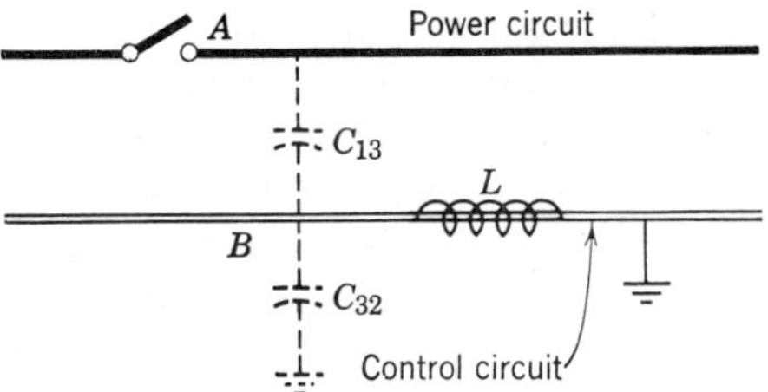

Fig. 8.2. Capacitive coupling between power circuit and control circuit.

of the potential between 1 and 2 will appear on conductor 3. Physical separation, having 3 much closer to 2 than to 1, will help in keeping this fraction small. Sometimes it is possible to artificially increase C_{31}, which achieves the same end. An example of this is discussed in Section 11.7 as it relates to the electrostatic coupling of surges between the windings of a transformer.

Frequently, the control circuit, relay circuit, or what have you, will be grounded at some point. This does not preclude its picking up spurious signals in the manner described if the point of induction is remote from the point of grounding. This is illustrated by Fig. 8.2. A sudden change of potential at A will bring about a change at B, because B is not directly grounded but grounded through L.

The other approach to solving this problem is through electrostatic shielding. A well-grounded conductor is interposed between the two circuits. This may take the form of a conduit or trough through which the control circuit is routed, or it may be a bulkhead or cubicle wall. The overhead ground wires on transmission towers help to provide this function. It is necessary to have good grounding on this shield for the reason mentioned in the last paragraph. Unfortunately, multiple grounding can sometimes create problems of a different kind, as we shall see later in this chapter.

8.3 A Review of Electromagnetic Induction and Related Topics

The fundamental law of electromagnetic induction was postulated by Faraday. It states that whenever the magnetic flux linking with a closed path changes, an emf is induced around the path proportional to the negative rate of change of the flux. The word "path" is used advisedly, since it is not necessarily a circuit. It could be in space, in dielectrics, along conductors, or it could be constituted from any combination of these. The emf is defined as the line integral of the field vector **E** around the path:

$$\text{emf} = \oint \mathbf{E} \cdot dl \tag{8.3.1}$$

Magnetic fields are produced by current flow. This is formalized in the law of Biot and Savart, which states that the integral of the magnetic force vector **H** around any closed path is equal to the current enclosed by the path. Whenever currents are changing, magnetic fields are also changing and are therefore creating electric fields in their vicinity (Faraday). Where these fields or emfs can be applied to a closed circuit, they will cause currents to flow. Of course, we constantly make use of this principle in the transformer; the changing magnetic flux of the primary current induces an emf in the secondary circuit, with which the primary is tightly coupled. Our concern here is not with such intentionally conceived devices, but with the countless inadvertently devised "transformers" that unintentionally, yet inescapably exist in power systems, and which couple undesired signals from one circuit to another, often with unfortunate consequences. These are of special concern under transient conditions when the inducing currents may be high and or changing very rapidly, for then the induced emfs can be considerable.

Faraday's law can be written in different ways according to whether it is the circuit aspect or the field aspect that one wishes to consider. From the circuit viewpoint, we may state

$$\oint \mathbf{E} \cdot dl = -\frac{\partial}{\partial t}\int_S \mathbf{B} \cdot d\mathbf{S} \tag{8.3.2}$$

In Eq. 8.3.2 the left-hand side of this equation is the emf, the right-hand side represents the negative time rate of change of flux, being the integral of the flux density vector **B** over any surface **S** bounded by the path. The partial derivative is used to indicate that **B** may also vary with position and it is the time variation which gives rise to the emf. The negative sign accords with the convention of considering the line integral positive when it is taken in a clockwise circulation about the area through which the flux passes. Thus, if one imagines magnetic flux proceeding outward through the door of the room where one is sitting, and that this flux is increasing, an electric field would be established in a counterclockwise direction around the door frame. Since the clockwise circulation is considered positive, this emf is considered negative.

If we wish to emphasize the field aspect of Faraday's law, it is written

$$\text{curl}\,\mathbf{E} \qquad \text{or} \quad \nabla \times \mathbf{E} = -\frac{\partial \mathbf{B}}{\partial t} \tag{8.3.3}$$

Since it is the author's experience that power system engineers are more comfortable and more adept at dealing with circuits than with fields, we will use Eq. 8.3.2 rather than Eq. 8.3.3 for the most part.

With a similar physical picture the law of Biot and Savart can be expressed:

$$\oint \mathbf{H} \cdot dl = \int_S \mathbf{J} \cdot d\mathbf{S} \tag{8.3.4}$$

where $\mathbf{J}$ is the current density vector, which in this case is integrated over the surface $\mathbf{S}$ to give the total current. In vector form this becomes

$$\nabla \times \mathbf{H} = \mathbf{J} \tag{8.3.5}$$

Actually, these equations are incomplete in that they do not include the displacement current (1), (2) arising from the changing electric flux that may pass through the surface S. A term

$$\frac{\partial}{\partial t}\int_S \mathbf{D} \cdot d\mathbf{S} \qquad \text{or} \qquad \frac{\partial \mathbf{D}}{\partial t}$$

as the case may be, should be added to the right-hand sides. This can be an important term in free space, but is usually negligible in the applications we shall consider.

Since magnetic flux lines form closed loops and currents flow in closed circuits, it is apparent from Eq. 8.3.3 that the two link each other mutually and that the flux linkage is proportional to the current. The proportionality factor is the *inductance* of the circuit, that is, the inductance of a circuit is the flux linkage with it per unit current flowing in it:

$$L = \frac{\Phi}{I} \tag{8.3.6}$$

In practical units Φ would be measured in weber turns, I is amperes, and L in henries.

Oliver Heaviside once remarked that the best way to measure inductance is with a tape measure. He was drawing attention to the fact that inductance is essentially a geometrical property of a circuit. The more extensive a circuit is, the more flux will it embrace when it supports a given current, and therefore the greater will be the inductance of the circuit.

With a simple loop of wire, or even with a long transmission line, the concept of magnetic flux linking the circuit is not difficult to grasp. We can visualize the closed loops of flux and the closed loop of the circuit mutually embracing like the links of a chain. With bus work that may meander all over the place, or with machine windings, it is more difficult to obtain such a visual image, although Carter (1) makes some interesting suggestions as to how this may be done.

The definition (8.3.6) is quite adequate for many determinations, but where conductors are closely spaced, as busbars sometimes are, a significant

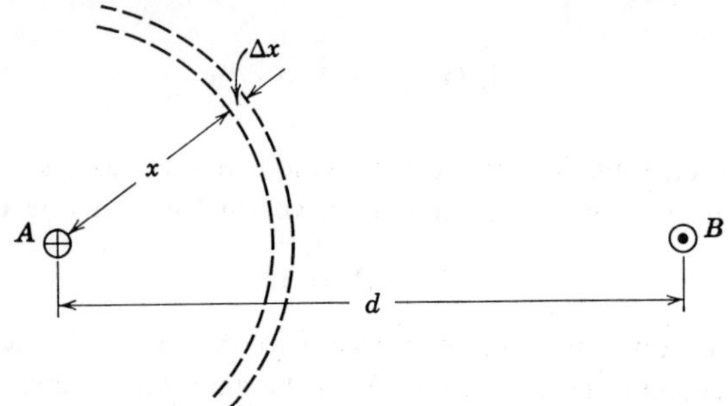

Fig. 8.3. Determining the inductance of a single-phase transmission line.

part of the total flux is within the conductors, and this links with only a part of the current. In this situation, the definition is modified as follows:

$$\left.\begin{aligned} L &= \frac{\sum(\Phi\, dI)}{I^2} \\ &\text{or} \\ L &= \frac{\sum(I\, d\Phi)}{I^2} \end{aligned}\right\} \tag{8.3.7}$$

depending upon whether attention is focused on the flux and how much current each flux element links, or on the current and how much flux each current element links. Where all the flux links all of the current Eqs. 8.3.7 reduce to Eq. 8.3.6.

To illustrate these points we now calculate the inductance of a single-phase transmission line comprising a pair of conductors of circular cross section. This has practical significance for Chapter 9. Such a circuit is a very long, narrow loop, each meter of which contributes to its inductance. It is convenient to think in terms of the inductance/meter for such a circuit. A circuit of this kind is shown in Fig. 8.3.

In the absence of any iron that would render the system nonlinear, the principle of superposition can be invoked, that is, we can determine what flux would link the circuit due to the current in conductor A and then due to the current in conductor B. When both carry current, the total flux will be the sum of these contributions. For the current in a single isolated conductor the magnetic flux lines are circles, concentric with the conductor, $\mathbf{H}$ being constant along such a circular path. When the conductor is carrying a current I, Eq. 8.3.4, for $\mathbf{H}$ at a radius x, gives

$$\begin{gathered} 2\pi \mathbf{H} x = I \\ \text{or} \\ \mathbf{H} = \frac{I}{2\pi x} \end{gathered} \tag{8.3.8}$$

It follows that

$$\mathbf{B} = \mu_0 \mathbf{H} = \frac{\mu_0 I}{2\pi x} \tag{8.3.9}$$

where μ_0 is the permeability of free space ($= 4\pi \times 10^{-7}$ in the MKS rationalized system of units).

The flux crossing an elementary area Δx wide and 1 meter long in the direction of the conductors (Fig. 8.3) will be

$$\Delta\Phi = \frac{\mu_0 I\, \Delta x}{2\pi x}$$

To determine the flux passing between conductors A and B due to the current in A (and therefore linking with their circuit) we must let Δx shrink to the infinitesimal dx and $\Delta\Phi$ to the infinitesimal $d\Phi$, and then integrate from $x = r$ to $x = (d - r)$:

$$\begin{aligned}\Phi_A &= \int_r^{d-r} \frac{\mu_0 I\, dx}{2\pi x} = \frac{\mu_0 I}{2\pi} \int_r^{d-r} \frac{dx}{x} \\ &= \frac{\mu_0 I}{2\pi} [\ln x]_r^{d-r} \\ &= \frac{\mu_0 I}{2\pi} \ln \frac{d-r}{r} \end{aligned} \tag{8.3.10}$$

By treating conductor B in precisely the same manner, it is clear that it will contribute an identical amount of flux. The total flux is therefore

$$\Phi = \frac{\mu_0 I}{\pi} \ln \frac{d-r}{r} \tag{8.3.11}$$

From Eq. 8.3.6 it is apparent that

$$L = \frac{\mu_0}{\pi} \ln \frac{d-r}{r} \tag{8.3.12}$$

If $d \gg r$, this reduces to

$$L = \frac{\mu_0}{\pi} \ln \frac{d}{r} \tag{8.3.13}$$

This inductance is due to the flux between the conductors. Flux within the conductors ($x < r$) will also contribute to the total inductance. How much this contribution will be can be calculated from Eq. 8.3.7. Referring to Fig. 8.4, the current enclosed by a flux line at radius x, assuming uniform current distribution over the cross section, is $(x/r)^2 I$. Thus, at this radius,

$$\mathbf{H} = \frac{Ix}{2\pi r^2} \quad \text{and} \quad \mathbf{B} = \frac{\mu_0 I x}{2\pi r^2}$$

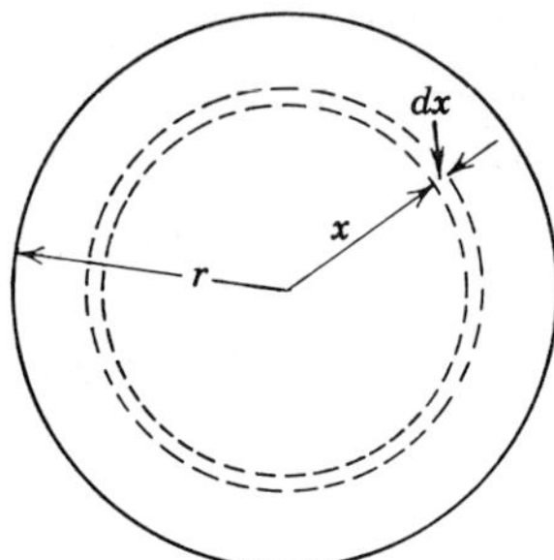

Fig. 8.4. Magnetic flux within a conductor.

An element of flux $d\Phi$ at this location contains $\mathbf{B}\,dx$ webers and links $(x/r)^2 I$ of the current, thus

$$I d\Phi = \frac{\mu_0 I^2 x^3}{2\pi r^4}\, dx$$

If we substitute this in Eq. 8.3.7, we obtain for the inductance due to the flux within the conductor

$$L_i = \frac{1}{I^2}\int_0^r \frac{\mu_0 I^2 x^3}{2\pi r^4}\, dx$$

$$= \frac{\mu_0}{8\pi} \tag{8.3.14}$$

This amount will be contributed by both conductors, so that the total inductance for the circuit, from Eq. 8.3.12, will be

$$L = \frac{\mu_0}{\pi}\left[\ln \frac{d - r}{r} + \frac{1}{4}\right] \tag{8.3.15}$$

When two circuits are adjacent, some of the magnetic flux from the current in one may link the other. This so-called mutual flux defines the mutual inductance between the two circuits. Thus M, or L_{AB}, is the flux linkage with circuit B due to unit current in circuit A. Mutual inductance is a reciprocal property of circuits; it can be equally well defined as the flux linkage with circuit A due to unit current in circuit B:

$$L_{AB} = \frac{\Phi_B}{I_A} = M = \frac{\Phi_A}{I_B} = L_{BA} \tag{8.3.16}$$

For some simple circuits the mutual inductance can be calculated readily in a manner similar to that used previously for the self-inductance of a transmission line. An example in Fig. 8.5 shows a power transmission line and a parallel communication circuit. It can be shown easily that to a close approximation the mutual inductance between these circuits is given by

$$L_{AB} = \frac{\mu_0}{2\pi}\left[\ln\frac{d_2}{d_1} - \ln\frac{d_4}{d_3}\right] = \frac{\mu_0}{2\pi}\ln\frac{d_2 d_3}{d_1 d_4} \tag{8.3.17}$$

When the current changes in a circuit it is evident that the flux linkage with the circuit changes (Eq. 8.3.6), and so does the flux linkage with neighboring circuits (Eq. 8.3.16). By Faraday's law this will generate emfs in the circuits involved, as follows:

$$\left.\begin{aligned} \text{the self induced emf} &= -L\frac{dI}{dt} \\ \text{the mutually induced emf} &= -M\frac{dI}{dt} \end{aligned}\right\} \tag{8.3.18}$$

Hence voltages are coupled from one circuit to another, such as the circuits in Fig. 8.5, especially under transient conditions. Currents induced in relay circuits by this means can cause false tripping. Errors can enter measuring circuits for the same reason. Voltages similarly induced in ground loops can imperil personnel. Examples of this will be presented later in the text (see Section 12.6).

In the power circuit itself undesirable phenomena can develop as a consequence of mutual coupling between circuits. On a tower system carrying two three-phase circuits, problems can arise when one of the circuits is taken out

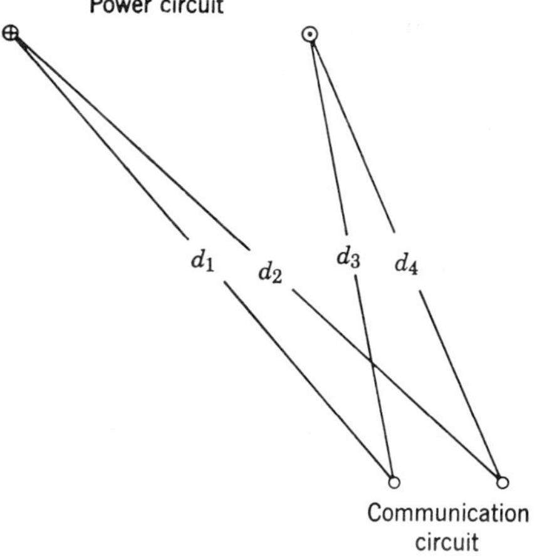

Fig. 8.5. Mutually coupled power and communication circuits.

of service for maintenance, for example. Notwithstanding that this circuit be grounded, it can still have significant voltages induced in it from the adjacent energized circuit. A similar, but accentuated condition is described by Pickett et al. (4) and Hesse and Wilson (5), in which a 138-kV circuit coupled into a neighboring 345-kV circuit. The latter had compensating shunt reactors installed at both ends, and the excessive voltages were observed across these reactors when the 345-kV breakers were opened. Apparently, the induction from the lower voltage circuit excited a near-resonance in the higher voltage circuit. This is perhaps not surprising because of the very nature of the compensating function the reactors are intended to provide under normal energized conditions. What has just been described falls in the category of dynamic overvoltages, which are persistent as long as the condition prevails. This distinguishes them from the transient overvoltages, which are the principal subject of this book. However, transients can be induced in the same manner, as discussed in Sections 9.7 and 13.4. Also, it should be recognized that transients can occur superimposed on these already abnormal conditions.

The solutions available for minimizing electromagnetic pickup are similar to those described in the last section for minimizing electrostatic pickup: separation and shielding. We will take up the subject of separation here; shielding will be considered later.

It is apparent from Fig. 8.5 and Eq. 8.3.17 that the greater the physical separation existing between two circuits, the less the mutual inductance and therefore the less does one circuit experience events taking place in the other. It is also evident that *small* separation between the individual conductors of a circuit promotes the best conditions for avoiding pickup. The secondary circuit thereby presents a small area through which the flux from the primary can thread. The twisted pair of conductors is particularly useful for control, relaying, and other auxiliary circuits, since as it twists, about half of the stray magnetic flux from adjacent circuits links it in one direction, the other half in the opposite direction. The net flux linkage is therefore small. Close spacing of the primary conductors (usually the power circuit) is desirable, since the closer they are, the more do they appear to any neighboring circuit like one conductor with equal and opposite currents. Such a conductor would produce no magnetic field.

This last observation prompts reference to a particular kind of power conductor—the isolated phase bus. In this equipment the conductors are concentric. If, therefore, one chooses a path around which to integrate $\mathbf{H}$, external to the bus, whether or not it embraces the conductors, the integral will be zero, because the path surrounds no net current provided the outer conductor is used for the return current. This does not necessarily mean that the magnetic field is zero at all points external to the bus. This is true if the bus comprises two perfectly concentric cylinders. Imperfections, lack of concentricity, will give rise to local external fields, which cancel out when one takes a complete closed path. For the most past, however, the magnetic flux is confined within the bus structure, making this a good conductor arrangement as far as the present problem is concerned.

For the same reason, where a number of control cables must run together in the same duct, the concentric type of cable is advantageous since it effectively limits both electric and magnetic fields to the confines of the outer conductor. On the other hand, this allows the possibility of a magnetic field from an external circuit linking loops formed by the outer conductors or sheaths of these concentric cables, which may cause currents to flow in them and thereby introduce spurious signals.

Before considering shielding and related topics, there is another concept that we will have occasion to use in later chapters, which can be conveniently discussed at this point. This is the method of images. Figure 9.2 shows the electric and magnetic field lines around a pair of parallel cylindrical conductors forming an isolated transmission line. These systems of lines form an orthogonal set, that is, they everywhere intersect at right angles. In an electric field the equipotentials also form an orthogonal set with the electric field lines; thus the magnetic field lines in Fig. 9.2 also serve as electric equipotentials. Conversely, the electric lines of force could equally well represent magnetic equipotentials. Thinking first of Fig. 9.2 as a representation of the electric field between the two conductors, it is apparent that the field possesses a symmetry about a vertical plane midway between the conductors and perpendicular to the plane in which they lie, the field on one side is a mirror image of the field on the other. If a conducting sheet were placed in this plane, either conductor could be removed without disturbing the field between this plane and the other conductor. The reverse of this is also true. If we have a single conductor running parallel to a conducting plane, the electric field between the conductor and the plane will be undisturbed if the plane is removed and replaced by an image conductor, symmetrically located on the other side of the plane. This artifice is used to plot the electric field of transmission line conductors, strung above a conducting ground. The method is applicable to multiconductor arrangements, so a three-phase line and ground can be replaced by six conductors, the three conductors of the line and their images in the ground. Strictly speaking, the representation is only precise when the ground is a perfect conductor, but the error introduced by usual soil conductivities is not great.

In like manner, for the magnetic field, ground can be replaced by an image conductor carrying an equal and opposite current. But here the fact that the earth is not a perfect conductor is much more significant. The poorer the conductivity of the soil, the deeper must the image conductor be put. In particular,

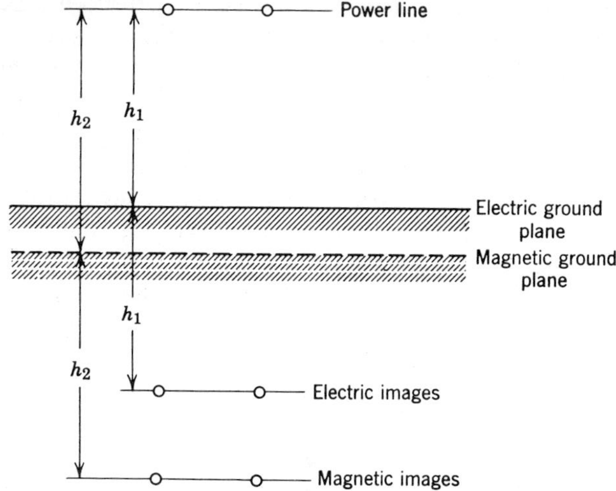

Fig. 8.6. Electric and magnetic image conductors for a single-phase transmission line.

the images for the magnetic field do not coincide with those for the electric field. In Chapter 9 it will be shown that this has some important consequences. An example for a single-phase transmission line is presented in Fig. 8.6. It shows the equivalent ground planes for the electric and magnetic fields do not coincide.

8.4 Steady-State Penetration of Magnetic Flux and Current into Conductors

The changing magnetic field of a changing current may link other metallic structures besides conventional circuits. These may be the magnetic cores of transformers or reactors, cubicle walls, or even structural steel work. Another very important conductor for this linkage is the conductor in which the current itself is flowing. Whatever the structure may be, the changing flux will induce emfs, which will drive currents in the conducting material. These currents, in turn, will set up their own magnetizing force, which will tend to oppose the inducing flux. This has important consequences of considerable practical significance. It is this phenomenon that retards or prevents the penetration of flux into conductors and which can be useful for shielding; but it is quite undesirable if the conductor is a magnetic core into which we wish to drive the flux. It is also responsible for the so-called "skin effect," whereby alternating currents tend to flow in the outer skin of a conductor rather than uniformly over the cross section. Induced or eddy currents caused by this effect are a source of loss, which can be both expensive and troublesome under some circumstances. Our concern in this section will be, for the most part, for the transient state, that is, with transient current distributions and transient losses. However, it is convenient to start with a hypothetical steady-state problem as an introductory illustration, since the mathematics is much simpler than for the transient case, yet it gives a real feeling for the important factors involved.

Suppose that everywhere to the right of AB in Fig. 8.7 is conducting material and everywhere to the left is free space; this is sometimes referred to as a semi-infinite plane. Let us further suppose that somewhere to the left there are electric currents flowing at some frequency $\omega/2\pi$, which produce an alternating magnetic field in that region, whose field is parallel to the conducting plane at the interface and has a value $\mathbf{H}_0$ in that location. Presumably this field will extend into the conducting material, and, linking with its substance, cause emfs to be generated and currents to flow therein. How will the field and currents vary with time and distance within the conductor? This is a very idealized situation, but one from which a great deal can be learned. The geometry is simple so that it does not complicate or obscure the fundamental phenomena taking place. Later, other geometries can be explored and transient conditions, in which fields are suddenly applied, can be examined.

According to Faraday's Law (Eq. 8.3.2), the induced emf and the current which it drives forms an anticlockwise circulation about an increasing magnetic field. It follows that if the flux $\mathbf{B}$ produced by $\mathbf{H}$ is parallel to the

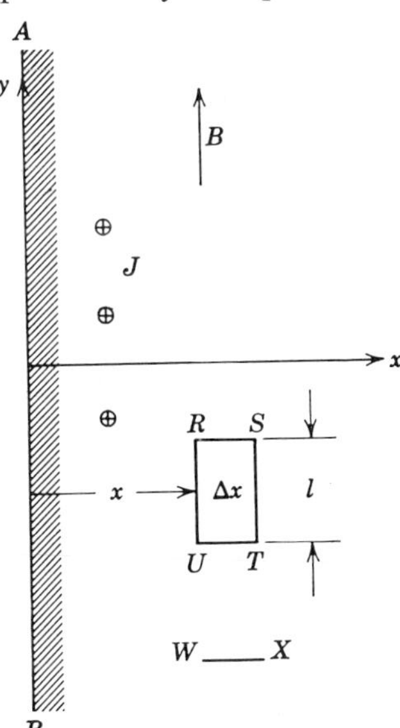

Fig. 8.7. Flux and current induced in a semi-infinite solid conductor.

plane of the diagram in Fig. 8.7, the currents will be perpendicular to the page. This is indicated by the symbols adjacent to the current density notation **J** in the diagram. Consider the small path $RSTU$, in which $RS = UT = \Delta x$, and $UR = TS = l$. The integral of **H** around this path is equal to the current enclosed:

$$\int_C \mathbf{H} \cdot dl = I$$

(right-hand rotation about the direction of current)

or

$$\mathbf{H} \cdot l - \left(\mathbf{H} + \frac{d\mathbf{H}}{dx}\Delta x\right) l = \mathbf{J} \cdot l \, \Delta x$$

(no component of H along x)

or

$$\frac{d\mathbf{H}}{dx} = -\mathbf{J} \tag{8.4.1}$$

Consider now a similar small circuit $WXYZ$ perpendicular to the paper, with only the edge WX exposed in the diagram. Faraday's law for the flux linkage with this circuit may be written

$$\int_C \mathbf{E} \cdot ds = -\frac{d\Phi}{dt} \tag{8.4.2}$$

where **E** is the electric field vector. Since Φ is varying sinusoidally with the frequency of **H**,

$$\frac{d\Phi}{dt} = j\omega\Phi \tag{8.4.3}$$

Taking Eqs. 8.4.2 and 8.4.3 together, and observing that $\mathbf{E} = \rho\mathbf{J}$, where ρ is the resistivity of the material, we may write

$$+\rho\mathbf{J}l - \rho\left[\mathbf{J} + \frac{d\mathbf{J}}{dx}\Delta x\right] l = -j\omega\Phi = -j\omega\mathbf{B}l\,\Delta x$$

or

$$-\rho\frac{d\mathbf{J}}{dx} = -j\omega\mathbf{B} \tag{8.4.4}$$

A third relationship is defined by the magnetic properties of the material,

$$\mathbf{B} = \mu\mu_0\mathbf{H} \tag{8.4.5}$$

where μ_0 is the permeability of free space and μ the relative permeability of the material.

Equations 8.4.1, 8.4.4, and 8.4.5 are three relationships between **J**, **B**, and **H**. They can be solved simultaneously to give an expression for any one of these quantities. Let us determine **B**. From Eqs. 8.4.1 and 8.4.4,

$$\mathbf{B} = \frac{j\rho}{\omega} \frac{d^2\mathbf{H}}{dx^2} \tag{8.4.6}$$

Combining Eqs. 8.4.5 and 8.4.6,

$$\mathbf{B} = \frac{j\rho}{\mu\mu_0\omega} \frac{d^2\mathbf{B}}{dx^2}$$

or

$$\frac{d^2\mathbf{B}}{dx^2} - \frac{j\mu\mu_0\omega\mathbf{B}}{\rho} = 0 \tag{8.4.7}$$

For reasons which will soon be apparent, it is convenient to make the substitution

$$\left[\frac{\rho}{\mu\mu_0\omega}\right]^{1/2} = d \tag{8.4.8}$$

then

$$\frac{d^2\mathbf{B}}{dx^2} - \frac{j}{d^2}\mathbf{B} = 0 \tag{8.4.9}$$

The solution to Eq. 8.4.9 is of the general form

$$\mathbf{B} = B_1 \exp \sqrt{j}\,\frac{x}{d} + B_2 \exp\left(-\sqrt{j}\,\frac{x}{d}\right) \tag{8.4.10}$$

as can be checked by substitution in Eq. 8.4.9. Here B_1 and B_2 are constants which must be evaluated from the boundary conditions of the problem.

Since

$$j = \cos\frac{\pi}{2} + j\sin\frac{\pi}{2} = \epsilon^{j\pi/2}$$

it is apparent that

$$\sqrt{j} = \epsilon^{j\pi/4} = \cos\frac{\pi}{4} + j\sin\frac{\pi}{4}$$

$$= \frac{1+j}{\sqrt{2}} \tag{8.4.11}$$

Substituting from Eq. 8.4.11 in Eq. 8.4.10,

$$\mathbf{B} = B_1 \exp\frac{1+j}{\sqrt{2}\,d}\,x + B_2 \exp\left(-\frac{1+j}{\sqrt{2}\,d}\,x\right)$$

$$= B_1 \exp\frac{x}{\sqrt{2}\,d} \exp\frac{jx}{\sqrt{2}\,d} + B_2 \exp\left(-\frac{x}{\sqrt{2}\,d}\right) \exp\frac{jx}{\sqrt{2}\,d} \tag{8.4.12}$$

It is evident that if B_1 is finite, the first term will increase without limit as x increases. This is physically unrealizable even in the idealized context being considered. We therefore conclude that $B_1 = 0$. On the other hand, when $x = 0$, that is, at the surface, $\mathbf{B} = B_0$, the value corresponding to $\mathbf{H}_0$, thus Eq. 8.4.12 reduces to

$$\mathbf{B} = B_0 \exp\left(-\frac{x}{\sqrt{2}\,d}\right) \exp\left(-\frac{jx}{\sqrt{2}\,d}\right) \tag{8.4.13}$$

The real exponent in this expression indicates that $\mathbf{B}$ declines exponentially as one penetrates deeper into the conductor. At $x = \sqrt{2}\,d$, the flux density is B_0/ϵ; $\sqrt{2}\,d$ might therefore be described as a space constant for the flux. At a depth of three space constants, $\mathbf{B}$ is only 5% of the surface value. The imaginary exponent in Eq. 8.4.13 indicates that the flux density changes continuously in phase as one proceeds into the material. At a depth of one space constant the flux lags the surface flux by 1 radian.

We chose to derive an expression for the flux density. Had we elected instead to examine the current density, we would have found that it was described by an identical expression.

From the expression for d (Eq. 8.4.8), it is instructive to calculate the value of this dimension for different materials at different frequencies. At 60 Hz, for example, for copper, $d = 5.5$ mm, and for steel $d = 0.5$ mm. It is no coincidence that these figures are of the order of the thickness of copper buses for power conductors and steel laminations for power transformers. At 1 kHz these dimensions shrink to 1.35 mm and 0.12 mm. At 1 MHz they are down to 0.0425 mm and 0.0039 mm. These numbers have some interesting implications.

If the conductor is no longer infinite but has a finite thickness in the x direction, and if the field is applied at both surfaces in the xy plane, flux will penetrate from both sides. This is a condition that would approximately apply, for example, to laminations being magnetized by a surrounding winding, or to a bus bar or a piece of structural steelwork in the magnetic field of the current in an adjacent busbar. This problem can be attacked in a similar manner to the semi-infinite solid. Equation 8.4.12 remains valid, only the boundary conditions must be changed. By shifting the y axis to the center of the conductor and setting $B = B_0$ at both surfaces (at $x = \pm b$, say, where $2b$ is the thickness of the plate), it is easily shown (1) that

$$\mathbf{B} = B_0\left\{\frac{\epsilon^{\sqrt{j}(x/d)} + \epsilon^{-\sqrt{j}(x/d)}}{\epsilon^{\sqrt{j}(b/d)} + \epsilon^{-\sqrt{j}(b/d)}}\right\}$$

or (8.4.14)

$$\mathbf{B} = B_0\left\{\frac{\cosh\,[(1+j)/\sqrt{2}](x/d)}{\cosh\,[(1+j)/\sqrt{2}](b/d)}\right\}$$

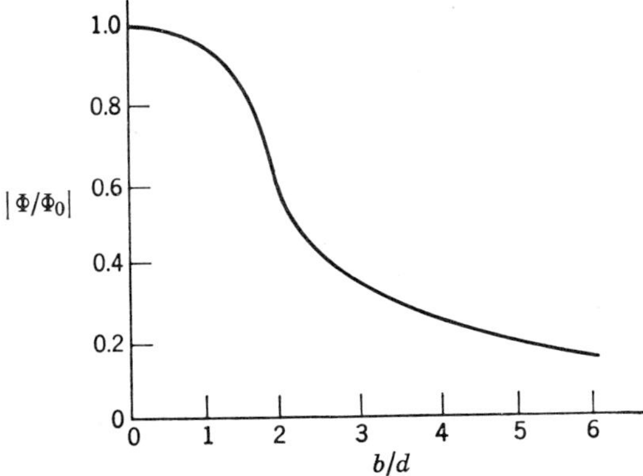

Fig. 8.8. The reduction in the effective permeability of a lamination by reason of eddy currents.

The flux density falls off as one penetrates the plate from either side, thus, at the center, where $x = 0$, the minimum value of B is

$$\mathbf{B}_{\min} = \frac{B_0}{\cosh\,[(1 + j)/\sqrt{2}](b/d)} \tag{8.4.15}$$

which gets less and less as b increases relative to d. This means that the magnetic properties of the material are not being fully utilized. Carter (1) derives the curve of Fig. 8.8 to show how the ratio Φ/Φ_0 varies with b/d. Here Φ is the total flux in the plate or lamination and Φ_0 the value it would have if the surface flux density were maintained through the section, that is, $\Phi_0 = 2B_0b$.

When a conductor is carrying an alternating current, it itself links the flux which its current produces. This causes the distribution of current over the cross section of the conductor to be nonuniform for the same reasons that affected the block and plate discussed above. The current tends to flow in greatest density in the outer layers of the conductor or the skin. Ramo and Whinnery (2) present a derivation of the current distribution for conductors of circular cross section, and show plots of the current density as a function of radius. A nonuniform distribution of this kind is responsible for an increase in the effective resistance of the conductor. Under d-c conditions resistance is

$$R_{dc} = \frac{\rho l}{A} \tag{8.4.16}$$

whereas under a-c conditions for a circular conductor it is

$$R_{ac} = \frac{2\pi\rho l \displaystyle\int_0^a J^2 r\, dr}{I^2} \tag{8.4.17}$$

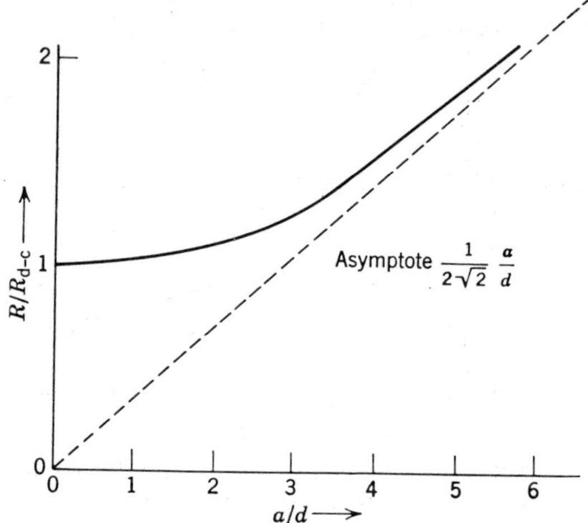

Fig. 8.9. Effective resistance of conducting rod as a function of the depth of penetration.

where A is the cross-sectional area ($= \pi a^2$ for the circular conductor), l the length of the conductor, and a its radius. Resistance is here defined in terms of the ohmic loss, which, when multiplied by I^2, gives the loss in the conductor.

Any current distribution can be divided into a uniform distribution and a nonuniform distribution, which, when superimposed, give the original distribution. Carter (1) shows that the losses associated with these two distributions can be added to give the correct total loss:

$$W = W_0 + \int \rho J_1^{\,2}\, dS \tag{8.4.18}$$

where W_0 is the loss due to the uniform distribution $\rho J_0^{\,2} S$, for a cross section S and uniform current density J_0, while J_1 is the nonuniform current density. For any set of conditions it allows one to see the effect of the distribution on the losses and the corresponding resistance. For the circular conductor mentioned above, the resistance ratio R/R_{dc} (the effective resistance to the d-c value) is shown in Fig. 8.9. The dependence on the depth of penetration d defined in Eq. 8.4.8 is evident. For small values of a/d the ratio approaches unity, for large values it is asymptotic to $a\ 2/\sqrt{2}\ d$.

This enhancement of resistance and losses increases the damping in circuits when transient high-frequency currents occur. The damping curves of Chapter 4 are still relevant, of course, but the R to be used must be modified to comply with the d of the conductors involved. Obviously, one does not go into a detailed analysis of every part of a circuit. What we are pointing out is that here is a factor to be considered when calculating damping.

The effect is not confined to the circuit in which the primary current is flowing. Losses will occur in adjacent conductors where eddy currents are induced by the primary current and these losses must be satisfied at the expense of energy from the primary circuit.

8.5 Transient Penetration of Magnetic Flux and Current into Conductors

We now turn to the more difficult problem of transient current distribution, transient resistance, and transient losses in conductors subjected to an applied magnetic field. In the last section it was shown that alternating fields give rise to nonuniform current distributions in conductors which they link, and this in turn results in an enhancement of the resistance of the conductors and the losses that accrue therefrom. In the transient period before the steady state is reached, in the first half cycle of a transient oscillation, for example, the current distribution is even more nonuniform, and the resistance losses are greater than those we have considered already. In what follows a determined attempt has been made to distill the lengthy analyses of original papers and from them cull such significant expressions, curves and other numerical data, as the engineer will find useful. At the same time it is hoped that the physical picture of what is taking place has been preserved or even clarified.

This subject of transient current and flux penetration has been studied in some detail by Miller (6) and by Tuohy et al. (7). Miller's approach was to apply a step function of current to a conductor. Initially, in this idealized situation, the current flows entirely on the conductor surface; for zero time the current density is infinite. Thereafter, the current proceeds to soak into the conductor in a manner exactly analogous to the penetration of heat into a conductor. In fact, the classical diffusion equation applies in both instances. For the cylindrical conductor this equation is

$$\frac{\partial^2 \mathbf{J}}{\partial x^2} + \frac{1}{x}\frac{\partial \mathbf{J}}{\partial x} = \frac{\mu\mu_0}{\rho}\frac{\partial \mathbf{J}}{\partial t} \tag{8.5.1}$$

and for parallel flow transverse to the plane surface of a semi-infinite solid it is simply

$$\frac{\partial^2 \mathbf{J}}{\partial x^2} = \frac{\mu\mu_0}{\rho}\frac{\partial \mathbf{J}}{\partial t} = \frac{1}{h^2}\frac{\partial \mathbf{J}}{\partial t} \tag{8.5.2}$$

where $\mathbf{J}$ is the current density as before and h has the dimensions of meters per square root second. The boundary conditions are such that the total current is zero until time zero, and constant, I, thereafter, so that the integral of $\mathbf{J}$ over the cross section is constant. These equations are very similar to the steady-state equations developed in Section 8.4 (compare Eqs. 8.4.7 and 8.5.2, for example). The difference is the substituting of $\partial/\partial t$ for $j\omega$.

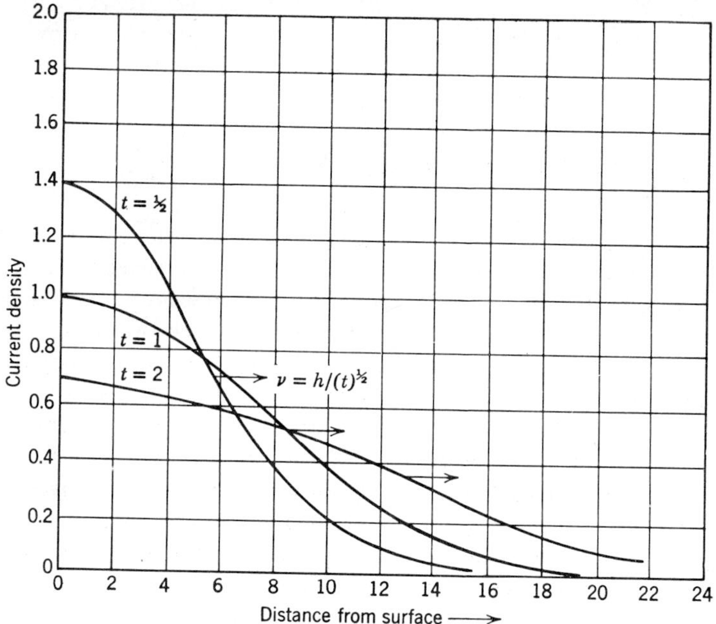

Fig. 8.10. Parallel current penetration from a plane surface.

The solutions to Eqs. 8.5.1 and 8.5.2 are quite complex, involving infinite series of Bessel functions. For a circular rod, Eq. 8.5.1 has the solution

$$\mathbf{J} = \frac{I}{\pi b^2}\left[1 + \sum_{n=1}^{n=\infty} \frac{J_0(\alpha_n x)}{J_0(\alpha_n b)} \exp\left(-h^2 \alpha_n^2 t\right)\right] \tag{8.5.3}$$

where the arguments are determined from the equation $J_1(\alpha_n b) = 0$. Here J_0 and J_1 are Bessel functions of the first kind, and of zero and first order, and are not to be confused with the **J** for current density.

Simpler solutions can be found for times so short that the depth of penetration is small compared with the dimensions of the conductor. Where this condition prevails the solution for Eq. 8.5.3 becomes

$$\mathbf{J} = I\left(\frac{\mu\mu_0}{\pi\rho t}\right)^{\frac{1}{2}} \exp\left(-\frac{\mu\mu_0 x^2}{4\rho t}\right) = \frac{I}{h(\pi t)^{\frac{1}{2}}} \exp\left(-\frac{x^2}{4h^2 t}\right) \tag{8.5.4}$$

which the reader may recognize as the *error function* from probability theory.

The curves of Fig. 8.10 are presented to give a clearer impression of this phenomenon than the equations convey. These plots, which are of Eq. 8.5.4, show the progress of the current or current density into the conductor. The scales are arbitrary, but the time intervals are chosen in the ratio $\frac{1}{2}$:1:2. Two significant facts are apparent. The surface current density declines at a rate inversely proportional to $t^{1/2}$. The velocity of penetration, characterized by the progress of the half amplitude point for each curve, is similarly proportional to $t^{-1/2}$, in fact,

$$v = \frac{h}{t^{1/2}} \tag{8.5.5}$$

It is worth reviewing once again the physical situation we are describing. It is the application of a step of current to the conductor. The surface current density is initially infinite, and there is no current within the conductor. The integral of current density across the section of the conductor at any time (the area under the curves of Fig. 8.10) gives the initial step of current. As pointed out, this is exactly analogous to the instantaneous application of a quantity of heat at the surface of a conductor. The curves of Fig. 8.10 would represent temperature profiles. No losses are assumed. The heat penetrates into the conductor, causing the surface temperature to fall and temperature within the conductor to rise.

In order to graph these functions in a way that makes them quite general in their application, it is desirable to plot them in a dimensionless form, much as was done for the generalized damping curves in Chapter 4. This requires the following substitutions:

$$x = b\bar{x}, \qquad t = \left(\frac{b}{h}\right)^2 \bar{t}, \qquad \alpha_n = \frac{\bar{\alpha}_n}{b}$$

or (8.5.6)

$$\bar{x} = \frac{x}{b}, \qquad \bar{t} = \left(\frac{h}{b}\right)^2 t, \qquad \bar{\alpha}_n = b\alpha_n$$

where b is a dimension of the system being studied, the radius of a rod, for example. When the equations are transformed in this way, they become functions of the pure numeric quantities $\bar{x}$, $\bar{t}$, and $\bar{\alpha}_n$ only, which, except for the scale factor, are each directly proportional to distance, time, and roots, respectively. Final numerical results tabulated or plotted as curves in terms of these pure numeric variables can be interpreted easily in terms of actual dimensions and materials for any specific problem by reference to Eqs. 8.5.6.

Equation 8.5.4, for the current distribution in a solid rod, has been plotted in this way in Fig. 8.11. It will be seen that as time progresses the current density distribution, which starts off nonuniform, gradually subsides to a uniform value. To show how these curves can be used, consider a copper rod 2 cm in diameter. For copper $\mu = 1$, $\rho = 1.72 \times 10^{-8}$ ohm meter and $\mu_0 = 4\pi \times 10^{-7}$. Whence,

$$h = \left[\frac{\rho}{\mu\mu_0}\right]^{1/2} = 0.117$$

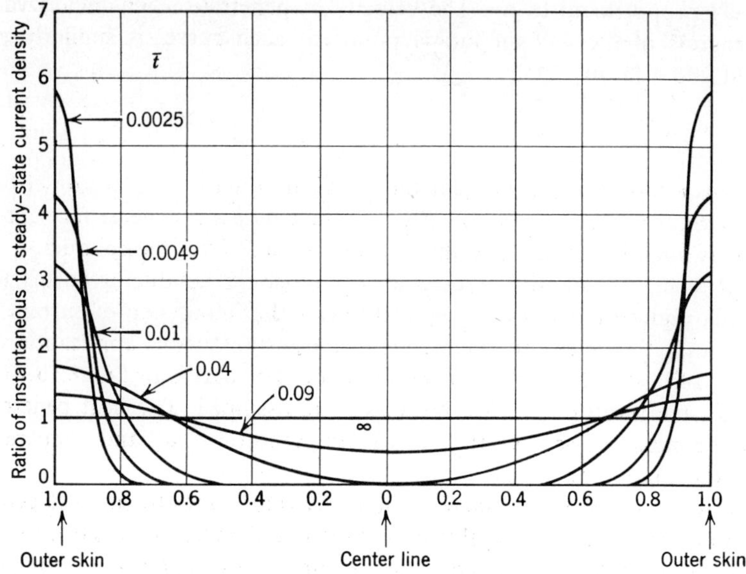

Fig. 8.11. Transient current distribution in a solid rod after the application of a step of current at the surface.

$b = 0.01$ meter,

$$\bar{t} = \left[\frac{h}{b}\right]^2 t = 137t$$

It is apparent from Fig. 8.11 that after $0.01/137 = 73$ μsec the surface current density is still three times the final steady value.

These results will vary with material, because h depends upon the material properties. Thus for aluminum $h = 0.130$, because the resistivity is higher. This causes the current to penetrate faster.

What practical situations are there in which a step of current is applied to a conductor? The implication is that the impedance of the conductor is small compared with other impedances in the circuit, and that these other impedances regulate the current rather than the conductor itself. We need not, of course, restrict ourselves to a step of current; it is simply a convenient starting point. It was shown in Section 2:6 how other waveforms can be derived from the step function, by applying Duhamel's integral. With this background, we may say that elements which typically experience current drives are fuses, shunts, and sometimes lengths of buswork.

Shunts are an interesting example. This subject will be treated in more detail in Chapter 16, but a few remarks are in order here. We often naively assume that the voltage developed across a shunt is directly proportional to the current flowing through it. When the current is constant or varying slowly, this is true, but the analysis above indicates that under transient conditions or steady-state a-c conditions, it is not so, since the nonuniform distribution of current in the shunt element must affect its apparent resistance, to say nothing of its inductance. If we pass 100 A through a shunt, and momentarily the current is forced to flow in a thin surface layer, the voltage developed across the shunt will be considerably higher than the value developed by a current of 100 A distributed uniformly over the cross section of the shunt conductor. Miller (6) defines this varying resistance in terms of the losses the current generates in the conductor much as we did for the a-c resistance in Section 8.4 (Eq. 8.4.17). The transient resistance ratio is not a fixed quantity but varies with time after initiation of the rectangular current wave, because the current distribution varies with time. Since the resistance ratio starts with an infinite and rapidly diminishing value, it is useful to define a cumulative time average ratio as well. The instantaneous power dissipation can be defined as

$$W = \int_A J^2 \rho \, dA \tag{8.5.7}$$

where the integral is performed over the entire cross section A. The average value of W up to a time t, after initiation, is

$$W_{av} = \frac{1}{t} \int_0^t W \, dt \tag{8.5.8}$$

From these two expressions can be derived the instantaneous transient resistance ratio,

$$K = \frac{W}{I^2 R_{dc}} \tag{8.5.9}$$

and the cumulative, or average, transient resient resistance ratio,

$$K_{av} = \frac{W_{av}}{I^2 R_{dc}} \tag{8.5.10}$$

In these definitions R_{dc} is the d-c resistance of the conductor. Graphs of both K and K_{av}, for solid rods and for thick-walled tubes, taken from Miller's paper (6) are given in Figs. 8.12 and 8.13. They are presented in dimensionless form through the transformation of Eqs. 8.5.6.

We noted earlier that for a 2-cm diameter copper rod, $h = 0.117$ and $\bar{t} = 137t$. After 10 μsec, $\bar{t} = 1.37 \times 10^{-3}$. Figures 8.12 and 8.13 indicate that at this time $K = 6$ and $K_{av} = 12$.

Although it may seem an easy extension to go from the current distribution in a cylindrical conductor to the distribution in a flat or square conductor,

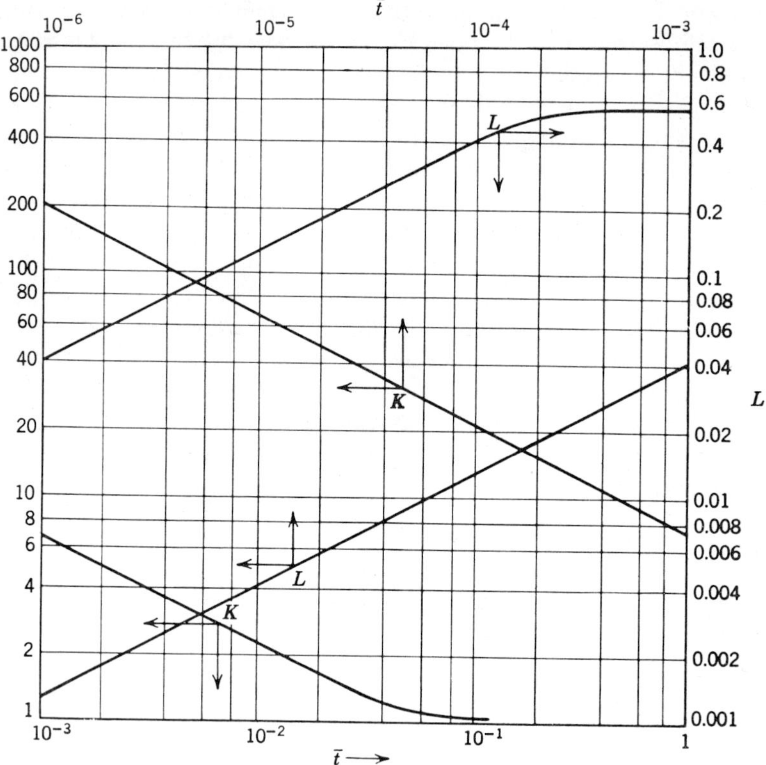

Fig. 8.12. L and K for solid rods ($K = W/I^2R_{dc}$).

this is in fact not so, because of some important differences. In a cylinder the fields are truly one-dimensional. At any radius within the conductor, the flux density is always tangential and the current density is always axial. In a flat conductor of finite thickness, the flux density will vary along the surface, as shown in Fig. 8.14, and will have both x and y components. Note that the flux lines cut the surface of the conductor, and that they are not perpendicular to the surface. Poritsky (8) has pointed out a very interesting fact for the limiting case of a very thin strip, wherein the width $2a$ is very much greater than the thickness b (Fig. 8.15). He showed that the initial current density is given by

$$\mathbf{J} = \frac{I}{\pi b}\left[\frac{1}{(a^2 - x^2)^{\frac{1}{2}}}\right] \tag{8.5.11}$$

The d-c density is $I/2ab$, thus at $x = 0$ the initial current density ratio is

$$\frac{J}{J_{dc}} = \frac{2}{\pi} = 0.638$$

This means that the transient resistance is no more than 1.57 times the d-c value.

Reference was made above to the transient inductance of a conductor. Bearing in mind the definition of inductance given in Section 8.3—the flux linkage per ampere—it is evident that the nonuniform and varying current density through the cross section of a conductor, and the distorted magnetic field, compared with the d-c condition, will materially change the inductance contributed by the internal flux. For a circular conductor, Miller (6) gives the following expression for this inductance:

$$L = \mu\mu_0\left\{\frac{1}{2} - 4\bar{t}\sum_{n=1}^{n=\infty}\frac{\epsilon^{-\bar{\alpha}_n^2\bar{t}}(2 - \epsilon^{-\bar{\alpha}_n^2\bar{t}})}{\alpha_n^2\bar{t}}\right\} \tag{8.5.12}$$

This is plotted in Fig. 8.12.

An interesting application where both the transient resistance and transient inductance can play an important role occurs where, for whatever reason,

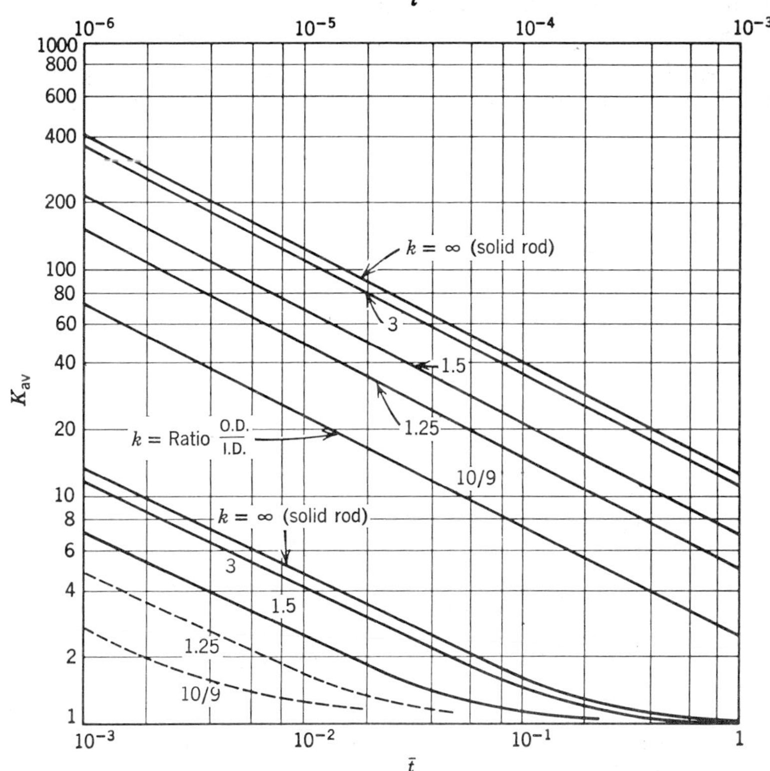

Fig. 8.13. $K_{av} = W_{av}/I^2R_{dc}$ for rods and thick tubes.

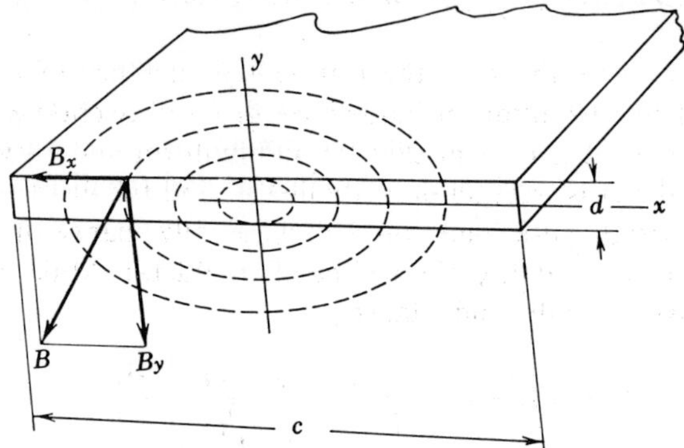

Fig. 8.14. Flux of the current in a flat strip.

current is conducted through a number of parallel paths. This occurs, for example, in rectifier equipments, where the individual rectifier elements have insufficient capacity to carry the entire current, necessitating the paralleling of a number of devices. It is important that these parallel paths share the load under steady-state and transient conditions in order that the design avails itself of the maximum potential of the devices. This is done by seeing to it

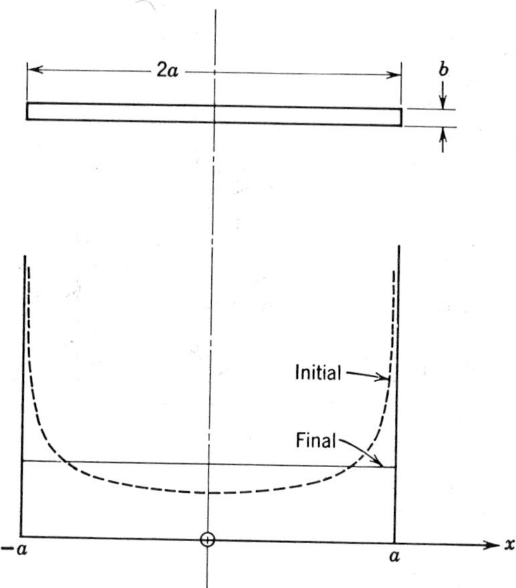

Fig. 8.15. Distribution of current in a thin strip conductor following the application of a current step [after Poritsky (8)].

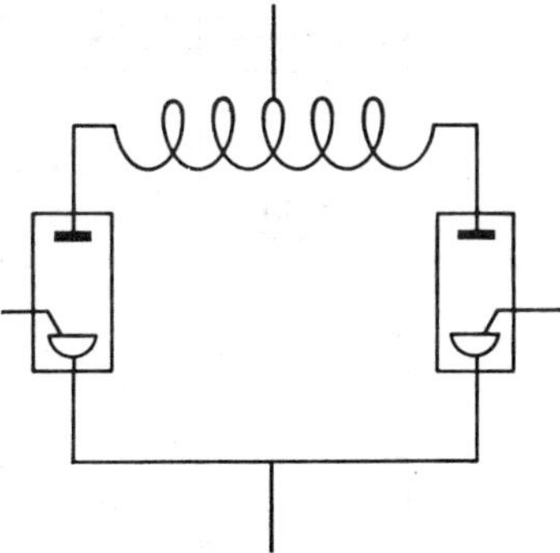

Fig. 8.16. Current-sharing reactors for ignitrons.

that as far as possible the paths have the same inductance and resistance. If just two paths are involved a balancing reactor may be used as shown in Fig. 8.16. The two halves are so coupled that if the current attempts to increase in either half, the voltage induced in it tends to oppose the increase in the usual manner, while the voltage induced in the other half tends to increase its current. Where many rectifying elements are used the balancing objective is achieved by disposing these elements in a circle with radial connections to a coaxial bus at the center. Where the rectifiers are controlled by a grid, gate, or ignitor, any failure to turn them on simultaneously will cause an unbalance to be created, which tends to persist. Whatever restoring emfs there may be attempting to restore balance, they must work against the impedance of the loop circuits around the paralleled rectifier elements, each of which has its L/R time constant. In this situation the transient Ls and transient Rs must be considered. These are changing with time, L tending to increase and R tending to decrease. Thus the time constant, at first short, increases with time. This reduces the persistance of severe unbalance, but may mean that true balance is never achieved before commutation takes place.

A quite different application for the theory outlined in this section occurs when a transmission line is struck by lightning. As explained in Chapter 10, this is essentially a situation wherein a current, often a very high current, is injected into the transmission circuit. The current divides, half flowing in one direction and half in the other direction along the line. Undoubtedly near the front of these traveling waves the effective resistance is much higher than the steady-state value. One would suppose that it diminishes with time at a particular point on the line as the wave passes by, or, put another way, at any instant the front of the wave is encountering a higher resistance than portions of the wave further back. The higher resistance again indicates a greater loss, which must be supplied at the expense of the traveling wave. This will increase the attenuation, especially at the front of the wave, and this in turn will lead to distortion of the wave. More will be said on this subject in Chapter 9.

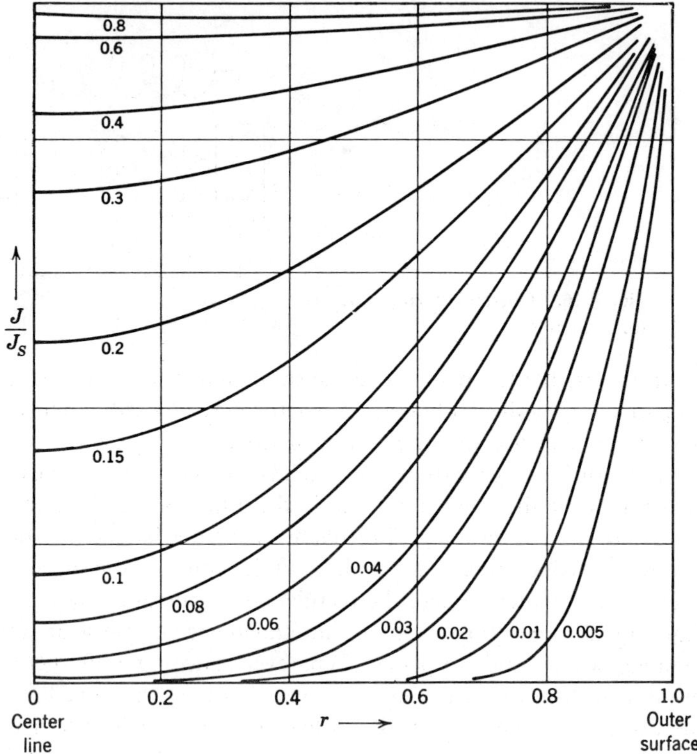

Fig. 8.17. Transient current distribution in a solid rod after the application of a step of electric field or current density at the surface.

In some circumstances it is more appropriate to consider the sudden application of a step of voltage to a circuit or a step of electric field to a conductor. This would be true if a transmission line is energized by closing a switch. Since

$$\mathbf{E} = \rho \mathbf{J}$$

this is tantamount to applying a step of current density to the surface of a conductor, rather than a step of current. Reverting to the thermal analogy once more, instead of applying to a conductor a finite quantity of heat in zero time, and seeing how this heat soaks in, or penetrates (the analogy with the step of current), it is as if one exposed the conductor surface to a fixed temperature and observes the heat penetration. The current is initially zero, but it builds up subsequently from the surface inward. The surface density is always $\mathbf{E}/\rho$. Points within the conductor gradually build up to this value.

Tuohy et al. (7) have made an analysis from this point of view. They derive an expression for the current distribution in a rod which is almost identical in form to Eq. 8.5.3. Data calculated from their expression are plotted in Fig. 8.17. It is interesting to compare these curves with those of Fig. 8.11. The difference in criteria is obvious. In Fig. 8.17 the surface current density is fixed and always equal to the final uniform current density. In Fig. 8.11 the total current is fixed, so that extremely high surface current densities are evident initially. The transient resistance derived from Fig. 8.17 is plotted in Fig. 8.18 using the same dimensionless scales. The writer finds this to be so close to Miller's data in Fig. 8.12 as to be considered identical.

8.6 Electromagnetic Shielding

The basic question being posed here is how to effectively shield one circuit from the magnetic fields produced by others, especially under transient conditions. The general answer is very simple; interpose between the circuits a barrier of conducting material. When the magnetic flux attempts to penetrate this barrier, it will induce therein eddy currents whose field will oppose the field of the inducing circuit. On the remote side of the barrier, only the net field will be experienced, which will be much less than the field without the barrier. However, to formulate the problem so that we can calculate the degree of shielding in any specific situation is very complicated. The principle is the same as that for the last two sections. It is necessary to solve the Laplace equation in the free space regions on either side of the barrier and the diffusion equation within the material of the barrier. Solutions must then be matched at the boundaries. This procedure is difficult even for simple geometrics and

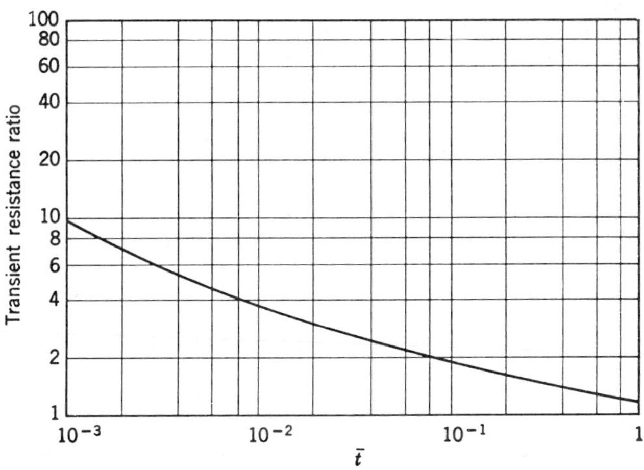

Fig. 8.18. Transient resistance ratio for a rod (suddenly applied electric field or step of current density).

becomes extremely difficult, if not impossible, for more complicated arrays. This does not mean that the task is hopeless for most practical systems. We will show that we can obtain quite a good idea of the effectiveness of any given shielding, and we will be able to prescribe barriers for practical purposes to give any desired degree of shielding. This will be done in some instances by good approximations, in others by considering various limiting cases. The material will be presented much as it was in the last section, as a distillation of original papers and texts on the subject. We will also follow the earlier pattern of considering steady-state conditions first, for these have practical significance in transient situation when the frequency is high. Later we will consider the pure transient.

Wilson and Mankoff (9) have studied this subject as it relates to isolated phase buses. This is the popular type of construction, mentioned earlier in this chapter, whereby high-voltage phase conductors are enclosed in concentric metal sheaths, which are almost invariably grounded. The primary purpose of such equipment is to protect personnel and guard the buses from disruptive phase-to-phase faults. The sheaths, incidentally, drastically reduce the magnetic fields established by the currents in the conductors in the vicinity of their neighboring conductors. This has the very useful effect of greatly reducing the mechanical forces on the phase conductors when short circuits occur. It is this aspect to which Wilson and Mankoff (9) address themselves.

They consider the shielding effect of a thin-walled cylindrical conducting tube placed in a uniform magnetic field, which is pulsating at a frequency f. The axis of the cylinder is perpendicular to the field as shown in Fig. 8.19*a*. The field, attempting to intersect the conducting material of the tube, induces emfs in that structure which cause eddy currents to flow. These currents are found to distribute sinusoidally around the circumference of the tube according to the expression

$$J = -\frac{2H_0\alpha}{(1+\alpha^2)}\sin\theta \tag{8.6.1}$$

Here, J is the rms current density, and it is expressed as a function of the angle θ in Fig. 8.19*a*; H_0 is the undisturbed magnetizing force remote from the cylinder; and α is given by

$$\alpha = \frac{\pi f \mu\mu_0\, dD}{2\rho} \qquad \left(= \frac{\pi f}{2} h^2\, dD \text{ in our former notation}\right) \tag{8.6.2}$$

where d = thickness of the metal sheath and D = diameter of the sheath. The other terms have their usual significance. This distribution of the induced currents is represented symbolically in Fig. 8.19*b*. The presence of these currents modifies the magnetic field both inside and outside the sheath. A

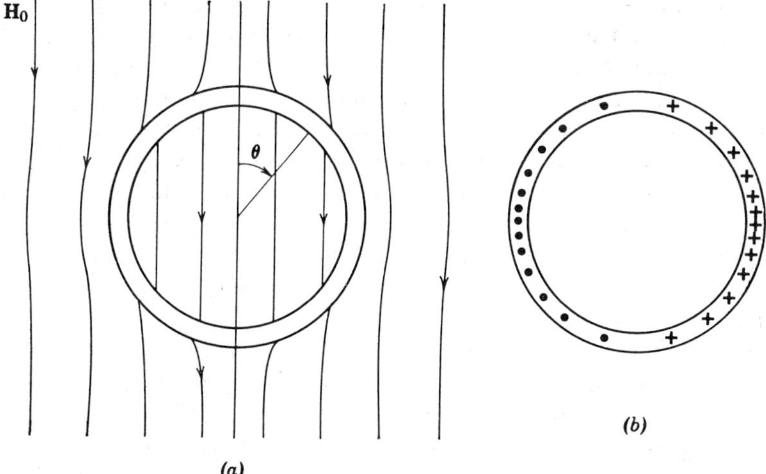

Fig. 8.19. (*a*) Field pattern—conducting cylinder in pulsating field. (*b*) Symbolic representation of eddy currents.

surprising feature is that the field inside the cylinder is uniform. For the flux density within Wilson and Mankoff give

$$\mathbf{B}_i = \frac{\mathbf{B}_0}{(1 + j\alpha)} \tag{8.6.3}$$

which indicates a substantial shielding effect when α is large. It also shows that there is a phase difference between the inside and outside flux density. Again, when $\alpha \gg 1$, this phase angle approaches 90°. The scalar value of B_i is given by

$$B_i = \frac{B_0}{(1 + \alpha^2)^{1/2}} \tag{8.6.4}$$

which leads to the concept of a steady-state a-c shielding factor, k_a, defined as the ratio of the flux density B_i at the axis of the sheath to the flux density B_0 that would appear at the same point if the sheath were not present:

$$k_a = \frac{B_i}{B_0} = \frac{1}{(1 + \alpha^2)^{1/2}} = \frac{1}{[1 + (2\pi f T_e)^2]^{1/2}} \tag{8.6.4}$$

where

$$T_e = \frac{\mu\mu_0 \, dD}{4\rho} \tag{8.6.5}$$

Here T_e has the dimensions of time, and is in fact the time constant with which sheath eddy currents decrease following the sudden application of a constant magnetic field.

We can make some observations at once on the effectiveness of the shielding. It is apparent that for a perfectly conducting cylinder, shielding would be complete. This is perhaps not surprising. Any attempt on the part of the flux to penetrate the conductor will be met with an immediate reaction in the form of eddy currents, which require no emf to drive them since the conductor is perfect, and which effectievly neutralize the applied field. For imperfect conductors, the shielding improves with the conductivity and also with the dimension d and D. To put this on a quantitative basis, we will substitute some numerical values. Suppose the cylinder is made of aluminum, for which $\rho = 2.83 \times 10^{-8}$ ohm meter, $D = 1$ meter, and $d = 1$ cm. From Eq. 8.6.5, $T_e = 0.111$ sec. At 60 Hz, $\alpha = 34.8$; thus, from Eq. 8.6.4, $k_a = 0.0286$, that is, the field within the sheath is but 2.86% of the field outside. This percentage will diminish directly as the frequency increases; thus at 1 kHz, $k_a = 1.72 \times 10^{-3}$ and at 10 kHz, $k_a = 1.72 \times 10^{-4}$.

It is interesting to observe how the shielding is influenced by the size of the tube, that is, by d and D. For a 1-cm diameter aluminum tube, with a 1-mm thick wall, the shielding at 60 Hz is very poor. Since $\alpha = 0.35$, the shielding may be considered negligible.

It is apparent from the expression for α (Eq. 8.6.2) that high conductivity and high permeability are desirable features in a shielding material. But the variations in conductivity between various likely metallic shield materials is not nearly as great as the variations in permeability between the ferromagnetic and nonferromagnetic materials. This indicates that much is to be gained by using a ferromagnetic material provided it does not saturate.

Fault currents are frequently asymmetrical, as was pointed out in Chapter 3. This means that they have a d-c component that decays with the L/R time constant of the circuit, which we will call T_d. If the fault is close to the generators, as it may very well be where isolated phase bus is applied, the a-c component of current may also have a decrement because of the changing reactance of the machines feeding the fault. These effects will be manifest in the magnetic flux produced by the current. Wilson and Mankoff (9) analyzed this condition; a feature of this analysis is interesting from the point of view of transient shielding. The magnetomotive force associated with the d-c component will follow this current instantaneously, so the variations of the uniform field with time will be as follows:

$$b_0 = B_0 \epsilon^{-t/T_d} \tag{8.6.6}$$

where b_0 = instantaneous value of uniform field strength and B_0 = initial value of uniform field strength. Because b_0 appears instantaneously, the eddy currents in the sheath must appear instantaneously to maintain zero flux linkages through the eddy current loops. The magnetic field intensity inside the sheath will therefore start from zero, build up gradually as the opposing eddy currents decay, and then gradually reduce to zero as the entire field vanishes. The following equation is given for b_i, the instantaneous field strength inside the sheath:

$$b_i = B_0\left[\frac{\epsilon^{-t/T_d} - \epsilon^{-t/T_e}}{1 - T_e/T_d}\right] \tag{8.6.7}$$

From a transient point of view, it is convenient to have the response to a simpler stimulus, namely, the sudden application of a step field with no decrement. Other forms of transient fields can be studied from this by various techniques such as Duhamel's integral (Chapter 2). The step function field response is formed by simply setting $T_d = \infty$ in Eq. 8.6.7. This gives

$$b_i = B_0[1 - \epsilon^{-t/T_e}] \tag{8.6.8}$$

This shows that after several time constants, T_e, the transient field will have penetrated the shielding cylinder. But if the shield is removed quickly, very little may penetrate inside. Now T_e can be calculated for any cylinder and for any material from Eq. 8.6.5.

Some shielding enclosures are not cylindrical. By means of the electrolytic field analyzer (see Chapter 14) it has been found that a square cross section sheath of a given thickness has the same shielding factor as a circular cross section sheath of the same thickness with the same perimeter. Thus a square with a side equal to the diameter of a cylinder will be approximately 25% better as a shield.

A somewhat similar shielding problem has been analyzed by Tuohy (10). He inquired into the effectiveness of a thin cylindrical tubular sheath in shielding its interior from the magnetic field of the current in an adjacent parallel conductor (Fig. 8.20). In this case the tube is assumed to have perfect

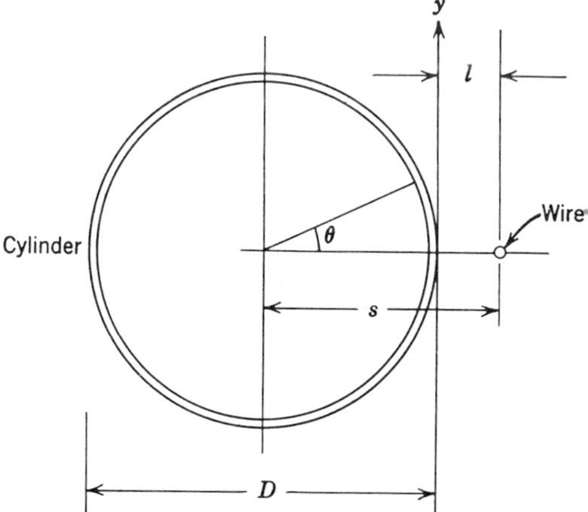

Fig. 8.20. Thin-walled cylinder and parallel wire.

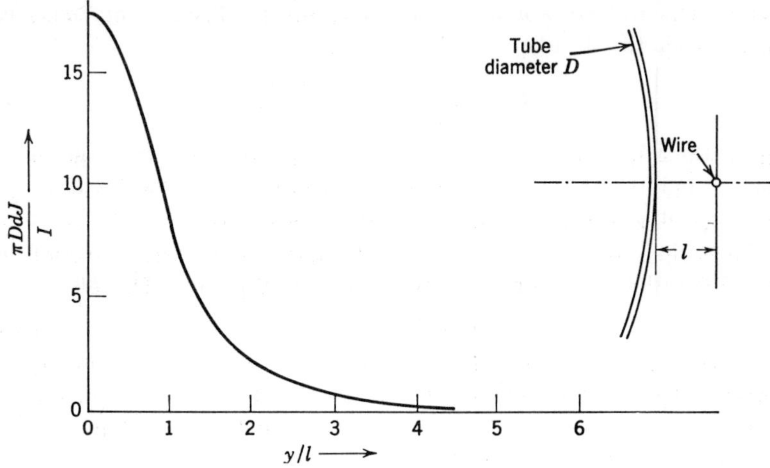

Fig. 8.21. Eddy current density in a thin conducting tube near a wire carrying a 60-Hz current.

conductivity. Tuohy's solution for shielding, which is reproduced here, makes use of an expression derived by Dwight (11) for current density distribution in a thin tube, due to an alternating current in a wire outside the tube. As the radius of the tube is increased, the tube appears like a flat plate to the wire. This current density expression has been used by Mankoff et al. (12), who noted that at 60 Hz the current density reduces to much simpler form than the infinite series presented by Dwight. The eddy current density in the sheet, although varying in magnitude, is always approximately 180° out of phase with the current in the wire. This fact suggests a relatively simple method for determining the magnetic field strength at any point beyond the sheet.

When $D/2s$ approaches unity, the expression for current density becomes

$$\mathbf{J}_x = \left\{\frac{D/s(\cos\theta - D/2s)}{1 + (D/2s)^2 + (D/s)\cos\theta}\right\} \tag{8.6.9}$$

where the symbols have the significance shown in Fig. 8.19. For $D/2s = 0.9$, the tube can be assumed flat for $\theta = \pm 13°$, and we may write

$$\frac{y}{l} = 9\theta \tag{8.6.10}$$

Substituting Eq. 8.6.10 in Eq. 8.6.9, the expression for J_x is given approximately by

$$\mathbf{J}_x = \frac{I}{\pi Dd}\left[\frac{\cos(y/9l) - 0.9}{1 - \cos(y/9l)}\right] \tag{8.6.11}$$

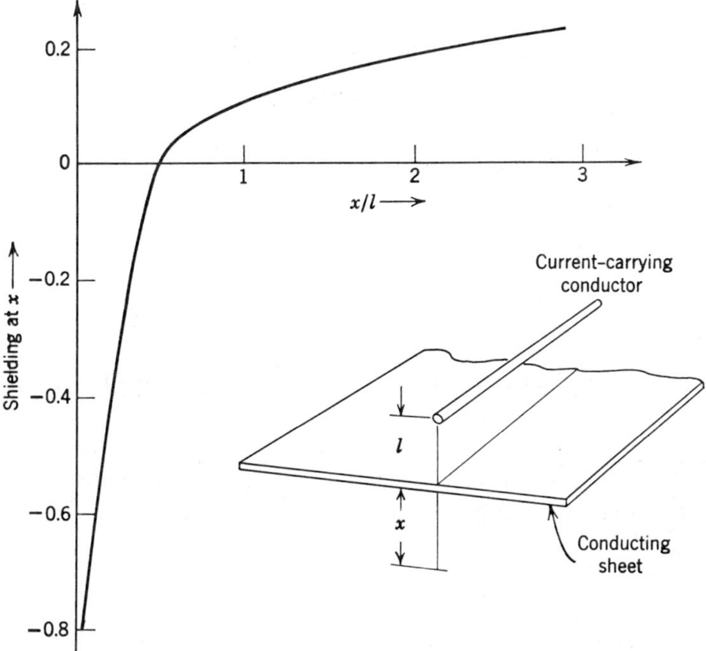

Fig. 8.22. Shielding of a magnetic field by a thin flat sheet of aluminum.

This is plotted in a convenient way in Fig. 8.21, which shows that most of the current flows in a section of the tube $\pm 4.5l$ on each side of the line between the wire and the tube centre. Thus, if this piece of the tube is divided into nine strips each of width l, and the current in each strip is assumed proportional to the current density at its center, the magnetic field produced by these nine currents can be found. When the magnetic field of the wire is superimposed, the net magnetic field is obtained. The field so obtained can be compared with the field of the wire above, without the sheet. This has been done in Fig. 8.22 using the data of Fig. 8.21. Some interesting information is contained in Fig. 8.22. It shows that about 90% of the magnetic field strength is shielded by the sheet at a point the same distance as the conductor from the sheet, but on the opposite side. There is very little shielding for points close to the sheet, and the eddy currents actually increase the field in the region between the conductor and the sheet.

Perhaps the most interesting analysis of the shielding problem is that given by Smythe (13), which appears to be based on earlier work by Jeans (14). This also examines the shielding effect of an infinite thin sheet of conducting material (see Fig. 8.23). Using the concept of vector potential, they show that the field on the side of the sheet remote from the conductor can be

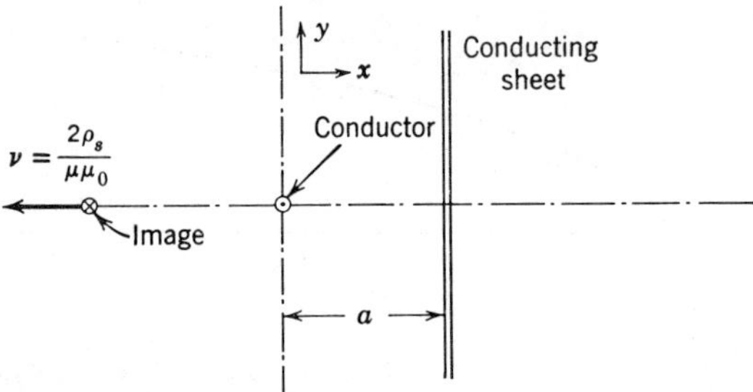

Fig. 8.23. Transient shielding effect of a conducting sheet using a retreating image.

determined by considering the field of the wire itself, and the field of an image conductor. At time zero, the image, whose current is precisely opposite to that in the wire, is located in the same position as the wire itself. The fields of the two therefore cancel at all points at this instant. Subsequently, the image moves in a direction perpendicular to the plane with a velocity,

$$v = \frac{2\rho_s}{\mu\mu_0} \tag{8.6.12}$$

where ρ_s is the surface resistivity of the sheet, that is, the resistance between opposite edges of a square with 1-meter side, and $\mu\mu_0$ is the absolute permeability of the sheet. As the image retreats, its influence will be felt less and less, so that, in the limit, the sheet will offer no shielding at all. Tuohy (10) has applied this analysis to the transient shielding problem.

For a step function of current I applied to the wire, the flux density in the plane of the wire on the remote side of the sheet will be

$$\mathbf{B} = \frac{\mu_0 I}{2\pi}\left[\frac{1}{x} - \frac{1}{x + vt}\right] \tag{8.6.13}$$

and will be parallel with the sheet. In this coordinate system, the origin is at the wire, x is measured perpendicular to the sheet, and y is parallel with that plane. The first term is the flux due to the wire, the second is that due to the image. The expression (8.6.13) is only valid for $x > a$, where a is the distance from the wire to the sheet.

For points not in the plane of the wire, the field will have both x and y components. Manipulation reveals that

$$\begin{aligned}\mathbf{B}_x &= \frac{\mu\mu_0 I}{2\pi}\left\{\frac{y}{x^2 + y^2} - \frac{y}{(x + vt)^2 + y^2}\right\}\\ \mathbf{B}_y &= \frac{\mu\mu_0 I}{2\pi}\left\{\frac{x}{x^2 + y^2} - \frac{x + vt}{(x + vt)^2 + y^2}\right\}\end{aligned} \tag{8.6.14}$$

If this field is compared with the undisturbed field of the wire, we can once more specify a field ratio or shielding factor. This is

$$\mathbf{B}_{\text{ratio}} = \left[(x^2 + y^2)\left\{\left[\frac{x}{x^2 + y^2} - \frac{x + \nu t}{(x + \nu t)^2 + y^2}\right]^2 + \left[\frac{y}{x^2 + y^2} - \frac{y}{(x + \nu t)^2 + y^2}\right]^2\right\}\right]^{1/2} \tag{8.6.15}$$

Again, this is only true for $x > a$. As time increases, this ratio approaches the value

$$\left\{(x^2 + y^2)\left[\left(\frac{x}{x^2 + y^2}\right)^2 + \left(\frac{y}{x^2 + y^2}\right)^2\right]\right\}^{1/2} \equiv 1$$

as would be expected.

For a sheet of copper 1 mm thick, $\rho_s = 1.72 \times 10^{-5}\ \Omega/\text{m}^2$, $(\mu\mu_0) = 4\pi \times 10^{-7}$, whence $\nu = 27.5$ m/sec. Using these data, let us examine the field 50 cm beyond such a barrier, from a conductor 50 cm away from the conductor on the other side. Let us assume that the stimulus is a square pulse of current, which rises instantaneously to a value I, retains that value for 1 msec, and then returns to zero.

If we limit consideration to points in the plane of the wire, Eq. 8.6.15 reduces to the much simpler expression,

$$[B_{\text{ratio}}]_{y=0} = \left[\frac{\nu t}{x + \nu t}\right] \tag{8.6.16}$$

Substituting $\nu = 27.5$, $x = 1$, gives

$$[B_{\text{ratio}}]_{y=0} = \frac{27.5t}{1 + 27.5t} \tag{8.6.17}$$

This must be plotted for 1 msec, at which time an identical term of opposite sign is added to acknowledge the termination of the current pulse. That is, we apply superposition. Equation 8.6.17 has been plotted in Fig. 8.24*a*. We see that the field reaches a maximum at the end of this pulse, but that the shielding is more than 97% effective. After 1 msec, although the current has ceased to flow in the conductor, a field will remain from the decaying eddy currents in the shield. Strictly speaking, we cannot consider a shielding ratio after this time, for in the absence of the shield, no field at all would exist once the conductor current has disappeared. However, the model of the wire and the retreating image (now extended, by superposition, to include the negative step which cuts off the pulse, and its retreating image) can be used to determine how the field due to the eddy currents lingers. Because copper is a good conductor, we find this decaying process is very slow. This is indicated

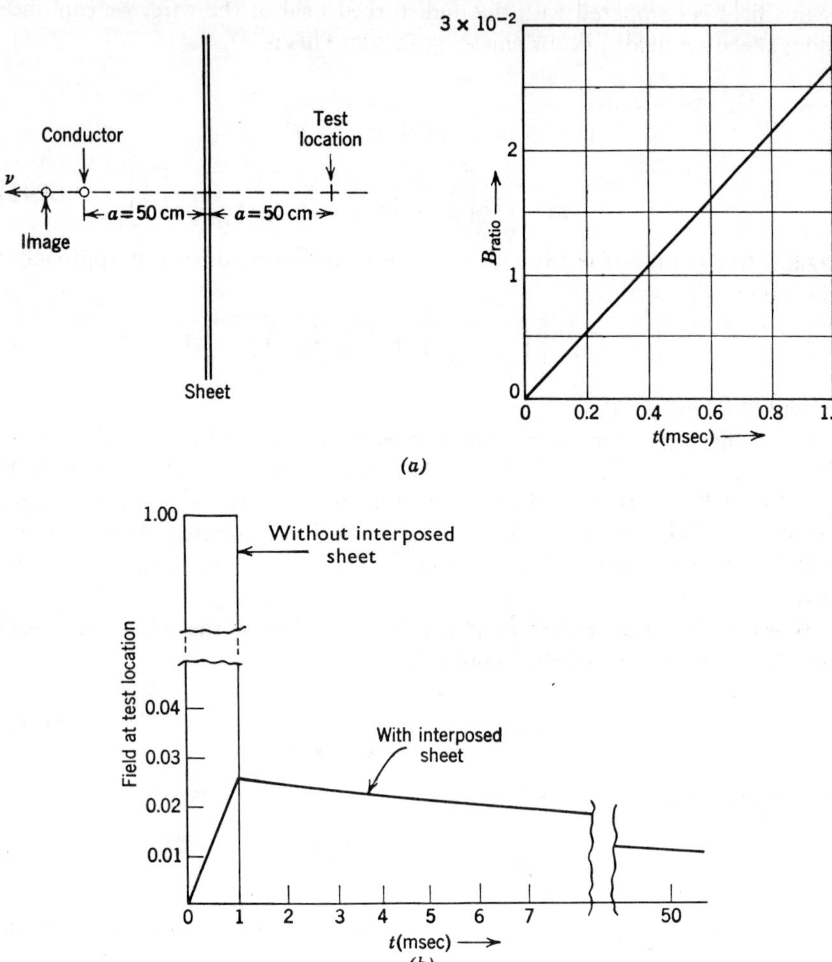

Fig. 8.24. (*a*) Magnetic shielding ratio for a 1-mm sheet of copper as a function of time for a 1 msec current pulse. (*b*) Field at the test location, with and without the interposed, copper sheet.

in Fig. 8.24*b*, which shows the field without the sheet (severely truncated), and the long-tailed field with the sheet interposed. After 50 msec the field still shows 44% of its peak value.

8.7 Implications of Electromagnetic Effects for Cryogenic Systems

One way to improve the efficiency of power systems is to operate the equipment at extremely low temperature, because at such temperatures the resistivity of conductors is greatly reduced. For pure copper, the resistivity diminishes by a factor of 10 at 77°K, the boiling point of nitrogen, and by about 500 at the boiling point of hydrogen (20°K). The changes are less striking for commercial grades of conducting material, but a ratio of 100 is still found at the hydrogen temperature. At still lower temperatures the resistance of conductors disappears altogether when they become superconductors.

Obviously, there are formidable problems in maintaining the required refrigeration, and the cost of this as well as the energy required to operate it must be figured into any meaningful assessment of the economics of cryogenic operation. The rewards promised by such an approach are such that investigations are being made for motors (15), and transformers (16), and in a recent paper (17) Neal considers the application for cables.

The topics we have been discussing in this chapter have some interesting implications for cryogenic systems. In steady-state a-c application, for example, the much lower resistivity markedly affects the skin depth. For copper at 60 Hz it is less than 0.4 mm. Neal (17) therefore proposes a thin-walled pipe for the cable conductor rather than a cylinder. Time constants for the penetration of flux and currents under transient conditions are significantly prolonged. Perhaps the most important concern would be the effect of transient currents. It seems possible that, being localized near the surface of conductors, they could generate sufficient loss to raise the local conductor temperature and momentarily increase the resistance. This could simply result in extra damping, which might be desirable. But if it should lead to vaporization of the coolant, with an attendant sharp deterioration of its cooling capability, it is possible that a disruptive pressure could be developed.

REFERENCES

1. G. W. Carter, *The Electromagnetic Field in Its Engineering Aspects*, Longmans, Green and Co., London, New York and Toronto (1954).
2. S. Ramo and J. R. Whinnery, *Fields and Waves in Modern Radio*, John Wiley & Sons, New York and London (1953).
3. J. A. Stratton, *Electromagnetic Theory*, McGraw-Hill Book Co., New York and London (1941).
4. M. J. Pickett, H. L. Manning, and H. N. VanGeen, "Near Resonant Coupling on E.H.V. Circuits: I—Field Investigations," *Trans. IEEE*, Vol. 87, Part III (1968), p. 322.
5. M. H. Hesse and D. D. Wilson, "Near Resonant Coupling on E.H.V. Circuits: II—Methods of Analysis," *Trans. IEEE*, Vol. 57, Part III (1968), p. 326.
6. K. W. Miller, "Diffusion of Electric Currents into Rods, Tubes and Flat Surfaces," *Trans. AIEE*, Vol. 66 (1947), p. 1496.

7. E. J. Tuohy, T. H. Lee, and H. P. Fullerton, "Transient Resistance of Conductors," *Trans. IEEE,* Vol. 87, Part III (1968), p. 455.
8. H. Poritsky, unpublished report.
9. W. R. Wilson and L. L. Mankoff, "Short-Circuit Forces on Isolated-Phase Busses," *Trans. AIEE,* Vol. 73, Part III (1954), p. 382.
10. E. J. Tuohy, private communication.
11. H. B. Dwight, *Electrical Coils and Conductors,* McGraw-Hill Book Co., New York and London (1945).
12. L. L. Mankoff, N. Swerdlow, and W. R. Wilson, "An Analog Method for Determining Losses in Isolated Phase Bus Enclosures," *Trans. AIEE,* Vol. 82 (1963), p. 535.
13. W. R. Smythe, *Static and Dynamic Electricity,* McGraw-Hill Book Co., New York and London (1950).
14. Sir James Jeans, *The Mathematical Theory of Electricity and Magnetism,* 5th ed., Cambridge University Press (1960).
15. A. D. Appleton, "Super Conducting Motor," *JIEE Electronics and Power,* Vol. 14 (1968), p. 114.
16. P. Burnier, "Domaines d'Application des Cryomachines Electrique," *Rev. Gen. d'Elec.,* Vol. 74 (1965), p. 623.
17. S. Neal, "Cryogenic Transmission in the Power Industry of the Future," *American Power Conference* (1968).

9 Traveling Waves on Transmission Lines

9.1 Circuits with Distributed Constants

The first eight chapters of this book have been concerned with circuits in which the constants R, L, and C have been lumped or concentrated, or where they could be so approximated. We realize that these parameters are really distributed in any circuit or piece of equipment, so it is perhaps surprising and gratifying to find how accurately the transient behavior of such circuits can be calculated on the basis of lumped circuit analysis. However, there are important parts of a power system where this approach is inadequate because the approximations are too great. The most obvious example is the transmission line. Here each meter of its length is much like every other meter. It possesses a certain inductance, capacitance, and resistance, so that these quantities are truly distributed over tens or hundreds of miles. A less obvious example is the machine winding. Heretofore, a transformer or rotating machine has been represented by an inductance, sometimes with capacitance at its terminals. Such a representation is adequate if the focus is on the action of the winding as a whole, as a component in the circuit. If, on the other hand, we are concerned with phenomena occurring within the winding, we must take cognizance of the distributed nature of the winding inductance, capacitance, and resistance.

A distinguishing feature of the circuit with distributed constants is its ability to support traveling waves of current and voltage. How this occurs will be presented first in a qualitative way, in order to establish a physical picture of what is taking place. Consider the two-wire circuit shown in Fig. 9.1*a* in which, by closing the switch S, the transmission line is connected to a source of voltage V, which will be assumed, for the time being, to have zero impedance. The action of closing the switch can be likened to opening a sluice or valve at the end of a channel or pipe, thereby admitting water to

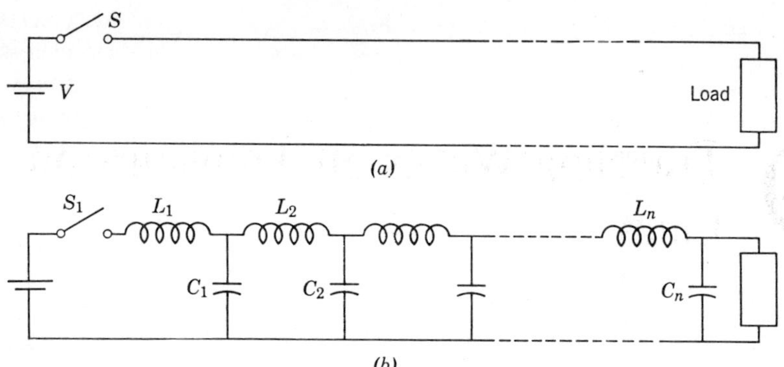

Fig. 9.1. (*a*) Two-wire transmission line. (*b*) "Lumpy" representation of a two-wire-line.

the pipe or channel from some reservoir behind. When the sluice is opened, the channel does not instantly fill with water. What we observe instead is a wave of water advancing down the channel. At any instant the channel ahead of the wavefront is dry, while that behind is filled with water to capacity. Contrast this with what happens when the switch is closed in the circuit of Fig. 9.1*b*. This figure is a very plausible representation of a transmission line; however, it has some inadequacies. The line is here divided into a large number of sections, each section being associated with a certain inductance L_1 and a certain capacitance C_1.

When S_1 is closed, current starts from zero and flows through the first inductance to charge the first capacitance. But, as soon as this capacitor has acquired any charge whatsoever, there is a voltage across the next section of the line and current commences to flow through the second L_2 to charge the second C_2. This argument can be carried to the third section, the fourth section, and so on. Thus, with a lumped circuit like Fig. 9.1*b*, some disturbance, no matter how slight, is felt in the nth section of the line when S_1 is closed. In Fig. 9.1*a* we assert that this is not so. The disturbance created by closing S on the smooth line travels down the line with a finite velocity and is felt only at remote points after a finite interval determined by the speed of propagation of electromagnetic waves in the medium surrounding the line. The different behavior of the circuit of Fig. 9.1*b* is a direct consequence of its lumpiness. This is not to say that this representation of the line is of no value. So-called "artificial lines" of this kind are very useful for several purposes. A discussion of this subject is given in Chapter 17. The more sections into which the line is divided, the more closely will it resemble the smooth line inasmuch as the response of the circuit to any stimulus will become progressively more like the response of the true transmission line.

In abandoning the lumpy approach, one must also abandon certain cherished notions about the behavior of circuits; for example, the idea that current is univalued at all points in a circuit and that voltage cannot be discontinuous. Such concepts have no place in transmission line theory. In describing what does happen when the switch S closes, it is helpful to consider events from the field standpoint. With the energizing of the circuit a current commences to flow from the source into the line charging the capacitance of the line to the source voltage as it does so. Let it be assumed that after a time Δt a length Δx of the line has been so charged. If the capacitance of the line is C farads per meter, a charge

$$Q = CV\,\Delta x \tag{9.1.1}$$

will have been imparted to the line. The consequences of this are twofold: an electric field is created between the conductors of the first Δx meters of line and a magnetic field is similarly created around the conductors by virtue of the currents flowing in them. The current, of course, is determined from the rate at which charge flows into and out of the line, that is, $I = dQ/dt$, thus, from 9.1.1,

$$I = VC\frac{\Delta x}{\Delta t} = CV\frac{dx}{dt} \quad \text{in the limit} \tag{9.1.2}$$

Now dx/dt is the rate at which the disturbance is propagating along the line. Let this be designated v. Then

$$I = CVv \tag{9.1.3}$$

The progressive establishment of magnetic flux linking the lines means that an emf equal to the rate of change of flux linkage is being induced in the loop formed by the conductors and the wavefront. The flux linkage can be determined from first principles or defined in terms of the line inductance. Taking the latter approach and supposing the line to have an inductance of L henries per meter, when the current has penetrated Δx meters, the flux linkage Φ is given by

$$\Phi = L\,\Delta x I = L\,\Delta x CVv \tag{9.1.4}$$

The emf induced is

$$\frac{d\Phi}{dt} = LCVv\frac{\Delta x}{\Delta t}$$

$$= LCVv^2 \quad \text{in the limit} \tag{9.1.5}$$

Thus

$$V = LCVv^2$$

or

$$v = \frac{1}{(LC)^{1/2}} \tag{9.1.6}$$

The velocity of propagation of these waves of current and voltage along the line appears to depend upon the line geometry and the electromagnetic properties of the surrounding medium, for they are these factors which prescribe L and C.

It is enlightening to enter some figures at this point. If the spacing d between the lines is large compared with the radius of the conductors r, the flux within the conductors can be neglected. A good approximation for the inductance is then

$$L = \frac{\mu_0}{\pi} \ln \frac{d}{r} \text{ H/m} \tag{9.1.7}$$

With the same assumptions, the capacitance can be written

$$C = \frac{\pi \epsilon_0}{\ln d/r} \text{ F/m} \tag{9.1.8}$$

Substituting these values in Eq. 9.1.6 gives

$$LC = \mu_0 \epsilon_0$$

or

$$v = \frac{1}{(\mu_0 \epsilon_0)^{1/2}} \tag{9.1.9}$$

which is the velocity of electromagnetic waves in free space, that is, the velocity of light, usually designated c. It is independent of line geometry. Actually, L and C are somewhat less than the approximations quoted in Eqs. 9.1.7 and 9.1.8 so that v is somewhat less than c.

In some practical cases the ambient medium is not free space or air; then the velocity of propagation is different. For example, in a cable the permittivity of the dielectric is $k\epsilon_0$, where k may be 3 to 5 or even more. Thus v, being reduced by a factor of $1/k^{1/2}$, may be less than half the velocity for an air-insulated line. Similarly, where a conductor is buried in steel, as are the conductors of most rotating machines, the permeability μ may greatly exceed μ_0. This will effect a change in velocity in the ratio $(\mu_0/\mu)^{1/2}$, once the flux has penetrated the steel.

Reduced to practical units, the wave velocity on an air-insulated line is approximately 1000 ft μsec or, put another way, it takes about 5.4 μsec for a wave to travel 1 mile.

Equation 9.1.3 can be alternatively stated as

$$I = CVv = \frac{CV}{(LC)^{1/2}}$$

or

$$\frac{V}{I} = \left(\frac{L}{C}\right)^{1/2} \tag{9.1.10}$$

Equation 9.1.10 expresses the ratio of the amplitudes of the voltage and current waves. It therefore has the dimensions of impedance and is designated

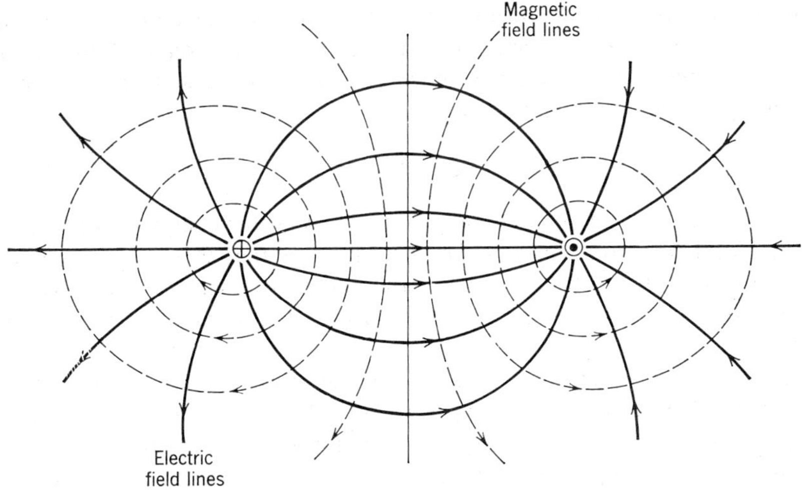

Fig. 9.2. Electromagnetic field of a two-conductor transmission line.

the characteristic impedance of the line. Because it is identical with the surge impedance of an LC circuit, we use for it the same symbol, Z_0:

$$Z_0 = \left(\frac{L}{C}\right)^{1/2} \tag{9.1.11}$$

Unlike the velocity, v, the characteristic impedance clearly depends upon the line geometry. For a typical overhead transmission line Z_0 is about 400 Ω. For a cable it is in the range 30–80 Ω because the closer spacing makes C larger and L smaller. The capacitance is additionally augmented by the higher permittivity of the cable dielectric.

Summarizing, it can be said that when a line is connected to a source of electric energy, a traveling wave of voltage passes along the line at a velocity approaching the speed of light, establishing an electric field between the conductors as it does so. The voltage wave is accompanied by a current wave of amplitude V/Z_0, which in turn creates a magnetic field in the surrounding space. The field patterns are shown in Fig. 9.2.

It is evident that energy is being supplied to the line at a rate of VI watts. In the absence of losses this energy must be stored in the electromagnetic field. Simple manipulation shows that it is distributed equally between the electric and magnetic fields. To the electric field is supplied $\frac{1}{2}CV^2v$ watts, to the magnetic field $\frac{1}{2}LI^2v$ watts. With the substitutions

$$\frac{V}{I} = \left(\frac{L}{C}\right)^{1/2} \quad \text{and} \quad v = \frac{1}{(LC)^{1/2}}$$

both these expressions reduce to $\frac{1}{2}VI$. Thus the phenomenon can be looked upon as a wave of energy spreading down the line. The reader familiar with electromagnetic theory will recognize that this accords with the concept of Poynting's vector for energy flow:

$$\mathbf{P} = \mathbf{E} \times \mathbf{H} \tag{9.1.12}$$

The vector **P** gives the magnitude and direction of the energy flow at a point in a field. The vector equation (9.1.12) indicates that **P** is perpendicular to both **E** and **H**. From this standpoint the line may be viewed as a waveguide transmitting energy through the space surrounding it rather than along the conductors themselves.

Any practical conductor must have some resistance. To overcome this resistance there must be a component of **E**, no matter how small, in the direction of the current flow. Therefore **P** must have a component perpendicular to this component of **E**, that is, directed inward to the centers of the conductors. This part of the energy is spent within the conductor as losses.

9.2 The Wave Equation

Thus far we have been concerned with a rather simple proposition—what happens when a uniform two-wire line is connected to a voltage source of zero impedance. A more general proposition will now be formalized by the development of the Wave Equation. It will be shown that any unbound system of charge of arbitrary form, which is on a transmission line, must travel along the line. We shall again consider a two-wire line and assume that it is loss-free. The consequence of losses will be considered later. At this point to introduce resistance and leakage greatly increases the complexity of the algebra without clarifying the picture.

Figure 9.3 shows a small element of a transmission line. If the line has an inductance of L henries per meter and C farads per meter, an elementary

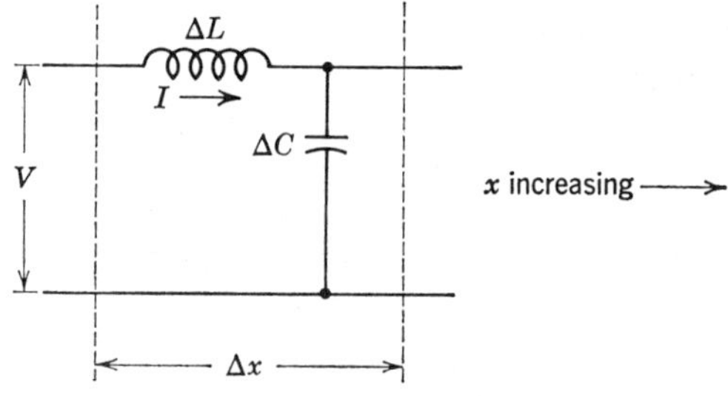

Fig. 9.3. Small element of a transmission line.

length Δx will have inductance and capacitance $L\,\Delta x$ and $C\,\Delta x$ as shown. The voltage across this element will be

$$-\Delta V = L\,\Delta x \frac{\partial I}{\partial t}$$

which in the limit as the element shrinks in length to the infinitesimal dx can be written

$$\frac{\partial V}{\partial x} = -L\frac{\partial I}{\partial t} \tag{9.2.1}$$

We use partial derivatives here because V and I are function of both position and time.

The current to charge the elementary capacitance ΔC is given by

$$-\Delta I = C\Delta x \frac{\partial V}{\partial t}$$

which in the limit becomes

$$\frac{\partial I}{\partial x} = -C\frac{\partial V}{\partial t} \tag{9.2.2}$$

The negative signs in Eqs. 9.2.1 and 9.2.2 arise from the convention being used. Figure 9.3 shows x increasing to the right. With the current flowing in the manner indicated, both V and I will diminish with increasing x.

Now I can be eliminated from the pair of simultaneous equations by differentiating Eq. 9.2.1 with respect to x and Eq. 9.2.2 with respect to t:

$$\frac{\partial^2 V}{\partial x^2} = -L\frac{\partial^2 I}{\partial x\,\partial t}$$

$$\frac{\partial^2 I}{\partial x\,\partial t} = -C\frac{\partial^2 V}{\partial t^2}$$

Eliminating $\partial^2 I/\partial x\,\partial t$ and rearranging the terms,

$$\frac{\partial^2 V}{\partial x^2} = LC\frac{\partial^2 V}{\partial t^2} \tag{9.2.3}$$

Solving Eqs. 9.2.1 and 9.2.2 for I instead of V leads to an equation of identical form for the current:

$$\frac{\partial^2 I}{\partial x^2} = LC\frac{\partial^2 I}{\partial t^2} \tag{9.2.4}$$

Equations 9.2.3 and 9.2.4 constitute the transmission line wave equations. It is possible to solve them by the Laplace transform method, taking two steps. The first transformation converts them from partial to ordinary differential equations. These can be transformed again, in the usual way, into algebraic equations. At this point we simply assert that the voltage equation can be satisfied by a solution of the form

$$V = f[t \pm (LC)^{1/2}x] \tag{9.2.5}$$

or, alternatively,

$$V = f\left(x \pm \frac{t}{(LC)^{1/2}}\right) \tag{9.2.6}$$

It will be recalled from Section 9.1 that $(LC)^{-1/2}$ is a velocity, so that dimensionally the first of these equations (9.2.5) states that V is a function of time, while the second specifies that V is a function of distance. They are different but equivalent ways of describing the same phenomenon.

One is useful if we are turning our attention to a particular location and determining how the voltage varies as a function of time. The other is applied when our interest is fixed on a particular time, and we seek a purview of all points at that time. Rewriting Eq. 9.2.6 in terms of v,

$$V = f_1(x + vt) + f_2(x - vt) \tag{9.2.6a}$$

That this is in fact a solution to the wave equation can be shown by direct substitution. We differentiate 9.2.6a twice, first with respect to x then with respect to t:

$$\left.\begin{aligned} \frac{\partial V}{\partial x} &= f_1'(x + vt) + f_2'(x - vt) \\ \frac{\partial^2 V}{\partial x^2} &= f_1''(x + vt) + f_2''(x - vt) \end{aligned}\right\} \tag{9.2.7}$$

$$\frac{\partial^2 V}{\partial t} = v[f_1'(x + vt) - f_2'(x - vt)]$$

$$\frac{\partial^2 V}{\partial t^2} = v^2[f_1''(x + vt) + f_2''(x - vt)]$$

$$= \frac{1}{LC}[f_1''(x + vt) + f_2''(x - vt)] \tag{9.2.8}$$

The functions $f'(x \pm vt)$ and $f''(x \pm vt)$ are derivatives with respect to the entire variables $f(x \pm vt)$. It is evident that Eqs. 9.2.7 and 9.2.8 satisfy Eq. 9.2.3. The point to note is that the functions of x and t are unspecified beyond the fact that they are differentiable.

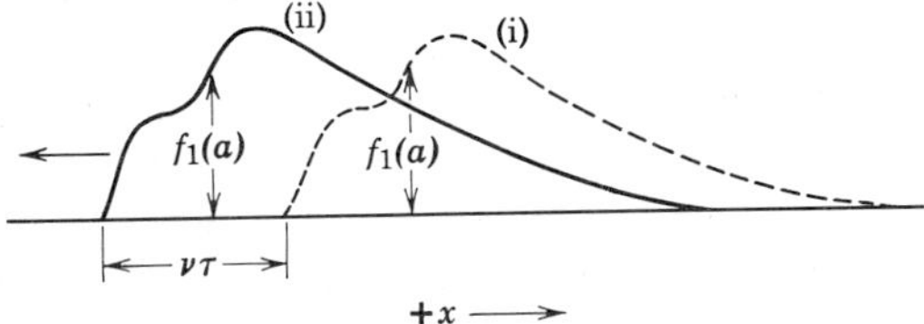

Fig. 9.4. The function $(x + vt)$ at (i) $t = 0$, and (ii) $t = \tau$.

What are the physical implications of this solution? Consider the function $f_1(x + vt)$. At $t = 0$, it has a spatial distribution $f_1(x)$ and a value at $x = a$ of $f_1(a)$. At any subsequent time, τ, it has the same value at $x = (a - v\tau)$ as it formerly had at $x = a$, which says that the voltage distribution has moved intact a distance $v\tau$ in the direction of minus x. This is illustrated in Fig. 9.4. Similarly, the function $f_2(x - vt)$ represents a voltage distribution moving in the direction of plus x with a velocity v. We conclude that to satisfy the wave equation, any unbound system of charge forming a voltage distribution must be moving along the line with a velocity $v = (LC)^{-1/2}$.

The current waves that are implied by the motion of the charge, and which therefore accompany the voltage waves, can be derived from Eq. 9.2.1:

$$\begin{aligned}\frac{\partial I}{\partial t} &= -\frac{1}{L}\frac{\partial V}{\partial x} \\ &= \frac{1}{L}[f_1'(x + vt) + f_2'(x - vt)] \qquad (9.2.9)\end{aligned}$$

Integrating both sides of Eq. 9.2.9 with respect to t,

$$\begin{aligned}I &= \frac{1}{Lv}[f_1(x + vt) - f_2(x - vt)] \\ &= \left(\frac{C}{L}\right)^{1/2}[f_1(x + vt) - f_2(x - vt)] \\ &= \frac{V}{Z_0} \qquad (9.2.10)\end{aligned}$$

Equation 9.2.10 states that the current waves are directly proportional to the voltage waves, the proportionality factor being the characteristic impedance Z_0. However, we note that the current wave traveling in the direction of minus x has the opposite sign to the voltage wave.

We have devoted ourselves so far to the solution of the wave equation given by Eq. 9.2.6 or 9.2.6a. This focuses attention on a point on the voltage distribution and shows how it travels with time up or down the line. Had we considered Eq. 9.2.5 instead, we would have focused on a particular point on the line and noticed that regardless of the point we chose, sooner or later, the voltage distribution would pass by, moving with a velocity v. As we have said already, this is just two different ways of looking at the same phenomenon.

Recapitulating, we come to the important conclusion that on loss-free transmission lines the current and voltage waves have the same shape, being related by the characteristic impedance of the line, and travel undistorted. A current wave in what has been arbitrarily chosen as the positive direction of x has the same sign as the voltage wave with which it is associated. Current waves traveling in the opposite direction have their sign reversed with respect to their voltage waves.

The concept of a negative current wave presents some latitude for misunderstanding. It is most readily comprehended by considering what is physically occurring when a wavefront travels down a line. At the wavefront there is a discontinuity. If the line is initially dead, the electrons in the conductor ahead of the front are at rest, or more correctly engaged in random motion as in any other piece of conducting material. Immediately behind the front they are experiencing a net drift along the conductor in a direction which may be the same as or opposite to the direction of the wavefront. If their drift is in the direction which has been arbitrarily chosen as positive for x, by definition their motion constitutes a negative current, since they are negative charges. If they are drifting in the negative direction, they constitute a positive current, regardless of the direction of travel of the wavefront. Figure 9.5 shows some combinations of current and voltage waves. We note

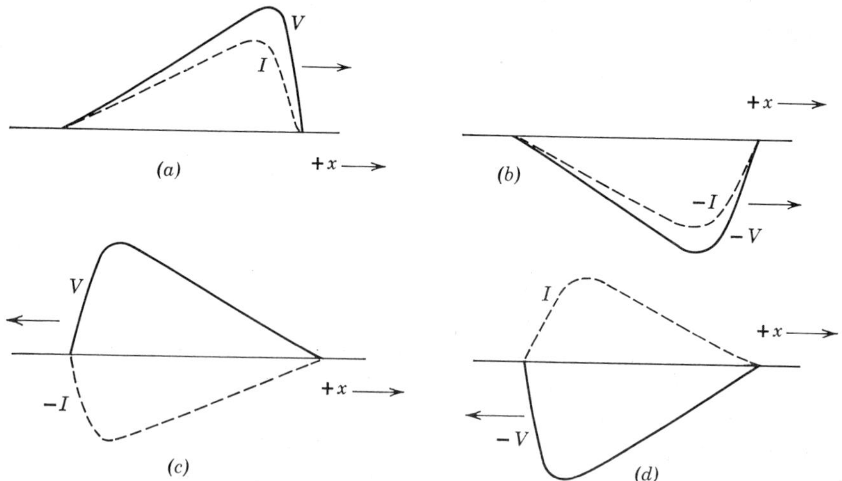

Fig. 9.5. Various combinations of voltage and current waves.

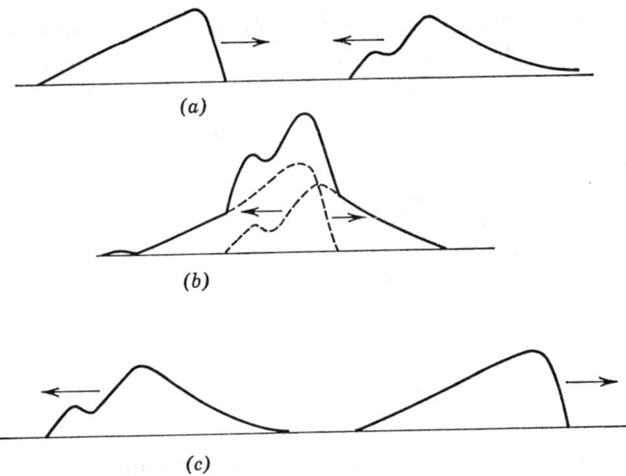

Fig. 9.6. Opposing waves (*a*) approaching, (*b*) combining, and (*c*) passing.

that V and I have the same sign when traveling to the right and opposite signs when traveling to the left. Figure 9.5*c* contains a negative current wave traveling in the negative direction, while Fig. 9.5*d* shows a positive current wave traveling in this direction. When two waves traveling in opposite directions meet, they add algebraically as they pass through each other. This is illustrated in Fig. 9.6.

9.3 Reflection and Refraction of Traveling Waves

We have seen that there is a strict proportionality between voltage waves on transmission lines and their associated current waves. The proportionality factor is the characteristic impedance Z_0 of the line. When a wave arrives at a discontinuity in a line, where the characteristic impedance of the line changes, some adjustment must occur if this proportionality is not to be violated. This adjustment takes the form of the initiation of two new wave pairs. The reflected voltage wave and its companion current wave travel back down the line superimposed on the incident wave. The refracted wave penetrates beyond the discontinuity. The amplitudes of the reflected and refracted waves are such that the voltage to current proportionalities are preserved for each, as demanded by the characteristic impedances of the lines on which they are traveling; current and voltages at the line discontinuity are themselves continuous, and energy is conserved. It will be found that energy conservation is automatically satisfied if the other two requirements are met.

Consider the junction between lines of characteristic impedances Z_A and Z_B and let us suppose that $Z_A > Z_B$. For example, this might be the junction between an overhead line and a cable. Suppose that a voltage surge of step function form and amplitude V_1 approaches the junction along the overhead line. The current wave will have the same shape and an amplitude

$$I_1 = \frac{V_1}{Z_A} \tag{9.3.1}$$

Let the reflected and refracted voltage waves be V_2 and V_3, respectively, so that their currents will be

$$I_2 = -\frac{V_2}{Z_A} \tag{9.3.2}$$

$$I_3 = \frac{V_3}{Z_B} \tag{9.3.3}$$

Note that I_2, because it is traveling in the direction of minus x, has a sign opposite to V_2. If voltage and current are to be continuous at the junction, it follows that

$$V_1 + V_2 = V_3 \tag{9.3.4}$$

$$I_1 + I_2 = I_3 \tag{9.3.5}$$

Equation 9.3.5 can be rewritten by substituting Eqs. 9.3.2 and 9.3.3:

$$\frac{V_1}{Z_A} - \frac{V_2}{Z_A} = \frac{V_3}{Z_3} \tag{9.3.6}$$

From Eqs. 9.3.4 and 9.3.6 it is possible to write expressions for the reflected and refracted waves in terms of the incident wave:

$$V_2\left[\frac{Z_B + Z_A}{Z_B Z_A}\right] = V_1\left[\frac{Z_B - Z_A}{Z_A Z_B}\right]$$

$$V_2 = \left[\frac{Z_B - Z_A}{Z_B + Z_A}\right] V_1 \tag{9.3.7}$$

The quantity $[(Z_B - Z_A)/(Z_B + Z_A)]$ is called the reflection coefficient and is designated a. It will be seen that a can be positive or negative depending upon the relative values of Z_A and Z_B. In the example we have taken, it is negative.

The refracted wave is obtained by eliminating V_2 between Eqs. 9.3.4 and 9.3.6:

$$V_3\left[\frac{Z_B + Z_A}{Z_B Z_A}\right] = \frac{2V_1}{Z_A}$$

$$V_3 = \left[\frac{2Z_B}{Z_B + Z_A}\right] V_1 \tag{9.3.8}$$

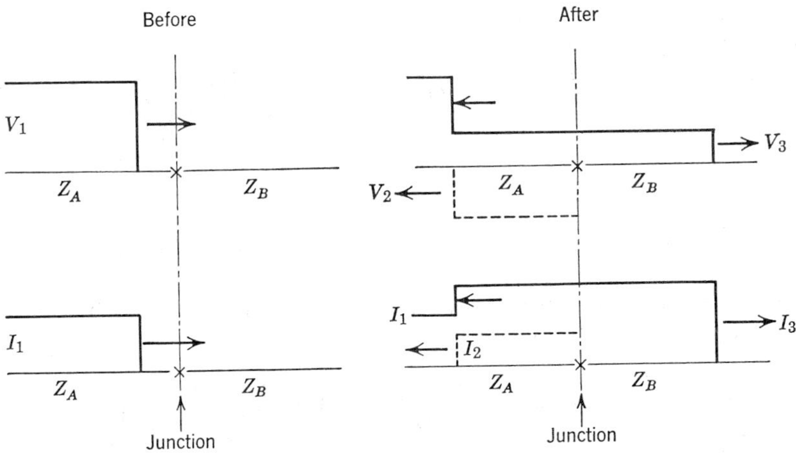

Fig. 9.7. Voltage and current waves being reflected and refracted at a junction between two lines.

This defines the refraction coefficient, $b = 2Z_B/(Z_B + Z_A)$, which varies between zero and two depending upon the relative values of Z_A and Z_B. What has just been described mathematically is shown in Fig. 9.7.

Suppose that $Z_A = 400\ \Omega$ and $Z_B = 50\ \Omega$ and that $V_1 = 300$ kV. We find that $I_1 = 750$ A. The reflection and refraction coefficients are, respectively, -0.78 and 0.22. Thus the wave that penetrates into the cable has a voltage of 66 kV and a current of 1320 A. The marked reduction of the incident wave as it penetrates the cable is sometimes utilized in power systems to protect terminal equipment from surges approaching down connected lines. A section of cable is inserted between the overhead line and the terminal equipment. This situation will be discussed in more detail in Section 13.3, where it is shown that the scheme is not as effective as might first appear.

That energy is being conserved by the initiation of the secondary waves when the incident wave reaches the junction can be demonstrated quite simply. Let us suppose that the incident wave was a consequence of the closing of a switch which connected the line A to source of V_1 volts. The source proceeds to supply the line at a rate of

$$I_1 V_1 = \frac{V_1^2}{Z_A} \text{ watts} \tag{9.3.9}$$

It will continue to do this until signalled to do otherwise by the arrival back at the source of a reflected wave from the discontinuity. When the incident wave arrives at the junction, energy will be reaching the junction at the same

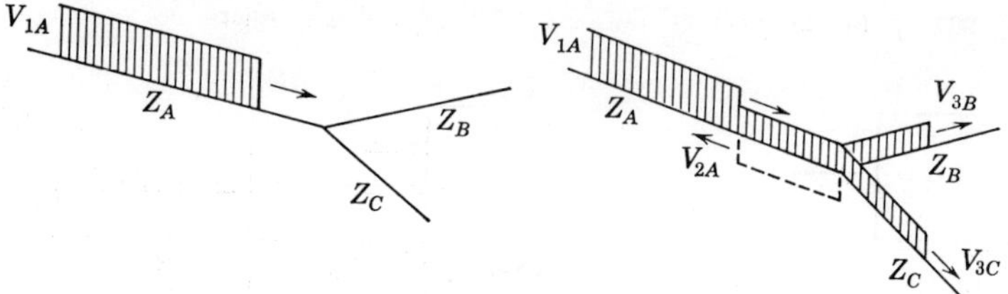

Fig. 9.8. Traveling wave of voltage encountering a line bifurcation.

rate, V_1^2/Z_A watts. It is being dispatched from the junction by the reflected and refracted waves at a rate

$$\left[\frac{V_2^2}{Z_A} + \frac{V_3^2}{Z_B}\right] \text{watts} = V_1^2\left[\frac{1}{Z_A}\left(\frac{Z_B - Z_A}{Z_B + Z_A}\right)^2 + \frac{1}{Z_B}\left(\frac{2Z_B}{Z_B + Z_A}\right)^2\right] \text{watts}$$

$$= \frac{V_1^2}{Z_A}\left[\left(\frac{Z_B - Z_A}{Z_B + Z_A}\right)^2 + \frac{4Z_AZ_B}{(Z_B + Z_A)^2}\right] \text{watts}$$

The quantity inside the square bracket reduces to unity, so that the reflected and refracted waves contain the same energy as the incident wave given in Eq. 9.3.9.

We have been considering what happens when a wave traveling on a transmission line reaches a point where the line is joined to a second line of different characteristic impedance. All that has been said could be equally well applied to a point where a line joins two or more other lines. A bifurcation of this kind is shown in Fig. 9.8; only voltage waves are represented. For the general situation where a line divides into n other lines, we have for the refracted waves

$$I_{3B} = \frac{V_{3B}}{Z_B}, \qquad I_{3C} = \frac{V_{3C}}{Z_C}, \ldots, I_{3N} = \frac{V_{3N}}{Z_N} \tag{9.3.10}$$

for the reflected wave

$$I_{2A} = -\frac{V_{2A}}{Z_A} \tag{9.3.11}$$

for continuity of voltage

$$V_{1A} + V_{2A} = V_{3B} = V_{3C} \cdots = V_{3N} \tag{9.3.12}$$

and for continuity of current

$$I_{1A} + I_{2A} = I_{3B} + I_{3C} + \cdots I_{3N} \tag{9.3.13}$$

Equations 9.3.10 to 9.3.13 are sufficient to specify the reflected wave and all the refracted waves in terms of the incident wave V_{A1} and the characteristic impedances of the lines.

9.4 Behavior of Traveling Waves at Line Terminations

Another obvious type of line discontinuity is the line termination, which, in impedance, can run the entire gamut from short circuit to open circuit. We will consider these two extremes first, as special cases.

a. Short Circuit. The unique characteristic of the short circuit is that it is impossible to develop any voltage across it. Thus, when a traveling wave of voltage reaches a short circuit, the reflected voltage wave must precisely cancel out the incident wave so that the refracted wave is zero. If the incident voltage wave is V_1 and the incident current wave I_1, the reflected voltage wave will be $-V_1$ and the reflected current wave $+I_2(= I_1)$. This is illustrated in Fig. 9.9.

The reflected wave of voltage annihilates the incident wave as it returns, while the reflected current wave augments the incident current wave, doubling the current flowing in the line.

This pattern of events can be understood in terms of energy. It was pointed out in Section 9.1 that the energy imparted to a line by a traveling wave as it passes down the line resides half in the electric field of the line capacitance and half in the magnetic field of the line inductance. When the reflected voltage wave from a short circuit forces the line voltage to zero, the electric energy is released. In the absence of any loss mechanism, this energy cannot be dissipated. It is therefore converted into magnetic energy. The magnetic field must contain the energy of the incident and reflected waves. This is accomplished by doubling the current, which quadruples the energy stored in the magnetic field (the energy is proportional to I^2).

Let us now examine what happens when a short circuit is applied to a transmission line fed by a voltage source, which, for the sake of simplicity, we will assume to have zero impedance and to provide a constant voltage V. We might suppose that if we neglect the line resistance, the fault current will increase indefinitely at a rate of V/lL, where lL is the inductance to the

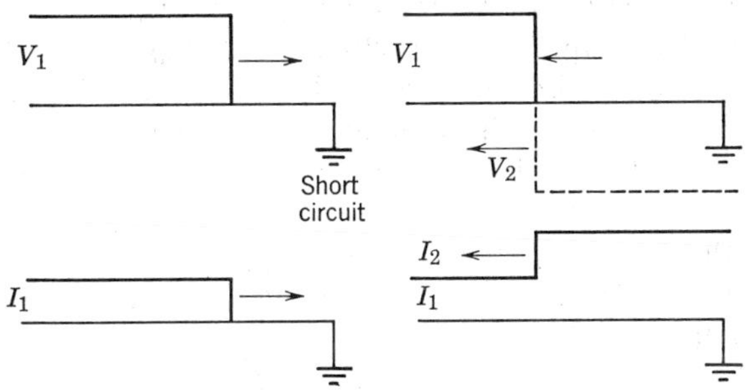

Fig. 9.9. Reflection of a traveling wave from a short circuit.

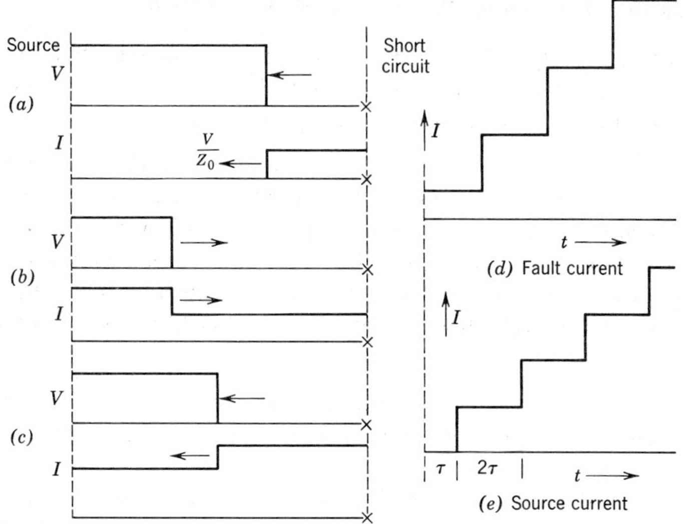

Fig. 9.10. Buildup of current when a short circuit occurs on a transmission line.

point of fault l meters away. We will see that this is only approximately true, for as long as we are dealing with a line, traveling wave phenomena are involved. Our specification of the problem implies certain boundary conditions. These are that at the short circuit the voltage is always zero, but at the source it is V at all times. To satisfy the first of these conditions when the short circuit is applied, a wave of voltage of amplitude $-V$ travels toward the source, reducing the line voltage to zero. Since this is the minus x direction, the accompanying current wave is $+V/Z_0$. This is illustrated in Fig. 9.10*a*. When this wave reaches the source, the boundary condition there demands the initiation of a new wave of voltage $+V$, which, because of its direction, is associated with a current $+V/Z_0$. This is illustrated in Fig. 9.10*b*. These waves in due course reach the short circuit, whereupon the cycle repeats (Fig. 9.10*c*), so that the short circuit current as seen at the fault or at the source increases in discrete steps as shown in Fig. 9.10*d* and 9.10*e*. At the source the effect of the short circuit is not felt until a time τ after its application. This is the time for the initial wave to travel from the fault to the source. Thereafter the current increases in steps of $2V/Z_0$ at intervals of 2τ. Suppose that the line has a capacitance of C farads/meter. Then the current increases in steps of $2V(C/L)^{1/2}$ every $2\tau = 2l(LC)^{1/2}$ sec. The average rise of current is therefore

$$2V\left(\frac{C}{L}\right)^{1/2}\frac{1}{2l(LC)^{1/2}} = \frac{V}{lL}$$

which is what we expected neglecting the wave effects.

In an a-c circuit the source voltage would vary with time. This affects what we have just discussed only to the extent that the voltage waves issuing from the source must always be such as to maintain equality between the line voltage and instantaneous source voltage whatever that may be. Between these discrete events, when waves from the short circuit arrive at the source, the source itself is generating a continuous traveling wave by virtue of its time-varying voltage.

b. Open Circuit. An open circuit at the end of a transmission line demands that the current at that point be zero at all times. Thus when a current wave of $+I$ arrives at the open circuit, a current wave of $-I$ is at once initiated to satisfy the boundary condition. This will travel toward the source in company with a voltage wave of $+V$. A current wave of $-I$ incident on the open circuit would be reflected as $+I$ and be associated with $-V$. What happens when an open-circuited line is energized from a source of V volts is shown in Fig. 9.11. In this situation the magnetic energy associated with the current disappears when the current is brought to zero at the open circuit. It reappears as electric energy, which manifests itself in the doubling of the voltage. If the travel time for the wave along the line is τ seconds, by the time the initial current and voltage waves reach the distant end, an amount of energy $VI\tau$ joules will have been fed to the line from the source. Half of this will reside

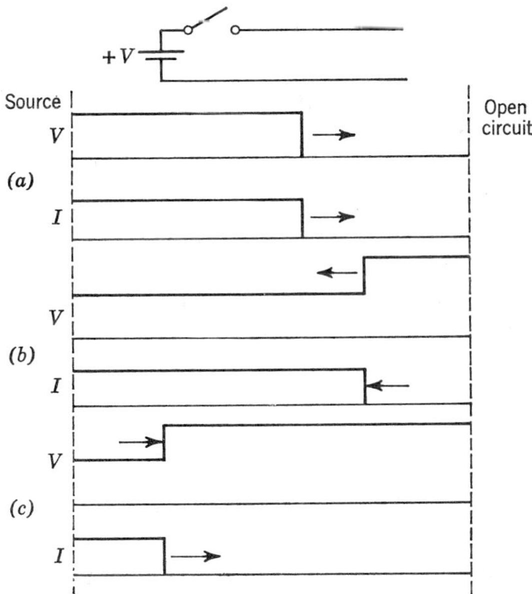

Fig. 9.11. Traveling waves initiated by energizing an open-circuited line.

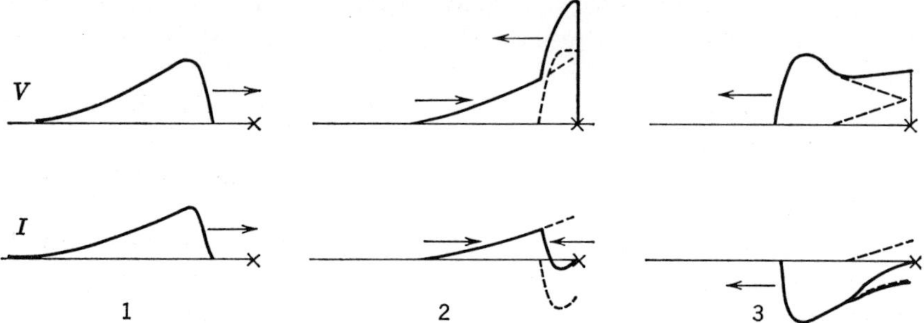

Fig. 9.12. Surge of double exponential form reflected from an open circuit.

in the electric field, and half in the magnetic field. τ seconds later, $2VI\tau$ joules will have been imparted to the line, but this is now all in the electric field. There is now four times the energy in the electric field that there was at time τ, the voltage along the line must be twice as high, since energy varies directly as the square of the voltage. Voltage surges can have very destructive effects on terminal equipment. The doubling effect just described is therefore of considerable practical importance, for it doubles the potency of the wave.

So far we have been concerned only with traveling waves of step-function form. To restrict our study to these alone is unrealistic and can be misleading. A surge waveform which is closely approximated by many practical surges is the double exponential,

$$V = V_0(\epsilon^{-\alpha t} - \epsilon^{-\beta t})$$

Figure 9.12 shows how such a wave is reflected from an open circuit.

c. General Termination. Transmission lines are most frequently terminated in some type of equipment, usually one or more transformers. To find out what happens when a traveling wave reaches such a termination we make use of the reflection and refraction coefficients computed in Eqs. 9.3.7 and 9.3.8. These are reproduced here for convenience.

$$\text{reflection coefficient} = a = \frac{Z_B - Z_A}{Z_B + Z_A} \tag{9.4.1}$$

$$\text{refraction coefficient} = b = \frac{2Z_B}{Z_B + Z_A} \tag{9.4.2}$$

The wave approaches down the transmission line of surge impedance Z_A and impinges on the terminal equipment which has an impedance Z_B. We write these quantities as operational impedances as described in Section 2.5.

In this form Z_A behaves as a resistor independent of s, since it is $(Ls/Cs)^{1/2}$. Suppose now that the terminal equipment is a capacitor C_1, then

$$Z_B(s) = \frac{1}{C_1 s}$$

and the reflection and refraction coefficients are

$$a = \frac{1/C_1 s - Z_A}{1/C_1 s + Z_A} \tag{9.4.3}$$

$$b = \frac{2/C_1 s}{1/C_1 s + Z_A} \tag{9.4.4}$$

If the traveling wave is a step function of amplitude V_1, its transform will be $v_1(s) = V_1/s$. Thus the transform of the reflected wave will be

$$v_2(s) = av_1(s)$$

which with appropriate substitution from Eq. 9.4.3 gives

$$\begin{aligned} v_2(s) &= \frac{V_1}{s}\left[\frac{1/C_1 s - Z_A}{1/C_1 s + Z_A}\right] \\ &= \frac{V_1}{s}\left[\frac{1/C_1 Z_A - s}{1/C_1 Z_A + s}\right] \end{aligned} \tag{9.4.5}$$

Remembering that Z_A has the dimensions of a resistance, $C_1 Z_A$ is a time constant, in fact the time constant to charge C_1 through the characteristic impedance Z_A of the line. Let

$$\frac{1}{C_1 Z_A} = \alpha$$

then Eq. 9.4.5 can be rewritten as

$$v_2(s) = V_1\left[\frac{\alpha}{s(s+\alpha)} - \frac{1}{(s+\alpha)}\right] \tag{9.4.6}$$

These are two simple, recognizable transforms:

$$\begin{aligned} V_2(t) &= V_1(1 - \epsilon^{-\alpha t} - \epsilon^{-\alpha t}] \\ &= V_1[1 - 2\epsilon^{-\alpha t}] \end{aligned} \tag{9.4.7}$$

This is the wave that will travel back down the line superimposing itself on the incident wave. Inasmuch as the sum of the incident and reflected waves must equal the refracted wave, it is evident from Eq. 9.4.7 that

$$V_3(t) = V_1[2 - 2\epsilon^{-\alpha t}] \tag{9.4.8}$$

This can be confirmed by deriving V_3 from the refraction coefficient:

$$\begin{aligned} v_3(s) &= bv_1(s) \\ &= \frac{V_1}{s}\left[\frac{2/C_1 s}{1/C_1 s + Z_A}\right] \\ &= \frac{V_1}{s}\left[\frac{2/C_1 Z_A}{1/C_1 Z_A + s}\right] \\ &= V_1\left[\frac{2\alpha}{s(s+\alpha)}\right] \end{aligned} \tag{9.4.9}$$

The inverse transform of Eq. 9.4.9 is indeed the expression given in Eq. 9.4.8. In this example Z_B is not a transmission line but a capacitance; $V_3(t)$ is therefore not a traveling wave but the voltage that will be impressed across the capacitor. Three instants during the reflection process for this example are illustrated in Figs. 9.13*a*, 9.13*b*, and 9.13*c*, the first before and the other two after the incident wave reaches the termination. As in previous diagrams the full line represents the net voltage distribution on the line at the instants shown. The dotted lines are the individual incident and reflected waves. The

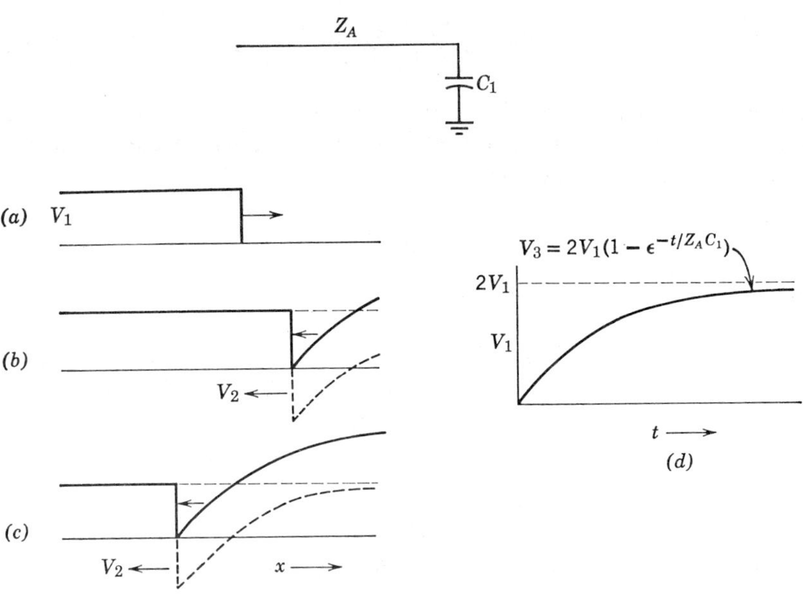

Fig. 9.13. Traveling wave on a line with capacitive termination. (*a*), (*b*) and (*c*) Disposition of the wave at different instants. (*d*) Capacitor terminal voltage as a function of time after the initial wave arrives.

voltage profile generated at the capacitor, $V_3(t)$, is plotted separately in Fig. 9.13*d*. It is exponential in shape, starting at zero and rising toward an asymptote of $2V_1$, with a time constant of $Z_A C_1$. This is surely what we would expect from a physical point of view. When the incident wave reaches the capacitor C_1, it cannot instantaneously change its potential.

Momentarily, the capacitor behaves like a short circuit; the front of the reflected wave cancels out the incident wave as we observed with a short circuit. Ultimately, the capacitor presents no path for direct current. In this respect it is like an open circuit, which we have seen results in voltage doubling. The transient voltage satisfies these initial and final conditions and passes from one to the other in a continuous manner by way of the familiar exponential function.

The inductive termination is the dual of the capacitive, for with an inductance, no current can penetrate instantaneously. When the wave arrives at its terminals, momentarily it appears like an open circuit. In due course the pure inductance presents no impedance to a direct current, so that ultimately it appears as a short circuit. If we insert $Z_B = L_1 s$ in Eqs. 9.4.1 and 9.4.2, we obtain for the reflected and refracted waves:

$$v_2(s) = \frac{V_1}{s}\left[\frac{L_1 s - Z_A}{L_1 s + Z_A}\right] = \frac{V_1}{s}\left[\frac{s - Z_A/L_1}{s + Z_A/L_1}\right] \tag{9.4.10}$$

$$v_3(s) = \frac{V_1}{s}\left[\frac{2L_1 s}{L_1 s + Z_A}\right] = \frac{V_1}{s}\left[\frac{2s}{s + Z_A/L_1}\right] \tag{9.4.11}$$

For the time constant we write $L_1/Z_A = 1/\beta$. This gives

$$v_2(s) = V_1\left[\frac{1}{s+\beta} - \frac{\beta}{s(s+\beta)}\right] \tag{9.4.12}$$

and

$$v_3(s) = V_1\left[\frac{2}{s+\beta}\right] \tag{9.4.13}$$

These expressions involve the simple transforms we encountered with the capacitive termination. The inverse transformations yield

$$V_2(t) = -V_1(1 - 2\epsilon^{-\beta t}) \tag{9.4.14}$$

and

$$V_3(t) = 2V_1\epsilon^{-\beta t} \tag{9.4.15}$$

Equations 9.4.14 and 9.4.15 are shown in Fig. 9.14.

Thévenin's theorem can be very effectively applied to the calculation of reflected and refracted waves at a line termination. One version of this theorem states that to find the current flowing in any branch Z_B of a network the branch is first removed and the voltage, V_0, say, appearing across the points x and y to which the branch was formerly connected, is noted. Next,

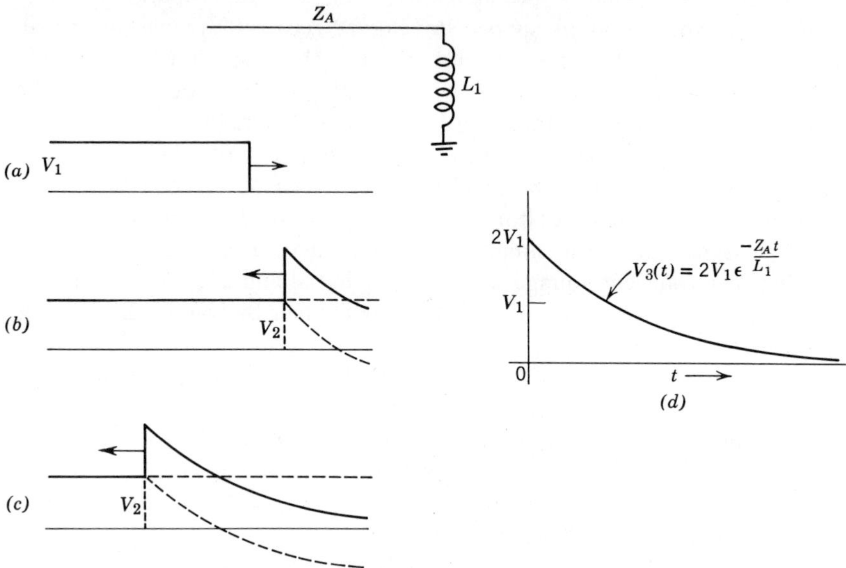

Fig. 9.14. Traveling wave on a line with inductive termination. (*a*), (*b*), and (*c*) Disposition of the wave at different instants. (*d*) Inductor terminal voltage as a function of time after the initial wave arrives.

all sources in the network are shorted and replaced by their internal impedances. Looking in at x and y impedance Z_A is calculated. The current through Z_B is then

$$I = \frac{V_0}{Z_A + Z_B}$$

It is clear from this that voltage across Z_B will be

$$IZ_B = V_0 \frac{Z_B}{Z_B + Z_A}$$

Now this same technique can be applied to a transmission line of surge impedance Z_A terminated in an impedance Z_B, as shown in Fig. 9.15. After the wave arrives at the termination, if Z_B is removed from the circuit, the voltage between x and y is $2V$, since the line now terminates in an open circuit. Looking into the line at x and y, with the source shorted, we see the characteristic impedance of the line, Z_A. According to Thévenin's theorem, Fig. 9.15*b* is the equivalent circuit for obtaining the refracted wave. The amplitude of this wave is the voltage developed across Z_B, which is

$$V_3 = \frac{2VZ_B}{Z_A + Z_B} \tag{9.4.16}$$

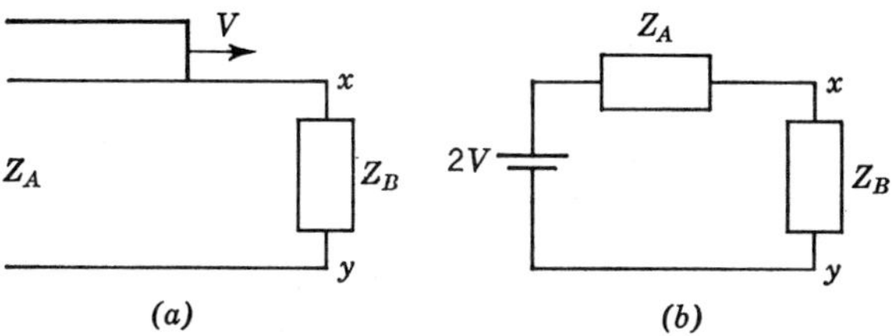

Fig. 9.15. Thévenin's theorem applied to a transmission line and a traveling wave. (*a*) Schematic of circuit. (*b*) Equivalent circuit for termination.

We recognize in Eq. 9.4.16 that the refracted wave is the incident wave V multiplied by the refraction coefficient.

There is a special case of interest, which has not been discussed so far. Suppose Z_B is a resistor numerically equal to the line characteristic impedance. The voltage developed across it, according to Eq. 9.4.16, will be V. This means that the incident wave is completely absorbed. There is no reflected wave, as can be verified from the reflection coefficient. This is of little practical significance in power circuits but is very important in communication circuits and also in measuring circuits, as will be shown later in this book.

A resistive termination dissipates energy. When a traveling wave reaches such a resistor, the energy that cannot be absorbed is reflected back. This constitutes the return or reflected wave. In the special case where the terminating resistor is equal to the characteristic impedance of the line, the wave brings energy to the termination at a rate such that it is completely dissipated, leaving nothing for reflection.

9.5 Lattice Diagrams

In an extensive network with many interconnected lines and a variety of terminations it is clear that the number of traveling waves initiated by a single incident wave will mushroom at a considerable rate as the wave splits and divides and as multiple reflections occur. It is true that we can have no more energy than that possessed by the initial surge and that this will be shared by the many waves that the initial wave gives rise to. Nevertheless, it is possible for the voltage to build up at certain points by the reinforcing action of several waves. In order to study such effects some systematic method of bookkeeping must be used. We will show in a later chapter that the digital computer can be used for this purpose. Before the days of digital computers, Bewley (1) proposed a scheme of space time diagrams, called lattice diagrams, which has received wide acceptance. To introduce the lattice diagram we will consider a single line terminating at the remote end in an impedance Z and having itself a surge impedance Z_1. Let us suppose that at the near end, the line can be connected to a bus whose surge impedance is so low as to be

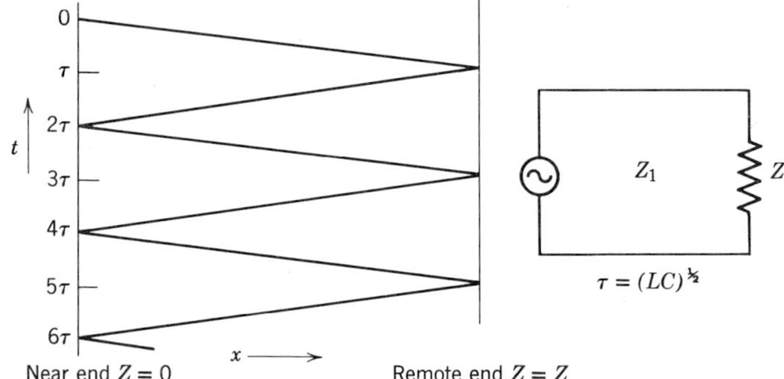

Fig. 9.16. Lattice diagram for a single transmission line terminated in an impedance Z.

considered a short circuit. The connection is made through a circuit breaker. We will study the sequence of events following the closing of the switch when the instantaneous bus voltage is V. In many instances the variation of this voltage at mains frequency can be neglected during the interval of interest, in which case the generator can be replaced by a battery of fixed voltage. If the line is long so that the transit time of traveling waves is not insignificant, then variations of the supply voltage must be taken into account.

We start by laying off a vertical time scale which can be either in seconds (or more likely microseconds) or in units equal to the transit time of a wave from end to end of the line, that is, $(LC)^{1/2}$ where L = total line inductance and C the total line capacitance in Fig. 9.16,

$$
\begin{aligned}
V_1 = V &= \text{the incident wave} \\
V_1' = a_1V_1 = a_1V &= \text{the reflection of } V_1 \text{ at the remote end} \\
a_1 &= \text{the reflection coefficient at the remote end} \\
V_2 = a_2V_1' = a_1a_2V &= \text{the reflection of } V_1' \text{ at the near end} \\
a_2 &= \text{the reflection coefficient at the near end} \\
V_2' = a_1V_2 = a_1^2a_2V &= \text{the reflection of } V_2 \text{ at the remote end} \\
&\cdot \\
&\cdot \\
&\cdot
\end{aligned}
$$

The voltage at any point on the line can be obtained at any instant by adding all the waves that have passed the point in both directions, up to the instant under review. This is an algebraic sum, so special attention must be paid to the signs of the various waves.

If we are concerned with the voltage at a line termination, the same procedure could be followed, or, alternatively, the number of waves could be reduced by using refracted waves rather than reflected waves. In the example just discussed, the refraction coefficient at the remote end is $b_1 = 2Z/(Z + Z_1)$, at the near end, $b_2 = 0$. Thus at the remote termination the following voltage fluctuations occur:

$$
\begin{aligned}
t < \tau \quad \text{voltage} &= 0 \\
\tau < t < 3\tau \quad &= b_1 V_1 = b_1 V \\
3\tau < t < 5\tau \quad &= b_1(V_1 + V_2) = b_1(1 + a_1 a_2)V \\
5\tau < t < 7\tau \quad &= b_1(V_1 + V_2 + V_3) = b_1(1 + a_1 a_2 + a_1^2 a_2^2)V
\end{aligned}
$$

Lattice diagrams can be drawn equally well for current waves. The current waves are readily derived from the corresponding voltage wave through the relationship

$$I = V/\pm Z_0$$

The choice of signs for Z_0 will depend upon the direction of the wave, as explained earlier in the analysis. With a short circuit at the remote end $a_1 = -1$, thus $V_1' = -V_1' = -V$. This wave will be associated with a current of opposite sign since it is traveling in the negative direction. Following the same convention for currents as for voltages, we have

$$I_1' = \frac{V_1'}{Z_1} = \frac{V}{Z_1}$$

This wave adds to the incident wave so that the current is doubled at the first reflection. However, not until this reflected wave arrives back at the near end will the source have any intimation of the load conditions. For a time 2τ, the line looks like a resistor of value Z_1 regardless of the termination. At the near end, where the line is again effectively short circuited, events repeat themselves in like fashion, and the current takes another jump of V/Z_1. Indeed the process keeps repeating at each reflection. A plot of the current at a point distance x along the line as a function of time is shown in Fig. 9.17 together with the voltage at this point.

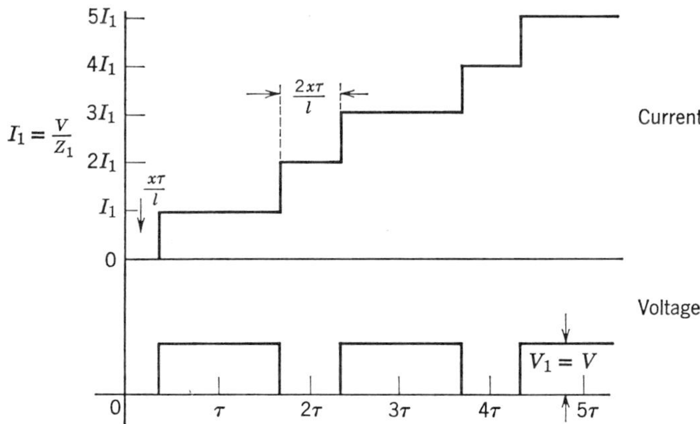

Fig. 9.17. Current and voltage at a point distant from the sending end of a short-circuited line following the application of a voltage V_1.

9.6 Attenuation and Distortion of Traveling Waves

Although it has not been explicitly stated, discussion so far in this chapter has been confined to simple twin-conductor transmission lines in which the conductors have formed either a concentric cable or a parallel pair so far removed from ground and all other lines or structures as to be entirely uninfluenced by them. It has also been assumed that the lines were completely free of losses. These limitations are very restricting and do not accord with practical systems. But the behavior of traveling waves on multiconductor transmission lines, adjacent to ground, and with attendant losses, is extremely complex. It was necessary to study the somewhat idealized conditions first, so that the basic principles would be understood. With this as a background, we can proceed to more realistic situations. In this section we will take up the matter of losses, and leave to the next the study of multiconductor systems and multiple velocity waves that they give rise to. Distortion, the changing of the shape of a surge or wave as it progresses, is a feature of both sections.

An important source of loss is the resistance of the line, and we saw in Chapter 8 how, on a transient basis, resistance can be very high. This effect will be felt most particularly at the front of the traveling wave where the current in the conductors is attempting to change most rapidly. The ground is rarely used as a conductor except in HVDC systems. However, currents flow through ground under fault conditions and currents can be induced in the ground. Depending upon soil conditions, the ground can introduce considerable resistance. Losses also arise from leakage through the finite resistance of the line insulation. Another very potent loss mechanism is corona.

The presence of resistance and leakance in a practical power line means that losses are incurred whenever a current flows in the line (I^2R), or a voltage is established between its conductors (V^2G). For a traveling wave the losses must be accommodated at the expense of the wave itself. The wave therefore is reduced in magnitude, or attenuated, as it progresses along the line. To handle resistance and leakage in a rigorous manner is complicated, for the wave equations (9.2.3 and 9.2.4) written in terms of L and C only, become considerably more difficult to manipulate when R and G are also included. Moreover, in a practical overhead power line, our representation of R and G is an approximation at best, for resistance is complicated by high-frequency effects as already mentioned and leakage, which is due mainly to the imperfection of the supporting insulator structure, appears in lumps of uncertain value at the supporting towers. Because of this, the approach here will be qualitative. We will establish certain useful concepts and then show by illustrations how they apply in practice.

Except in a very special case, which does not occur on practical power transmission lines, attenuation of traveling waves will be accompanied by distortion. This is evident from the following considerations. As a traveling wave of voltage passes along a line, we have seen that it charges the line capacitance, imparting $\frac{1}{2}CV^2$ joules to each meter of length. It is easiest to think in terms of a "square wave" surge such as would arise from the sudden application of steady d-c voltage, but the principle is in no way upset by waves of different form. Electrical energy is therefore being supplied to the line as a whole at a rate of $\frac{1}{2}CV^2v$ watts. If the circuit possesses leakance, each meter of energized line is dissipating energy through this cause at a rate GV^2v watts. This continuous attrition is responsible for the attenuation mentioned above. Inasmuch as the rate of supply of energy and the rate of loss are both proportional to V^2, this attenuation takes on an exponential form, the exponential being the function whose rate of change is proportional to the function itself. The front of the wave decays exponentially as it proceeds along the line. We may write

$$V = V_0 \exp\left(-\frac{G}{C}\frac{x}{v}\right) = V_0 \exp\left(-\frac{G}{C}t\right) \tag{9.6.1}$$

where V_0 is the amplitude of the voltage wave at its source.

In a similar way a wave of current supplies $\frac{1}{2}LI^2$ joules to each meter of line in the form of magnetic energy. But if the line possesses resistance, I^2R of this energy is being spent within the conductor. Again we observe that the rate of supply and the loss of energy are both proportional to the same quantity, in this case I^2. We would therefore expect that the current wave would also attenuate exponentially as it propagates down the line. This may be expressed as

$$I = I_0 \exp\left(-\frac{R}{L}\frac{x}{v}\right) = I_0 \exp\left(-\frac{R}{L}t\right) \tag{9.6.2}$$

where I_0 is the amplitude of the current wave at its source.

In the preceding treatment voltage and current waves have been considered separately. In fact we know that they travel together, and more specifically that the equations of the line demand a strict proportionality, at least at the wavefront:

$$\frac{V}{I} = Z_0 = \left(\frac{L}{C}\right)^{1/2}$$

Reviewing the points thus far made, we expect that the voltage and current waves will attenuate exponentially as they travel, but that at the wavefront the two waves must preserve a strict proportionality. These conditions can be satisfied simultaneously only if the exponents on the current and voltage waves are numerically equal. From Eqs. 9.6.1 and 9.6.2 this requires that

$$\frac{R}{L} = \frac{G}{C} \tag{9.6.3}$$

But this is clearly a very special situation. What happens where it does not apply? This question is best understood by first considering the situation where Eq. 9.6.3 is valid.

We can rearrange Eq. 9.6.3 in the following manner:

$$\frac{R}{G} = \frac{L}{C} = Z_0^2 = \frac{V^2}{I^2} \tag{9.6.4}$$

which says that in this special case

$$I^2R = V^2G \tag{9.6.5}$$

The rate of loss due to line resistance is exactly equal to the rate of loss from line leakance. We showed in Section 9.1 that when a source is connected to a transmission line, it imparts energy at equal rates to the electric and magnetic fields established by the traveling waves of voltage and current. Thus in this special case energy is being brought to each point on the line in equal proportions by the current and voltage waves, and equal fractions are being dissipated in line resistance and leakance losses. It follows that where Eq. 9.6.3 is *not* valid there is an excess of energy, either electric or magnetic, when the losses have been satisfied, if the ratio of V/I is to be preserved.

This situation is not new. We have encountered it at line discontinuities. Energy was conserved in those situations by initiating new waves; more especially, energy was reflected back from the discontinuity by the voltage or current wave or by both. This is precisely what happens in the situation being studied. A *continuous process of energy reflection* is taking place from each point on the line as it is encountered by the wavefront. This allows the incident waves to maintain their strict proportionality at the wavefront (the only place where the wave equations demand it) and also furnish the losses as required. The consequences of this is that waves change their shape as they travel as well as suffer attenuation. From what has been said one would suppose that the center and rear of a surge would gain energy at the expense of the front, that steep fronts would become less steep, and tails would become longer. This is in fact what is observed.

In the special case where $R/L = G/C$, the losses can be supplied as required and the proportionality of V and I preserved at the wavefront without the continuous reflection of energy. Indeed $V/I = Z_0$ applies for each point on the wave, so that the wave, although attenuated, travels undistorted along the line. Distortionless lines are obviously very desirable in communication circuits and steps are taken to obtain this condition. But in power transmission lines it is a matter of small consequence. Here resistive losses are always much greater than those due to leakage. A distortionless condition is no incentive to artificially increase the leakage losses.

There are other phenomena that contribute to attenuation and distortion, of which the most potent is corona. When the electric field gradient in air exceeds a critical value, the air becomes partially ionized and loses its insulating properties. As the potential of a transmission line conductor is progressively raised, a level is reached at which the critical gradient is attained at the surface of the conductor. If the potential is raised still further, the phenomenon just described occurs, that is, a corona discharge is established. It may be audible and, in the dark, is clearly visible as a bluish glow. It appears first at asperities on the surface where the electrical gradient is locally enhanced and spreads to give an increasing coverage as the voltage mounts. It requires a significant amount of energy to ionize air. This must be supplied by the power system. A corona current will appear superimposed on the load current whenever corona is present. Normally, lines are operated well below the corona threshold, but corona can be observed in wet weather when the corona voltage is lower.

Voltage surges on transmission lines are unaffected by corona until their potential exceeds the corona threshold. Above this point the effect is quite dramatic, as is shown in experimental evidence obtained by Lacey (2) and reproduced in Fig. 9.18. Here we see the shapes of surges as they originated and at intervals as they traveled down a 132-kV transmission line. Note how corona has removed the initial peak by the time the wave has traveled only $1\frac{1}{2}$ miles. Thereafter attenuation continues but at a reduced rate. The kind of distortion that builds the center and tail of the wave at the expense of the front is quite evident.

9.7 Multiconductor Systems and Multivelocity Waves

This topic will be approached in what at first may seem to be a rather circuitous route. It will be shown that waves travel in more than one mode on multiconductor lines and that these different modes can have different velocities, although they need not necessarily do so. As in other places in this book, the subject will be introduced through a very simple system, with the objective of establishing the fundamental ideas. Later it will be shown how the approach can be extended to more complicated systems.

We start the simple system by adding a ground plane to the two conductor line studied earlier. A cross section through the line is shown in Fig. 9.19. The parameter values are per meter of line. It will be seen that, in addition to the mutual capacitance C_2 between the conductors, they possess a capacitance C_1 farad/meter to ground. Similarly, each circuit has a self-inductance L henries/meter associated with the flux of its own current, the return path for these currents being through ground. The circuits are additionally coupled by a mutual inductance M. We will analyze these twin circuits in a rather general way, much as the single circuit was analyzed in Section 9.2.

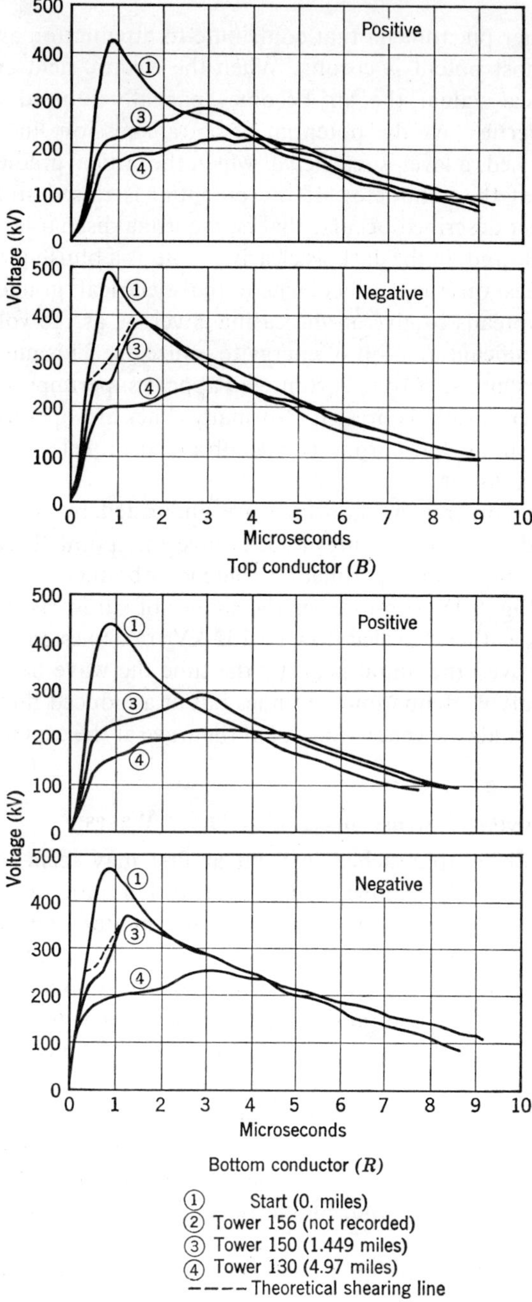

Fig. 9.18. Modification of surges with distance traveled.

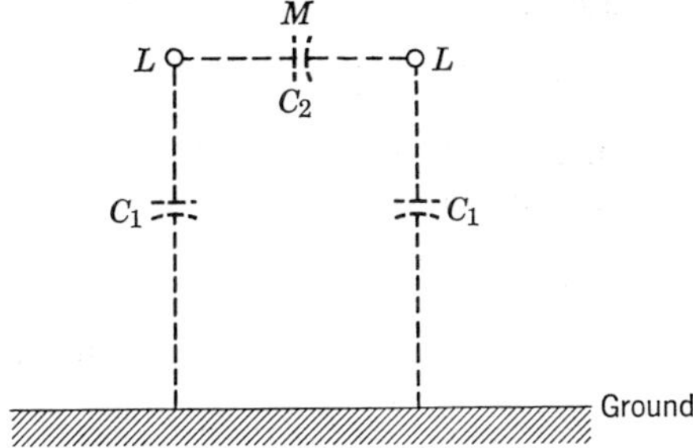

Fig. 9.19. Capacitances and inductances of a single-phase line in the presence of ground.

For circuit 1:

$$-\frac{\partial V_1}{\partial x} = L\frac{\partial I_1}{\partial t} + M\frac{\partial I_2}{\partial t}$$

It is helpful to adopt the following notation: let

$$\frac{\partial}{\partial x} = q \qquad \text{and} \qquad \frac{\partial}{\partial t} = p$$

The preceding equation may then be written

$$-qV_1 = LpI_1 + MpI_2 \tag{9.7.1}$$

For current:

$$-\frac{\partial I_1}{\partial x} = C_1\frac{dV_1}{\partial t} + C_2\frac{\partial}{\partial t}(V_1 - V_2)$$

$$\text{or} \quad -qI_1 = (C_1 + C_2)pV_1 - C_2pV_2 \tag{9.7.2}$$

Similar equations can be written for circuit 2:

$$-qV_2 = LpI_2 + MpI_1 \tag{9.7.3}$$

$$-qI_2 = (C_1 + C_2)pV_2 - C_2pV_1 \tag{9.7.4}$$

Equations 9.7.1 to 9.7.4 are four simultaneous partial differential equations with four unknowns, V_1, V_2, I_1 and I_2. It is possible to obtain an equation for any one of these by eliminating the other three.

Eliminating I_1 from Eqs. 9.7.1 and 9.7.2 gives

$$[L(C_1 + C_2)p^2 - q^2]V_1 = MpqI_2 - LC_2p^2V_2 \tag{9.7.5}$$

Similarly eliminating I_1 from Eqs. 9.7.2 and 9.7.3 gives

$$M(C_1 + C_2)p^2V_1 = LpqI_2 + (MC_2p^2 + q^2)V_2 \tag{9.7.6}$$

It is now possible to eliminate I_2 from Eqs. 9.7.5 and 9.7.6 to obtain

$$\{[L(C_1 + C_2) - MC_2]p^2 - q^2\}V_1 + [M(C_1 + C_2) - LC_2]p^2V_2 = 0 \quad (9.7.7)$$

By similar manipulations, but involving Eq. 9.7.4, an independent relationship between V_1 and V_2 is found:

$$[M(C_1 + C_2) - LC_2]p^2V_1 + \{[L(C_1 + C_2) - MC_2]p^2 - q^2\}V_2 = 0 \quad (9.7.8)$$

Further progress is made easier if we make the substitutions

$$\left.\begin{aligned} a &= [L(C_1 + C_2) - MC_2] \\ b &= [M(C_1 + C_2) - LC_2] \end{aligned}\right\} \quad (9.7.9)$$

Equations 9.7.7 and 9.7.8 can now be simplified to read

$$[ap^2 - q^2]V_1 + bp^2V_2 = 0 \quad (9.7.7a)$$

and

$$bp^2V_1 + [ap^2 - q^2]V_2 = 0 \quad (9.7.8a)$$

Either V_1 or V_2 can be eliminated from these two equations to provide identical equations for V_1 or V_2:

$$[(a^2 - b^2)p^4 - 2ap^2q^2 + q^4]V_1 = 0 \quad (9.7.10)$$

or

$$(a^2 - b^2)\frac{\partial^4 V_1}{\partial t^4} - 2a\frac{\partial^4 V_1}{\partial t^2\,\partial x^2} + \frac{\partial^4 V_1}{\partial x^4} = 0 \quad (9.7.11)$$

There is a striking similarity between the two Eqs. 9.7.7a and 9.7.8a and the wave equation (9.2.3), and it is not surprising to find that they are satisfied by solutions of the same kind. For V_1 we can write

$$V_1 = f_1(x + vt) \quad (9.7.12)$$

and for V_2

$$V_2 = f_2(x + vt) \quad (9.7.13)$$

If these tentative solutions are substituted in Eqs. 9.7.7a and 9.7.8a, we obtain

$$(av^2 - 1)f_1''(x + vt) + bv^2f_2''(x + vt) = 0 \quad (9.7.14)$$

and

$$bv^2f_1''(x + vt) + (av^2 - 1)f_2''(x + vt) = 0 \quad (9.7.15)$$

Integrating Eqs. 9.7.14 and 9.7.15 twice with respect to x,

$$\left.\begin{aligned} (av^2 - 1)f_1(x + vt) &= -bv^2f_2(x + vt) \\ \text{or} \qquad f_2(x + vt) &= \left[\frac{1 - av^2}{bv^2}\right]f_1(x + vt) \end{aligned}\right\} \quad (9.7.16)$$

and

$$\left.\begin{aligned} bv^2 f_1(x + vt) &= -(av^2 - 1) f_2(x + vt) \\ \text{or} \qquad f_2(x + vt) &= \left[\frac{bv^2}{1 - av^2}\right] f_1(x + vt) \end{aligned}\right\} \tag{9.7.17}$$

This is a rather interesting result; it says that

$$\frac{1 - av^2}{bv^2} = \frac{bv^2}{1 - av^2}$$

or

$$v^4 - \frac{2av^2}{a^2 - b^2} + \frac{1}{a^2 - b^2} = 0$$

which is a quadratic in v^2, with solutions

$$v^2 = \frac{a \pm b}{a^2 - b^2} = \frac{1}{a - b} \quad \text{or} \quad \frac{1}{a + b}$$

whence

$$v = \frac{\pm 1}{(a - b)^{1/2}} \quad \text{or} \quad \frac{\pm 1}{(a + b)^{1/2}} \tag{9.7.18}$$

This suggests that the line can support pairs of traveling waves of two distinct velocities, in contrast with the isolated twin conductor line of Section 9.2 where there was one velocity, $\pm(LC)^{1/2}$. What is the physical significance of this? We are encountering here two modes of propagation for waves. For a single pair of conductors in complete isolation, there is but one mode, current flows in one direction on one conductor returns in the opposite direction on the other. This mode is possible with the circuit in Fig. 9.19, but, in addition, current can go down either line and return in the ground. This is a second mode and it appears that it has a different velocity. If we had three conductors and ground, we would have three modes, the third being that which goes on one conductor and returns on the other two. With n conductors there would be n modes, and so on. This is akin to phase sequences in polyphase lumped circuits. More will be said about this later. For the moment there is another point to consider. Suppose the constant b, defined in Eq. 9.7.6, should be zero. This would be true if $M/L = C_2/(C_1 + C_2)$. Physically, this requires that the potential coupled electrostatically, or capacitively, to conductor 2 as a consequence of a charge on conductor 1 be the same as that coupled electromagnetically, or inductively, by the current which the movement of that charge constitutes. Whether or not this is so depends upon the ground conditions. In Section 8.3 it was shown that for a conductor over a perfectly conducting ground, the ground could be replaced by an image conductor

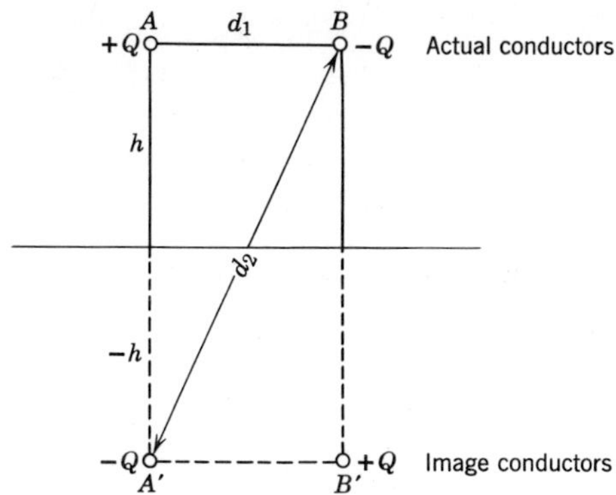

Fig. 9.20. Single-phase transmission line over a perfect ground plane and the image conductors.

placed below the ground at a depth equal to the conductor height above ground, without affecting either the electric or the magnetic field between the conductor and ground. By superposition this can be extended to any number of conductors, so that the field configuration for the conductors in Fig. 9.20 will be the same as that in Fig. 9.19. It was also explained in Chapter 8 that where the ground conductivity is poor, the image conductors, especially for the magnetic field, must be put at a considerably greater depth. The effect of this on traveling waves has been studied by Carson (3). In general, under these conditions, the image conductors for the electric and magnetic fields will not coincide. This is a likely practical situation. In Fig. 9.19 we have indicated the actual ground plane. We have not stated whether or not the L and C_1 used in our analysis are or are not related to this plane. If the ground is a perfect conductor, the equivalent configuration of Fig. 9.20 can be substituted, but, more specifically, it will now be shown that $b = 0$, which means there is only one velocity of propagation.

Referring to Fig. 9.20, let the charge density on conductor 1 at A be $+Q$ coulombs/meter. On the image conductor it will therefore be $-Q$ coulombs/meter. As a consequence of their charges the potential at point B,

$$V_B = \frac{Q}{2\pi\epsilon_0} \ln \frac{d_2}{d_1} \tag{9.7.19}$$

This is the potential, relative to ground that an isolated conductor would assume, if placed in this location.

Now consider that the charges $\pm Q$ are propagating as a wave down the line. They will constitute currents $\pm I$, given by

$$I = Qv = \frac{Q}{(\mu_0\epsilon_0)^{1/2}} \tag{9.7.20}$$

where v is the velocity of the wave, in meters/second. These currents will set up a magnetic field that will link with the circuit formed by the conductor 2 at B and its image at B'. The combined flux from I at A and $-I$ at A' will be

$$\Phi = \frac{2\mu_0 I}{2\pi} \ln \frac{d_2}{d_1} \tag{9.7.21}$$

The rate of change of Φ as I advances, and therefore the emf electromagnetically induced between A and A', is

$$\Phi v = \frac{\Phi}{(\mu_0 \epsilon_0)^{1/2}} = \frac{2}{2\pi} \left(\frac{\mu_0}{\epsilon_0}\right)^{1/2} I \ln \frac{d_2}{d_1}$$

and the potential to ground will be half this value, that is,

$$\frac{1}{2\pi} \left(\frac{\mu_0}{\epsilon_0}\right)^{1/2} I \ln \frac{d_2}{d_1}$$

If we substitute for I from Eq. 9.7.20, the electromagnetically induced potential is the same as the electrostatically induced potential given in Eq. 9.7.19. It is clear that where, because of ground conductivity, the image conductors do not coincide, the relationship just described does not apply, $b \neq 0$, and there are two distinct propagation velocities for the waves.

If a voltage is impressed on conductor 1 of Fig. 9.19, as a consequence of the line being struck by lightning, for example, it will result in the initiation of traveling waves of voltage on that line which will propagate in both directions along the line. Voltage waves, E_2, will be induced on the neighboring conductor, 2. Inducing and induced voltage waves will be associated with current waves, again of the same form, and we would suppose that the pairs of waves E_1 and I_1, and E_2 and I_2, are related through surge impedances. However, it has just been shown that, except for a rather special condition of a perfectly conducting ground, the waves are capable of two modes of propagation with different velocities. This means that there are two inducing and two induced voltage waves emanating in both directions from the point of initiation, each associated wtih its current wave. The question then arises, how do the applied and induced waves proportion themselves between the two modes?

Nature always has these situations neatly cut and dried; she always supplies just enough pieces for the jigsaw puzzle and no more. In our analysis of lumped circuit transients, we always found that there were just enough initial conditions to specify all the constants of integration that appeared in our equations. A similar situation arises with the traveling waves. There are conditions in the circuits, which once specified will cause the proportionality between the modes to fall into place. Consider the example cited, wherein line 1 in Fig. 9.19 is struck by lightning. The current in conductor 1 is supplied by the lightning discharge. There is no corresponding current injected into conductor 2. Conditions therefore demand that at the point of initiation of the surge, the net induced current in the second conductor must be zero. For this particular example, this is the datum that completes the specification. This will become clearer as we proceed.

We will concern ourselves with the forward-traveling waves only, since those traveling in the negative direction are in all ways similar. The subscripts F and S will be used to designate waves of the fast and slow modes. Thus for the inducing voltage waves on conductor 1 and the induced waves on conductor 2, we have

$$\left.\begin{aligned} E_1 &= f_F(x - v_F t) + f_S(x - v_S t) \\ E_2 &= k_F f_F(x - v_F t) + k_S f_S(x - v_S t) \end{aligned}\right\} \tag{9.7.22}$$

where

$$v_F = \frac{1}{(a-b)^{1/2}} = \frac{1}{[(L-M)(C_1 + 2C_2)]^{1/2}}$$

$$v_S = \frac{1}{(a+b)^{1/2}} = \frac{1}{[C_1(L+M)]^{1/2}}$$

and k_F and k_S are coupling factors between the circuits. They are obtained from Eq. 9.7.16 or 9.7.17,

$$\left.\begin{aligned} k_F &= \left[\frac{1 - a v_F^{\,2}}{b v_F^{\,2}}\right] = -1 \\ k_S &= \left[\frac{1 - a v_S^{\,2}}{b v_S^{\,2}}\right] = +1 \end{aligned}\right\} \tag{9.7.23}$$

This is not generally true, but it is true for the circuit we are considering. It simplifies the expressions (9.7.22) for the voltages on the line:

$$\left.\begin{aligned} E_1 &= f_F(x - v_F t) + f_S(x - v_S t) \\ E_2 &= -f_F(x - v_F t) + f_S(x - v_S t) \end{aligned}\right\} \tag{9.7.22a}$$

Thus the waves that travel in the fast mode are equal in magnitude and opposite in sign. Those in the slow mode are identical as regards amplitude and sign.

It remains to determine how the surge energy divides between the modes. If we substitute for E_1 and E_2 from Eq. 9.7.22a in Eq. 9.7.4 we obtain

$$\begin{aligned} -\frac{\partial I_2}{\partial x} &= (C_1 + C_2)[v_F f'_F(x - v_F t) - v_S f'_S(x - v_S t)] \\ &\quad - C_2[-v_F f'_F(x - v_F t) - v_S f'_S(x - v_S t)] \end{aligned} \tag{9.7.24}$$

where, as before, the prime represents the derivative of the function. Integrating Eq. 9.7.24 with respect to x gives

$$-I_2 = (C_1 + 2C_2)v_F f_F(x - v_F t) - C_1 v_S f_S(x - v_S t) \qquad (9.7.25)$$

This shows the fast and slow components of the current wave on conductor 2 and infers the surge impedance for the two modes. However, it was pointed out earlier that, at least at the time of initiation of the surge, $t = 0$, there can be no net current on this conductor, since it was conductor 1 that was struck. Accordingly, I_2 must be zero when $t = 0$, or, from Eq. 9.7.25,

$$f_S(x) = \left[\frac{C_1 + 2C_2}{C_1}\right]\frac{v_F}{v_s} f_F(x)$$

Substituting for v_F and v_S, this gives

$$f_S(x) = \left\{\frac{(L + M)[C_1 + 2C_2]}{(L - M)C_1}\right\}^{1/2} f_F(x) \qquad (9.7.26)$$

This indicates that the slow component is greater than the fast component.

The algebraic manipulations of this section would serve little purpose if we did not consider the physical consequences of the results obtained. The most striking feature perhaps is the distortion of the total surge that the multi-velocity components introduce, as they separate in time. An example will make this clear. Let us again suppose that conductor 1 of a line like that shown in Fig. 7.19 is struck by lightning. Figure 9.21 gives a picture of the voltage waves thereby initiated on the conductors. These are the waves traveling in the positive direction; identical waves would travel in the negative direction. The waves v_1 and v_2 are shown resolved into their fast and slow

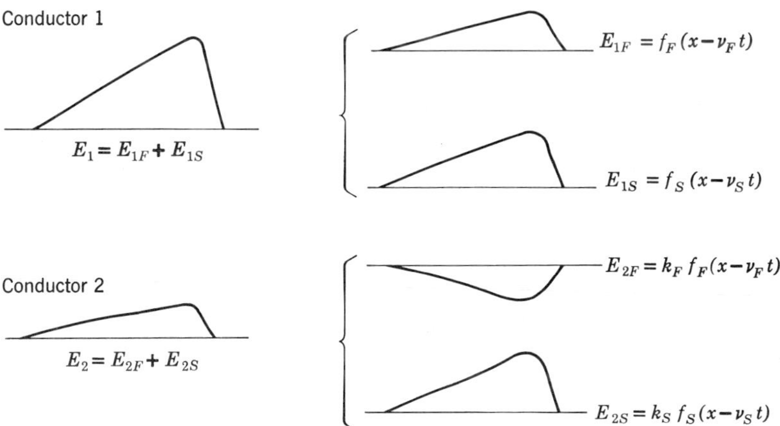

Fig. 9.21. Fast and slow components of induced and inducing waves.

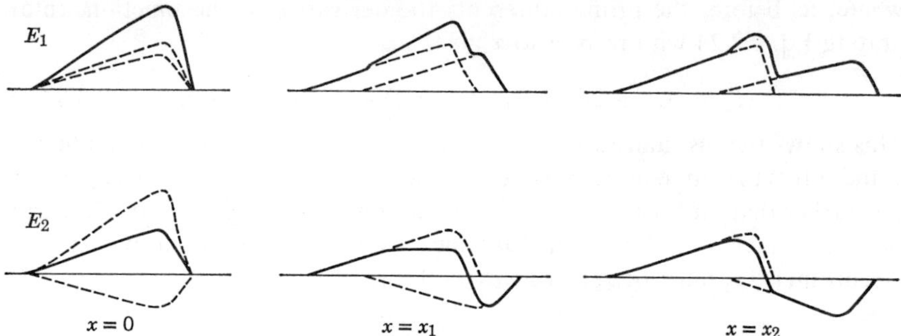

Fig. 9.22. Separation of fast and slow components as wave progresses.

components according to Eq. 9.7.26, and with coupling factors ± 1 from Eq. 9.7.23. The total or composite waveshapes can be considered as the time variations of voltage at the point of initiation, or, fixing attention on any one component, it can be as the spatial distribution of potential on the conductor. As the waves progress at their peculiar velocities, the components separate so that at different distances along the line the total surges appear as in Fig. 9.22. The result is considerable distortion, and in the case of conductor 1 the result appears to be attenuation. No true attenuation has been factored into Fig. 9.22, that is, no losses have been included. The apparent attenuation is simply another aspect of the separation of the fast and slow component waves.

We have considered only the single circuit in the presence of ground, shown in Fig. 9.19. This contains the essential elements for describing the behavior of multiconductor systems. Nor is it without practical significance, for single-phase rural distribution are sometimes of this type, as are some HVDC lines. However, it is more usual to find three-phase circuits with three conductors and ground. Sometimes there are two such circuits on the same line of towers. Not infrequently, one or more ground wires are strung above the conductors. The approach we have taken for the single-phase circuit can be broadened to encompass n conductors. This has been done by Bewley (9). There are n voltage and n current equations describing the system and they are correspondingly increased in their numbers of terms to accommodate the couplings between the conductors. The simultaneous solution of these equations reveals n modes of propagation with, in the general case, each having its own velocity. Where the conductors of a line are continuously transposed, or transposed frequently enough to approximate this condition, there are still n modes of propagation, but $(n - 1)$ of them are equal, so that there are only two velocities on the system. We present here an analysis by McElroy and Smith (4). Figure 9.23 shows an n conductor system parallel to the earth's surface. The image conductors below the earth's surface are also shown. Instead of Eq. 9.7.1 we use the following equations:

$$\left.\begin{aligned} -\frac{\partial E_1}{\partial x} &= L_{11}\frac{\partial I_1}{\partial t} + L_{12}\frac{\partial I_2}{\partial t} + \cdots L_{1n}\frac{\partial I_n}{\partial t} \\ -\frac{\partial E_2}{\partial x} &= L_{21}\frac{\partial I_1}{\partial t} + L_{22}\frac{\partial I_2}{\partial t} + \cdots L_{2n}\frac{\partial I_n}{\partial t} \\ -\frac{\partial E_n}{\partial x} &= L_{n1}\frac{\partial I_1}{\partial t} + L_{n2}\frac{\partial I_2}{\partial t} + \cdots L_{nn}\frac{\partial I_n}{\partial t} \end{aligned}\right\} \tag{9.7.26}$$

wherein the Ls are the self-inductance of, or mutual inductances between, the many circuits which the conductors and their images comprise (see Section 8.3). Equations 9.7.26 are more conveniently written in matrix form:

$$-\left.\frac{\partial E_l}{\partial x}\right] = [L_{ls}]\left.\frac{\partial I_s}{\partial t}\right] \tag{9.7.27}$$

where $l = 1, 2, \ldots n$ and $s = 1, 2, \ldots n$ and L_{ls} are the coefficients of partial inductance of the circuit. These Ls, of course, depend upon the location of the ground images, which in turn depends upon the earth conductivity. We would also expect from Chapter 8 that frequency would be involved. This is indeed the case, as Carson (3) has shown. This matter is treated in a brief and qualitative way at the end of this section. Here we simply quote from Carson that

$$L_{ls} = 2.303\frac{\mu_0}{2\pi}[A_{ls} + 0.8684Q_{ls}] \text{ henry/meter} \tag{9.7.28}$$

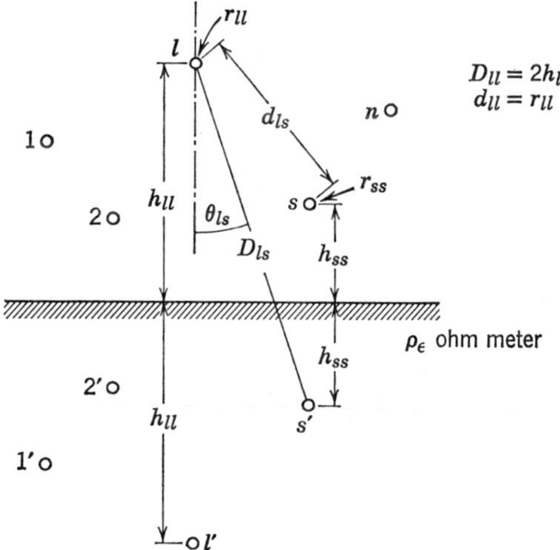

Fig. 9.23. System of n conductors and their images.

where internal inductance of the conductors is neglected.

$$A_{ls} = \log \frac{D_{ls}}{d_{ls}} \tag{9.7.29}$$

(see Fig. 9.23) and where Q_{ls}, which represents the effect of earth for a uniform resistivity, ρ_E, is a function of θ_{ls} (also indicated in Fig. 9.23) and,

$$\rho_{ls} = 0.8565 \times 10^{-3}\, D_{ls}\left[\frac{f}{\rho_E}\right] \qquad (D_{ls} \text{ in feet}) \tag{9.7.30}$$

Corresponding to Eq. 9.7.2 for the single circuit line there will be for the multiple conductor configuration a series of equations for current that can be similarly written in matrix form,

$$\left.\frac{\partial I_l}{\partial x}\right] = [C_{ls}]\left.\frac{\partial E_s}{\partial t}\right] \tag{9.7.31}$$

where C_{ls} are the partial coefficients of capacitance between the pairs of conductors and the conductors and their images. This is written

$$C_{ls} = \frac{2\pi\epsilon_0}{2.303}\,[K_{ls}] \tag{9.7.32}$$

where $[K_{ls}] = [A_{ls}]^{-1}$. This procedure can be made somewhat neater by letting

$$\mathfrak{L}_{ls} = A_{ls} + 0.8684Q_{ls} \tag{9.7.33}$$

Thus Eqs. 9.7.27 and 9.7.31 can be rewritten

$$-\left.\frac{\partial E_l}{\partial x}\right] = 2.303\,\frac{\mu_0}{\pi}\,[\mathfrak{L}_{ls}]\left.\frac{\partial I_s}{\partial t}\right] \tag{9.7.34}$$

and

$$-\left.\frac{\partial I_l}{\partial x}\right] = \frac{2\pi\epsilon_0}{2.303}\,[K_{ls}]\left.\frac{\partial E_s}{\partial t}\right] \tag{9.7.35}$$

Adopting the same procedure for Eqs. 9.7.34 and 9.7.35 that we used for the corresponding equation for the single circuit, we differentiate and combine:

$$\left.\frac{\partial^2 E_l}{\partial x^2}\right] = \frac{1}{C^2}\,[K_{lk}][\mathfrak{L}_{ks}]\left.\frac{\partial^2 E_s}{\partial t^2}\right] \tag{9.7.36}$$

and

$$\left.\frac{\partial^2 I_L}{\partial l^2}\right] = \frac{1}{C^2}\,[K_{lk}][\mathfrak{L}_{ks}]\left.\frac{\partial^2 I_s}{\partial t^2}\right] \tag{9.7.37}$$

where $k = 1, 2, \ldots n$, and $c = 1/(\epsilon_0\mu_0)^{1/2} = 3 \times 10^8$ m/sec.

McElroy and Smith (4) go on to show that it is possible to normalize Eqs. 9.7.34 to 9.7.37 and thereby replace them by a similar set of equations free of mutual terms. The problem then reduces to the solution of n single-phase equations of the same general form. Just as the solution of the single-circuit wave equation leads to the mode of propagation and relationship between current and voltage (velocity and surge impedance) for the voltage and current waves on the single circuit, so the solution of these equations yields the n modes of propagation for the multiple conductor system. It is obviously very difficult to keep track of this multiplicity of waves. Indeed on anything but a very simple system the problem becomes quite intractable without the help of a fair-sized computer (Chapter 14).

This last example brings us to the threshold of *Modal Analysis*, which is a systematized approach to the current topic. This is beyond the scope of the present book, but the reader who wishes to pursue the subject further will find Hedman's paper (5) very helpful in itself and as a bibliography.

We have said little up to this point about the effects of the ground, other than the fact that it can be simulated by image conductors and that if the ground resistivity is high, the image conductors for electric and magnetic fields will not coincide. In our example, the two velocities are a consequence of this. There is another important role that ground plays. This was touched on briefly in Chapter 8. The current associated with the ground wave, flowing through the finite resistance of the ground, must dissipate energy, which must be supplied by the wave. This inevitably results in attenuation of the wave. This is most pronounced at the front of the wave, for the current density is highest there. Typical current paths are shown in Fig. 9.24. The current takes time to penetrate into the ground and establish a more or less horizontal flow. The wavefront in this figure is shown sloping backwards, indicating that

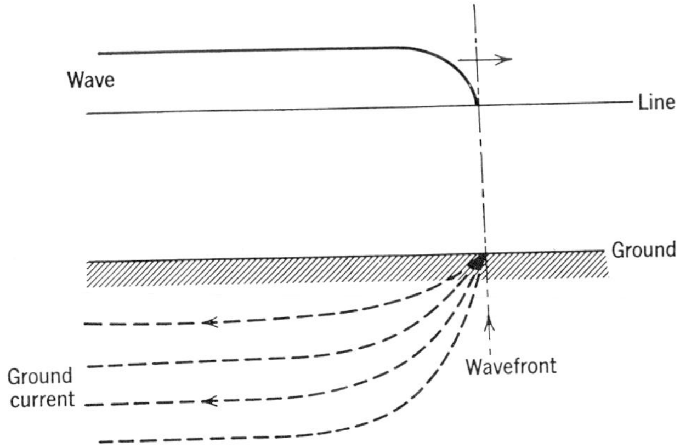

Fig. 9.24. Traveling wave proceeding above a ground plane.

although it may have started with a steep front, the loss of energy in ground resistance renders it progressively less steep. As long as we neglected losses we could visualize traveling waves proceeding without distortion, but once resistance is introduced either in the conductors or in the ground, the advancing waves must charge the line capacitance through this resistance. In these circumstances there can be no abrupt steps in voltage at this wavefront or anywhere else.

The effect just described is intensified by a compounding phenomenon in the conductor. As explained in Chapter 8, it takes time for the surge current to penetrate from the skin into the conductor. At the wavefront the effective resistance is the highest and so therefore are the losses.

REFERENCES

1. L. V. Bewley, *Traveling Waves on Transmission Systems*, John Wiley & Sons, New York, (1951).
2. H. M. Lacey, "The Lightning Protection of High-Voltage Overhead Transmission and Distribution Systems," *Proc. IEE*, Vol. 96 (1949), p. 287.
3. J. R. Carson, "Wave Propagation in Overhead Wires with Ground Return," *Bell System Tech. Journ.*, Vol. 5 (1926), p. 539.
4. A. J. McElroy and H. M. Smith, "Propagation of Switching Surge Wavefronts on EHV Transmission Lines," *Trans. AIEE*, Vol. 81, Part III (1962), p. 983.
5. D. E. Hedman, "Propagation Overhead Transmission Lines I—Theory of Modal Analysis; II—Earth Conduction Effects and Practical Results," *Trans. IEEE*, Vol. PAS 84 (1965), p. 200.

10 Lightning

10.1 The Scope of the Lightning Problem

That there is ample justification for studying lightning and for devoting space to it in this text is evident if one reads the report issued by a Joint IEEE-EEI Committee studying outages on E.H.V. Lines (1). In it we find the statement that "· · · lightning is the greatest single cause of outages . . . , accounting for about 26% of the outages on 230-kV circuits and about 65% of the outages on 345 kV." This conclusion was reached after analyzing data obtained over a 14-year period from some 42 operating companies in the United States and Canada. The data were gathered from 386 high-voltage circuits, representing 25,499 miles of transmission line. A similar study has been made of utility distribution systems in Britain (2), also over a 14-year period. Over 50,000 fault reports were analyzed on circuits up to and including 33 kV. It was found that 47% of all these incidents, including those of unknown origin, were caused by lightning. Many earlier reports could be cited that convey essentially the same message.

In Section 1.1 we asserted that electrical transients are the outward manifestation of a sudden change in circuit condition. How do we relate this to lightning? The physical phenomenon of lightning, which we will look into in Section 10.2, shows that clouds acquire charge or at least become polarized, so that electric fields of considerable strength are created within the cloud and between the cloud and adjacent masses such as earth and other clouds. When these fields become excessive, to the extent that the dielectric of the intervening space can no longer support the electrical stress, a breakdown or lightning flash occurs; this is usually a high-current discharge. The lightning strokes that create problems for the utility engineer are those that terminate on or near to power lines. These close a "grand switch" between the cloud and the power line or adjacent earth, so we are indeed changing circuit conditions. This is either a direct connection to the line or it completes a circuit with close mutual coupling to the line. Frequently the line will be raised to such a potential that further flashes will occur to grounded structures. Or the grounded structures may be raised to such a potential that they flash over to the line. In either event, yet another path is introduced for current and yet another source for transients is created. We will give these matters our consideration in Section 10.3.

The material of Chapter 9 is important in studying lightning because any disturbance on a power line involves traveling waves. If a lightning current is injected into a transmission line, we can expect that it will flow in both directions from the point of injection, creating voltage waves of $I/2Z_0$, where I is the lightning current and Z_0 the surge impedance of the transmission line. In this way the effects of lightning are conducted to remote parts of the system where surges impinge on terminal equipment and even enter industrial plants, commercial buildings, or private homes.

As engineers and scientists have learned more about lightning and have been able to study its effects, a growing and continuing effort has been made to improve the performance of power systems during thunderstorms. This work has led to the development and application of ground wires, counterpoise, lightning arresters, and so on. A sequence of presentation somewhat parallelling this historical development has been used in this chapter. We start by considering the physical processes involved in the lightning discharge.

10.2 The Physical Phenomenon of Lightning

Sometimes intriguing, more often terrifying, lightning has been a source of wonder to mankind for thousands of years. In his book *The Flight of Thunderbolts* (3), Schonland gives an interesting history of the subject. He points out that it was not until the eighteenth century, when Franklin's curiosity was aroused by lightning, that any real scientific inquiry was made into the phenomenon. In fact, a systematic study of lightning was not really undertaken until this century, and it was largely stimulated by the matter that concerns us here, that is, the hazard that lightning presents to the electric supply industry.

The turmoil that is apparent inside a thundercloud is most impressive to the viewer. Its shape changes continually, and one notes especially the development of the towering "thunderhead." It is very easy to imagine the fierce updrafts within the cloud and the downdrafts near its surface, which are matters of practical experience for aviators. It is now generally accepted that the updraft is responsible for charge separation within the cloud, like some gigantic electrostatic generator, which leads to the creation of electric fields within and around the cloud and ultimately to the electric breakdown that we call lightning. The precise details of the charge separation mechanism are not completely understood, although there are a number of theories, more than one of which may be active in different locations within the cloud. Wilson (4) has postulated that raindrops falling in the electric field that normally exists in the earth's atmosphere become polarized, bottom positive and top negative, which is the normal direction of this field, and that subsequently they attract negative charges to them. Clusters of ions of both signs are to be found in the atmosphere at all times. The raindrops apparently capture these selectively, acquiring a negative charge themselves and leaving a preponderance of positive charge in the air. The updrafts tend to carry up to the top of the cloud the positive air and smaller drops that the wind can blow against gravity. Meanwhile, the falling heavier raindrops bring negative charge to the base of the cloud. It is an observable fact that most thunderclouds are polarized this way, and it has been demonstrated (5) that falling waterdrops can acquire charge in the manner described.

Conditions within the thundercloud appear to be more complicated than can be accounted for by this mechanism alone. Simpson and Scrase (6) have made some interesting measurements and evolved a plausible theory to explain their observations. This is based on the observed fact that when water droplets break up, they acquire a positive charge and in the process negative charge is imparted to the air around. They have been able to obtain a reasonable assessment of the charge distribution in thunderclouds. This has been done by floating balloons equipped to measure the electric field, through thunderclouds, and also by measuring the electric field at the earth's surface beneath a thundercloud as it passed overhead. Figure 10.1 is reproduced from their paper. The electric flux lines, of course, start on a positive charge and terminate on a negative charge. The charge may be within the cloud or on the ground below. The field pattern on the ground and in the cloud accord with the distribution of charge shown in Fig. 10.1. The field measurements in planes z_1 and z_2 are represented by the columns shown in those locations, the width of the columns being proportional to the field. The potential gradient pattern on the ground, which is associated with bound charges on the earth's surface or structures thereon, moves as the cloud moves and changes as the charge distribution inside the cloud changes.

Figure 10.2, also from Simpson and Scrase's (6) paper, shows what these investigators believe is happening in the cloud. The progress of the cloud is right to left, and the air currents are indicated by the solid lines. It will be seen that these account for the high winds that appear as the storm approaches. The highest velocities are where the flow lines are close together. Thus near the front of the cloud there is considerable upward current, so powerful in fact that it precludes raindrops falling through it. Drops falling in this region are broken up with the charge separation described. They tend to coalesce at a somewhat higher altitude, and the heavier ones fall as the familiar heavy

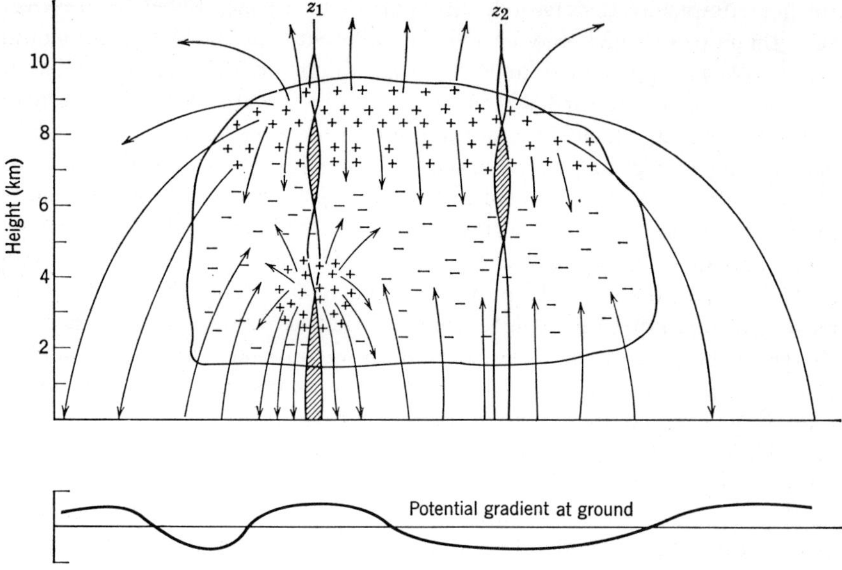

Fig. 10.1. Hypothetical case of a cloud with a positive charge in the upper part, a negative charge in the lower part, and a small region of strong positive charge near the base.

drops that characterize the beginning of a thundershower. Smaller spray and the negatively charged air is driven further upward into regions of sub-zero temperatures, where the moisture freezes into ice crystals. It has been observed that, by what must be a frictional charging process, such ice crystals acquire a negative charge when blown by air. The air near the top of the cloud tends to become positively charged, while the negatively charged ice crystals drift gently down to melt and recycle or fall as rain. We see therefore that charge separation occurs and charge patterns of quite complicated and ever-mobile forms are set up within the cloud. Conditions are now propitious for a lightning stroke.

Perhaps the best description of lightning itself is to be found in a series of five papers by Schonland et al. [(7) to (11)]. As the charges build up within the cloud and by induction on the earth below, a point is reached locally within the cloud where the electric field is sufficient to cause breakdown. In air at atmospheric pressure this is approximately 30 kV/cm. However, in the cloud with the high moisture content at a lower pressure because of its altitude, it is conjectured that local breakdown may occur at a lower gradient, perhaps as low as 10 kV/cm. Photographic evidence indicates that this initiates the development of a *stepped leader* stroke, a discharge which progresses in a somewhat random manner by short steps from the cloud

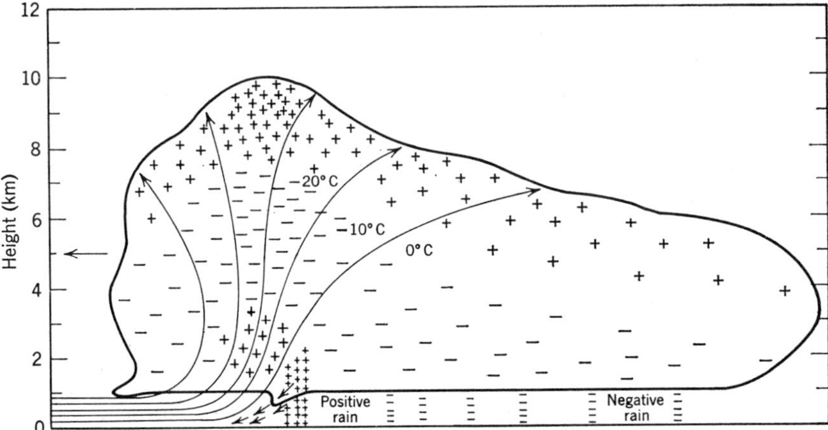

Fig. 10.2. Generalized diagram showing air currents and distribution of electricity in a typical heat thunderstorm.

toward the earth. These observations have been made with what is known as a Boys camera after C. V. Boys, who invented it. Such a camera produces a time-resolved image of the phenomenon. Actually, it produces two images simultaneously for it has two lenses which are fixed with respect to each other but rotate at a high speed with respect to the film. From a knowledge of the speed and direction of rotation it is possible to chart the progress of the stepped leader from cloud to ground. A good description of the Boys camera and some fine examples of the photographs it produces are to be found in reference (3). The leader steps are of the order of 50 meters in length, and are accomplished in about a microsecond. However, the pauses between these sallies, and the irregular path pursued, usually results in the leader taking several milliseconds to reach earth. Branches on either side are also evident. The precise mechanism whereby the stepped leader proceeds is not fully understood, though it would seem that at each pause the electric gradient at the head of the leader has become reduced below the critical value for cumulative ionization at this locality. Progress is resumed when this situation is changed, perhaps by transfer of more charge down the leader from the cloud. Griscom (12) in his so-called prestrike theory proposed a possible stepping mechanism. He postulated that the head of the leader has a pestle shape and that it contains far more charge than originally supposed. He believes that before the leader reaches the ground, a discharge somewhat similar to it rises from ground to meet it. He questions the validity of the interpretation of some earlier photographic data. The author is impressed with similarities between the lightning stroke in its formative stages and phenomena observed in negative point-to-plane corona. Trichel (13) showed that this is a pulsating discharge; current, which is choked off by field distortion produced by negative ions, develops again as the ions disperse. But it has also been shown (14) that at high fields a pulse-free mode can exist. Loeb (15) has given a comprehensive review of this subject.

Once the stepped leader has established a connection to earth, there is general agreement that a power return stroke moves swiftly up the ionized channel prepared by the leader. This is very intense and travels much more quickly than the leader. What we are witnessing is the neutralization of the charge in the leader channel, or the progressive discharge of the channel to ground. The rate of propagation and the quantity of charge involved is of great practical concern when the ground termination of the stroke resides on a transmission line tower, conductor, or ground wire, for it gives a measure of current involved. We will take up this matter in the next section.

When the leader commences its progress towards earth, other leaders may well push out within the cloud to other centers of charge so that the sequence of events just described is frequently not confined to one location in a thundercloud. Alternatively, the discharge to ground may disturb the potential balance within the cloud to the point that an internal discharge is triggered. The consequence of either of these events is that quite often they will precipitate further strokes to ground down the original channel. It is well established that what appears to the eye as a single flash of lightning is frequently the integrated effect of several strokes, at intervals of 50 msec or so. In fact the majority of flashes probably involve more than 1 stroke, and as many as 30 or 40 have been recorded.

10.3 Interaction between Lightning and the Power System

Thus far we have been concerned for the most part with lightning as a phenomenon. In this section we will consider it in relation to the power system. A vast number of studies have been made on this subject in both the United States and Europe. Some have been concerned simply with collecting statistics, some with making measurements, still others with evaluating the efficacy of the various means that have been employed to reduce the outages due to lightning, and others have tried to probe more deeply into the physical nature of the problem. The ultimate objective in all cases has been to improve our understanding, so that the incidence of service interruption due to lightning will be reduced. It would not be practical or appropriate to review this voluminous literature here, although references (16) to (24) are some of the more recent studies. Instead, we will attempt to point up the underlying principles involved at our present stage of understanding and show how these affect concepts of design.

A good place to start is with a model for the lightning stroke, something that can be expressed in terms of circuit elements and can therefore be used in conjunction with the circuit elements of the power system itself to study the interaction. Unfortunately, a simple, meaningful model is not easy to devise. Bewley (25) has attempted a model which includes the power system, the stroke channel, and even the distributed charge centers in the cloud. Conceptually, this may be helpful, but it is doubtful whether it can be used for a quantitative assessment of tower top potentials. Griscom et al. (12), (24) have concerned themselves with a model for the stepped leader and the return stroke, which attempts to fit in with the physical aspects of the discharge on the one hand and the circuit aspects on the other.

It is apparent from what has been said already that when lightning strikes a power line a current is injected into the power system. This is a very useful concept. What voltages this current will give rise to depend upon its waveshape and the impedances through which it flows. For example, if it is a tower that is struck, the impedance of the tower will be of concern. The voltage drop down the tower will appear across the line insulation. If this is excessive, flashover of the insulation will occur and a fault will be placed on the system. In an attempt to avoid this, one or more ground wires are often strung on the towers above the line conductors. These serve several purposes. They shield the phase conductors, that is, frequently they are themselves struck instead of the phase conductors. In the event of a stroke to a tower, the tower impedance is paralleled by the surge impedance of the ground wires, which extend away in both directions, so that the total impedance is reduced and the tower top potential is correspondingly less. Finally, there is considerable electric and magnetic coupling between the ground wires and the power line proper, which tends to limit the voltage that can be established between them, thereby reducing the chance of insulation flashover in the manner described. It is of small concern if the top of the distant tower flies up to a high potential if the line conductors do likewise. It is true that these disturbances travel as waves along the line and must ultimately reach a terminal where they will be impressed across the equipment at that point. However, attenuation plays an important role, as pointed out in Chapter 9, so that damage to terminal equipment becomes likely only when the lightning strikes close to the terminal.

The tower current must of course flow in the ground; therefore the ground impedance or tower footing resistance must be taken into consideration. More will be said about this and counterpoise in Chapter 12.

But how is the circuit completed? We have postulated current coming into a tower by lightning stroke from a cloud, and then disappearing into the ground; where does it finally return? Perhaps a useful concept here is to think of the cloud and earth as forming a vast capacitor which is being

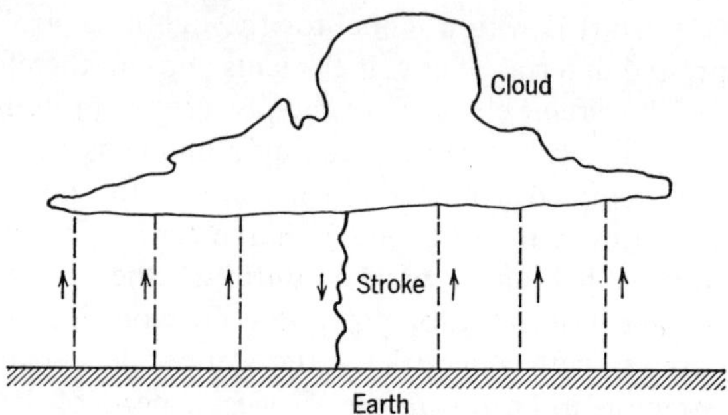

Fig. 10.3. Lightning stroke from cloud to earth discharges a vast capacitor.

discharged by the stroke. The return circuit would then be completed by displacement current in the electric field. This is suggested by Fig. 10.3. But this conceptualization only partially suffices, for it leaves open the question of how to represent the stroke path, shown in Fig. 10.3 as if it were a solid conductor. Bewley (25) calculates the inductance of this path by computing the flux linkage per ampere with it. He then assumes that the electric field is in free space and that wave velocities appropriate to free space prevail. In this way he calculates a capacitance. Thence, treating the path as a transmission line, he deduces its surge impedance from L and C. However, it should be possible to calculate the capacitance from the size of the cloud and its height above the ground and find a surge impedance looking upon the assemblage as a lumped circuit. An example will show that such a simple computation can yield results not entirely at odds with observable phenomena, if we recognize that damping will be present.

Suppose we represent the cloud and earth as a capacitor having parallel plates of circular shape with a radius of 1 km and a separation (height of cloud) of 1 km, and let the stroke be 20 cm in diameter. Then

$$L = 2 \times 10^{-7} \int_{r_1}^{r_2} \left(1 - \frac{x^2}{r^2}\right) \frac{dx}{x} \quad \text{H/m}$$

$$= 2 \times 10^{-7} \ln \frac{r_2}{r_1} \quad \text{H/m, approximately}$$

for $r_1 = 10$ cm, $r_2 = 1$ km, and a path 1 km long, the total inductance is 2.18 mH. The capacitance is given by

$$C = \frac{\epsilon_0 A}{d} = \frac{8.854 \times 10^{-12} \times \pi \times 10^6}{10^3 \times 4} = 6.95 \times 10^{-9}\ \text{F}$$

This gives a surge impedance of 540 Ω and a period, if the current would oscillate, of about 24 μsec. We would expect, however, that the resistance of the ionized path would damp the current. If this resistance amounted to 5000 Ω, the current would approximate to a 1.5:30 μsec wave.

Although this approach may give us an insight into the current characteristics of the return stroke and conform to the kind of waveforms suggested by the AIEE Committee Report (23) for a method of estimating lightning performance of transmission lines, it is clearly a very crude approximation for describing the phenomenon as a whole. A more accurate representation must take the leader stroke and prestrike into account.

On the ground under a thundercloud, and on objects on the ground, such as power transmission lines, charge is induced by the proximity of the charged cloud, as suggested in Fig. 10.1. Whenever the charges in the cloud move or redistribute, the charges on the ground also move and redistribute. Such motion represents current flow, and momentarily potential differences are established in ground elements. These movements in the cloud are usually relatively slow, unless a discharge occurs; consequently the ground currents are very small. However, when a leader stroke commences to make stepped progress toward the ground, charge is lowered from the cloud. Sudden movements occur in the induced ground charges which become more localized as the leader approaches earth. But even here, the induced currents are relatively small. However, it is apparent that our simple picture of a parallel plate capacitor is beginning to look a little inadequate.

Griscom (12) postulates that the head of the leader stroke contains considerable charge, most of which becomes immediately available to discharge into the line or tower with which contact has been made, when the prestrike occurs. The capacitance being discharged in this case is something smaller than we have been considering, but something much more closely coupled to the power line. Griscom (12) constructs quite a sophisticated model to represent this situation. The charge involved is a small fraction of that residing in the cloud, but it is capable of being delivered rapidly to the object struck. This implies a high current, albeit one of short duration, which can generate voltages of sufficient magnitude to cause line flashovers.

Figures 10.4*a* and 10.4*b* are taken from the paper by Griscom, Skooglund, and Hileman (4); they are two possible equivalent circuits for the elements involved in the prestrike. In Fig. 10.4*a* the situation is based upon traveling waves, so that the main plasma channel D_1 is viewed as a transmission line of prescribed length and surge impedance Z. The extending plasma channel from the tower D_2, is similarly depicted with surge impedance Z_2, and likewise the tower Z_T, and the ground wire Z_{GW}. The prestrike is initiated when the switch SW is closed, which causes the charge in the leader head (charge residing on the capacitor $C_{H'}$ in the figure) to circulate in the circuit, by

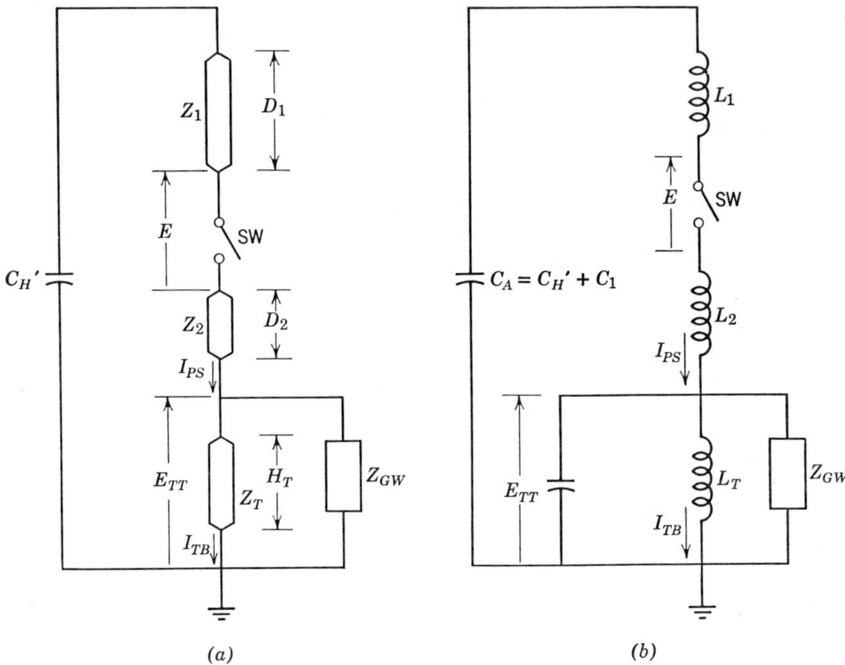

Fig. 10.4. Equivalent circuits for lightning stroke to tower. (*a*) Traveling wave equivalent. (*b*) Lumped parameter equivalent.

traveling waves of current and voltage spreading in both directions from the switch. The waves will be reflected and refracted at each discontinuity, so that a fairly complicated pattern will be set up, from which, however, the tower top potential can be determined. The circuit is simplified somewhat in Fig. 10.4*b*, where the distributed representations of the arc channels and tower are replaced by lumped circuit constants. The lengths of components we are describing are relatively short, of the order of 100 m, thus the lines involved are very short. Most of what is important is over in the first microsecond. The leader head charge and potential or capacitance are such that currents of tens or hundreds of thousands of amperes are believed to flow, which, with impedances on the order of hundreds of ohms, create megavolts or tens of megavolts. A tower, for example, with a surge impedance of 125 Ω, paralleled by two ground wires, might have an effective surge impedance of 75 Ω. Currents of the order of 50,000 A lead to voltages of the order of 4 MV.

Unfortunately, there is a dearth of current data, although more are being collected all the time (17). It is difficult to collect such data, especially on current waveforms, because the instrumentation is complex, and of course it must be in the right place at the right time. Some discussion of measuring equipment and measuring methods for lightning investigations are given in Chapters 16 and 17. Most of the available data probably relate to the return stroke. It will be recalled that once the leader has established a conducting arc channel to ground, the return stroke represents a process of progressive neutralization of the charges of this channel. The neutralizing front moves up the channel with a speed about one-third the velocity of light. This rate and the quantity and disposition of the charge to be neutralized determine the magnitude and shape of the current. Thus, if the channel contains 1 mC/m and the stroke travels at 100 m/μsec, the current involved would be $10^{-3} \times 10^{8} = 10^{5}$ A.

It has been suggested that in the initial stages all the neutralizing charge does not flow from the object struck but from local space charge in the air around the prestrike channel. For this reason the initial rise of current in the return stroke through the tower is perhaps not as fast as we might suppose. The actual tower top current probably can be represented with reasonable accuracy by the double exponential form rising to its peak in 1 to 10 μsec and diminishing to half value in 20 to 100 μsec.

Some reference has been made to electrostatically induced lightning surges where bound charges on a line are suddenly released. These are relatively innocuous. Much more serious are electromagnetically induced surges which arise from lightning striking close to the line, but not actually striking the line itself. Chowdhuri and Gross (26) have made a considerable study of this phenomenon. They conclude that surges produced in this way can be high enough to cause flashover.

The distribution of peak amplitudes of lightning currents is shown in Fig. 10.5, which is taken from a paper by Lacey (27). This represents the percentage of strokes exceeding the value shown on the abscissa. Note that only currents in excess of 5000 A are included.

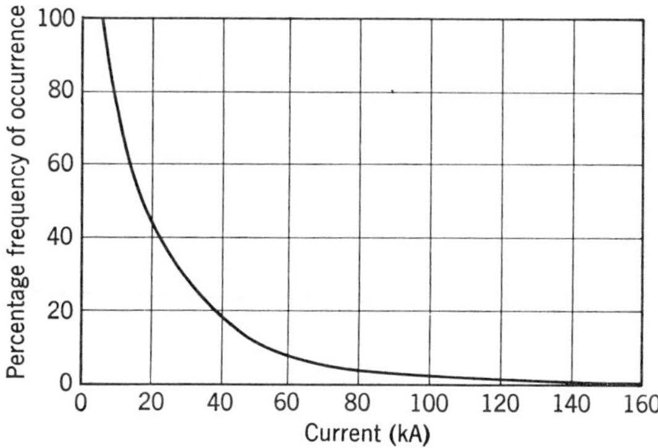

Fig. 10.5. Frequency of occurrence of lightning currents of 5-kA or greater.

The number of strokes per year/mile of line depends upon a number of factors, which are discussed in reference (16). The incidence of lightning in the areas through which the line runs is obviously important. This is the so-called isokeraunic level, which states on how many days of the year thunder is heard in a particular location. This varies from almost zero in arctic regions to 100 or more in some tropical areas. In the United States, Florida is most severely affected, having an isokeraunic level between 70 and 90 depending upon location (22). But, as pointed out by Golde (16), the isokeraunic level can be expected to be no more than an indicator. A particular storm on one day can create more incidents than several lesser storms on several other days.

System voltage plays an important role in the incidence of lightning problems on overhead lines. Generally the number of incidents diminishes as the voltage goes up (16) because of the improved insulation. One would suppose that at extra high voltages insulation levels would be set by switching surge levels rather than by consideration of the lightning problem. There are some exceptions to this trend, however, as the following figures show. It has been found in recent extensive study in the United States (17) that whereas the number of lightning outages/100 miles of line at 220–240 kV and 287–300 kV were 0.409 and 0.262, respectively; at 345–360 kV, the figure was 3.040. This points up the very important factor of the design. This is discussed in detail in reference (16) among others. We can only treat it briefly here.

10.4 Factors Contributing to Good Line Design

In order to reduce the hazard that lightning poses to the power system, certain factors that determine line performance must be understood. The object of good line design is to reduce the number of outages caused by lightning. First we try to keep the incidence of strokes to the system to a minimum, and then we try to minimize the effects of those strokes that do terminate on the system. As in all design work, requirements in one area often conflict with needs in another, not least the economic needs. Lightning problems would be all but eliminated if all transmission was through tunnels at least 20 ft under the ground. But the cost of this reliable service could scarcely be justified. A judicious compromise can be made only when the influential factors can be fully appreciated and the risks involved accurately assessed. Such an evaluation is frequently quite difficult.

It is a matter of experience that lightning seeks out tall objects, thus tall towers are more vulnerable than low goalpost-like structures. A compromise must enter here, however, inasmuch as adequate clearances must be provided. These cannot be maintained with a low construction unless the spans are short, which means frequent supporting structures and their attendant cost.

Can anything be done to prevent those lightning strokes that must strike the line from causing an outage? The answer is yes, a well conceived ground wire system can dramatically reduce the number of outages. As explained earlier, the first function of ground wires is to shield the phase conductors, that is, to serve in lieu of those conductors as the termination of the lightning stroke. The degree of protection afforded in this way depends upon the disposition of the ground wires with respect to the conductors proper. Lacey (27) states that a ground can be regarded as providing adequate protection to any conductor lying below a quarter circle drawn with its center at the height of the ground wire and with its radius equal to the height of the ground wire above the ground. If two or more such wires are provided, the vulnerable area between two adjacent wires can be taken as a semicircle having as its diameter a line connecting the two ground wires. As an example, the area protected by two ground wires is shown in Fig. 10.6, which is taken from Lacey's paper (27). Ground wires inevitably increase the number of strokes likely to terminate somewhere on the line, but this should not increase the number of outages.

The lightning current propagates in both directions along the ground wire, and we can expect that traveling waves will be induced on the phase conductors below in the manner described in Chapter 9 (28). For the ground wires to be effective, the potential difference created between them and the phase conductors should not be great enough to cause flashover between the two. For if this should occur, a line-to-ground fault is established which must be cleared by the circuit breakers at the end of the line. This will result in at least a momentary outage. Mid-span flashovers rarely occur. Breakdowns, when they happen, are usually across the insulators at a tower. The tower presents a discontinuity to the traveling waves of current and voltage on the ground wires. The waves undergo reflection and refraction, the reflected wave returning toward the point where the lightning struck, while

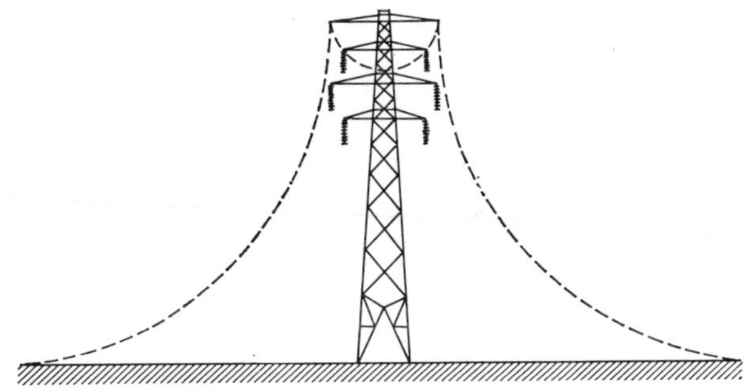

Fig. 10.6. Protection afforded by ground wires.

two refracted waves are generated. One will proceed into the next span of the ground wire, the other will pass down the tower to ground. Much will depend upon the ground impedance the latter encounters. If it is low, the wave reflected back up the tower will largely remove the potential existing from the incident wave. In this way the chance of flashover is eliminated. If, on the other hand, the incident wave encounters a high ground impedance, it may be reflected with the same sign, that is, the reflected wave will reinforce the incident wave. This will greatly increase the chance of flashover.

It is apparent that a considerable number of waves are quickly brought into being, which makes the problem difficult to study analytically. Some success has been achieved by means of models. Although any attempt to rigorously analyze such an incident is unlikely to lead to very accurate quantitative results because of uncertainties in parameter values, some qualitative conclusions can be drawn that are helpful in designing the system. It is clear that lightning currents with high rate of rise are to be feared most, since the voltages they give rise to will reach disastrous levels before reflected waves can ameliorate the situation. High surge impedances in the ground wire circuits, the tower structures, and above all high ground impedance or tower footing resistance, are to be avoided. Finally, close coupling between the ground circuit and the power circuits is to be sought, inasmuch as this minimizes the voltage that can be established between them. Golde (16) discusses how these considerations influence the design of distribution systems. Similar discussion as relating to transmission circuits will be found in reference (23).

In this chapter we have not considered the impact of lightning on terminal equipment nor how we protect such equipment against lightning. These topics are taken up in Chapters 11 and 12, respectively. Further relevant information on the subject of grounding will be found in Chapter 12.

REFERENCES

1. Report of Joint IEEE-EEI Subject Committee on EHV Line Outages, *Trans. IEEE.*, Vol. 86 (1967), p. 547.
2. R. A. W. Connor and R. A. Parkins, "Operations Statistics in the Management of Large Distribution System," *Proc. IEE*, Vol. 113 (1966), p. 1823.
3. B. F. J. Schonland, *The Flight of Thunderbolts*, Oxford University Press (1950).
4. C. T. R. Wilson, "Investigations on Lightning Discharges and on the Electrical Field of Thunderstorms," *Phil. Trans. Royal Soc.*, Series A, Vol. 221 (1920), p. 73.
5. J. P. Golt, "On the Electrical Charge Collected by Water Drops," *Proc. Royal Soc.*, Series A., Vol. 151 (1935), p. 665.
6. G. B. Simpson and F. J. Scrase, "The Distribution of Electricity in Thunder-clouds," *Proc. Royal Soc.*, Series A, Vol. 161 (1937), p. 309.

7. B. F. J. Schonland and H. Collens, "Progressive Lightning," *Proc. Royal Soc.*, Series A, Vol. 143 (1934), p. 654.
8. B. F. J. Schonland, D. G. Malan and H. Collens, "Progressive Lightning II," *Proc. Royal Soc.*, Series A, Vol. 152 (1935), p. 595.
9. D. J. Malan and H. Collens, "Progressive Lightning—III—The Line Structure of Return Lightning Strokes," *Proc. Royal Soc.*, Series A, Vol. 162 (1937), p. 175.
10. B. F. J. Schonland, "Progressive Lightning—IV—The Discharge Mechanism," *Proc. Royal Soc.*, Series A, Vol. 164 (1938), p. 132.
11. B. F. J. Schonland, D. B. Hodges and H. Collens, "Progressive Lightning—V—A Comparison of Photographic and Electrical Studies of the Discharge," *Proc. Royal Soc.*, Series A, Vol. 166 (1938), p. 56.
12. S. B. Griscom, "The Prestrike Theory and Other Effects in the Lightning Stroke," *Trans. IEE.*, Vol. 77 (1958), p. 919.
13. G. W. Trichel, "The Mechanism of the Negative Point-to-Plane Corona Near Onset," *Phys. Rev.*, Vol. 54 (1938), p. 1078.
14. A. N. Greenwood, "Pulse-Free Discharges in Negative Point-to-Plane Corona," *Phys. Rev.*, Vol. 88 (1952), p. 91.
15. L. B. Loeb, *Electrical Coronas*, University of California Press, Berkeley and Los Angeles (1965).
16. R. H. Golde, "Lightning Performance of British High Voltage Distribution System," *Proc. IEE*, Vol. 113 (1966), p. 601.
17. S. B. Griscom, R. W. Caswell, R. E. Graham, H. R. McNutt, R. H. Schlomann and J. K. Thornton, "Five-Year Field Investigation of Lightning Effects on Transmission Lines I—Data Sumary and Suggestions for Improved Line Performance," *Trans. IEEE*, Vol. 84 (1965), p. 257.
18. S. B. Griscom, R. W. Caswell, R. E. Graham, H. R. McNutt, R. H. Schlomann and J. K. Thornton, "II—Instrumentation and Installations," *Trans. IEEE*, Vol. 84 (1965), p. 261.
19. S. B. Griscom, R. W. Caswell, R. E. Graham, H. R. McNutt, R. H. Schlomann and J. K. Thornton, "III—Lightning Stroke Characteristics and their Effects on Transmission Lines," *Trans. IEEE*, Vol. 84 (1965), p. 268.
20. G. M. Clayton and F. S. Young, "Estimating Lightning Performance of Transmission Lines," *Trans. IEEE*, Vol. 83 (1964), p. 1102.
21. C. F. Wagner, "A New Approach to the Calculation of Lightning Performance of Transmission Lines," *Trans. AIEE*, Vol. 75 (1956), p. 1233.
22. G. S. Forrest, "The Performance of the British Grid System in Thunderstorms," *Proc. IEE*, Vol. 97, Part II (1950), p. 345.
23. AIEE Committee Report, "A Method of Estimating Lightning Performance on Transmission Lines," *Trans. AIEE*, Vol. 63 (1950), p. 1187.
24. S. B. Griscom, J. W. Skooglund and A. R. Hileman, "The Influence of the Prestrike on Transmission-Line Lightning Performance," *Trans. AIEE*, Vol. 77, Part III (1958) p. 933.
25. L. V. Bewley, *Traveling Waves on Transmission Systems*, 2nd ed., John Wiley & Sons, New York (1951).

26. P. Chowdhuri and E. T. B. Gross, "Voltage Surges Induced on Overhead Lines by Lightning Strokes," *Proc. IEE*, Vol. 114, No. 12 (1967), p. 1899.
27. H. M. Lacey, "The Lightning Protection of High-Voltage Overhead Transmission and Distribution Systems," *Proc. IEE*, Vol. 96 (1949), p. 287.
28. P. Chowdhuri and E. T. B. Gross, "Voltage Induced on Overhead Multiconductor Lines by Lightning Strokes," *Proc. IEE.*, Vol. 116, No. 4 (1969), p. 561.

11 The Behavior of Windings under Transient Conditions

11.1 Introduction

The response of transformer and machine windings to the impact of surges has engaged the interest of engineers for many years. Indeed, it is probably true to say that no single topic in the area of electrical transients has such a profuse literature. Abetti (1), (2), (3), in three bibliographies, lists 1263 papers relating to this subject. More have appeared since these bibliographies were published, and doubtlessly more will be written.

Obviously, it would be impossible to attempt to cover such a mass of literature within the confines of one chapter of this book. Instead, we attempt to delineate the principles involved and present the results of analyses and tests. With this for a background, the reader who wishes to delve more deeply into the subject should have no difficulty in doing so.

There are two broad avenues along which the problem has been attacked; one uses the concept of traveling waves, the other, first described in the classical paper of Blume and Boyajian (4), is based on standing waves in the winding. Rudenburg (5) has pointed out the equivalence of these two approaches. We give considerable attention to his analysis.

In detail, there are two other areas in which different authors have introduced variations, these are the approximations made regarding the physical parameters of the winding, and in the method by which they are expressed. What are these parameters? Generally speaking, each turn of a winding possesses a self inductance attributable to the flux of its own current with which it is linked. It is also linked with the flux of neighboring coils and therefore has a mutual inductance with these coils. If we consider eddy currents in the core, another mutual coupling is involved. Regarding capacitance, the turn or coil has capacitance to ground, ground in this context being the core, the tank, or perhaps another winding. In addition, it has some capacitive coupling with the other coils of the winding, this mutual capacitance being greatest with those other coils in closest proximity to it. Finally, there will be losses in the resistance of the windings, in the leakance of the insulation, and in the core, all of which will dissipate energy at the expense of the surge.

To take account of all of these items would be a monumental task, even with a large computer, and it is extremely doubtful whether the results obtained would warrant the effort expended. Inevitably approximations are made. Losses are almost invariably neglected. In the case of alternator windings where the conductors are almost buried in the steel, the only mutual capacitance considered is that to conductors in the same slot. Such simplifications can greatly reduce the analysis without seriously impairing its accuracy. The point we wish to make here is that in reading any dissertation on this subject, it is important to realize what approximations are being made, and also to have a grasp of the physical significance of the parameters being retained. Authors do not always call the same thing by the same name. Here, as elsewhere in this text, we try to define the terms used and put some physical interpretation on them.

We start by studying the response of a uniform single transformer winding to a step function of voltage. This is not because such surges are commonplace in practice, although the chopped wave comes close to this, but rather because it represents the ultimate in severity. Also, it is possible to construct any waveform, to a close approximation, by a succession of step functions, positive and negative, and then, by superposition, determine the response to the waveform by adding the individual responses to the steps. This may sound tedious, but with computational aids it is quite manageable. Transformers will be given the greatest attention because they are much more exposed to surges, especially those caused by lightning.

It is helpful to divide the time after the surge strikes the winding into three time intervals. The first of these periods is extremely short, usually a fraction of a microsecond. During this time no significant current can penetrate the winding proper because of its inductance. Currents flow only as displacement currents in the capacitance of the winding. This gives rise to an initial voltage distribution. In the third period we assume that the voltage distribution in the winding has reached some steady state. During the intervening or second period the voltage pattern goes through many intricate contortions as it relaxes from the initial to this final state.

11.2 Initial Voltage Distribution when a Step Function Voltage Surge Strikes a Uniform Transformer Winding

The winding is represented by the circuit shown in Fig. 11.1. This figure shows a finite number of meshes. In fact, we now assume that the capacitance

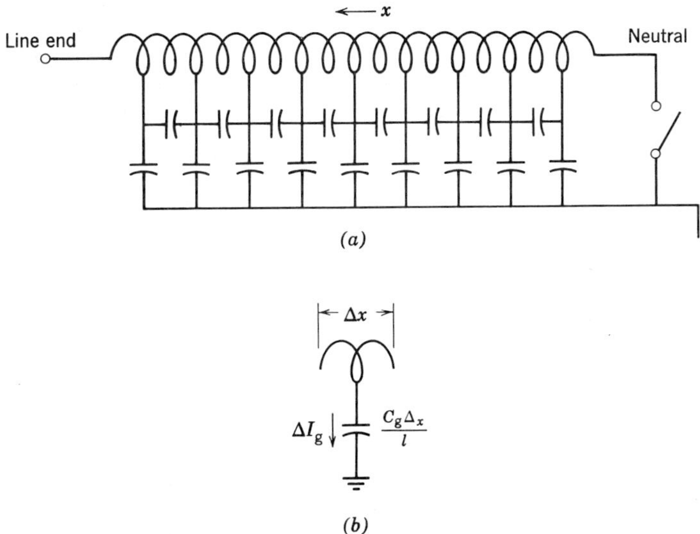

Fig. 11.1. (*a*) Schematic representation of a tranformer winding. (*b*) Elementary length Δx of the winding.

is uniformly distributed along the winding. Each short length of conductor will have a capacitance to ground and to the adjacent turns. We are contriving to represent these capacitances in Fig. 11.1*a*. The neutral of the winding may or may not be grounded. Whatever the condition, it can be satisfied by the position of the switch, shown at the neutral.

In deriving the initial voltage distribution we assume that no current has penetrated the winding proper. It is therefore the capacitance network only that is effective at this time, and its response to one frequency will be the same as its response to any other. Let the capacitance to ground of the entire winding be C_g and let the capacitance from end to end of the winding be C_s. Then

$$\frac{\text{ground capacitance}}{\text{unit length}} = C_g/l$$

$$\frac{\text{series capacitance}}{\text{unit length}} = C_s l$$

where l is the length of the winding. Let

E = voltage to ground at any point in the winding
I_g = total current in the ground capacitance
I_s = current in the series capacitance

Traditionally, the positive direction for x has been from the neutral to the terminal. We follow this convention here.

Consider an elementary length Δx in Fig. 11.1*b*. Its ground capacitance will be $(C_g/l)\,\Delta x$ and the ground current through this element is given by

$$\Delta I_g = \frac{C_g\,\Delta x \omega E}{l} \tag{11.2.1}$$

Here ω is the frequency of the voltage E. But

$$\Delta I_g = \frac{dI_s}{dx}\Delta x \tag{11.2.2}$$

Combining Eqs. 11.2.1 and 11.2.2,

$$\frac{dI_s}{dx} = \frac{C_g \omega E}{l} \tag{11.2.3}$$

Now

$$\text{series capacitance of the elementary length} = \frac{lC_s}{\Delta x} \tag{11.2.4}$$

and

$$\text{voltage across the element} = \frac{dE}{dx}\Delta x \tag{11.2.5}$$

Inasmuch as the series capacitive current $= C\,dE/dt$, from Eqs. 11.2.4 and 11.2.5 we can write

$$I_s = lC_s \omega \frac{dE}{dx} \tag{11.2.6}$$

Differentiating Eq. 11.2.6,

$$\frac{dI_s}{dx} = lC_s \omega \frac{d^2E}{dx^2} \tag{11.2.7}$$

Equating Eqs. 11.2.3 and 11.2.7 gives

$$\frac{C_g \omega E}{l} = lC_g \omega \frac{d^2E}{dx^2}$$

or

$$\frac{d^2E}{dx^2} - \frac{1}{l^2}\frac{C_g}{C_s}E = 0 \tag{11.2.8}$$

As we anticipated, this equation is independent of ω. It is familiar in form as is the solution:

$$E = A_1 \epsilon^{px} + A_2 \epsilon^{-px} \tag{11.2.9}$$

where

$$p = \frac{1}{l}\left[\frac{C_g}{C_s}\right]^{1/2}$$

The constants A_1 and A_2 will depend upon the boundary conditions of the problem. Suppose that the neutral is grounded; then

$$\text{at } x = 0, \qquad E = 0$$
$$\text{at } x = l, \qquad E = V$$

where V is the amplitude of the step function surge. Inserting these conditions in Eq. 11.2.9 gives

$$\left.\begin{aligned} A_1 + A_2 &= 0 \\ A_1\epsilon^{pl} + A_2\epsilon^{-pl} &= V \\ A_1 = -A_2 &= \frac{V}{\epsilon^{pl} - \epsilon^{-pl}} \end{aligned}\right\} \tag{11.2.10}$$

Let

$$pl = \left[\frac{C_g}{C_s}\right]^{1/2} = \alpha$$

then

$$A_1 = -A_2 = \frac{V}{2\sinh\alpha} \tag{11.2.11}$$

If we substitute Eq. 11.2.11 in 11.2.9, we obtain

$$\begin{aligned} E &= \frac{V}{2\sinh\alpha}\left[\epsilon^{px} - \epsilon^{-px}\right] \\ &= V\frac{\sinh(\alpha x/l)}{\sinh\alpha} \end{aligned} \tag{11.2.12}$$

If the neutral were isolated, the boundary conditions would be

$$\text{at } x = l, \qquad E = V$$

$$\text{at } x = 0, \qquad I_s = 0 \qquad \text{or} \qquad \frac{dE}{dx} = 0$$

Thus we have

$$p(A_1 - A_2) = 0$$
$$A_1\epsilon^{pl} + A_2\epsilon^{-pl} = V$$

from which we establish that

$$A_1 = A_2 = \frac{V}{2\cosh pl} = \frac{V}{2\cosh\alpha} \tag{11.2.13}$$

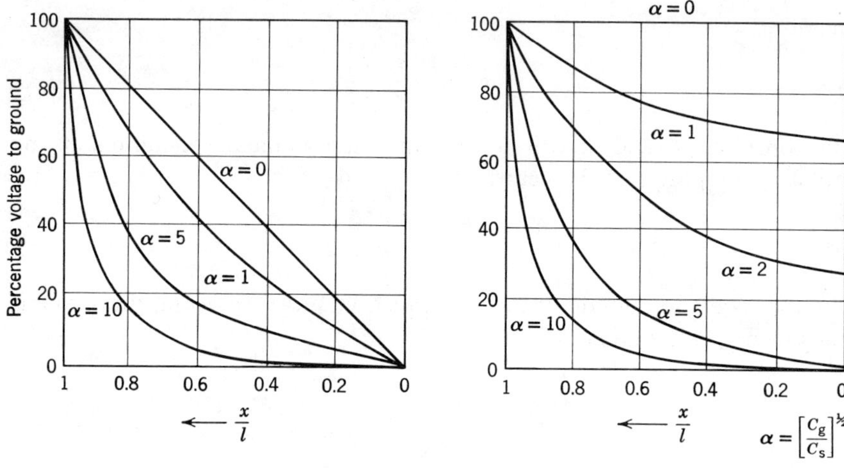

Fig. 11.2. Initial voltage distribution down a uniform winding in response to a step-function voltage surge. (*a*) Neutral grounded. (*b*) Neutral isolated.

Substituting Eq. 11.2.13 in 11.2.9 gives, for the isolated neutral case,

$$E = \frac{V}{2\cosh\alpha}[\epsilon^{px} + \epsilon^{-px}]$$

$$= V\frac{\cosh(\alpha x/l)}{\cosh\alpha} \tag{11.2.14}$$

Equations 11.2.12 and 11.2.14 describe the initial voltage distributions for winding struck by a step function surge voltage. Observe how both expressions depend upon α, the ratio $[C_g/C_s]^{1/2}$. Figure 11.2 shows a plot of these results.

We observe at once how very nonuniform the voltage distribution can be as α increases. When $\alpha = 10$, 60% of the voltage is developed across the first 10% of the winding at the line end. The precise gradient at this terminal can be obtained by differentiating Eq. 11.2.12 or 11.2.14:

(*a*) Grounded neutral:

$$\frac{dE}{dx} = \frac{\alpha}{l}V\frac{\cosh(\alpha x/l)}{\sinh\alpha}$$

$$\text{if } x = l, \qquad \frac{dE}{dx} = \frac{\alpha V}{l}\coth\alpha \tag{11.2.15}$$

(*b*) Isolated neutral:

$$\frac{dE}{dx} = \frac{\alpha}{l}V\frac{\sinh(\alpha x/l)}{\cosh\alpha}$$

$$\text{if } x = l, \qquad \frac{dE}{dx} = \frac{\alpha}{l}V\tanh\alpha \tag{11.2.16}$$

For large values of α, $\coth \alpha \approx \tanh \alpha \approx 1$. Thus in both instances $dE/dx = \alpha V/l$, which is α times the mean gradient.

Under surge conditions the insulation between turns at the line end of the winding is very severely stressed. When this was first discovered, it explained why many transformers had failed by breakdown of their insulation in this location.

The capacitance of a winding is strictly a matter of the winding geometry, that is, the physical disposition of the turns and coils relative to each other and to grounded parts of the structure. It is not surprising therefore that different styles of winding have different αs and as a consequence perform differently when steep-fronted voltage surges impinge upon them. There are two principal types of winding, the layer winding and the disk or barrel winding. Sketches of these two types of winding are to be found in Fig. 11.3*a* and 11.3*b*. As the name implies, the disk winding is made up disks or pancakes. The turns are concentric and the winding progresses outwards on one disk, drops down one on the stack, and progresses inward on the disk below, and so on, down the winding. In general this design has a higher α than the barrel winding, which is made up of a number of concentric cylinders, each of which comprises many turns.

The first remedial step taken to protect transformer windings against the excessive electrical stresses was to strengthen the insulation on the end turns. In most instances this failed to correct the problem, because it affected the capacitance distribution in the direction of making the voltage distribution yet more nonuniform.

The most common method now practiced of grading the surge voltage down windings is to place metallic shields in strategic positions adjacent to the coils. The manner in which they achieve the desired end can be understood if one appreciates how the voltage distribution becomes nonuniform in the first place. Consider Fig. 11.1*a*. If an alternating voltage is applied between the line terminal and ground, it is clear that the current flowing through the series capacitance will diminish progressively as one proceeds down the winding, since current is being shunted off to ground through the ground capacitance. If we assume the capacitance distribution is uniform, the voltage across any series section will be proportional to the current through it. The grading shields are so placed and so connected to the winding that the capacitance current flowing between them and the winding tends to compensate for the ground capacitance currents, thereby making the current through the series capacitance much more uniform. A typical electrostatic shield system is shown in Fig. 11.3*c*.

Another interesting method, widely used, has been described by Chadwick et al. (6). The capacitance distribution is adjusted by the method in which the coils of the winding are interconnected. The electrical sequence in which

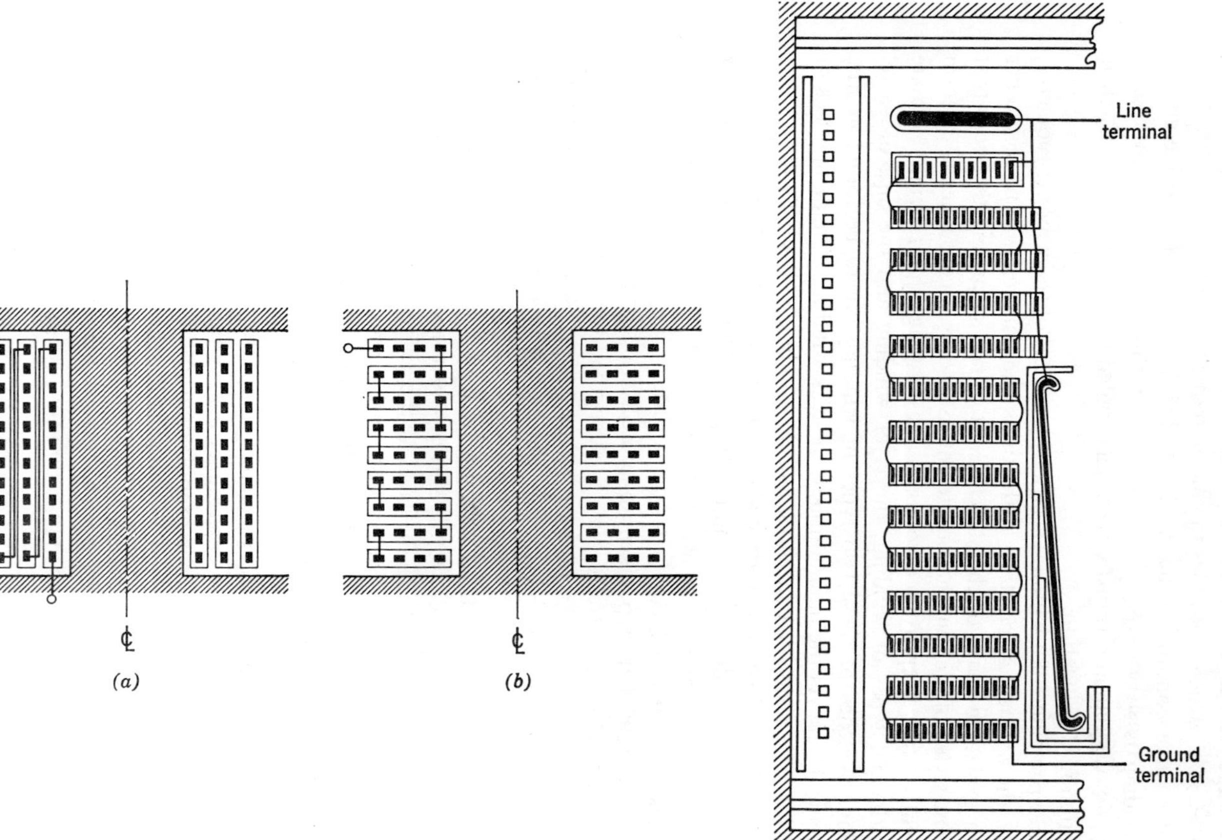

Fig. 11.3. Schematic diagrams of (*a*) layer type winding and (*b*) disk type winding. (*c*) Electrostatic shield for grading a disk type transformer winding.

the disks are joined does not coincide with their geometrical sequence. A reasonably uniform distribution of voltage is achieved at the expense of a somewhat more complicated assembly process.

So far the transformer winding has been considered in isolation; in fact, we have considered only the capacitive attributes of the winding. We should not overlook the fact that the surge must have arrived from somewhere, and that there must be some interaction between the winding and this external circuit. In all probability the surge will have reached the transformer along a transmission line to which the transformer is connected. The transformer represents a particular termination for traveling waves. So far we have chosen to represent this as capacitive, a type of termination studied in Section 9.4. It was shown there that a square wave of voltage traveling along a line and reaching a capacitor C at the end of the line proceeds to charge the capacitor to a voltage $2V$ at a rate determined by the time constant Z_0C, where Z_0 is the characteristic impedance of the line. This is described by Eq. 9.4.9 and illustrated in Fig. 9.13. It will be shown later that the effective capacitance of the winding is $(C_sC_g)^{1/2}$. Inasmuch as C_s is typically in the range 10^{-5}–10^{-4} μF and C_g is in the range 10^{-3}–10^{-2} μF, the effective capacitance will be $10^{-4} < C_{\text{eff}} < 10^{-3}$ μF. Thus the time constant for a 400-Ω line will be a fraction of a microsecond. This is the time to establish the initial voltage distributions. The other important feature disclosed by Eq. 9.4.9 is the doubling effect of the surge voltage. If a square surge of voltage V arrives at the transformer, the winding will see it as having an amplitude $2V$, but also having its front sloped somewhat by the transformer capacitance in the exponential manner described.

11.3 Winding Oscillations

The initial distribution quickly gives way to a very complicated system of oscillations within the winding. The distribution writhes in a most tormented way. We are familiar with the idea of the potential of a point starting a transient at some initial value, oscillating about some final value, and in the process swinging almost as far above the steady value as it starts below it. The potential of points along the winding behave in a somewhat similar manner. Consider one of the curves in Fig. 11.2*a*, $\alpha = 10$, for example. With the neutral grounded, the final voltage distribution will be uniform, corresponding to $\alpha = 0$ in the figure. The excursions of any point on the winding during the oscillation almost always lie within an envelope formed by reflecting the initial distribution in the final distribution. This is shown in Fig. 11.4. This figure is not to be construed as showing that the entire initial distribution oscillates as a whole about the steady-state line in the manner of a jump rope. It merely indicates that the topmost curve is a limit which at some time every point may expect to approach. A typical example

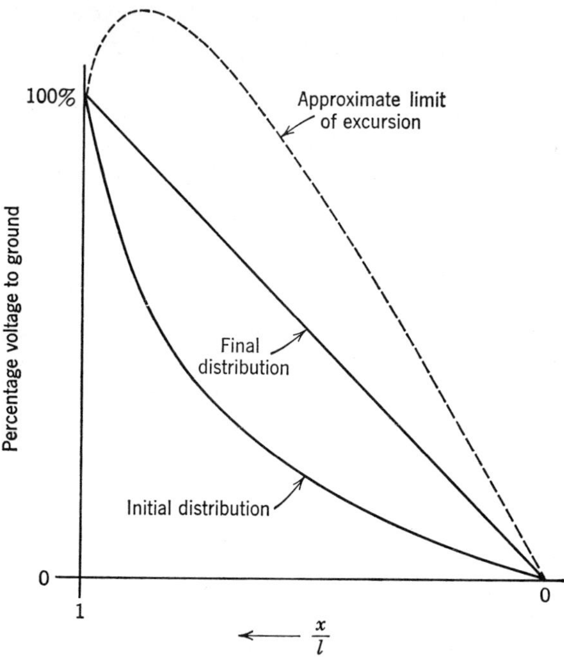

Fig. 11.4. Voltage oscillations in a transformer winding—initial distribution and approximate limit of excursion, neutral grounded.

of a winding transient in a transformer is shown in Fig. 11.5. This is taken from a paper by Lacey (7).

It has been mentioned that strengthening the end line insulation, which may improve the ability of these turns to accept excessive voltage stress, may also subject this insulation to higher stresses because of the change in capacitance distribution. This is well illustrated in Fig. 11.5*b* also taken from Lacey's paper. There are two ways of analyzing such disturbances, one in terms of standing waves, the other in terms of traveling waves. We will consider both in the order stated.

The standing wave approach starts by recognizing the initial distribution as the sum of the steady-state distribution and a transient distribution. For Fig. 11.4, these are, respectively,

$$V_1 = \frac{Vx}{l} \tag{11.3.1}$$

and

$$V_2 = V\left[\frac{x}{l} - \frac{\sinh(\alpha x/l)}{\sinh \alpha}\right] \tag{11.3.2}$$

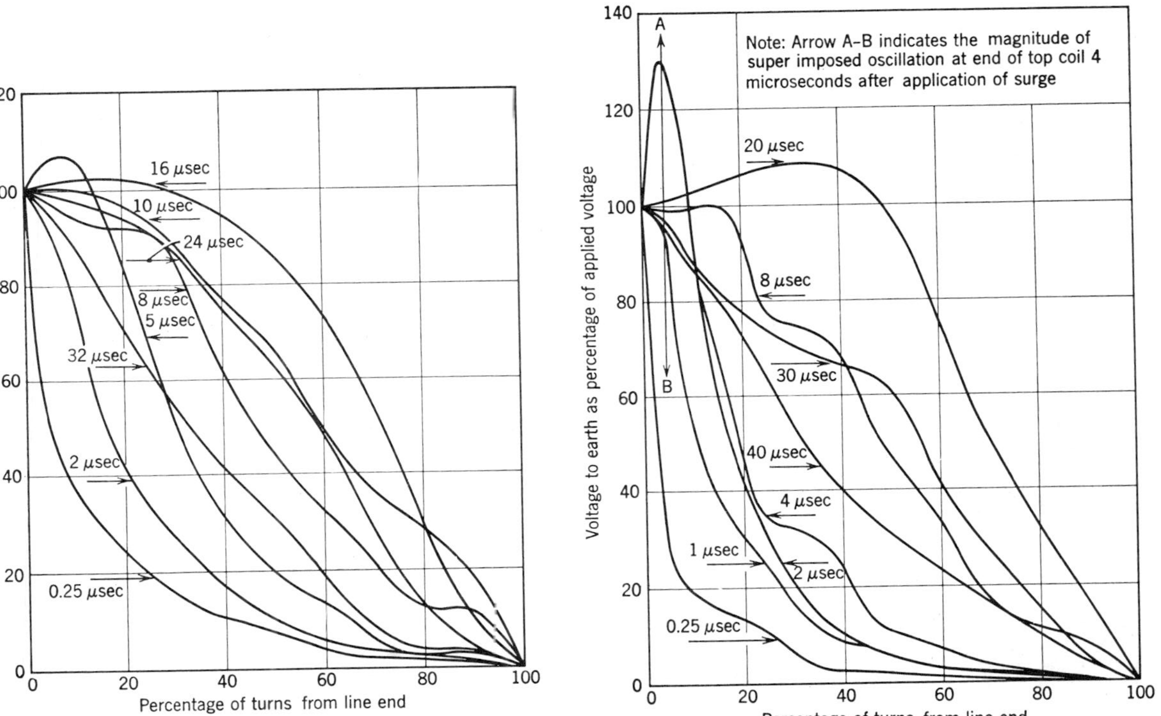

Fig. 11.5. Surge distribution in transformer windings as they vary with time. (*a*) Uniformly insulated winding. (*b*) Winding with reinforced end-turn insulation.

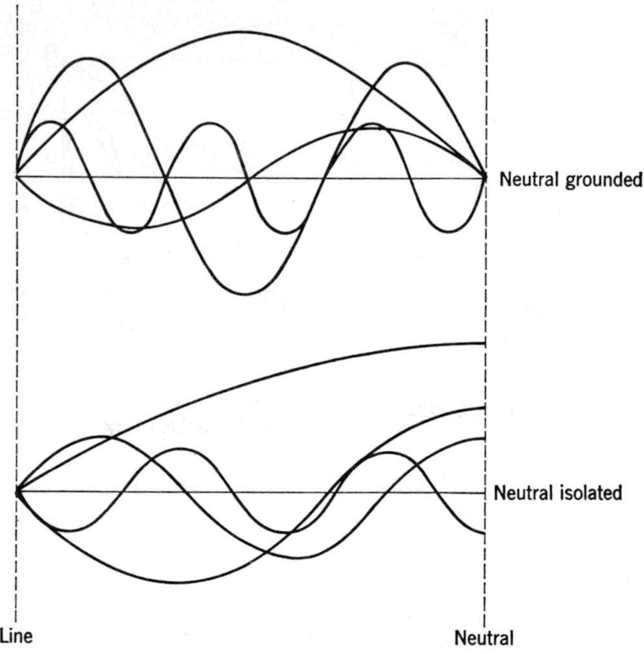

Fig. 11.6. Standing wave patterns in transformer windings.

The method then proceeds to analyze V_2 into its space harmonics by equating it with an infinite series:

$$V\left\{\frac{x}{l} - \frac{\sinh(\alpha x/l)}{\sinh}\right\} = A_{1/2} \sin \frac{\pi x}{l} + A_1 \sin \frac{2\pi x}{l} + A_{3/2} \sin \frac{3\pi x}{l} \cdots \tag{11.3.3}$$

$$= \sum_{m=1}^{m=\infty} A_{m/2} \sin \frac{m\pi x}{l} \tag{11.3.4}$$

The coefficients are determined by the usual Fourier method. There are only certain standing waves that can exist on the winding. They are determined by the boundary conditions. When the winding is grounded at one end, the wavelength must be such that an integral number of half waves fit into the length of the winding. It will be seen that Eq. 11.3.4 complies with this. If the neutral of the winding is isolated, an odd number of quarter wave lengths must fit into the winding length, that is,

$$V_2 = \sum_{m=1}^{m=\infty} A_{m/4} \cos \frac{m\pi x}{2l} \tag{11.3.5}$$

Examples of some of these space harmonics are shown in Fig. 11.6 for both cases.

Each space harmonic oscillates in time at a frequency peculiar to itself. We shall show momentarily how these frequencies can be determined. After analyzing the initial distribution into its significant space harmonics, it is possible from a knowledge of the corresponding periods to predict the potential of any point on the winding or the entire distribution at all subsequent times.

The next step is to establish an equation describing the voltage distribution on a uniform winding. It is here that we must be quite clear about the physical meaning of the terms in which this equation will be written. Figure 11.7 will help with the physical picture. It shows three adjacent turns in the winding.

Suppose, as before, that the total capacitance to ground of the winding is C_g; then let the capacitance per unit length be C_1, defined by

$$C_1 = \frac{C_g}{l}$$

If the small shaded portion in Fig. 11.7 were of length Δx, its capacitance to all grounded parts of the structure would be $C_1\,\Delta x$. The element also has capacitance to adjacent sections of neighboring turns on both sides. Let this be C_2 per unit length or bC_2 for a turn of length b.

In addition, the coil will possess inductance. Each turn may be deemed to have a self-inductance due to the flux that links it alone; and it will have a mutual inductance with other turns due to the flux they mutually share. The sum total of all these inductances, for all the turns, will be the self-inductance of the complete winding. Let the self-inductance be L_1 per unit length, or $L_1\,\Delta x$ for the elementary length. For a turn of length b it will be bL_1.

The degree of coupling between turns will depend upon their proximity, being greatest when they are close together. Consider two turns A and B.

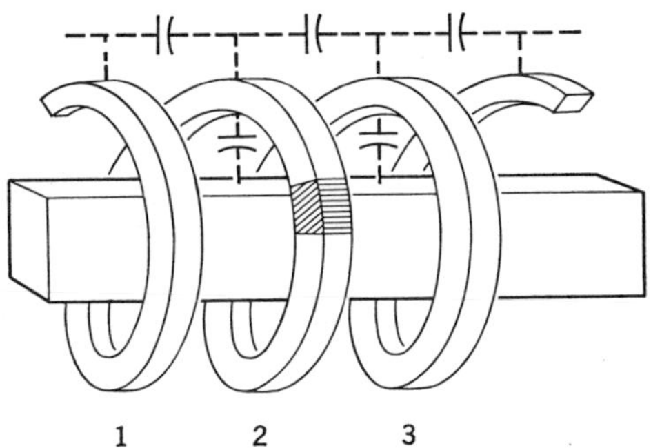

Fig. 11.7. Three adjacent turns on a winding.

The voltage induced in A due to a current in B will be

$$bL_{AB}\frac{dI_B}{dt}$$

Every other turn will similarly induce a voltage so that the total mutually induced voltage may be written

$$b\sum L_{AB}\frac{dI_B}{dt}+L_{AC}\frac{dI_C}{dt}+L_{AD}\frac{dI_D}{dt}+\cdots \tag{11.3.6}$$

The currents I_B, I_C, I_D, . . . , are not generally equal because the current, like the voltage, is not uniform. In some turns it will be greater than A and in others it will be less. If these currents are each written in the form $I_A + \Delta I_N$, where ΔI_N is the difference, positive or negative, between I_A and I_N, then the expression (11.3.6) can be rewritten as

$$b\frac{dI_A}{dt}\sum L_{AB}+L_{AC}+L_{AD}+\cdots L_{AN}+b\sum L_{AB}\frac{d\,\Delta I_B}{dt}+L_{AC}\frac{d\,\Delta I_C}{dt}+\cdots \tag{11.3.7}$$

The second part of Eq. 11.3.7 contains a summation of difference terms some positive and some negative. More careful analysis (5) shows that the net value of this summation is small. We therefore neglect this term, recognizing that by so doing we are introducing an approximation. Taking the remaining terms, together with the self-inductance term, the induced voltage in turn A will be

$$bL_1\frac{dI_A}{dt}+b\sum(L_{AB}+L_{AC}+L_{AD}+\cdots)\frac{dI_A}{dt}$$

$$=b[L_1+\sum L_{AN}]\frac{dI_A}{dt} \tag{11.3.8}$$

The quantity $(L_1 + \sum L_{AN})$ is what each unit length of turn A, or any other turn, contributes to the total inductance of the winding. This will be designated L. The element Δx will therefore have an inductance $L\,\Delta x$. Shrinking to the infinitesimal scale, the induced voltage can be written

$$\partial E = L\frac{\partial I}{\partial t}\,\partial x$$

or

$$\frac{\partial E}{\partial x} = L\frac{\partial I}{\partial t} \tag{11.3.9}$$

This is consistent with our convention of signs. We are considering the current flowing into the winding, but are measuring x positively from the neutral toward the terminal.

The reader will find in the literature several different ways of obtaining the leakage inductance of a winding and they will not all mean the same thing. The author finds the preceding approach, which is due to Rudenburg (5), fairly easy to comprehend. It differs from that of Blume and Boyajian (4), for example, and this is reflected in their analysis.

We turn now to the consideration of capacitance. As a consequence of voltage variations at any location on the winding, charging currents will flow in the capacitances. The capacitance current to ground from one turn will be

$$bC_1 \frac{\partial E}{\partial t} \tag{11.3.10}$$

Referring to Fig. 11.7 and using the same convention of signs, the capacitance current to turn 2 from adjacent turns will be, from turn 1,

$$\left.\begin{aligned} & bC_2 \frac{\partial}{\partial t}\left[\frac{\partial E}{\partial x}\right]_{1,2} b \\ \text{and from turn 3} \qquad & -bC_2 \frac{\partial}{\partial t}\left[\frac{\partial E}{\partial x}\right]_{2,3} b \end{aligned}\right\} \tag{11.3.11}$$

To a close approximation,

$$\left.\begin{aligned} & \left[\frac{\partial E}{\partial x}\right]_{1,2} = \left[\frac{\partial E}{\partial x}\right]_2 + \left[\frac{\partial^2 E}{\partial x^2}\right]_2 \frac{b}{2} \\ \text{and} \qquad & \left[\frac{\partial E}{\partial x}\right]_{2,3} = \left[\frac{\partial E}{\partial x}\right]_2 - \left[\frac{\partial^2 E}{\partial x^2}\right]_2 \frac{b}{2} \end{aligned}\right\} \tag{11.3.12}$$

$[\partial^2 E/\partial x^2]_2$ is the space rate of change of the voltage gradient at conductor 2. Substituting Eqs. 11.3.12 in 11.3.11, we obtain the net charging current from adjacent turns:

$$bC_2 \frac{\partial}{\partial t}\left[\left(\frac{\partial E}{\partial x}\right)_2 + \left(\frac{\partial^2 E}{\partial x^2}\right)_2 \frac{b}{2}\right] b - bC_2 \frac{\partial}{\partial t}\left[\left(\frac{\partial E}{\partial x}\right)_2 - \left[\frac{\partial^2 E}{\partial x^2}\right]_2 \frac{b}{2}\right] b = C_2 b^3 \frac{\partial^3 E}{\partial t\, \partial x^2} \tag{11.3.13}$$

If we subtract from this the capacitance current to ground (Eq. 11.3.10), we have the net rate of change of total current per turn length,

$$-b \frac{\partial I}{\partial x} = C_2 b^3 \frac{\partial^3 E}{\partial t\, \partial x^2} - C_1 b \frac{\partial E}{\partial t} \tag{11.3.14}$$

the negative sign because x is measured positively from neutral to the line end. Equations 11.3.9 and 11.3.14 describe the fundamental behavior of the winding. E or I can be eliminated to reach a single equation in an unknown. I is eliminated by differentiating Eq. 11.3.9 with respect to x and Eq. 11.3.14 with respect to t. From Eq. 11.3.9,

$$\frac{\partial^2 I}{\partial x\,\partial t} = \frac{1}{L}\frac{\partial^2 E}{\partial x^2} \tag{11.3.15}$$

From Eq. 11.3.14,

$$\frac{\partial^2 I}{\partial x\,\partial t} = -C_2 b^2 \frac{\partial^4 E}{\partial t^2\,\partial x^2} + C_1 \frac{\partial^2 E}{\partial t^2} \tag{11.3.16}$$

Equating Eqs. 11.3.15 and 11.3.16 and rearranging,

$$\frac{\partial^2 E}{\partial x^2} - LC_1 \frac{\partial^2 E}{\partial t^2} + LC_2 b^2 \frac{\partial^4 E}{\partial t^2\,\partial x^2} = 0 \tag{11.3.17}$$

If Eq. 11.3.17 is solved and the boundary conditions for any specific problem are inserted, the solution will give the value of E for any point at any time for that particular problem.

Equation 11.3.17 is similar to the transmission line equation for a smooth line (Eq. 9.2.3); the difference is in the third term. This term arises as a consequence of the mutual coupling, inductive and capacitive, along the winding, which is absent from lines. A similar equation can be derived for current.

The simplest form of solution for the winding equation is

$$E = V_\gamma \epsilon^{j\omega t} \epsilon^{j\gamma x} \tag{11.3.18}$$

according to which a voltage of amplitude V_γ oscillates in time and space along the extent of the conductor. These are the standing waves we have referred to. When solution (11.3.18) is substituted in Eq. 11.3.17, it is found that

$$-\gamma^2 + LC_1\omega^2 + LC_2 b^2 \gamma^2 \omega^2 = 0 \tag{11.3.19}$$

from which

$$\left.\begin{aligned} \gamma &= \left[\frac{LC_1\omega^2}{1 - LC_2 b^2 \omega^2}\right]^{1/2} \\ \omega &= \frac{\gamma}{(LC_1 + LC_2 b^2 \gamma^2)^{1/2}} \end{aligned}\right\} \tag{11.3.20}$$

showing that there is a definite relationship between the space harmonics and the frequencies at which they oscillate. An example will make this clear.

Consider a winding with grounded neutral with the following characteristics:

$$L = 2.5\ \text{mH}$$

$$\left.\begin{aligned} \text{no. of turns} &= 200 \\ b &= 1.5\ \text{m} \end{aligned}\right\} \quad \text{whence } l = 300\ \text{m}$$

$$\left.\begin{aligned} C_g &= 1000\ \text{pF} \\ C_2 &= 60\ \text{pF} \end{aligned}\right\} \quad \text{whence } C_1 = 3.33\ \text{pF}$$

According to Eq. 11.3.4,

$$\gamma = \frac{m\pi}{l}$$

where m takes on sequential positive integral values. Thus the following values of γ are possible in this winding:

$m =$	1	2	3	4	5	$\cdots$
$\gamma =$	0.0105	0.0211	0.0316	0.0421	0.0526	$\cdots$

If these values are entered in Eq. 11.3.20 with the given values of the constants, the corresponding frequencies can be found:

$\gamma =$	0.0105	0.0211	0.0316	0.0421	0.0527	$\cdots$
$\omega =$	1.21	2.40	3.57	4.68	5.73×10^5	$\cdots$

Apparently, the lowest natural frequency of this winding, corresponding to the half wave space harmonic, is about 19 kHz.

Having obtained the amplitudes of the various space harmonics by the Fourier analysis of Eq. 11.3.4 and determined the frequency at which each oscillates, we can derive the voltage distribution by synthesis, for any point at any time. In the precomputer era this method of calculating the response of a winding to a surge was criticized because of slow convergence of Eq. 11.3.4, which made it necessary to evaluate a considerable number of terms. With computational aids, this is less of a problem.

From the denominator in the expression for γ in Eq. 11.3.20, we note that if

$$LC_2b^2\omega^2 > 1$$

γ becomes imaginary. This corresponds to a critical value of ω, which will be designated ω_c,

$$\omega_c = \frac{1}{b(LC_2)^{1/2}} \tag{11.3.21}$$

As m assumes larger and larger values, the frequencies of the space harmonics become ever closer, eventually converging on ω_c. The winding apparently behaves like a filter, refusing to admit or support any frequencies above this critical value.

11.4 The Traveling Wave Solution

An equally valid solution to Eq. 11.3.17 is

$$E = V_\omega \epsilon^{j\omega(t+x/v)} \tag{11.4.1}$$

Comparing this with the standing wave solution (Eq. 11.3.18), the equivalence is at once apparent. The connection between γ and v is simply

$$\gamma = \frac{\omega}{v} \qquad \text{or} \quad \gamma^2 = \frac{\omega^2}{v^2} \tag{11.4.2}$$

Solution 11.4.1 lends itself to another interpretation; it represents a system of traveling waves, which oscillate in time at frequencies related to their velocities. This relationship can be deduced from Eq. 11.3.19 by substituting for γ^2 from Eq. 11.4.2:

$$-\frac{\omega^2}{v^2} + LC_1\omega^2 + LC_2 b^2 \frac{\omega^4}{v^2} = 0$$

or

$$v = \left[\frac{1}{LC_1} - \frac{C_2}{C_1} b^2\omega^2\right]^{1/2} \tag{11.4.3}$$

The physical interpretation of Eq. 11.4.3 is as follows. In reference (8) it is pointed out that a surge, being aperiodic, is represented by a continuous spectrum of frequencies, not a fundamental and finite, often small number of harmonics. When such a surge arrives at the terminals of a transformer the different frequencies penetrate the winding at different velocities, frequency and velocity being related by Eq. 11.4.3. From a knowledge of the surge in the frequency domain, it is possible to see how the winding responds to its particular frequency spectrum.

Examination of Eq. 11.4.3 shows that when ω is low, the velocity is given by

$$v = \frac{1}{(LC_1)^{1/2}} \text{ approx.} \tag{11.4.4}$$

This reminds us of the wave velocity on a transmission line, which is perhaps not surprising, for we would also arrive at this solution were $C_2 = 0$, that is, with no capacitive mutual coupling. As ω increases, the second term in Eq. 11.4.3 becomes progressively more important. The velocity diminishes progressively. When

$$\frac{C_2}{C_1} b^2\omega^2 = \frac{1}{(LC_1)} \tag{11.4.5}$$

the velocity is zero, for yet higher frequencies v becomes imaginary, whatever that may mean. The critical frequency here is the same frequency we observed in the standing wave solution and designated ω_c in Eq. 11.3.21. It has the same physical significance, that is, components of supercritical frequency cannot penetrate the winding. For the example studied in the last section the critical frequency is

$$\omega_c = \frac{1}{1.5(2.5 \times 10^{-3} \times 6 \times 10^{-11})^{1/2}} = 1.72 \times 10^6$$

or

$$f_c = 275 \text{ kHz}$$

We arrive at the same conclusion regarding the critical frequency from consideration of the winding surge impedance. Taking the winding as a whole, the change in current along its length is due solely to capacitance current flowing to ground. For an elementary length Δx this will amount to

$$\Delta I = C_1\,\Delta x \frac{dE}{dt} \tag{11.4.6}$$

Until a wave reaches the neutral, the total ground capacitance current will be the total current flowing in the winding. Thus, at the terminal,

$$I = \sum \Delta I = C_1 \int \frac{\partial E}{\partial t}\,dx \tag{11.4.7}$$

Substituting for E from Eq. 11.3.18,

$$\begin{aligned} I &= j\omega C_1 V_\gamma \epsilon^{j\omega t} \int \epsilon^{j\gamma x}\,dx \\ &= C_1 V_\gamma \epsilon^{j\omega t} \epsilon^{j\gamma x} \\ &= \frac{\omega}{\gamma} C_1 E \end{aligned}$$

Thus, the impedance

$$Z = \frac{E}{I} = \frac{\gamma}{\omega C_1} \tag{11.4.8}$$

From Eq. 11.3.20,

$$\frac{\gamma}{\omega C_1} = \left[\frac{L/C_1}{1 - LC_2 b^2 \omega^2}\right]^{1/2}$$

But $(L/C_1)^{1/2} = Z_0$ and $LC_2b^2 = 1/\omega_c^2$, so we may write

$$Z = \frac{Z_0}{[1 - (\omega^2/\omega_c^2)]^{1/2}} \tag{11.4.9}$$

For small values of ω, the surge impedance approaches the surge impedance of a smooth line:

$$\text{as } \omega \to 0, \qquad Z \to Z_0 = \left(\frac{L}{C_1}\right)^{1/2}$$

As ω increases, Z increases, until at the critical frequency ω_c, it becomes infinite. Above ω_c, Z becomes imaginary, and

$$\text{as } \omega \to \infty \qquad Z \to \frac{1}{jb\omega(C_1C_2)^{1/2}} \tag{11.4.10}$$

The winding behaves like a capacitance $b(C_1C_2)^{1/2}$, which we referred to earlier.

If supercritical frequencies cannot penetrate the winding, it is reasonable to suppose that they are reflected from its terminals. How then do they affect the voltage distribution within the winding? Since, for $\omega > \omega_c$, the velocity becomes imaginary; then we can write $v = j\omega/\beta$. Substituting this in the traveling wave solution for E (Eq. 11.4.1),

$$E = v_\omega \epsilon^{j\omega t} \epsilon^{\beta x} \tag{11.4.11}$$

where

$$\beta = \left[\frac{\omega^2}{LC^2b^2\omega^2 - 1}\right]^{1/2} \tag{11.4.12}$$

This is reminiscent of the standing wave solution Eq. 11.3.18, with β corresponding to $j\omega$. The part $\epsilon^{\beta x}$ represents a distribution in space, being greatest at the terminal, where $x = l$, and declining exponentially to a value of unity at the neutral, where $x = 0$. Each supercritical frequency gives rise to a distribution of this kind. In the case of the step function surge, which contains a continuous spectrum of all frequencies, each frequency contributes an infinitesimal contribution of this kind. In any event, at time zero, the summation or integral of all of these contributions is what gives rise to the initial voltage distribution down the winding, which we have already studied. At that time, of course, none of the subcritical components of the surge, which give rise to traveling waves, will have had time to penetrate the winding. The part $\epsilon^{j\omega t}$ in Eq. 11.4.11 indicates that as time passes this distribution oscillates as the traveling wave components pass up and down through it.

The components of supercritical frequency in any incident surge, since they cannot penetrate the winding, are reflected back down the line. Rudenburg (5) uses the Fourier integral to the subcritical and supercritical components of a square wave:

$$\text{subcritical (transmitted)} = \frac{E}{2} + \frac{E}{\pi}\int_0^{\omega_c} \frac{\sin \omega t}{\omega}\, d\omega$$

$$\text{supercritical (reflected)} = \frac{E}{\pi}\int_{\omega_c}^{\infty} \frac{\sin \omega t}{\omega}\, d\omega$$

The analyses of Sections 11.2 to 11.4 have been restricted to simple uniform windings, which are themselves complicated. However, windings in practice are often not uniform; for example, the successive layers of a high-voltage winding frequently diminish in axial length as they become progressively bigger in diameter. Attempts have been made to study non-uniform windings and these will be found in the literature. But it is a subject that goes beyond the scope of this book. As an alternative to such analysis it is possible to build analogues and models of transformer windings and study them experimentally. Some discussion of these techniques will be found in Chapters 14 and 17. Chapter 17 also considers the full-scale testing of windings.

11.5 Behavior of the Transformer Core under Surge Conditions

So far in our analyses we have almost ignored the transformer core, treating it simply as an equipotential ground surface to which the winding has a certain capacitance. This is a good first approximation, because eddy currents will initially preclude magnetic flux penetrating the iron. However, in Chapter 8 we pointed out that with the passage of time, flux does build up in iron, and when it does so in the transformer core, we can expect it to modify our earlier conclusions regarding the behavior of the winding under surge conditions. How quickly are these modifications felt? Christoffel (9) has addressed himself to this question.

As we showed in Chapter 8, a problem of this kind resolves itself into the solution of Maxwell's equations for the appropriate stimulus—a surge of voltage applied to the winding in this instance—consistent with the boundary conditions applied by the lamination thickness. Christoffel considers a surge of step function form and approaches it from the opposite extreme to that which we have so far used. That is to say, he considers only the inductance of the winding and ignores all capacitances. When a step function of voltage is applied to an inductance we would expect it to give rise to a linear ramp of current, for neglecting resistance, the inductive emf must balance the applied voltage. If the voltage is V_0, we would write

$$V_0 = L\frac{dI}{dt}$$

or

$$I(t) = \frac{V_0}{L}t \tag{11.5.1}$$

The question is, what is L? Initially L will have a low value since the flux is all airborne. It will correspond to the inductance we have been using in our work so far. But as time passes and the magnetic flux penetrates the laminations, the inductance will increase considerably to some much higher value, L_∞. In practice, of course, the core would ultimately saturate and the inductance would fall again. The expression that Christoffel derives for the current is an infinite series:

$$I(t) = \frac{V_0}{L_\infty}\left\{t + \frac{\pi^2}{3} T_1 \frac{[1 - \sum_{n=1}^{n=\infty}(1/n^2)\epsilon^{-n^2t/T_1}]}{\sum_{n=1}^{n=\infty}(1/n^2)}\right\} \tag{11.5.2}$$

T_1 is a time constant, which gives a measure of the rate at which the flux moves into the iron. It is given by

$$T_1 = \frac{\mu}{\rho}\left[\frac{d}{2\pi}\right] \tag{11.5.3}$$

where μ and ρ are the absolute permeability and resistivity of the core material, respectively, and d is the thickness of the laminations.

Equation 11.5.2 has been plotted in a dimensionless form in Fig. 11.8. It is seen that the current slope is initially high when the inductance is small but becomes asymptotic to a line of slope t/T_1 as time increases. Typically, the following values might apply for transformer steel:

$$d = 0.35 \text{ mm or } 3.5 \times 10^{-4} \text{ m}$$

$$\mu = 10^3 \times 4\pi \times 10^{-7} = 4\pi \times 10^{-4}$$

$$\rho = 6 \times 10^{-7} \text{ ohm meter}$$

Substituting these numbers in Eq. 11.5.3, $T_1 = 6.5\ \mu\text{sec}$, indicating that the core will influence the transient behavior of the winding in a relatively short time.

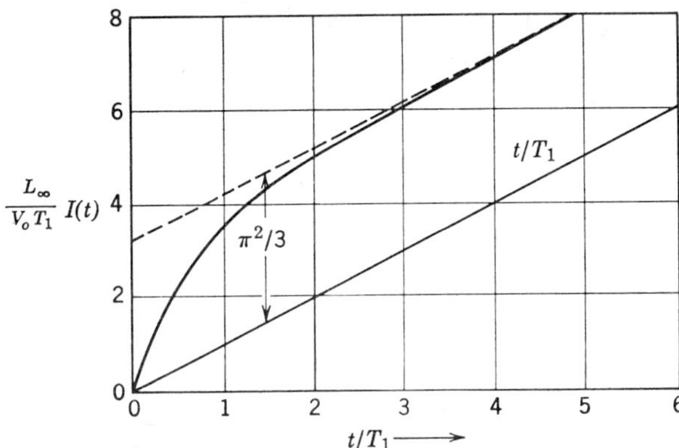

Fig. 11.8. Flux penetration into transformer core under surge conditions—dimensionless plot of Eq. 11.5.2.

11.6 Response of a Transformer Winding to a Current Stimulus

This topic has an important practical application in its own right, but it also provides a rather interesting extension of the material of Section 11.5. The practical application is the matter of current suppression. This subject was discussed in Section 5.2 but only as it affected a coil or winding as a whole. It was shown that abnormally high overvoltages can be generated in inductive circuits when current suppression occurs and the chopping of magnetizing current in the winding of a transformer was cited as an example. A question might be raised as to how this voltage distributes within the winding. This problem has been studied by Lee (10) and Greenwood (11). As mentioned above, this leads to an interesting variation as far as the material of the last section is concerned, for we are not concerned with the penetration of flux into the transformer core. Flux already exists in the core. However, when the current is suppressed, the collapse of this flux will be opposed by eddy currents in the core. In this respect the problem is similar.

Lee (10) treated the chopping of current as being equivalent to the injection into the winding of a negative current, which wiped out the existing current. To analyze the response to this current stimulus, he used an approach similar to that of Blume and Boyajian (4) for a voltage stimulus and showed that this method is better suited to the current problem than the voltage problem, since the series solution converges more rapidly. It was shown that for the transformer with a uniform winding and with its secondary winding open-circuited, the total voltage, $I_c Z_0$, (I_c is the current chopped, Z_0 is the effective surge impedance) could be determined by assuming that Z_0 is the square root of the quotient of the magnetizing inductance, L_m, and one-third of the total capacitance to ground of the winding or $C_g/3$. Likewise the natural frequency would be $1/2\pi(L_m C_g/3)^{1/2}$. These would be modified by damping effects due to hysteresis and eddy currents as explained in Chapter 5. For such a transformer it was also shown that *this voltage distributes uniformly down the length of the winding*. For loaded or short-circuited transformers Lee's analysis indicated that the voltage is much lower and is no longer uniformly distributed. These results were confirmed experimentally by Greenwood (11).

It was pointed out in Section 5.2 that the circuit breaker often cannot support the voltage which its chopping may generate and that it will reignite, sometimes many times. An example of this is shown in Fig. 5.5. Such reignition, of course, constitutes *voltage surges*, as far as the transformer is concerned, and therefore the transient voltages which they give rise to do not distribute uniformly but follow the procedures outlined in Sections 11.2 to 11.4.

11.7 Transmission of Voltage Surges through Transformers

In the earlier sections of this chapter attention has been focused on the winding subjected to the surge. We now consider how voltage surges are coupled to other windings. Regardless of the manner in which they normally function, for our purposes we shall regard the primary winding as that to which the surge is applied. The secondary winding is that to which the surge is subsequently communicated. There are two mechanisms whereby this transfer can be effected, the first is electrostatic, the second electromagnetic.

When a steep-fronted voltage surge first impinges on a transformer the characteristic of immediate concern is its capacitance, as we have pointed out earlier. Figure 11.9*a* is a simplified schematic representation of a two winding transformer which indicates pertinent capacitances between the windings and from the secondary to the core. As a first approximation this winding/core combination forms a coaxial system of capacitors, with capacitances C_1 and C_2 as indicated in Fig. 11.9*b*. This is indeed an approximation, since the neutral end of one or both windings may be connected to ground. Yet, initially, the system behaves in the manner described. Thus the windings and core constitute together a crude potential divider. When a step of voltage is applied to a series combination of capacitors, the voltage divides inversely as the capacitance. In this instance we may expect the potential of the secondary winding to attain a value

$$\frac{C_1}{C_1 + C_2} V \tag{11.7.1}$$

where V is the surge voltage. It is evident that C_1 and C_2 are determined by the geometry of the coil and core structure, so that the voltage (11.7.1) can represent a very significant fraction of the surge voltage. Specifically, it bears no relationship to the turns ratio of the transformer. When the primary winding is simply energized, especially if this occurs near the peak of the voltage wave, a considerable fraction of the primary voltage may appear on the secondary winding. Such electrostatically induced voltage surges are of very short duration because other attributes of the winding quickly come into play. However, they last long enough to cause concern, if not to the secondary winding itself, to equipment connected thereto. In recent years there has been an increasing practice of using high transformation ratios to bring power into industrial plants. In some instances the voltage may be transformed from 100 kV or higher down to a few hundred volts in one stage of transformation, and then it may be applied to a solid-state rectifier. Such installations are particularly vulnerable to the electrostatic phenomenon just described. The only limit to the surge voltage on the primary side is set by the breakover of arresters in the primary circuit, which is much higher than the system voltage. The rectifiers, on the other hand are particularly sensitive to overvoltages, as pointed out in Chapter 7.

Fortunately, this problem is quite readily solved in one of two ways. Perhaps the simpler way is to drastically upset the divider ratio indicated by Eq. 11.7.1. This is done by increasing C_2. Physically, this means connecting capacitors from the secondary terminals to ground in accordance with the practice described in Chapter 12. An alternative method is to insert a grounded shield between the windings to hide one winding from the other, thereby eliminating interwinding capacitive coupling to all intents and purposes. In rectifier installations of the kind described above, both types of protection are frequently provided.

To show how surges are electromagnetically coupled through a transformer from winding to winding, consider Fig. 11.10*a*. Here we have a transformer with transmission lines connected on either side with surge impedances Z_1 and Z_2, respectively. The surge is assumed to approach along line 1, and for convenience we will consider that it is a step function of amplitude V. The transformer represents a change of surge impedance. The question we are posing concerns the refraction coefficient (see Section 9.3).

We are concerned here with what are, comparatively speaking, the longer time effects of the surge, which follow after the electrostatically coupled phenomenon described in the last section has already subsided. Accordingly, we will omit capacitance in our representation of the transformer.

Figure 11.10*b* would be a circuit for this device. The leakage inductances of primary and secondary windings are L_{s1} and L_{s2}, while M is the mutual inductance. The surge impedances of the lines will be represented by resistors

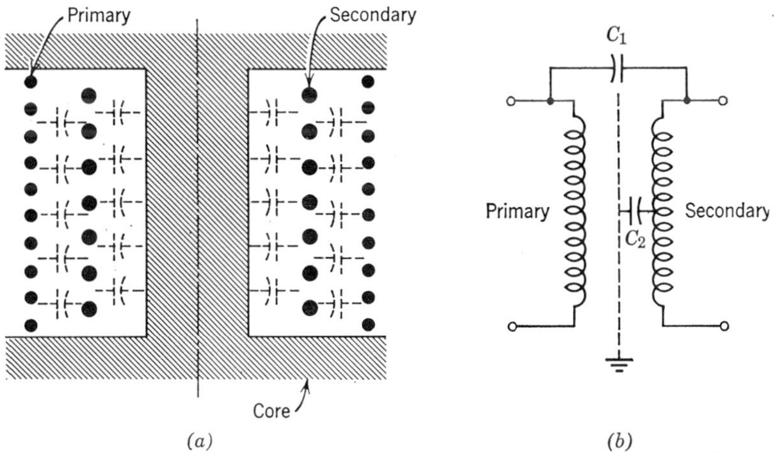

Fig. 11.9. Capacitive coupling between primary and secondary windings of a transformer.

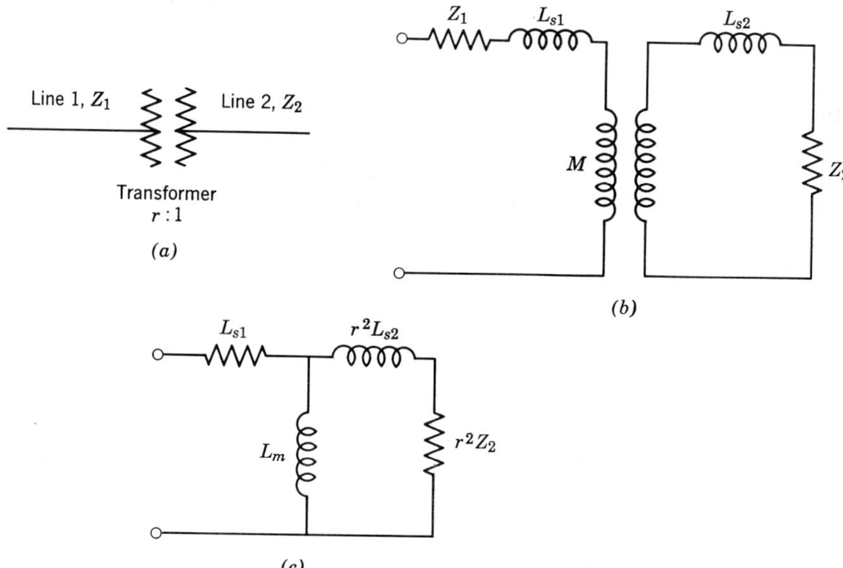

Fig. 11.10. Electromagnetic coupling of surges through a transformer—equivalent circuits.

of the same ohmic values as Z_1 and Z_2. It is possible to set down differential equations for this circuit and derive an expression for the voltage across Z_2, which is the wave that would penetrate the line on the secondary side. However, this involves a considerable amount of algebraic manipulation. A quite accurate idea of how the transformer behaves can be obtained much more simply by a less rigorous, more heuristic approach.

We start by considering that the transformer is ideal, and subsequently refine our thinking to take account of some of its nonideal attributes. With an ideal transformer, the impedance Z_2 can be reflected to the primary side as r^2Z_2, where r is the primary to secondary turns ratio. The refracted wave will then be

$$\frac{2r^2Z_2}{Z_1 + r^2Z_2}V \tag{11.7.2}$$

which, referred to the secondary side, will be

$$\frac{2rZ_2}{Z_1 + r^2Z_2}V \tag{11.7.3}$$

When the turns ratio is high, this approximates to $2V/r$, which, on a per unit basis, is twice the voltage of the incident wave. For a 1:1 transformer with $Z_2 = Z_1$, the voltage wave on line 2 is V_1, the same as the incident wave.

That is, the surge proceeds as if the transformer were not there, as we would expect.

The modifications wrought by the imperfections of the transformer are twofold: one arises as a consequence of its leakage inductance, the second is due to its magnetizing inductance. If secondary impedances are transferred to the primary by multiplying them by the square of the turns ratio, Fig. 11.10*c* is a reasonable equivalent circuit for the transformer/lines combination. Here L_m is the magnetizing inductance, which is very much greater than the leakage inductance L_s. It will be observed that the primary and secondary leakage inductances could be combined on one side of L_m. This is a convenience which introduces little error in the circumstances being studied.

When the surge is first impressed on the transformer terminals, we can safely neglect L_m, temporarily, since the current in L_m will increase very slowly, L_m being so high. The surge voltage will momentarily be impressed across L_s, for although this may be small, it is nevertheless an inductance, and the current cannot change in it instantaneously. The output voltage therefore does not immediately accord with (11.7.2) but only approaches this value after the current builds up in L_s. What we have here is no more than an *RL* circuit in which $R = Z_1 + r^2Z_2$. This is described in Chapter 2 and, as an inductive termination for a transmission line, in Section 9.4. It exhibits here, as there, its characteristic time constant $L_s/(Z_1 + r^2Z_2)$. The output voltage rises as the current rises, heading for (11.7.2) as an asymptote. In lieu of (11.7.2), we write

$$V_2' = \frac{2r^2Z_2}{Z_1 + r^2Z_2} V\left[1 - \exp - \frac{(Z_1 + r^2Z_2)t}{L_s}\right] \tag{11.7.4}$$

and for (11.7.3),

$$V_2 = \frac{2rZ_2}{Z_1 + r^2Z_2} V\left[1 - \exp - \frac{(Z_1 + r^2Z_2)t}{L_s}\right] \tag{11.7.5}$$

Thus far L_m has been ignored. We recognize that current will build up therein; indeed, in the limit, with a step function of voltage, it will offer no impedance to the associated direct current. The voltage across it, and therefore across Z_2, will decline to zero. In considering this eventuality, we can neglect L_s compared with L_m. From a circuit point of view, this places L_m in parallel with Z_1 and r^2Z_2. The time constant for the decline of the secondary voltage is $L_m(Z_1 + r^2Z_2)/Z_1r^2Z_2$, which is much slower than the rise. Assuming for the moment that the rise is instantaneous, that is, (11.7.2) and (11.7.3) are valid, these voltages would subsequently diminish exponentially with this time constant. Thus, for example, (11.7.3) would become

$$\frac{2rZ_2}{Z_1 + r^2Z_2} V \exp - \frac{Z_1r^2Z_2}{L_m(Z_1 + r^2Z_2)} t \tag{11.7.6}$$

If we take both L_s and L_m into account, (11.7.6) and (11.7.5) can be combined to give for the secondary voltage

$$V_2 = \frac{2rZ_2}{Z_1 + r^2Z_2} V\left\{\exp - \frac{Z_1r^2Z_2}{L_m(Z_1 + r^2Z_2)} t - \exp - \frac{Z_1 + r^2Z_2}{L_s} t\right\} \quad (11.7.7)$$

This is not rigorous, but it is accurate for the conditions described. The components of this expression and the composite waveshape are shown in Fig. 11.11.

We have considered only single-phase transformers in this discussion, as in fact we have elsewhere in this chapter. Multiwinding transformers have been treated by Bewley (12) and Hileman (13). The reader who wishes to proceed to such matters should be able to do so through this literature with the foundation layed in this text. It is perhaps worth pointing out that in some areas the extension to polyphase transformers is not a big step. For example, when a wye/delta transformer with grounded neutral experiences a surge on one phase of the wye winding, the secondary winding in the delta, to which it is most closely coupled, will have a line on each terminal. To a first approximation these can be treated as being in series. Thus Z_2 is replaced by $2Z_2$.

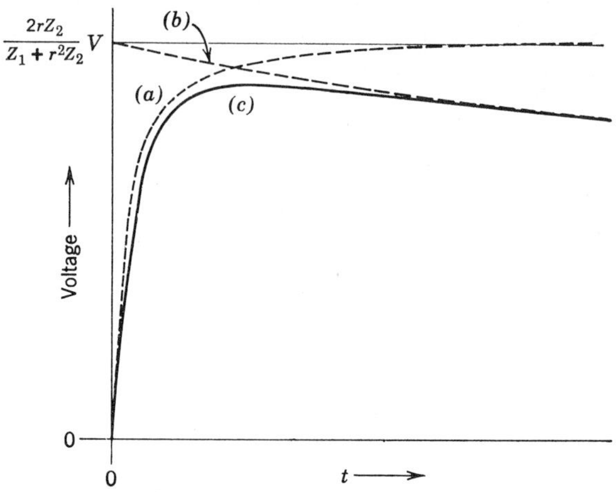

Fig. 11.11. Surge coupled electromagnetically to the secondary of a transformer.

(a) $\dfrac{2rZ_2}{Z_1 + r^2Z_2} V\left\{1 - \exp\left[-\dfrac{(Z_1 + r^2Z)t}{L_s}\right]\right\}$

(b) $\dfrac{2rZ_2}{Z_1 + r^2Z_2} V \exp\left[-\dfrac{Z_1r^2Z_2}{L_m(Z_1 + r^2Z_2)}\right]t$

(c) Resultant wave (Eq. 11.7.7).

In Section 11.6 we described the initial disturbance that is coupled electrostatically to the secondary winding. In this section the electromagnetically coupled component has been described. Between these in time, there are superimposed oscillations arising out of events described for the primary in Sections 11.3 and 11.4. However, the two components to which we have devoted most attention are usually the most important, since the electrostatic surge is capable of producing the highest voltages, and the electromagnetic surge conveys most of the energy.

11.8 Surges in Generators and Motors

The exposure of generators to high-voltage surges is much less than the exposure of transformers to them since such surges are rarely connected directly to overhead lines on which lightning may impinge. With a few exceptions, such as transportation and excavation equipment, this is also true of motors, although they may experience frequent switching surges. It is probably for this reason that the volume of literature on surges in rotating machines is not as extensive as that for transformers.

Robinson (14), (15) and White (16) discuss this subject as it relates to turboalternators. The biggest differences between such machines and transformers of comparable rating are that the turns are fewer but longer and are deeply buried in the stator steel. The capacitance to ground is comparatively high, but series capacitance effects are only significant where conductors occupy the same slot. The winding behaves more like a transmission line with a relatively high surge impedance, of the order of 1000 Ω. Because L and C are significantly greater than the corresponding quantities in a transmission line, the speed of propagation for surges in alternator windings is much lower than for the line. Typically, it is about 15–20 m/μsec. This means that for surges with fast fronts, much of the voltage will appear across the first turn as the surge enters the winding. Where two or three conductors occupy the same slot, surges are induced on these adjacent conductors much as they are induced on the other conductors of a transmission line, when one conductor is surged.

It is generally easier to insulate the windings of transformers than the windings of motors of comparable rating, and in fact the insulation codes for motors are less stringent. Motors are exposed to frequent switching surges when they are switched off and on, but these are relatively mild. One important reason for this is that motors represent a back emf load. Thus, when a motor is switched off, it continues to generate as it coasts down until the magnetic field of the rotor has had time to die down. On the scale of electrical switching transients this takes a long time. The time constant of the rotor circuit is often many cycles. The disconnecting contactor sees across its opening contacts the difference in voltage between the supply and the machine. This develops very slowly, relatively speaking, so that the chanced restrikes in the switch, which could lead to abnormal transient voltages, is minimal.

Large motors, like transformers, are almost invariably protected against surge voltages. This subject will be dealt with in Chapter 12.

REFERENCES

1. P. A. Abetti, "Bibliography on the Surge Performance of Transformers and Rotating Machines," *Trans. AIEE.*, Vol. 77 (1958), p. 1150.
2. P. A. Abetti, First supplement to reference 1, *Trans. AIEE*, Vol. 81 (1962), p. 213.
3. P. A. Abetti, Second supplement to reference 1, *Trans. IEEE*, Vol. 83 (1964), p. 855.
4. L. F. Blume and A. Boyajian, "Abnormal Voltages Within Transformers," *Trans. AIEE*, Vol. 38 (1919), p. 577.
5. R. Rudenburg, "Performance of Traveling Waves in Coils and Windings," *Trans. AIEE*, Vol. 59 (1940), p. 1031.
6. A. T. Chadwick, J. M. Ferguson and Y. F. Stearn, "Design of Power Transformers to Withstand Surges due to Lightning, with Special Reference to a New Type of Winding," *Journ. IEE*, Vol. 97 (1950), p. 737.
7. H. M. Lacey, "The Lightning Protection of High Voltage Overhead Transmission and Distribution Systems," *Proc. IEE*, Vol. 96 (1949), p. 287.
8. H. Tropper, *Electric Circuit Theory*, Longmans, Green and Co., London, New York and Toronto (1949).
9. M. Christoffel, "The Effect of Iron and Secondary Windings on the Current and Voltage during Surges in Transformer Windings," *The Brown Boveri Review*, Vol. 45, No. 6 (1958), p. 254.
10. T. H. Lee, "The Effect of Current Chopping in Circuit Breakers on Networks and Transformers, Part I. Theoretical Considerations," *Trans. AIEE*, Vol. 79 (1960), p. 535.
11. A. N. Greenwood, "The Effect of Current Chopping in Circuit Breakers on Networks and Transformers, Part II. Experimental Techniques and Investigations," *Trans. AIEE.*, Vol. 79 (1960), p. 545.
12. L. V. Bewley, *Traveling Waves on Transmission Systems*, 2nd ed., John Wiley & Sons, New York (1951).
13. A. B. Hileman, "Surge Transfer through 3-Phase Transformers," *Trans. AIEE*, Vol. 77, Part III (1958), p. 1543.
14. B. C. Robinson, "The Propagation of Surge Voltages Through High-Speed Turbo-Alternators with Single-Conductor Windings," *Proc. IEE.*, Vol. 100 (1953), p. 453.
15. B. C. Robinson, "The Propagation of Surge Voltages Through Large Turbo-Alternators with Two Parallel Windings," *Proc. IEE.*, Vol. 101 (1954), p. 335.
16. E. L. White, "Surge-Transference Characteristics of Generator/Transformer Installations," *Proc. IEE*, Vol. 116 (1969), p. 575.

12 The Protection of Systems and Equipment against Transient Overvoltages

12.1 Some Basic Ideas about Protection

Transient overvoltages are facts of life on power systems. Those due to switching operations, which are the most frequent kind, can be controlled to some extent, as we have seen. But there remain transients arising from abnormal conditions and the intrusion of lightning, which introduce relatively unpredictable disturbances. How one protects a power system against overvoltages is really a matter of economics. It would certainly not pay to so insulate the system that it could support any voltage that would ever appear. It would be equally unrealistic to insulate against normal system voltage and accept all failures caused by transients with the attendant cost of repair and replacement, to say nothing of the interruption of service. A judicious compromise must be sought, which makes a reasonable investment in reliable protective devices against those transients that cannot be controlled. This chapter will be concerned with such devices and their application.

Before attempting to describe surge suppressors, lightning arresters, and the like, it is important that we first study the requirements for such equipment and consider the philosophy of their application. We will treat these broad subjects at the beginning of the chapter and then move on to various embodiments. Finally, we will consider some specific protective systems to meet the needs of specific stituations.

Every insulation structure has a certain electrical strength. The function of the protective device is to see to it that this level is not exceeded at any time. Thus, if a surge attempted to raise the voltage above the insulation capability, the protector would have to exert a clamp or restraint to keep the voltage down to an acceptable level. This seems straightforward, at least in principle, but in fact it is complicated by our inability to define the voltage withstand

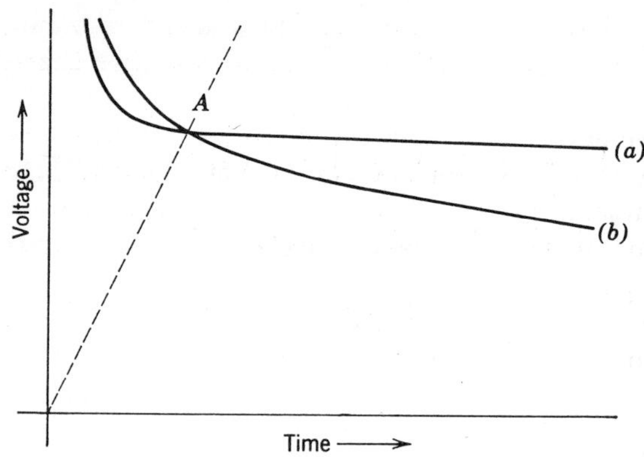

Fig. 12.1. Volt/time curves. (*a*) Insulation. (*b*) Rod gap.

capability of an insulation system, at least by a single number. This is because it is a function of time. Insulation can support a higher voltage sometimes a much higher voltage, if it is applied quickly than if it is applied more slowly. This leads to the concept of a volt-time characteristic. Such a relationship is shown as curve *a* of Fig. 12.1. It is seen that there is a significant turnup in the characteristic for short times. The degree varies with the kind of insulation involved.

Protective devices exhibit this same phenomenon to a greater or lesser extent. The volt-time curve of a simple rod gap is plotted as curve *b* in Fig. 12.1. The reason for the turnup varies. In the case of the rod gap, breakdown depends upon the formation of a spark, and perhaps subsequently an arc, between the rods. There is a formative time for a spark to propagate across the gap. Moreover, the avalanching process that this involves requires the initial presence of free electrons in the high field regions of the gap. There is a certain amount of random ionization of the air at any time, but it is a statistical matter as to how soon an initiating electron will find itself in a position to start an avalanche. Under normal atmospheric conditions this statistical time lag is very short. The formative time lag that follows is dependent on the voltage, being shorter the higher the voltage or field. This is primarily responsible for the turnup shown. Figure 12.1 is a qualitative comparison; Fig. 12.2 is included to give quantitative data for a specific rod gap. We might suppose that spread can be expected in the volt/time data from rod gaps, and this is surely the case. Rather than plot a curve, we should perhaps show a band. Where a line is drawn, it is usually the upper boundary of this band, for reasons we shall see in a moment.

Reverting to the requirement for a protective device stated at the beginning of this chapter—the protector must maintain the voltage across the device protected at a level below its withstand capability at all times—it is apparent that there must be coordination between the volt-time characteristics of the protecting and protected equipments. Specifically, the volt-time curve of the protective device must lie below the corresponding characteristic for the insulation system being protected, for all conditions. If these characteristics are bands rather than lines, then the highest value the protector can clamp must lie below the lowest value at which the insulation it is protecting can fail. Referring to Fig. 12.1, let us suppose that curve *a* represents the volt-time characteristic for the most vulnerable piece of insulation in a transformer, which the rod gap is striving to protect. The condition just mentioned is satisfied only by points to the right of *A*. Thus, if a surge should impinge on the insulation in such a manner as to cause the voltage to rise at a rate faster than the slope of the line *OA*, the insulation may well fail before the rod gap flashed over. This region of vulnerability could be shortened by reducing the rod gap setting, which would move curve *b* downward. The objective would then be achieved at the expense of more rod gap flashovers, which could be quite undesirable. This matter of coordination is important and basic to the understanding of overvoltage protection. It will be referred to many times in this chapter in the context just described and also as it relates to coordination between several protective devices in a system.

The next important idea regarding protection can perhaps best be described as impedance division. Consider the rod gap again; the thought behind its use as a protective device is as follows. Under normal circumstances, the open

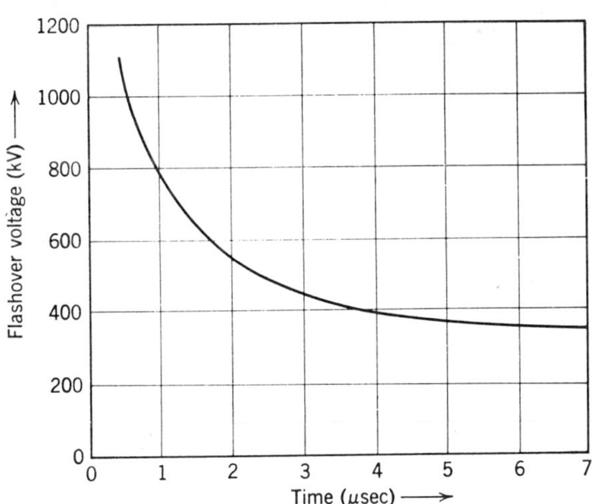

Fig. 12.2. Sparkover characteristic of 20-in. rod gap. Positive polarity.

gap represents a very high impedance, one might almost say an infinite impedance. When the gap flashes over, it switches to a low impedance mode, for the voltage of the arc that forms is probably a few hundred volts only, or a few thousand volts for a long gap on a high-voltage system. The voltage is low because the arc impedance is low compared with the impedance of the system, which is the source of the surge. Since the gap is in parallel with the equipment to be protected, it maintains a low voltage across that equipment also. Thus we note an effect whereby the surge voltage is divided between the system impedance and the protector impedance. This can be placed on a more rigorous basis by invoking Thévenin's theorem. One form of this theorem states that when a switch is closed in a circuit, the current that flows can be determined by dividing the voltage across the switch prior to closing, by the impedance seen looking into the circuit at the open switch contacts. In Fig. 12.3, suppose that Z_1 is the impedance of a system on which a surge is being generated and that Z_2 is a piece of equipment connected to the system. This latter is paralleled by a surge suppressing device of impedance Z_p, which is switched into the circuit when S is closed; S may be a gap of some kind that the surge causes to flash over. According to Thévenin, the current through S, when it is closed, is given by

$$\frac{V}{Z_p + Z_1Z_2/(Z_1 + Z_2)} = \frac{Z_1 + Z_2}{Z_pZ_1 + Z_pZ_2 + Z_1Z_2} V \qquad (12.1.1)$$

where V is the voltage across S prior to closing. Looking into the circuit at the switch we see the impedances Z_1 and Z_2 in parallel, and they in turn are in series with Z_p. The voltage V_1 across Z_p can be obtained simply from Eq. 12.1.1:

$$V_1 = \frac{Z_p(Z_1 + Z_2)}{Z_pZ_1 + Z_pZ_2 + Z_1Z_2} V \qquad (12.1.2)$$

For the surge suppressor to do a satisfactory job of protecting the equipment represented by the impedance Z_2, neither the voltage V, which it sees initially

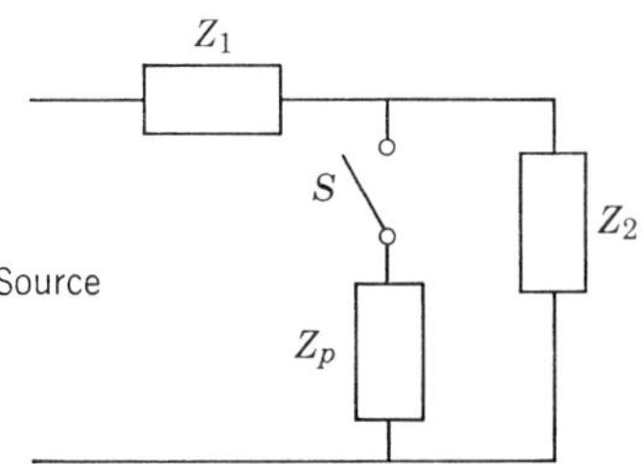

Fig. 12.3. Impedance dividing effect of protector and system. Z_1 = source impedance; Z_2 = equipment impedance; Z_p = protector impedance.

at the time S closes, nor the voltage V_1, given by Eq. 12.1.2, should exceed the voltage that the equipment is capable of sustaining. One would suppose that if V is satisfactory, V_1 would also be satisfactory since $V_1 < V$. Actually, this is not necessarily true; we have oversimplified the problem, as an example will show.

Suppose that Z_1 represents the surge impedance of a transmission line along which the surge approaches, and that Z_2 is the surge impedance of a transformer at its termination. Let us also suppose that the surge voltage peak, V_{peak}, is well in excess of the electrical strength of the transformer insulation. Without any protection this would be particularly serious, for, as was shown in Section 9.3, when $Z_2 \gg Z_1$, as would be the case here, the surge would almost double on reaching the transformer terminals. But in the situation before us, we have a suppressor that sparks over at some safe voltage V, well below V_p. When this occurs, the voltage will instantaneously drop to the value V_1 of Eq. 12.1.2. Subsequently, however, as the surge proceeds, the effective driving voltage will increase, and so therefore will the voltage across the transformer and suppressor.

Sometimes it is more appropriate to consider the surge as a source of current. Suppose, for example, that lightning strikes the transmission line injecting a current I into the circuit. In Section 10.3 it was explained how this develops traveling waves of voltage of amplitude $IZ_1/2$, which spread from the point of intrusion in both directions along the line, half the current flowing with each wave. When the suppressing device flashes over as a consequence of the voltage, we may think of the associated current as being partially diverted into the device, and thereby creating a voltage across it. The maximum voltage would be generated in this way if all of the current were so diverted. This condition would be approached if the line terminating impedance had a particularly high value. In this case the maximum voltage would be $IZ_p/2$. It is this voltage, generated by the follow current, that must not exceed the safe value set for the plant being protected.

The point we wish to make here is that the impedance of the protective device and the impedance of the system as well as the value of the surge itself, all play an important role in determining how effective the suppressor shall be. This topic has been expanded by Sakshaug and Hedman (1).

The third important factor regarding protective devices relates to their potential for storing and/or dissipating energy. When current is diverted into a surge suppressor of whatever kind, and voltage is generated across it, it is at the time absorbing energy. The amount of energy involved depends upon the magnitude and duration of the surge. The suppressor must be capable of handling this energy by storing or dissipating it, without damage to itself; it is not sufficient that it hold the voltage down to an acceptable level.

12.2 Lightning Arresters and Surge Suppressors

The rod gap was mentioned in the last section as a form of overvoltage protection. Such gaps have the virtue of being very cheap, but they have several disadvantages, the most important of which is that when they flash over they throw a fault on the circuit. It is true that this can be removed by opening a circuit breaker, and that power can be restored by reclosing the breaker after the rod gap has deionized. Certainly this is preferrable to damaging a costly piece of equipment and perhaps sustaining a lengthy outage. But a device that can effect the voltage limiting without creating a fault is obviously more attractive. The nonlinear resistor is such a device.

These resistors have the property that their resistance diminishes sharply as the voltage across them increases. This characteristic is usually expressed in the following way:

$$I = kV^n \tag{12.2.1}$$

The "constant" n is typically in the range $2 < n < 6$, which gives a measure of the nonlinear relationship between the current and the voltage. For a given material k is a geometrical factor, determined by the dimensions of the resistor. The material used for these resistors is marketed under a number of trade names, but each is basically silicon carbide. It is formed into different shapes such as rods and disks to arrive at a range of resistance values. Figure 12.4 shows typical V/I curves for such resistors. It will be seen that n is not strictly constant.

Nonlinear resistors of this kind are applied in circuits at all voltage levels, sometimes as minute devices to protect relay coils, sometimes in large chunks, under oil, across the windings of power transformers. The principle is the same wherever they are used. They are connected across the apparatus to be protected and so experience the system voltage under normal operating conditions. Their resistance is such that this results in a minimal power dissipation. On the incidence of a surge of voltage, the resistance falls rapidly as the voltage rises, thereby diverting much of the current and energy of the surge into the suppressor. All the factors mentioned in Section 12.1 regarding coordination, impedance, and energy dissipation are relevant. With the highest anticipated surge current flowing through the suppressor, the voltage must not exceed what the protected apparatus can support. As we have seen, this requires a knowledge of the surge source impedance. Energy dissipating capability depends upon the bulk of resistor element, since the energy is usually put in so quickly that there is little opportunity for heat to be transferred elsewhere until the surge has passed. It is apparent that protectors of this kind must be selected with care, after studying the circuit involved. It is also clear that circumstances will arise in which it is impossible to obtain the desired protection by this approach. For example, if the ceiling voltage to

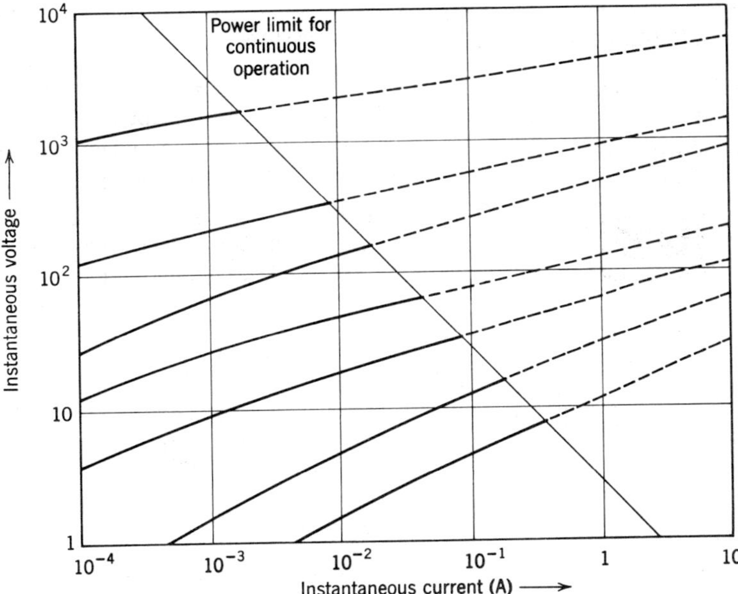

Fig. 12.4. Typical curves for nonlinear resistance material.

which surges must be limited is even three or four times the normal system voltage, it may be found that values of the resistance at system voltage would be so low they would make the steady-state losses prohibitive. The lower the permitted ceiling voltage, the more acute will this problem become.

The problem is overcome by suppressors of the next degree of complexity. These utilize the nonlinear resistor elements as before, but they have a gap or gaps in series with them. In this way the resistor is isolated from the circuit under normal conditions and is introduced when a surge appears by the flashover of the gap(s). With this principle it is possible to design the resistor element from the point of view of energy dissipation and voltage-limiting under surge conditions, without regard to steady-state power dissipation. However, a new consideration emerges; when the suppressor operates and an arc is established in the gap, this arc must be quenched when the surge is passed or the resistor will be destroyed by the current that would continue to flow once conditions are restored to normal. If this was not the case, there would be no need for the gap. To meet this new requirement various forms of arc control and quenching elements are introduced into the suppressor. We are now speaking of "lightning arresters." Lewis (2) provides a good general reference on this subject.

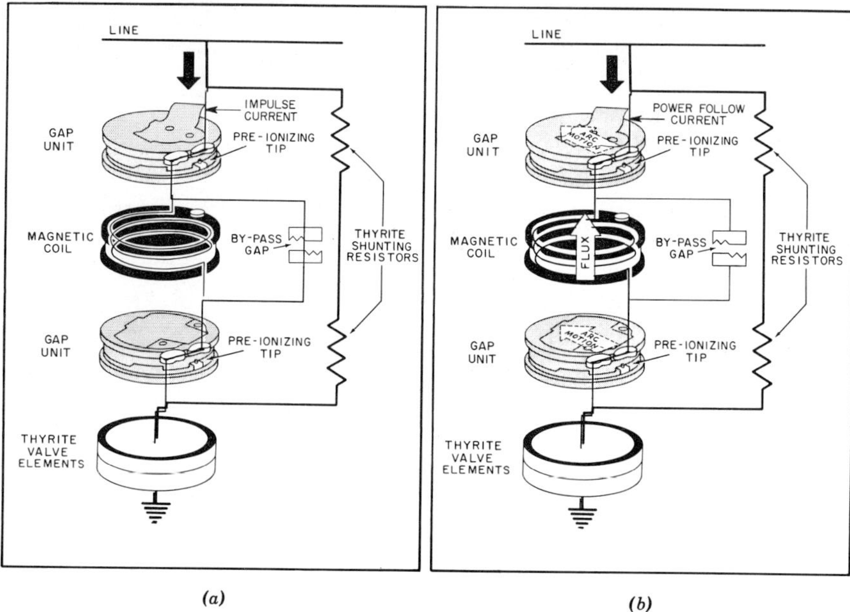

Fig. 12.5. Schematic diagram of valve type arrester showing path of (*a*) surge current. (*b*) Follow current. (Courtesy of the General Electric Co.)

Arresters vary in sophistication depending upon their voltage class and duty but they generally comprise gap units, coil units, and valve elements of the nonlinear resistance material. These are stacked in series and hermetically sealed in a porcelain housing. The principle of their operation can be understood by referring to Fig. 12.5, which shows schematically the circuit arrangement of the elements of an intermediate class of lightning arrester. The components shown in this figure comprise the requirement for a 6-kV arrester, that is, it will coordinate with the characteristics of power equipment typically found in a 6-kV circuit. For higher voltages, other similar sets of elements are connected in series. A station-type arrester for 96 kV is shown in Fig. 12.6.

The operation proceeds as follows. When a transient voltage, such as a lightning surge, reaches the arrester terminals, it causes the two gap units to flash over, thereby creating a path for the surge current through the coil and Thyrite* valve element. In the example shown, breakdown of the gaps is made more consistent by the inclusion of a "preionizing tip," which maintains a higher-than-normal number of electrons in the gap, ready to initiate a

* Proprietary name for the nonlinear resistance material marketed by the General Electric Company.

breakdown when an overvoltage appears. The surge current quickly develops across the coil a voltage sufficient to cause the bypass gap to flash over. This removes the coil from the circuit and leaves only the impedance of the valve element and connections in the circuit. During this phase the voltage developed is essentially the *IR* drop in the resistance element. Conditions at this time are those shown in Fig. 12.5*a*. When power frequency conditions are restored, following the passage of the surge current, the impedance of the coil is much lower. This causes the arc in the bypass gap to become unstable and extinguish. The current is transferred to the coil. This condition is shown in Fig. 12.5*b*. The magnetic field created by the follow current in the coil reacts upon this current in the arcs of the gap assemblies, causing them to be driven

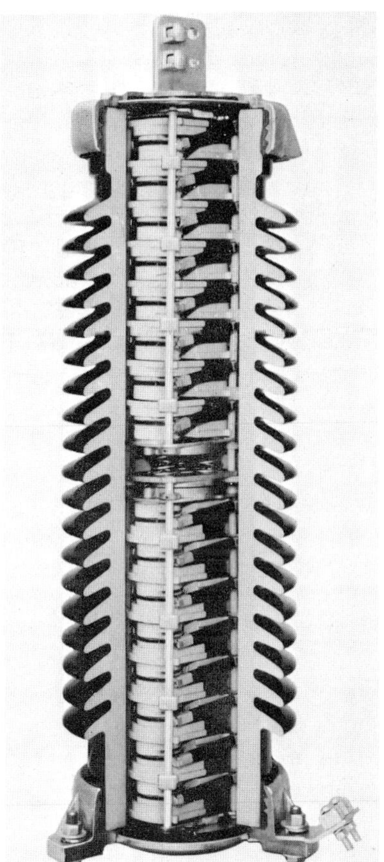

Fig. 12.6. Cutaway of 96-kV station type lighting arrester. (Courtesy of the General Electric Co.)

into arc quenching chambers, which are an integral part of the gap units. Arc extinction is brought about at the first current zero by elongating and cooling the arc, the arrester returns to its quiescent state and the surge voltage has been successfully limited.

As shown in Fig. 12.5, the main gaps are shunted by other nonlinear resistance elements that provide automatic self-regulation of the power frequency voltage across each each set of 6-kV gap elements. They also provide similar self-regulation of power voltage across the arrester gaps or complete units, throughout the line-to-ground stack, and reduce any electrical effect of external surface deposits or stray fields. This construction makes sparkover voltage practically independent of arrester location or type of arrester connection (whether the line terminal is connected to the top of the arrester with ground connected to the bottom, or vice versa).

The efficient performance characteristics of this type of arrester arc is demonstrated in Fig. 12.7 by cathode ray oscillograms of design tests conducted in accordance with IEC and USA (3) standards for lightning arresters. These oscillograms show repeated volt-time impulse sparkovers of intermediate class arresters rated at 12-kV maximum system voltage, using impulse waves of both positive and negative polarities. (With the numerous impulses applied to the arrester, starting from zero at slightly different points on the oscillograms, the zero of the time scale does not coincide with the beginning of any particular impulse, but does enable scaling the time to arrester sparkover.) From these similar data for arresters of other voltage classes, it is possible to construct volt-time characteristics for these devices. Examples are shown in Fig. 12.8. The "turnup" described in Section 12.1 is evident.

The performance of a 36-kV arrester of this type, discharging a 5000-A surge current of 10×20 μsec form, is shown in Fig. 12.9, which includes the voltage and current waves, together with 100-kHz timing waves. The initial spike of voltage indicates where flashover occurred. The voltage drops abruptly at this instant, but then rises again as current builds up. As explained in Section 12.1, it is essential that the sparkover voltage and the subsequent voltage generated by the follow current be well below the insulation strength of the equipment being protected.

A summary of protective characteristics of station-type arresters is given in Table 12.1. The "front-of-wave" data in the second column indicate the voltage at which flashover will occur with a very fast-rising voltage surge (3). The next column indicates the flashover voltage for the standard $1\frac{1}{2} \times 40$ μsec wave (that is a surge which reaches its peak in approximately $1\frac{1}{2}$ μsec and declines to one-half of this value in 40 μsec, more carefully defined in Chapter 17) sparkover occurring near the peak. The numbers in these columns, together with the 60-Hz data in the fifth column, give a measure

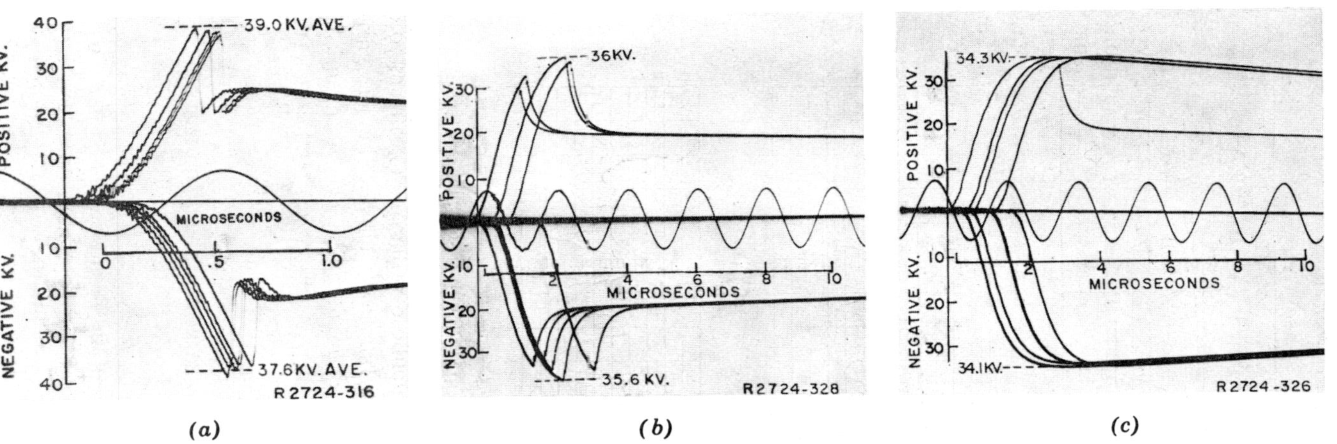

(a) (b) (c)

Fig. 12.7. Cathode ray oscillograms of repeated volt-time impulse sparkovers of 12 kV intermediate class arrester. (*a*) Average time to sparkover: 0.4 μsec. (*b*) Average time to sparkover: 1.8 μsec. (*c*) Repeated application of predominantly $1\frac{1}{2} \times 40$ μsec impulses. (Courtesy of the General Electric Co.)

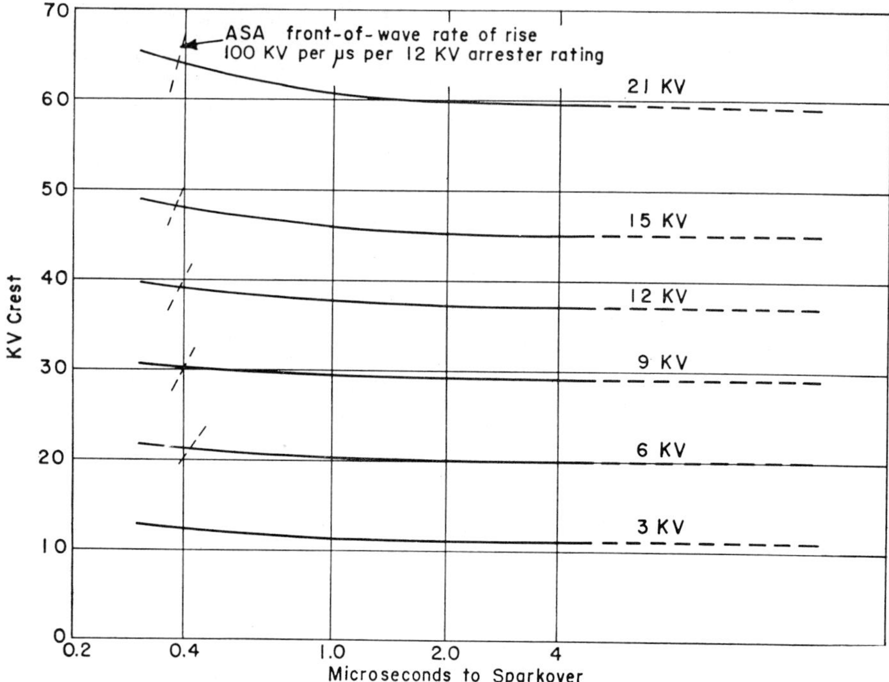

Fig. 12.8. Average impulse volt-time sparkover curves for intermediate class arresters. (Courtesy of the General Electric Co.)

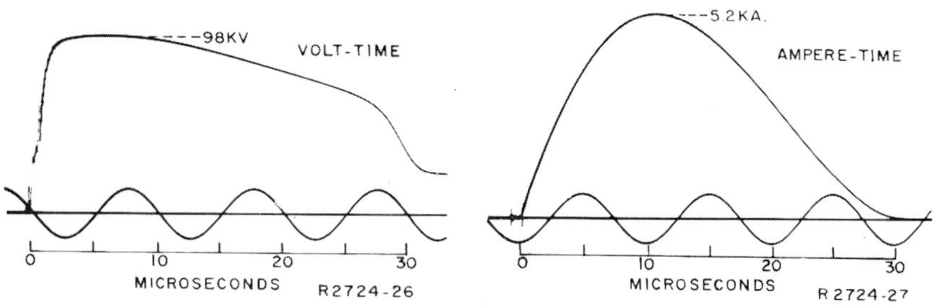

Fig. 12.9. Discharge voltage and current for 36 kV intermediate class arrester; current 5000 A, 8 × 20 μsec wave. (Courtesy of the General Electric Co.)

Table 12.1 Typical Protective Characteristics of Station-Type Lightning Arresters

Arrester Rating kV RMS	Maximum ASA Front of-wave Sparkover kV Crest	Maximum $1\frac{1}{2} \times 40$ Sparkover $1\frac{1}{2} \pm 0.2$ μsec Range kV Crest	Maximum Switching Surge Sparkover kV Crest	Maximum 60-Hz Sparkover kV RMS	Maximum Discharge Voltage (in Crest KV) at Indicated Impulse Current, 8×20 μsec					
					1.5 kA	3.0 kA	5.0 kA	10.0 kA	20.0 kA	40.0 kA
3	12	12	12	5.3	5.0	5.8	6.4	7.3	8.3	10.2
4.5	16	15	15	7.9	7.4	8.7	9.5	10.8	12.3	15.1
6	20	18	18	10.5	9.8	11.5	12.6	14.3	16.3	19.9
7.5	25	22	23	13.2	12.2	14.3	15.7	17.7	20.3	24.8
9	30	25	27	15.8	14.6	17.1	18.8	21.2	24.3	29.6
12	39	32	35	21.0	19.4	22.7	24.9	28.1	32.1	39.2
15	48	39	43	26.3	24.2	28.2	31.0	35.0	40.0	48.8
18	57	47	51	31.5	28.9	33.7	37.1	41.8	47.8	58.5
20	66	54	59	36.8	33.7	39.3	43.2	48.7	55.5	68.0
25	76	61	67	42.0	38.4	44.8	49.2	55.5	63.5	77.5
30	95	75	84	52.5	47.8	56.0	61.5	69.5	79.0	96.5
37	113	90	110	63.0	57.5	67.0	73.5	83.0	94.5	115
40	123	97	108	68.3	62.5	72.5	79.5	89.5	102	125
50	151	118	132	84.0	76	89.0	97.5	110	125	153
60	180	136	142	99	95	111	122	137	156	190
73	213	166	170	119	114	133	146	164	187	227
78	231	183	184	129	123	144	158	178	202	246
84	247	198	198	139	133	155	170	191	217	265
90	267	214	213	149	142	166	182	204	232	283
96	280	231	227	159	151	177	194	218	248	302
108	315	262	253	178	170	198	218	245	278	339
120	347	294	284	198	188	220	241	272	309	376
132	380	320	312	218	207	241	263	294	333	402
144	413	350	340	238	226	262	287	321	363	439
168	497	396	397	277	263	305	334	374	422	510
180	538	430	425	297	281	327	358	400	452	550
192	580	460	453	317	300	349	382	427	482	585
228	702	547	539	376	355	413	452	510	575	695
240	743	575	567	396	374	435	476	535	605	730
258	805	623	609	426	402	467	515	575	650	785
264	825	640	624	436	411	478	525	585	665	800
276	876	672	650	455	429	500	550	615	690	835
288	907	706	680	475	448	525	570	640	720	875
300	948	738	709	495	467	545	595	665	750	910
312	988	771	737	515	485	565	620	690	780	945

Courtesy of the General Electric Co.

of the turnup on the volt-time curve. The last six columns are self explanatory. They show the measure of protection available against high-surge follow currents.

Arresters of this type are frequently called upon to discharge transmission lines. Long lines store a considerable energy in the electric field associated with their capacitance, which the arrester must be able to absorb when it operates. How much line a station type arrester can discharge is given in Table 12.2. This chart assumes that the line is charged by a switching surge to 2.7 times the nominal system line-to-ground voltage. Lower levels of charge will allow longer lines to be discharged. References (4) to (7) give some helpful information on arresters and their application.

The usefulness of any protective device can be completely vitiated by improper application. Several important points should be observed in applying arresters. It is a good general rule that all protective devices be located as close as possible to the equipment they are to protect. This minimizes the chance of surges entering the circuit between the protecting and protected devices. But it also has another advantage in many instances. If a steep-fronted surge approaches a transformer that is protected by an arrester, the arrester will spark over when its protective level is reached. In doing so, it will let through a spike of voltage with a crest equal in magnitude to the sparkover voltage, which will travel on and impinge on the transformer terminals. This is shown in Figs. 12.10*a* and 12.10*b*. At the same time a

Table 12.2 Miles of Single Circuit Transmission Line That Can Be Successfully Discharged by Alugard* Arrester

Normal Circuit Voltage (kV)	Arrester Rating (kV)	Miles	Line Surge Impedance (ohms)
115	96	>300	450
115	120	>300	450
138	120	>300	450
138	144	>300	450
161	144	>300	400
161	168	>300	400
230	180	>300	400
230	192	>300	400
345	258	200	325
345	264	205	325
345	276	215	325
345	300	235	325
345	312	240	325

* General Electric Co. registered trade name.

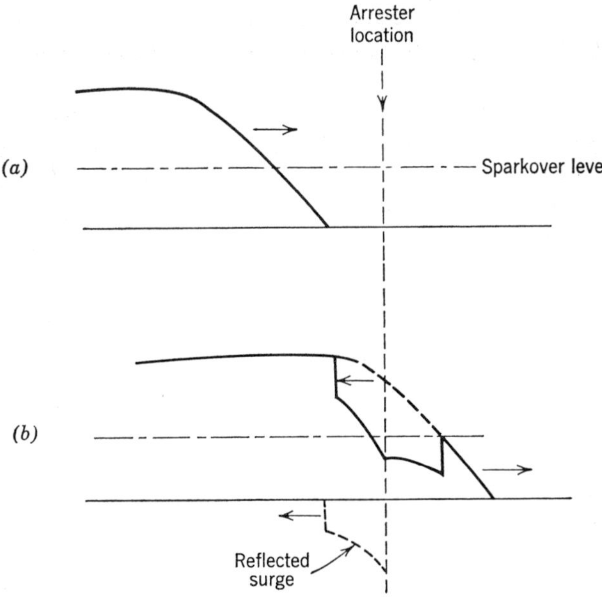

Fig. 12.10. (*a*) Surge traveling along a line approaching an arrester. (*b*) Refracted and reflected surges caused by arrester operation.

reflected wave will travel back down the line modifying the incident surge. The transformer will present a comparatively high impedance to the surge so that, as shown in Chapter 9, the reflected voltage wave from the termination will add to the incident surge. This results in the terminal voltage approaching twice the incident voltage, that is, the transformer may experience a surge twice as high as that admitted by the arrester. If the arrester is right at the transformer terminal, this could not occur, but where separation exists between the transformer and the arrester, it will persist until the reflected wave has returned to the arrester and a second reflected wave has propagated back from the arrester to the transformer.

The only place where some separation may be introduced between the arrester and the protected equipment is where a surge capacitor is applied in addition to the arrester, and the capacitor is located at the equipment terminals. Capacitors are sometimes used in conjunction with arresters for the purpose of "sloping" the front of the surge. We saw in Chapter 11 how steep-fronted voltage surges distribute in a very nonuniform manner across windings. When the winding is shunted by a capacitor, the voltage can rise across the winding only as quickly as the capacitor can be charged through the line or bus on which the surge enters. Thus the presence of the capacitor can have a salutary effect on the voltage distribution. When an arrester sparks over it does apply a negative step of voltage to the protected equipment, the amplitude being the difference between the sparkover voltage and the voltage generated by the follow current (see Fig. 12.7, for example). The addition of a capacitor at the winding terminals, with some separation between it and the arrester, will tend to soften this step. Where a number of motors are connected at points along a bus, a capacitor at each motor and one arrester on the bus are effective and at the same time economical in the use of protective equipment.

Another important point of application for both arresters and capacitors concerns the connections to line and ground. These should be as short as possible; otherwise, their inductance will render the arrester or capacitor ineffective for a long enough time for the surge to damage equipment. The connection of a capacitor as shown in Fig. 12.11*b* is to be preferred to that in Fig. 12.11*a* in that there is less mutual inductive coupling between the parts of the circuit.

In selecting an arrester to coordinate with basic insulation level of a particular plant, other considerations must not be overlooked, most notably the requirement for the arrester to seal-off against the existing system voltage, once the surge has been diverted through the arrester. An extreme case in point is the ungrounded system. Under normal circumstances the neutral of such a system, although it is isolated from ground, remains close to ground potential, being held there by the balanced capacitance of the three phases. If a flashover should occur on one phase, and virtually ground that phase, the neutral will shift in potential and the unfaulted phases will have line voltage, rather than phase voltage, to ground. If a disturbance on one of these phases should cause an arrester there to spark over at this time, the arrester would have to clear against this higher voltage once the surge was passed. If it failed to do so, it would be quickly destroyed by the follow current. The follow current is not necessarily very great, but it persists. Thus

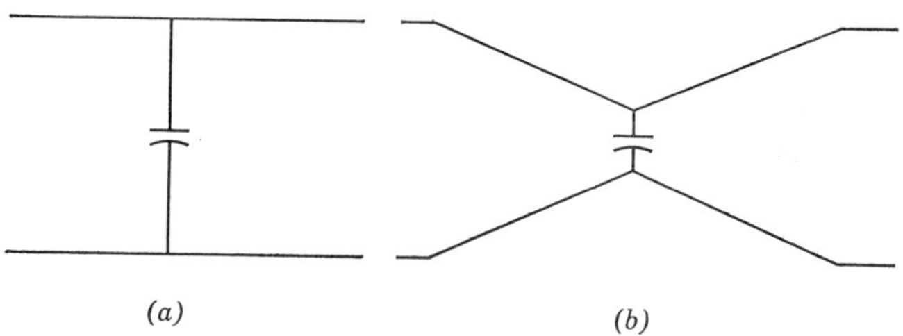

Fig. 12.11. Connections to surge capacitors should be short. (*b*) is preferred to (*a*).

arresters should be rated for the full line-to-line voltage, except where the neutral is effectively grounded. By definition (3), this means that neutral shift under ground fault conditions is slight.

The triggered vacuum gap (8) holds promise as a protective device. This is a fast-triggered gap, comprising a pair of main electrodes sealed into a vacuum enclosure. The trigger is a third electrode juxtaposed by one of the main electrodes, such that a modest voltage applied between the two will cause a discharge between them. This propagates plasma into the main gap, which in turn breaks down. Such a gap, in conjunction with a nonlinear resistor, would have the properties of an arrester, but because of the trigger, it would possess a new degree of freedom. Within limits the sparkover voltage would be controlled by the trigger, that is, the device could be switched into the circuit on command, or in response to a rising surge voltage, without waiting until the surge itself spilled over the main gap.

12.3 Surge Suppressors and Arresters for Direct-Current Circuits

Direct-current circuits are particularly difficult to protect against overvoltages, since any d-c surge suppressor must be equipped to interrupt a follow current which has no cyclical current zeros. This contrasts with the requirements for a-c arresters, where the follow current can be interrupted at the power frequency current zeros. A d-c surge suppressor must therefore incorporate what amounts to an interrupter for direct current.

A suppressor for relatively low voltage circuits has been described by Lee and Liao (9). It was developed primarily for the protection of silicon rectifier equipment for industrial drives and for electric transportation applications. These are situations in which the ratio of protective level to normal system voltage is unusually low, which places a further restriction on the protective device. A sketch of a suppressor of this type is shown in Fig. 12.12. It will be seen to have many of the same elements as a conventional arrester—Thyrite, arc quencher, and magnetic coils to drive the arc—but there are some important differences. When a surge is impressed across the terminals it appears across the main gap A and also across the much smaller auxiliary or trigger gap, B, to which it is coupled from the line through the large capacitor (1 μF) and the pulse transformer. The barium titanate distorts the field of gap B, causing it to fire almost instantaneously and creating, at the same time, enough local ionization to precipitate flashover of the main gap. The coils are so arranged as to drive the arc along the runners, which are sandwiched between arc quenching ceramic plates. Somewhere near the end of the runners the lengthened arc will extinguish because of its high arc voltage. But if the surge still persists, a reignition will occur at the narrow part of the gap, and a new arc will be swept around. This procedure repeats until the surge has been dissipated.

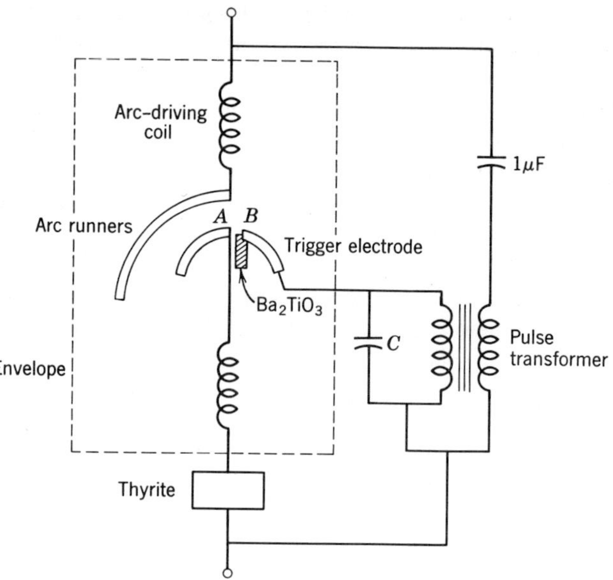

Fig. 12.12. Principal elements of a d-c hydrogen surge suppressor.

A very important factor in this device is the hydrogen filling. Hydrogen is an extremely good interrupting medium at low voltage because the lightness of its ions allows them to diffuse quickly, and also because its low atomic mass gives it a high thermal conductivity. With the help of the arc quencher that lengthens and cools the arc, and the Thyrite that keeps the value of the follow current down, it is possible for the hydrogen to interrupt this follow current though it has no natural current zero. Suppressors of this kind are able to dissipate several thousand joules at an operation and keep the voltage in the range of 1.6 to 2 times the system voltage.

The suppressor of Fig. 12.12 is not designed to handle lightning surges, but a variation of it (9) is made for this purpose. Of course, even these devices have a relatively low energy dissipating capability, but they can be successfully applied in coordination with protective gaps on the overhead wires of transportation systems, which are subject to lightning surges. They can protect terminal or car equipment in this way.

The triggered vacuum gap, mentioned in Section 12.2, appears to offer possibilities for protecting d-c systems. Although with direct current such gaps are not self-clearing, it would seem possible to commutate the follow current into a capacitor, perhaps by the use of a second triggered device.

Comparatively recent advances in converter technology have led to an expansion of the use of direct current for high-voltage transmission. Indications are that this trend will continue. Protection of HVDC lines against lightning presents considerable challenge because the problems already referred to are greatly magnified by the size of the system and the operating voltage. Again it is essential to seal against the system voltage without benefit of a cyclical current zero after the surge has been diverted. When the system voltage is 400 kV or more, the problem is a mean one. Energy levels in such systems are very high too. Typically, a 500-mile line carrying 1200 A has close to 1,000,000 joules stored in its magnetic field. Under fault conditions, the energy is considerably higher. Under certain conditions a significant fraction of this energy might have to be dissipated in an arrester.

A so-called "dual-cycle" arrester has been developed for use on HVDC systems. It has been described by Russell (10). The arrester consists of two columns of current-limiting gaps connected in parallel and the whole assembly connected in series with valve disks of the type developed for HVAC arresters and described in Section 12.2. When this arrester is connected to the transmission line, a voltage surge may cause one of the arrester gap columns to spark over. Since both columns are identical, either one may spark over first, depending upon many random factors. Initial current flow through the arrester will be limited only by the valve material and the resultant voltage drop across the arrester is established by design to meet the system and equipment needs. Within a few microseconds, the current-limiting gaps begin to rapidly build voltage, which reaches the sparkover of the other gap column in some 500 to 1000 μsec. The second column then sparks over, the voltage reverts to the valve drop, and column one goes out. This process is repeated until the transmission line is discharged to a voltage in the order of 80% of gap sparkover, at which point the arrester reseals.

12.4 Surge Capacitors and Surge Reactors

Very early in this book it was pointed out that one cannot suddenly change the voltage on a capacitor, because to do so would require the momentary flow of an infinite current. If a capacitor is therefore placed in parallel with a piece of equipment or a component, it will provide some measure of protection against voltage surges, for before the surge can impress a high voltage on the equipment or component, it must first charge the capacitor. This provides the sloping effect referred to in the Section 12.2. The effectiveness of the capacitor in holding down the voltage depends upon the energy conveyed by the surge and the capacitance of the capacitor. Such capacitors are particularly effective against voltage "spikes," that is, short duration voltage surges. The energy of these surges is such that it can be absorbed by the capacitor with little change of voltage. An example will make this clear.

Suppose a lightning surge rising at 200 kV/μsec enters a primary distribution circuit, which has an insulation flashover voltage of 180 kV. The surge will reach 180 kV in 0.9 μsec and then cause a flashover, limiting the peak voltage to this value. The remainder of the surge will be diverted to ground, while the sharp spike travels on down the line, much as is shown in Fig. 12.10. If the line surge impedance is 400 Ω, this voltage spike will be associated with a similar current surge with a peak of 450 A, which rises from zero at 500 A/μsec. The energy in this surge is

$$\int_0^t IV\,dt = \int_0^{9\times10^{-7}} 5 \times 10^8 \times 2 \times 10^{11} t^2\,dt$$

$$= 10^{20}\left[\frac{t^3}{3}\right]_0^{9\times10^{-7}}$$

$$= 24.3 \text{ joules}$$

Suppose the instantaneous voltage on the line prior to the appearance of the surge was 20 kV, and that a 0.25 μF surge capacitor on the line was charged to that value. The surge capacitor will have stored

$$\tfrac{1}{2}CV^2 = 1.25 \times 10^{-7} \times (2 \times 10^4)^2$$

$$= 50 \text{ joules}$$

A simple calculation indicates that the extra 24.3 joules can be accommodated in the capacitor with an increase in its voltage of only about 5 kV.

High-energy surges cannot be handled by capacitors. The required energy storage capability makes their physical size and cost prohibitive, as a calculation will show. Suppose the energy in a surge on a 250-V system amounts to 5000 joules, and it is desired to hold the voltage to 500 V, how big a capacitor would be needed?

$$\text{energy} = \tfrac{1}{2}CV^2 = \tfrac{1}{2}C(500^2 - 250^2) = 5000 \text{ joules}$$

whence

$$C = 53{,}000\ \mu\text{F}$$

Capacitors find their place in the protective scheme as absorbers of sharp spikes of surges, and as waveform modifiers to soften the front of surges. The combination of a capacitor and some other protective device, such as an arrester, can be very useful, since together they complement each other. Arresters have a significant turnup at short times in their volt-time curve, as discussed earlier in the chapter. This is of much less consequence if a capacitor is present, because this reduces the steepness of surge wavefronts, thereby eliminating the short time part of the volt-time curve from consideration. To illustrate this point, suppose that an arrester sparkover characteristic is reasonably flat after 1 μsec. If a capacitor is put in parallel with the arrester, such that even with the maximum surge conceivable, having a vertical wavefront, it is impossible to charge the capacitor, through the surge impedance of the connected line or bus, in excess of this sparkover value, in less than 1 μsec, then this combination will keep all surges down to the flat sparkover level. For example, suppose such a combination is used to protect a large 13.8-kV motor, which is a common and recommended practice, and suppose that the bus impedance is 100 Ω. It might be assessed that the highest voltage surge that could appear on this bus would be 120 kV; this figure would be determined from flashover characteristics of bus supports and so on. The arrester might typically have a sparkover of 39 kV. The surge current would be

$$I = \frac{V}{Z_0} = 1200 \text{ A}$$

$$\text{the rate of rise of capacitor voltage} = \frac{I}{C} = \frac{1200}{C} \text{ V/sec}$$

If the capacitor already is at peak voltage ($13.8/\sqrt{3} \times \sqrt{2} = 11$ kV) when the surge arrives, the voltage would not climb more than 28 kV in the first microsecond, if

$$\frac{1200}{C} \times 10^{-6} = 28000$$

or

$$C - 0.043 \; \mu\text{F}$$

Typical installations of this kind employ 0.25 μF surge capacitors.

This approach is particularly useful where semiconductor devices, such as silicon rectifier installations, are concerned, because the semiconductor components have very little turnup in their volt-time curve and therefore present a coordination problem. For reasons mentioned in Chapter 7, it is sometimes necessary to put a resistor in series with the capacitor in some applications in this area, to form an *RC* "snubber circuit." Any such resistor reduces the effectiveness of the capacitor in the context we have just discussed, since the surge current diverted into the capacitor will at once produce the *IR* drop of the series resistor. The situation is further aggravated if the resistor also possesses inductance. The purity of components for these applications, be they resistors or capacitors, cannot be overstressed.

Capacitors designed specifically for surge duty have minimal inductance and resistance. The so-called extended foil construction is used, with multiple connections to the foils. Electrolytic capacitors are not well suited for this task, but it is possible to capitalize on their cheapness and high capacitance-to-volume ratio by placing a small, high-quality surge capacitor in parallel with a bank of electrolytics. A useful scheme which allows the use of polarized electrolytic capacitors on a three-phase a-c circuit, and yet economizes in

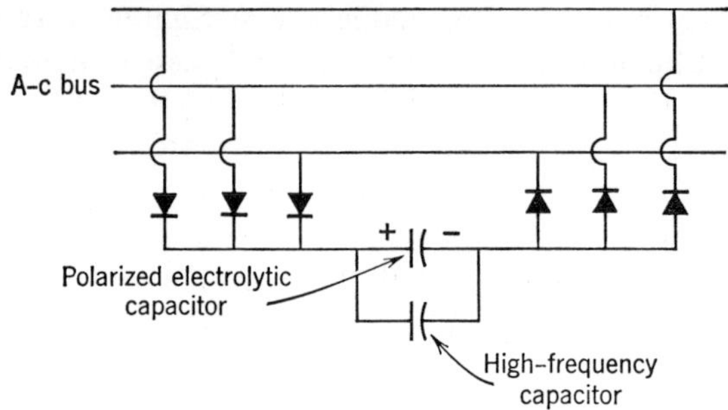

Fig. 12.13. Capacitive surge protection with a bridge of diodes.

their use, is shown in Fig. 12.13. If a surge appears between any pair of phases, regardless of its polarity, it will be communicated to the electrolytics with the polarity they can handle. It is advisable to shunt the capacitor with a high resistance to bleed off any excess charge acquired by them, which would otherwise remain trapped by the diodes.

Cable has a quite significant capacitance and can be used to advantage for surge protection. It was shown in Section 9.3 that when a surge reaches a transition point in a circuit where there is a change in surge impedance, the surge undergoes a change by the initiation of reflected and refracted waves from the transition point. More specifically, the refracted surge, which penetrates beyond the transition point, is related to the incident wave by Eq. 9.3.8:

$$\text{refracted wave} = V_3 = \left[\frac{2Z_2}{Z_1 + Z_2}\right] V_1$$

where V_1 is the incident surge or wave, and Z_1 and Z_2 are the surge impedances on either side of transition point. It is apparent that if line 1 is an overhead transmission line (Z_1 typically 400 Ω) and line 2 is a cable (Z_2 typically 50 Ω), only a fraction of the incident surge will penetrate the cable and travel to the terminal equipment. This presentation, however, is incomplete, for the surge that enters the cable will be reflected at the terminal equipment, return to the junction, and be reflected again. In fact it will travel up and down the cable many times, until it is attenuated by losses. During this time the voltage will build up at the terminal equipment. This will be demonstrated by an example.

Suppose that $Z_1 = 400\ \Omega$ and $Z_2 = 50\ \Omega$, and further suppose that the cable is 100 ft long and terminates at a transformer whose surge impedance is so high compared with that of the cable, or line, that it can be considered an open circuit. For the purpose of this example we assume that the surge has a vertical front and a flat top, that is a square wave of amplitude V_1. The wave that initially penetrates the cable will have an amplitude

$$V_3 = \frac{2 \times 50}{50 + 400} V_1 = 0.222 V_1$$

It will typically travel at 400 ft/μsec on the cable and will therefore reach the transformer terminals in 0.25 μsec. The reflected wave from this termination will be equal to the incident wave and will add to it, thus the arrival of the wave at the transformer will be signalled by the voltage there taking a jump of 0.444 V_1.

Then 0.25 μsec later the reflected wave will have returned to the junction point, where new reflected and refracted waves will be initiated. The reflected wave will be

$$\left[\frac{Z_1 - Z_2}{Z_1 + Z_2}\right] V_3 = 0.173 V_1$$

(This is derived from Eq. 9.3.7; Z_1 and Z_2 are interchanged because the wave is traveling on the cable towards the line.) This wave is also reflected from the transformer with sign unchanged and returns to the transition point after a round trip of 0.5 μsec to initiate another pair of waves. Sequential waves will be

$$\left[\frac{Z_1 - Z_2}{Z_1 + Z_2}\right]^2 V_3; \qquad \left[\frac{Z_1 - Z_2}{Z_1 + Z_2}\right]^3 V_3; \qquad \text{etc.}$$

Thus the voltage at the transformer terminal will rise in a staircase fashion, the steps being each 0.5 μsec long, but becoming progressively smaller. Letting

$$\left[\frac{Z_1 - Z_2}{Z_1 + Z_2}\right] = a$$

the transformer terminal voltage will be given by

$$2V_3[1 + a + a^2 + a^3 + \cdots a^n]$$

It will be found that this series converges on a value V_1/V_3, that is, the terminal voltage is asymptotic to $2V_1$. This is shown in Fig. 12.14*a*. This is the value it would have attained instantaneously, had the cable not been there, since the termination is effectively an open circuit. The value of the cable, therefore, is not in reducing the terminal voltage but in modifying its rate of rise as would a capacitor. Indeed, much the same result would be achieved with a capacitor having the same capacitance as the cable. In that case the voltage at the transformer would rise smoothly as a (1-exponential) curve, as shown in Fig. 9.13, rather than with the steps of Fig. 12.14*a*.

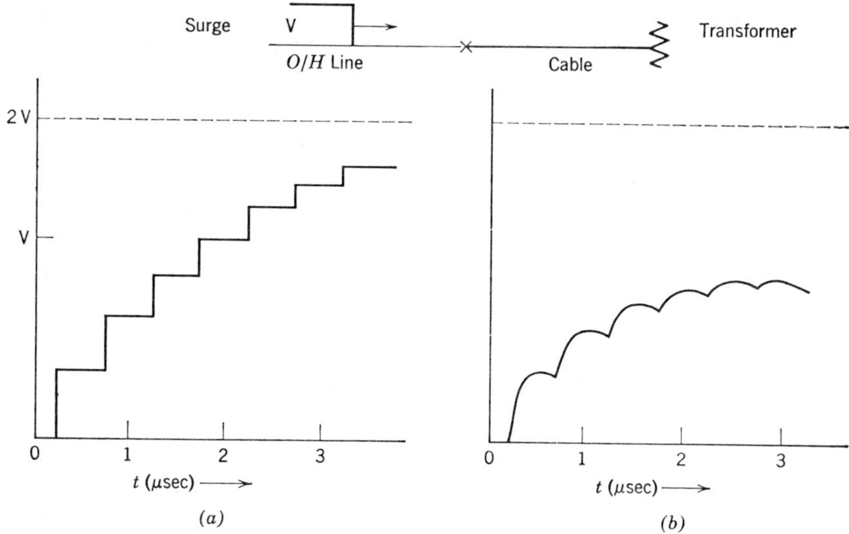

Fig. 12.14. Modification of a surge by a length of cable (voltage at transformer terminals). (*a*) Rectangular surge. (*b*) Double exponential surge.

An additional advantage will accrue for cable or capacitor if the surge has a finite front and tail, since the voltage will not have reached twice the peak value of incident wave before the surge declines. This is illustrated for the cable in Fig. 12.14*b*.

Reactors are sometimes used to protect against voltage surges, particularly within pieces of conversion equipment, in a manner that complements the use of surge capacitors. The capacitor parallels the equipment to be protected; the reactor is placed in series with the equipment. When a surge reaches such a series combination, the initial tendency is for the surge voltage to appear principally across the reactor, since this offers the highest impedance to the rapidly changing surge current. Put another way, the surge energy is initially absorbed in the magnetic field of current in the reactor. Like capacitors, the ability of reactors to store energy is limited, but again like capacitors they can take care of low-energy surges, or act temporarily to protect a device, until such time as a gap can flashover or an arrester operate.

When voltage is impressed across a reactor the following equation applies:

$$V = -\frac{d\Phi}{dt}$$

where Φ is the magnetic flux linkage of the reactor. Rearranging and integrating,

$$\int V\,dt = \Phi_2 - \Phi_1 \qquad (12.4.1)$$

The quantity on the left has the dimensions of volt seconds—how many volts the reactor can support, for how many seconds. To support the voltage the flux linkage must be changing and Eq. 12.4.1 indicates the change in flux linkage required for any desired number of volt seconds. This is particularly useful in determining the core cross section for reactors with a magnetic core. For example, suppose we wish to support a voltage of 20 kV for 5 μsec with a core material that saturates at 1.4 webers/meter^2 (or 14000 gauss). Let the cross section of the core be A (meter)^2 and let there be n turns on the reactor. Assuming all the flux links with all of the turns and that the core is initially demagnetized, the change in flux linkage to the point of saturation will be

$$\Phi_2 - \Phi_1 = \Phi_2 = 1.4 \times An$$

$$\text{the volt seconds called for} = 2 \times 10^4 \times 5 \times 10^{-6} = 0.1$$

whence

$$An = \frac{0.1}{1.4} = 0.071 \text{ m}^2 \text{ turns}$$

Thus a reactor with 20 turns would require a core cross section of 3.55×10^{-3} m^2 or 35.5 cm^2.

There are obviously combinations of A and n that would satisfy these particular requirements. The choice in any practical circumstances will be principally an economic one determined by the cost of the core versus the cost of the coil. The design is constrained by two other relationships,

$$\int \mathbf{H}\, dl = I$$

and the **B**/**H** curve of the core material.

We have been describing here a saturable reactor, one which presents a high inductance until it saturates, but thereafter presents the inductance of a comparable air cored device. They can absorb voltage, and therefore delay its appearance elsewhere, until saturation is reached.

The discussion so far has been idealized in one or two important ways that we should be aware of. It was pointed out in Section 8.5 that the penetration of flux into a conducting solid is opposed by eddy currents. The net flux is always the consequence of the applied ampere turns and those produced by these eddy currents. To minimize the effects of the eddy currents the core must be finely laminated, especially where a rapid change of flux is called for as with steep-fronted surges. If this is not done, it will not be possible to take advantage of the presence of the core. In extreme cases, powder cores or ferrites must be used.

Another important fact is that the core and winding structure of the reactor or choke represents a complicated system of capacitances, which effectively shunt the inductance. In the initial stages when the surge first reaches the equipment, these, in combination with similar capacitances in the apparatus being protected, are the determining parameters in the voltage distribution. Conditions are quite analogous to those discussed in Chapter 11. This points up the advantages to be gained by a combination of series reactor and shunt capacitor to form a surge filter in front of a piece of vulnerable equipment. Such combinations are used very effectively, especially in control circuits.

12.5 Transient Voltages and Grounding Practices

In this section we consider how good grounding practices contribute to more reliable and safer utility and industrial power systems. We are concerned with minimizing the damage to equipment and with false operations of equipment, under transient conditions, and also with the avoidance of shock hazard to personnel. Space precludes a very detailed discussion of grounding practices, although references presented (11) to (15) indicate where such data can be found. Instead the focus will be on the underlying principles.

Under steady-state, quiescent operating conditions in a power system, the integrity of equipment is preserved by adequate insulation, and safety of personnel is secured by keeping sufficient separation between people and high-voltage conductors. Preferably a grounded barrier in the form of a metal enclosure, wire fence, or the like, is placed between the circuit and any personnel. "Grounded" barrier is unambiguous; it refers to a physical equipotential surface, at the same potential as local ground and any other grounded objects around.

Under abnormal conditions, as when a fault occurs, this situation may no longer prevail; "ground" becomes a relative thing. To take a simple illustration, consider Fig. 12.15. Suppose a short circuit occurs across the conductors from the transformer, perhaps by a foreign body falling across them. This could be simulated by closing the switch in Fig. 12.15. We know that whatever

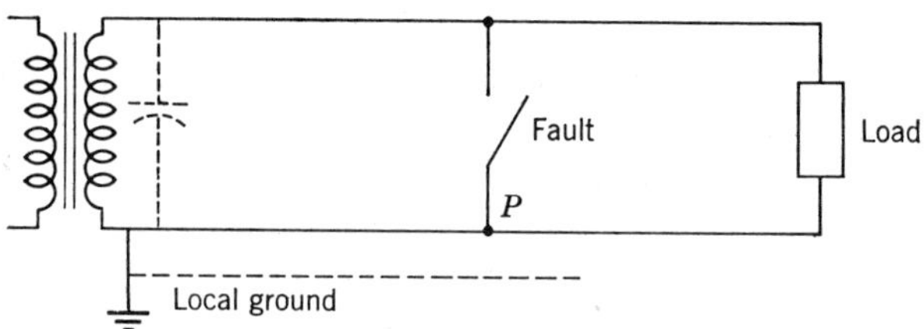

Fig. 12.15. Illustrating the shift of potential in a "grounded" conductor when a fault occurs.

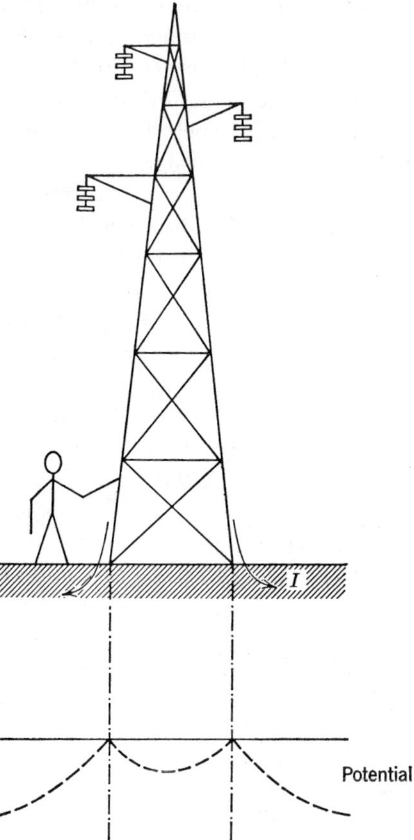

Fig. 12.16. Shock hazard when surge currents pass to ground through tower footings.

voltage existed across the switch before it was closed will be anihilated by the closing. But since the flux linkage with a circuit cannot change instantaneously, this voltage will reappear distributed around the circuit. Because of its capacitance, the transformer will tend to hold its voltage momentarily, so that considerable differences of potential will suddenly appear across the ends of short lengths of bus, as between point P and local ground. This becomes particularly disturbing when one of these conductors is part of the "ground" structure, as is often the case.

Another example is shown in Fig. 12.16. This relates to events occurring when a structure, such as a utility pole, becomes part of a circuit by a flashover from the line or perhaps by being itself struck by lightning. Some consideration has already been given to this in Section 10.4. It is apparent that a high surge current flowing down the structure will develop a considerable resistive volt drop, which an individual in the position shown will pick up. In addition, and this may be more important, the loop formed by the person and that part of the structure and ground that he spans, is mutually coupled with the circuit in which the surge current is flowing. Thus, as this current is building up, a considerable voltage, $M(dI/dt)$, can be induced in the loop. Indeed, the man may be involved without actually touching the structure, by simply standing beside it, for sufficient voltage could be induced in the incomplete loop to cause a flashover from the man to the structure. Again, the surge current will presumably flow into the ground and spread radially away from the pole. If high ground resistance should create a high radial voltage gradient, a person would be shocked by simply spanning a stride on such a gradient with his two feet. The unfortunate propensity of cattle for standing under trees during thunderstorms has had some sad consequences. But it has been observed on some occasions that those standing radially have been killed, while those standing circumferentially, and therefore having their feet approximately on an equipotential, have survived, when the tree was struck by lightning.

The preceding examples illustrate that under abnormal conditions, when transient currents, sometimes of high magnitude, take unusual paths, high transient voltages can appear between points that are normally at or close to the same potential. This can endanger personnel and, by causing further flashovers, can damage equipment, thereby adding considerably to the problem of the initial disturbance. Some bouncing around of potentials is unavoidable, even during normal switching operations on a system. But the adverse consequences of this and of abnormal voltage disturbances can be greatly reduced by correct design.

The object of this design is to reduce transient potential gradients as far as possible. It is of little consequence if even a safety guardrail around an equipment flies up to a high potential with respect to some remote ground, provided all other adjacent structures and circuits follow suit. This is achieved in a substation, vault, or where have you, by having a number of sturdy copper strips run around the equipment and exposed metal structural members of the room, to which these items are firmly bonded at frequent intervals. The periphery of the building or room should be circled by such a ground conductor, which would attach to structural steel work, water pipes, and the like. This would be connected within the room to cubicles, machine frames, transformer tanks, bus ways, and so on. This copper harness should itself be attached to the ground mat of the building. The ground mat may comprise the steel reinforcing in the concrete of the floor, or a mesh of carefully bonded copper cables specifically put down for the purpose. Such an arrangement provides a low impedance path for ground fault currents and therefore minimizes the potential difference that can exist between any pair of points that a man could reach. Grounding schemes of this kind are described in reference (11).

In an industrial plant, ground fault currents are usually confined to such networks and to cable sheaths. On utility systems true ground is often involved. As yet we have said nothing about connections from the ground mat to true ground, although such connections are clearly necessary. In a switchyard, the mesh just described will be buried in the ground. At its periphery it is frequently connected to metal rods driven into the ground, typically 5–10 ft long. The effectiveness of these rods is critically dependent on the resistivity of the soil. This varies widely with location and weather conditions, from a few thousand to hundreds of thousands of ohm centimeters. The presence of moisture greatly improves the conductivity, especially if certain minerals are also present. Dry sandy soils offer a high resistance, which in acute situations must be modified by adding salt. This eventually leaches away and must be replaced. It is also found that ground resistivity increases sharply when the ground freezes. The poorer the soil conditions, the closer the intervals at which ground rods must be placed. When a fault or surge current is dissipated to ground through a rod, the voltage gradient is highest adjacent to the rod, where the current density is highest. The closer together the rods are placed, the less will the local potential fluctuate. The subject of ground gradients is discussed in considerable detail in a 1958 AIEE Committee Report (12). The overall outcome of any ground mat design should be as low a resistance path for ground fault currents as is economically possible. Suggested maximum values are 1 Ω for a large station and 5 Ω for a small substation, where the fault current is likely to be lower. But it will be clear that these values can represent a significant part of the ground fault impedance, and therefore impress a significant fraction of the system voltage between station ground and true ground. In a survey of substation ground resistances (13) of a large number of U.S. utilities, it was found that only 80% of those reporting did in fact have ground resistance less than 5 Ω. This is indicated in Fig. 12.17, taken from the survey report.

Sometimes when we solve a problem, we do so only to have it reappear somewhere else, or in some other form. This is true in designing grounding systems. It is possible, for example, by the expedients described above, to have a site or building made secure from high potential gradients under the most adverse conditions, at the expense of introducing very high gradients in adjacent areas. The potential picture is of a flat-topped plateau on which the substation sits, but beyond its perimeter the gradient is very steep. Under these conditions, personnel within the station will be quite safe, but those close by outside would be imperiled. The solution here is to extend the

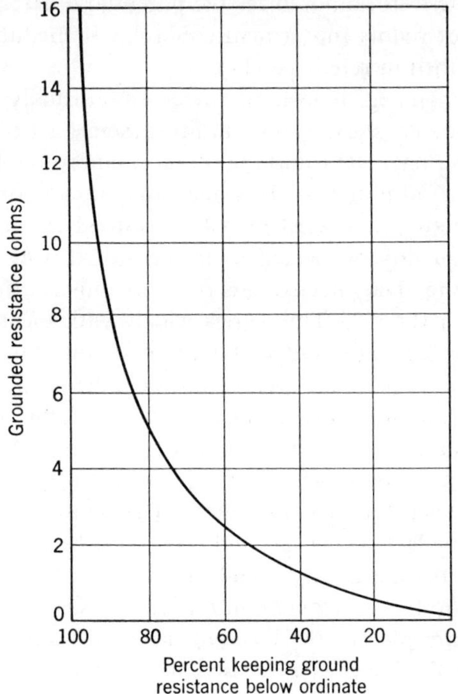

Fig. 12.17. Survey of a-c substation ground resistances.

grounding system beyond the perimeter of the station, tapering it to give an acceptable gradient.

Another way in which the benefits of a good grounding system can be vitiated is by bringing into the area, or taking from it, insulated grounds. This can be done inadvertently; for example, a water pipe coming into a building may have its closest good ground connection some distance from the building, so that when a disturbance causes building ground potential to change suddenly, a considerable voltage can appear between the water pipe and adjacent objects. This problem can be remedied by ensuring that the pipe is connected to the building ground mat, but it must then be recognized that the pipe has become part of that ground mat system.

In like manner, dangerous transient voltages can be taken out of an area by insulated conductors connected to ground in the area, which then pass to adjacent areas. The hazard is then in the adjacent areas. The most frequent offenders here are control and communication circuits. Sometimes it is not possible to connect such circuits directly to local grounds, in which case capacitors can be used to tie down the ends of the circuits against transient disturbances. However, this too may be unacceptable because of circulating currents that are introduced. It may then be necessary to install a fairly elaborate shielding scheme. These problems are discussed in Chapters 9 and 18.

Transmission line tower grounding has been discussed to some extent in Chapter 11. The emphasis with this type of structure is on the response of the tower and its ground footing to high surge currents due to lightning. If the surge enters on the ground wire at the top of the tower, the tower behaves like a rather complicated spur with a surge impedance. A wave travels down to ground and is reflected. In fact, waves will continue oscillating up and down the structure. A high ground impedance will do little to modify the initial surge; indeed it will reflect it with little modification, doubling the potential momentarily. In contrast, a low ground impedance will reduce the voltage on the structure and perhaps prevent back flashover to the phase conductors.

The subject of ground rod behavior under high surge current conditions has been studied and well documented by Bellaschi, Armington, and Snowden (14). They too observed that different soils offer quite different impedances to the flow of current. However, with high surge currents and high associated voltage gradients, it is observed that the resistance falls. Apparently, electrical breakdown of the soil occurs, which sometimes leads to its vitrification. Such vitrified curiosities are sometimes found on beaches and other sandy areas, where lightning has actually struck the beach. They are referred to as fulgurites.

Quite elaborate grounding methods must be applied where soil conditions are particularly adverse. This sometimes takes the form of "counterpoise," which is a form of continuous ground mat laid under the ground beneath the phase conductors.

12.6 Protection of Control Circuits

Following the old adage that prevention is better than cure, we consider first some principles for keeping transients out of control circuits. As mentioned, this is of vital importance for the correct operation of such circuits, aside from the need for their protection. Kotheimer (16), (17) has addressed himself to this problem, particularly as it affects relay equipment in high-voltage switchyards. These references are drawn on liberally in the following discussion.

The first point to recognize is that two general classes of electrical conductor systems coexist in every switching station:

1. The HV power circuits, including all of the HV buses and apparatus, primary circuits of instrument transformers, and all devices operating at high

potential such as line traps, arresters, and gaps. Also included are the station ground grid, all apparatus grounds, and *all bushing capacitances and stray capacitances to ground.*

2. Control circuits, including all potential and current transformer secondary circuits, all battery, d-c control, a-c auxiliary power, protective relaying, supervisory, alarm, and communication circuits.

During HV switching operations, high-magnitude transient currents circulate between the buses and the ground system through the various bushing capacitances and stray capacitances in the station. These currents are in close proximity to the various control circuits at the base of most HV apparatus and therefore favorably disposed to induce transients therein. The object of good design is to reduce the coupling between these circuits by layout and shielding.

Coupling can be electrostatic, electromagnetic, or conductive. The first two of these were discussed at some length in Sections 8.2 and 8.3. Conductive coupling exists when two circuits share a common conductor for part of their length. In the circuits being described, this is most likely to exist through sharing of common ground connections. The problem is best illustrated by specific examples.

One of the most prevalent sources of pickup is the switching of a section of EHV bus by an air break disconnect. Figure 12.18*a* shows a typical installation, with the area of interest circled. The physical arrangement of the apparatus can be understood from Fig. 12.18*b*. Consider the case when all the disconnect switches except *C* and *D* are open, the breakers are likewise open, and the transformer is being excited from the 161-kV side. Disconnect switch *C* is opened and a restriking arc occurs between its arms, making and breaking the charging current flowing into the series combination of the circuit breaker voltage dividing capacitance and the *CT* bushing capacitance. Reignitions and restrikes are to be expected until the disconnect reaches a considerable gap, which takes time to achieve since disconnects are slow-moving devices.

We saw in Section 5.4 that it is readily possible to have twice peak system voltage across the switch contacts when it reignites. It is possible to compute the inductance of the local discharge loop from its dimension and, knowing the capacitance involved, calculate the magnitude and frequency of the resulting current. An equivalent circuit which neglects all losses is shown in Fig. 12.18*c*. Here L_T, the inductance of the loop between the *CT*s, may be typically 120 μH. If the capacitance of each *CT* is 800 pF and the breaker is 125 pF,

$$C_N = \frac{C_A C_B}{C_A + C_B} = 108 \text{ pF}$$

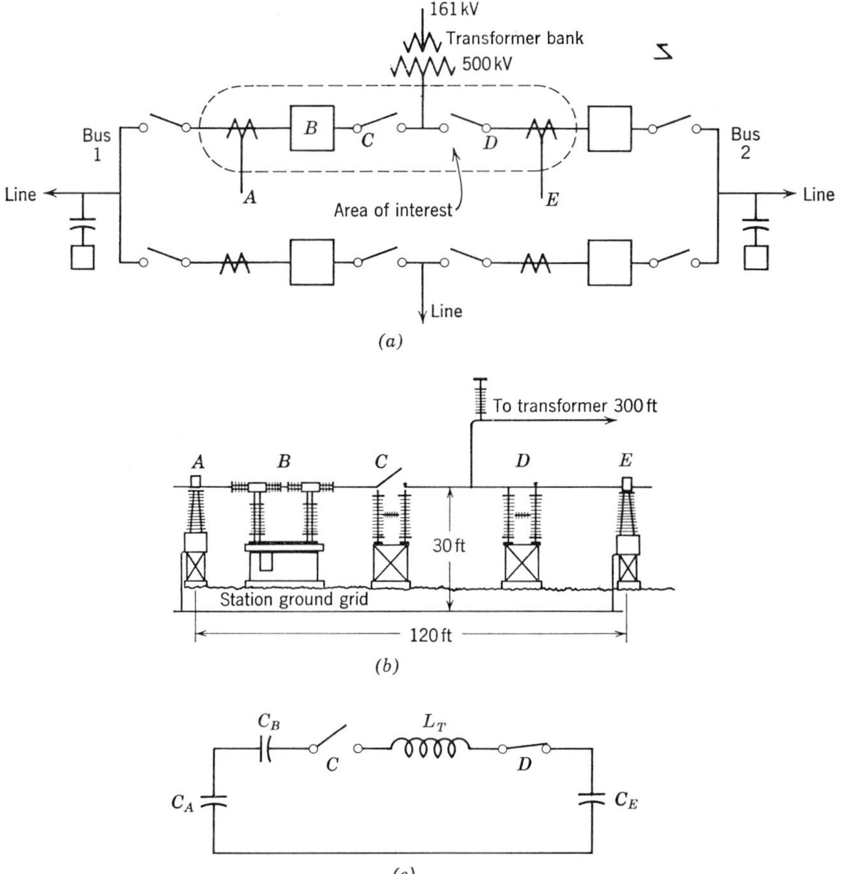

Fig. 12.18. (*a*) One-line diagram of a typical 500-kV switching station. (*b*) Elevation view of part of EHV station. (*c*) Equivalent circuit when disconnect *C* restrikes.

The effective capacitance involved in the circuit oscillation is

$$C_T = \frac{C_N C_E}{C_N + C_E} = 95 \text{ pF}$$

The surge impedance of the circuit will be

$$\left(\frac{L_T}{C_T}\right)^{1/2} = \left[\frac{125 \times 10^{-6}}{95 \times 10^{-12}}\right]^{1/2} = 1140\ \Omega$$

With an instantaneous voltage of 700 kV across the switch when reignition occurs, the current would be 620 A. The frequency is

$$f_0 = \frac{1}{2\pi(L_T C_T)^{1/2}} = 1.34\ \text{MHz}$$

These high-frequency currents flow not only in the buswork, but also in the *CT*s and their ground connections, and in the ground grid. This gives rise to capacitive and magnetic coupling to secondary and control wiring due to their close proximity to the ground conductors.

To obtain some idea of the magnitude of the induced voltages, consider Fig. 12.19*a*, which shows the high-frequency equivalent circuit of a high voltage *CT*. The capacitor C_1 represents the bushing capacitance (790 pF), C_2 represents the capacitance of the primary to the internal shield (about 50 pF), and C_3 is the distributed capacitance of the secondary winding to the shield, core and case (about 500 pF). The magnetic flux of the high frequency current in the ground conductor and ground grid links the loop formed by these conductors and the control circuit leads. The loop is completed by the capacitance C_3. By estimating some dimensions of this loop, it is possible to make a rough calculation of the voltage induced. Equation 8.3.10 is helpful

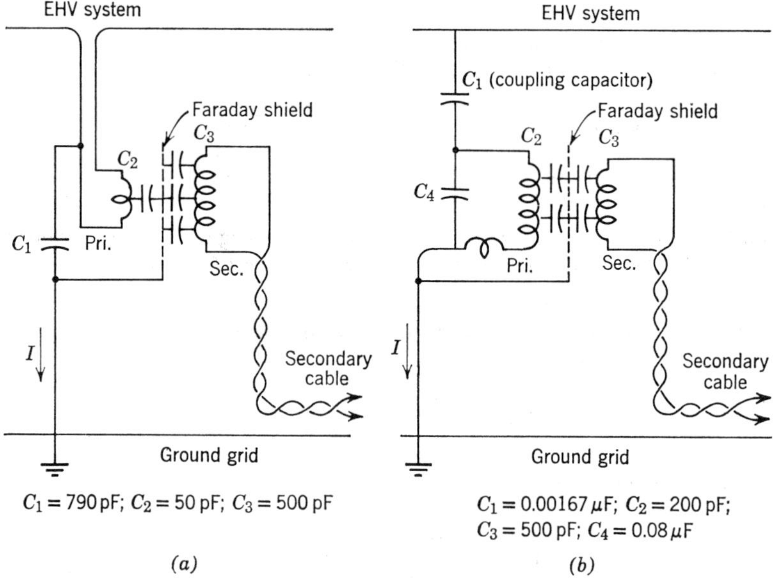

Fig. 12.19. (*a*) High-frequency equivalent circuit of EHV current transformer. (*b*) High-frequency equivalent circuit of EHV potential transformer.

in this regard; the flux linking with the loop is given by

$$\Phi = \frac{\mu_0 I}{2\pi} \ln \frac{d-r}{r} \text{ webers/meter} \qquad (8.3.10)$$

(this assumes that the ground conductor is part of a long line), r is the radius of the ground conductor (1 cm, say), and d is the width of the loop to the control circuit (1 m, say). The induced voltage will be

$$V = \frac{d\Phi}{dt} = \frac{\mu_0}{2\pi}\left[\ln \frac{d-r}{r}\right]\left(\frac{dI}{dt}\right) \text{ volts/meter}$$

which has a maximum value

$$V_{max} = \frac{4\pi \times 10^{-7}}{2\pi}\left[\ln \frac{d-r}{r}\right]\omega I_{max} \text{ volts/meter} \qquad (12.6.1)$$

Substituting the dimensions and setting $\omega = 2\pi \times 1.34 \times 10^6$ and $I_{max} =$ 620 A,

$$V_{max} = 4 \times 10^{-7}(\ln 100)2\pi \times 1.34 \times 10^6 \times 620 = 4800 \text{ V}$$

If the ground conductor is 2 m long, the peak induced voltage would be almost 10 kV.

This voltage appears on both secondary conductors with respect to ground, that is, as a common mode voltage. The control cable may be taken several hundred feet to the control house, forming a short transmission line with the ground mat. The induced voltage will be impressed on this "line" and travel down it to the control house.

A similar situation to that described can exist in coupling capacitor potential devices as indicated in Fig. 12.19*b*. The physical layout is clearer in Fig. 12.20*a*. Two schemes for drastically reducing the pickup are shown in Figs. 12.20*b* and 12.20*c*. In the first example the loop is eliminated; transient currents are conducted by a tubular ground conductor, while control leads pass down the field-free regions inside. In the second approach the loop is short-circuited by using a shielded secondary cable with the shield grounded at the base of the potential device and at the receiving end. The shield must be a good conductor.

Power systems abound with relay, control, and monitoring circuits of many kinds. To avoid damage to these circuits or misoperation from voltage surges, it is essential that they be protected. The need for protection has become significantly more acute in recent years by the transistorizing of such equipment. Such circuits operate at a lower voltage and energy level than formerly yet are more sensitive to overvoltage. Most frequently the surge voltage incursions are induced from neighboring power circuits, in the manner just described. The first step, therefore, is to minimize the coupling with

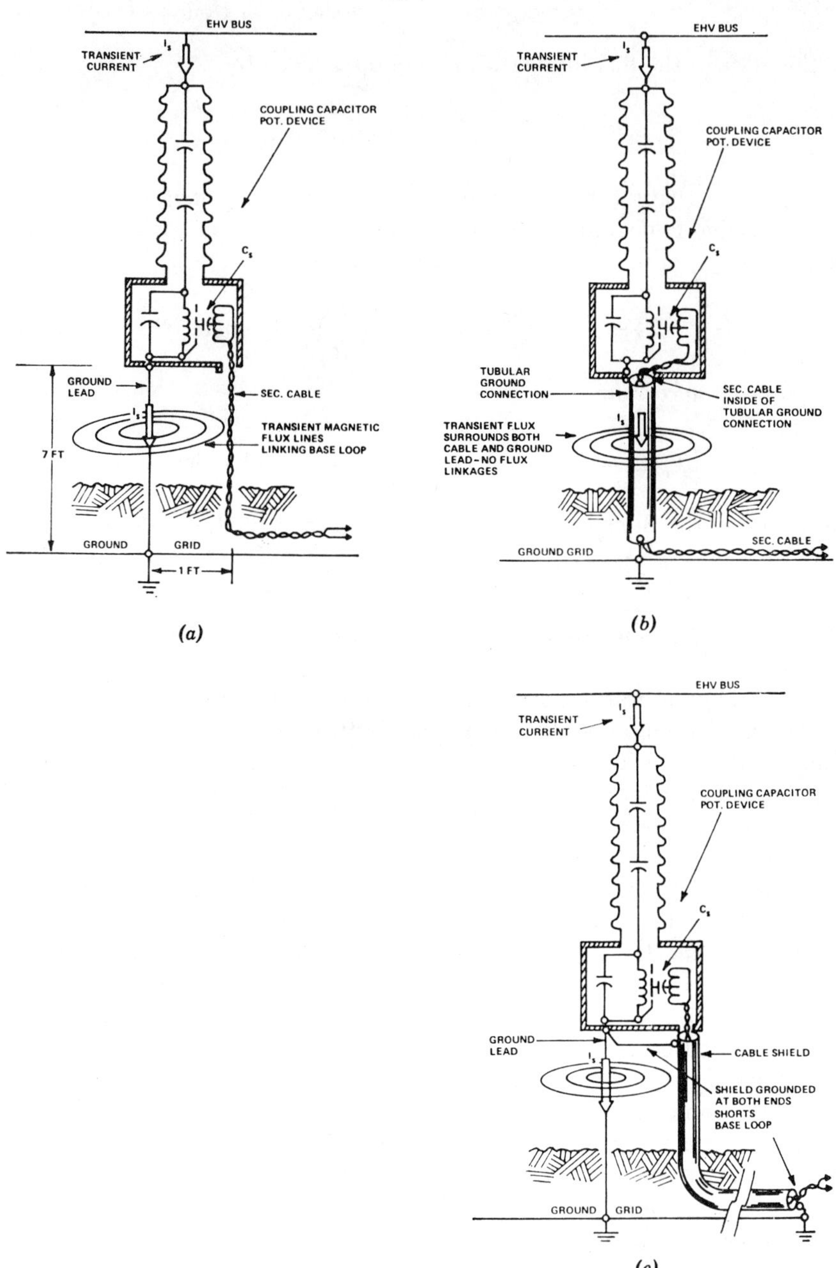

Fig. 12.20. Coupling capacitor potential device under transient conditions. (*a*) Source of magnetic coupling. (*b*) Elimination of the inductive loop by cylindrical ground conductor. (*c*) Reduction of coupling by shielding.

these circuits by effective shielding through the various methods mentioned above and described in Chapter 8. Even when this has been done, there sometimes remains a residue of voltage disturbances, which can still create serious problems.

The approach to protection for control circuits is basically the same as that for the power circuits themselves; we must consider coordination, impedance division, and energy. Except for those circuits that are exposed to lightning, the voltage and energy are much smaller than the voltage we associate with power system surges. But, as pointed out previously, the ability of the circuit components to cope with voltage and energy is also drastically reduced.

Nonlinear resistors are made in minute sizes for application in this area, but, applied alone, they are inadequate for many circuits, for the same reason that rendered them unacceptable in many power circuits, namely, if they hold the voltage to an acceptable level, their steady-state dissipation with normal voltage is too high. Surge capacitors can be used within their limitations, especially in conjunction with other devices. A surge capacitor, in this context, may be no bigger than a dime, since the surge energy level in many circuits is low. The small size of the components has the advantage of minimizing stray inductance. The capacitor, for example, can be tightly connected across the terminals of a relay card.

Since the operating voltage level of control circuits is low, it is very difficult to protect them with air gaps. The gap spacing required, typically a few thousandths of an inch, is very difficult to maintain. It will vary with temperature because of differential expansion of the parts in the gap structure. It will also vary as a consequence of operation, since the electrode surfaces are always disrupted to some extent by the discharge current. These effects are minimized by using graphite for the electrodes, but in spite of this, consistent operation within relatively narrow tolerances cannot be obtained. Another shortcoming of this type of gap is its considerable turnup for fast-voltage transients.

These characteristics have been dramatically improved in the several types of gas-filled gap that are now available (18). These comprise a pair of metal electrodes, sealed into an enclosure, usually made of glass, which is filled with an inert gas to some fraction of an atmosphere. The electrodes themselves are normally, though not always, made of tungsten or some other refractory metal. Gaps of this kind have a spacing typically ten times as great as an air gap of corresponding voltage rating. It is therefore not difficult to maintain consistent operation within about $\pm 15\%$ of nominal sparkover. The turnup of the volt-time curve is reduced, and by using a radioactive additive to assist in preionizing, it sets in at shorter times than with the air gap. This may be contained in the electrode or in the gas. Additionally, the burning voltage, once the gap is ignited, is relatively low; 40 V can be readily achieved, which is compatible with transistor circuits.

Gas gaps are usually operated in conjunction with a current-limiting resistor, although they are capable of passing considerable current. For example, units no bigger than a thimble exist, which are capable of discharging 5000-A peak current surges of 20 μsec duration. The series resistor helps to dissipate energy, but of course at the expense of an increase in total voltage. Where the circuit voltage subsequently falls below the sustaining voltage of the discharge, no further protection is required. When this is not the case, a series interrupter of some kind must be used. Fuses, reed switches, or other resetable circuit breaking devices are used for this purpose.

Close control of voltage can be obtained in light current circuits by the use of avalanche diodes or Zener diodes; the latter are common standard components in voltage regulating circuits. Under surge condition they usually lack the energy dissipating capability. However, if they are used in combination with a more rugged device, such as a gas-filled gap, one can compliment the other and thereby provide a very effective package. The gap might be shunted by a surge capacitor to slow the front of fast-rising surges. The combination would then be separated from the zener diode by some impedance across which the surge voltage might momentarily be impressed until reduced to safer proportions by the gap. In this way, the Zener diode can maintain the voltage within narrow limits despite the surge and at the same time not be unduly extended from an energy point of view.

The function of decoupling impedance just described can often be provided very satisfactorily by simply slipping one or more ferrite cores over the conductors between the gap and the circuit being protected. This is a very good way of introducing inductance but, more especially, inductance that will be effective at high frequency. Although the ferrite has a much lower permeability than laminated steels of various kinds, its structure, which minimizes eddy currents, renders its flux-supporting ability very accessible, even at high frequency. In this application we would like the reactor formed by the conductor and the linking ferrite to hold off or support voltage temporarily, while the situation is taken in hand by the gap. The ferrite can do this, as long as its flux is changing, and its flux will change until it saturates. Reactors for this purpose are analyzed in Section 12.4. In most applications of this kind the transient voltages are not very high and the time required to support them is short, so that only a modest cross section of ferrite is required.

Another protective device of modest energy capability, well-suited for low power and control circuits is the Selenium Thyrector* Diode. This is a

* Trade name of the General Electric Co.

specially manufactured selenium rectifier. The major difference between this device and the conventional selenium rectifier is that its reverse characteristic has been modified to provide a sharp knee in the *I* versus *V* trace. This characteristic is compared with the comparable characteristic of a standard selenium rectifier in Fig. 12.21. The cell's reverse resistance is nonlinear and decreases logarithmically with increasing voltage. Because of the shape of the Thyrector curve, the cell performs in a manner similar to a zener diode. When a spike of voltage appears across a circuit which one of these devices is protecting, it momentarily introduces a shunting low impedance and limits the voltage. A conventional selenium rectifier would fail dielectrically in similar circumstances. Protectors of this kind have an energy limit, of course, beyond which they fail thermally. They are made in different sizes, depending on the energy dissipation required, and by stacking disks in series, they are made for different voltage levels. For a–c applications the disks are arranged back-to-back.

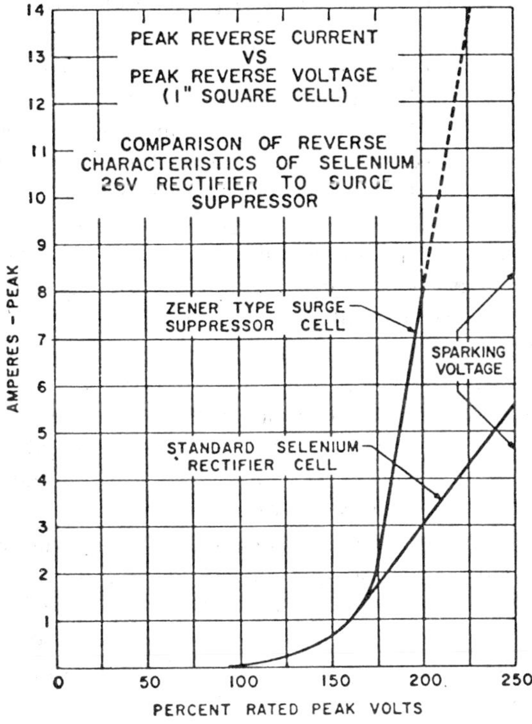

Fig. 12.21

12.7 Surge Protection Scheme for an Industrial Drive System

To pull together the many aspects of circuit protection that have been discussed in this chapter, an example will now be considered of an integrated protection scheme in which most of these aspects have to be considered. Protective devices described earlier are shown in a typical environment. The example selected is an industrial drive that might typically be found in a steel mill. It is shown schematically in Fig. 12.22 and in more detail in Fig. 12.23.

The motive power derives from a direct current motor whose speed is controlled by regulating the voltage of the bus to which it is connected. This regulation is achieved by a bridge rectifier. The particular drive shown is made reversible by having two sets of oppositely poled rectifiers, which allow the polarity of the bus to be reversed. It also permits dynamic breaking by regeneration to assure rapid reversal of the motor. The source of power is an a-c line which delivers energy to the rectifier through a stepdown transformer. Note that where one thyristor is shown it may represent a considerable number in series/parallel combination, depending upon the size and voltage of the motor.

Starting at the line, it is seen that arresters are located line-to-ground at the high-voltage terminals of the transformer. These would be either station or intermediate type, with a voltage rating appropriate to the rating of the transformer (15-kV class in this case). It would limit voltages from lightning or switching surges to their sparkover voltage. In most installations the transformers are located outside the building close to the wall. This outside area would usually have additional lightning protection in the form of overhead structural steelwork and/or ground wires, with lightning rods on the building above and strategically placed around the switchyard if that is extensive.

It was pointed out in Chapter 11 that short but potentially very high voltage transients can be electrostatically coupled through a transformer from one winding to another. These are effectively suppressed by putting capacitors on the low voltage side of the transformer connected from each phase to ground. These are shown in Fig. 12.22 as C_1. Where the transformation ratio is very high this might be supplemented by a grounded metal shield between the windings. Longer duration surges coupled electromagnetically through the transformer, as might appear from an induced lightning surge which just failed to spark over the arrester, are absorbed by the large capacitor C_2, which was described in Section 12.4. As an added safeguard, a surge suppressor (S.S.) is placed in parallel.

The remainder of the protective equipment is applied primarily to protect the thyristors from surges they themselves generate in the course of their normal operation. These have been discussed in Section 7.6, and will therefore

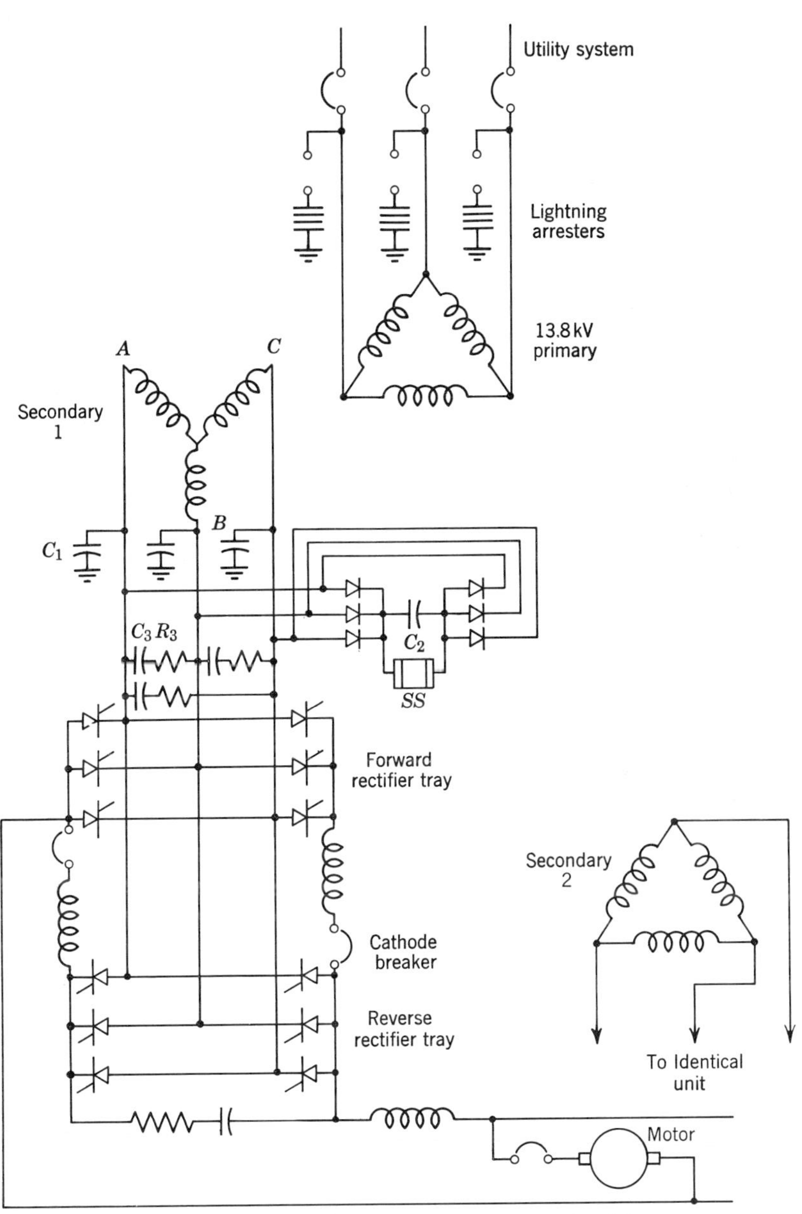

Fig. 12.22. Protective equipment on a reversing drive.

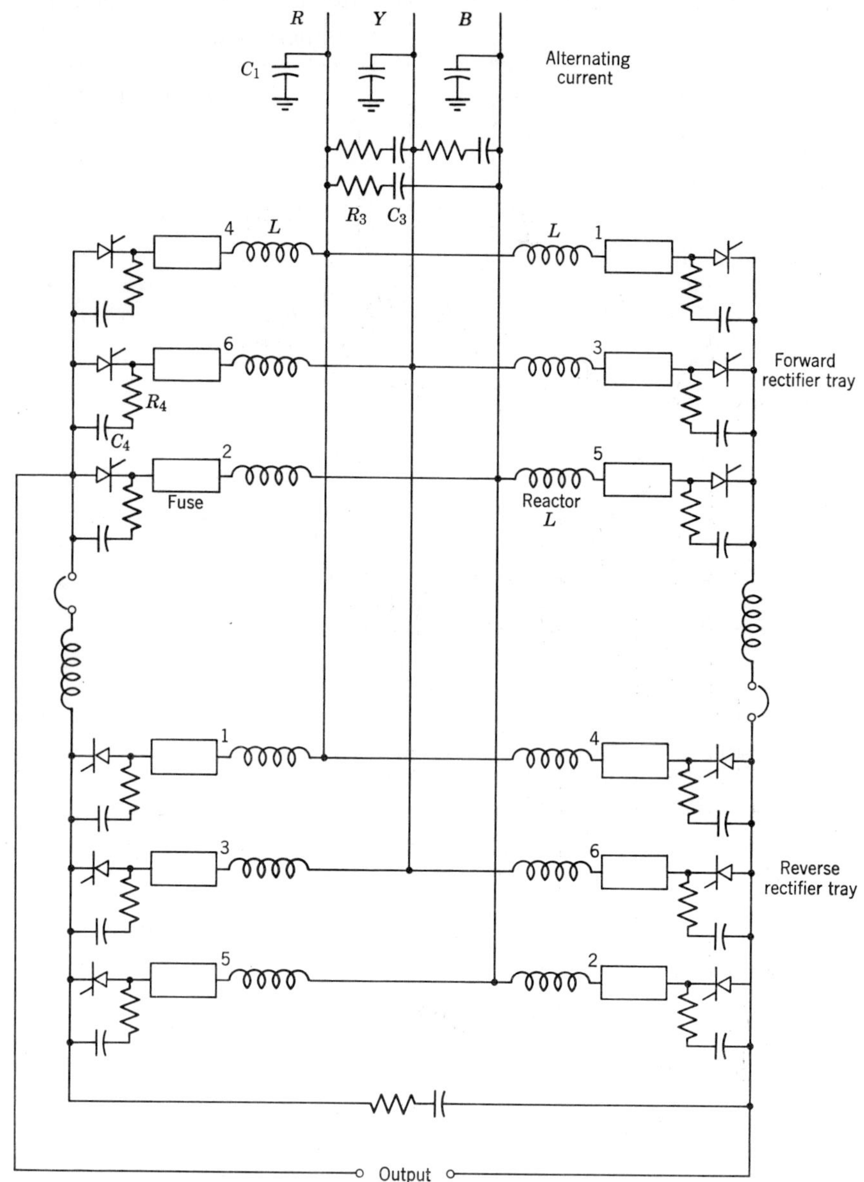

Fig. 12.23. Detail of protective equipment on reversing drive.

be reviewed here only briefly. We note in Fig. 12.23 inductors marked L, in series with the controlled rectifiers. These assure equitable division of current between the various strings of thyristors, but they perform an important function during commutation in the manner described in Section 12.4. When a string such as 6 is turned on, the voltage formerly across it is not suddenly transferred to 3. Instead it is absorbed momentarily by the Ls and then transferred in an orderly manner to C_4 with R_4 acting as a damper. At the time 6 is gated, 4 will already be conducting (see Fig. 7.16*b*), thus a momentary short is applied between phases R and Y. The protection capacitors C_1 will therefore discharge into each other, and once again the inductors, L, will modify the current. This transient will be damped by the presence of R_3 and C_3. At the end of commutation the transients evoked by the turnoff of the outgoing phase will be similarly modified.

An installation of this kind would be grounded according to practices described in Section 12.5. Outside the building the transformer tank, arrester, and circuit breaker framework would be connected by short, heavy copper braids to the ground mat. Inside the building a harness of copper strip would join the rectifier and control cubicles, the motor frame, and any exposed metallic structure of the building. This harness also would be connected at several points to the ground mat.

REFERENCES

1. E. C. Sakshaug and D. E. Hedman, "Trade-Offs Developed for Arrester Selection," *Electrical World* (November 1967).
2. W. W. Lewis, *The Protection of Transmission Systems Against Lightning*, Dover Publications, Inc., New York (1965).
3. U.S.A. Standard C62-1967, "Lightning Arresters" and AIEE Standard No. 32, Section 32-1.05 (1947).
4. IEE Working Group of the Lightning Protective Device Subcommittee "A Report on Protective Characteristics of Lightning Arresters," *Trans. IEEE*, Vol. PAS 82 (1963), p. 254.
5. Committee Report, "Investigation and Evaluation of Lightning Protective Methods for Distribution Circuits, Part I: Model Studies and Analysis," *Trans. IEEE*, Vol. PAS 88 (1969), p. 1234.
6. Committee Report, "Investigation and Evaluation of Lightning Protective Methods for Distribution Circuits, Part II: Application and Evaluation," *Trans. IEEE*, Vol. PAS 88 (1969), p. 1239.
7. Y. Ozaki, "Operating Duty of Lightning Arresters Against Switching Surges," *Trans. IEEE*, Vol. PAS 85 (1966), p. 831.
8. J. M. Lafferty, "Triggered Vacuum Gaps," *Proc. IEEE*, Vol. 54 (1966), p. 23.
9. T. W. Liao and T. H. Lee, "Surge Suppressors for the Protection of Solid State Devices," *Trans. IEEE*, Vol. IGA-2 (1966), p. 44.
10. K. E. Russell, "Overvoltage Protection of HVDC Converter Stations," *Transmission and Distribution* (September 1967).

11. "Grounding and Industrial Power Systems," AIEE Bulletin No. 953.
12. AIEE Committee Report, "Voltage Gradients through the Ground under Fault Conditions," *Trans. AIEE*, Vol. 73, IIIA (1954), p. 271.
13. AIEE Committee Report, "Application Guide on Methods of Substation Grounding," *Trans. AIEE*, Vol. 73, IIIA (1954), p. 271.
14. P. L. Bellaschi, R. E. Armington and A. E. Snowden, "Impulse and 60 Cycle Characteristics of Driven Grounds," Part I, *Trans. AIEE*, Vol. 60 (1941), p. 123; Part II, *Trans. AIEE*, Vol. 61 (1942), p. 349; Part III, *Trans. AIEE*, Vol. 62 (1943), p. 334.
15. E. C. Sunde, *Earth Conduction Effects in Transmission Systems*, D. Van Nostrand, New York, (1949).
16. W. C. Kotheimer, "Control Circuit Transients," *Power Engineering*, Vol. 73, No. 1 (1969).
17. W. C. Kotheimer, "Control Circuit Transients," *Power Engineering*, Vol. 73, No. 2 (1969).
18. V. W. Vodicka, "Voltage Transients Tamed by Spark-gap Arresters," *Electronics*, p. 109 (April 1966).

13 Transients in the Integrated Power System

13.1 Introduction

In the early chapters of this book we were investigating transients in relatively simple circuits with lumped parameters. In Chapter 9 the study was extended to the behavior of transmission lines under transient conditions, specifically to the study of traveling waves. We now address ourselves to the switching operations that produce these traveling waves and to consideration of the interaction between the lines and the lumped parameter terminal equipment. Moving further in this direction, we attempt to analyze the response of more extensive systems to switching stimuli.

We do this to reinforce the earlier contention that transients are system problems—a disturbance created in one location will permeate throughout the system, often causing difficulties at points quite remote from its origin. This is true of all systems, be they industrial distribution systems or EHV utility transmission systems. A second reason for a more careful study of switching transients in integrated systems relates specifically to the EHV field. For a long time the most serious problems from voltage surges in utility systems were the consequence of lightning, because lightning produced the highest voltages. This fact is well documented in Chapter 10. As system voltages continue to increase, switching transients, which in magnitude are geared to the system voltage, are assuming increasing importance. There is a very strong economic pressure to limit and control switching surges on EHV systems, because of the large slice that line insulation represents in the cost of such transmission systems. It is well known that because of poor distribution of voltage due to capacitive effects, the benefits obtained per unit as more insulators are added to a string diminishes progressively as the string becomes longer.

There are related problems concerned with the application of lightning arresters. It has also been shown that the sudden reversal of voltage polarity, which can occur with traveling waves from switching surges, causes a marked reduction in the flashover voltage of certain equipments.

These considerations are causing system planners to take a new look at the overall design of transmission networks, especially in the area of insulation coordination (1). In Sections 13.3 and 13.4 we look specifically at switching transients on transmission systems, at the voltages that are developed, and how these voltages can be controlled. A good general reference to this field is the 1961 AIEE Committee Report (2), which includes an extensive bibliography.

It will soon be apparent, if it is not so already, that any detailed investigation of this kind is very complicated and quantitatively quite intractable without some computational aids. This leads naturally to the subject matter of Chapter 14, where such aids are discussed. There are topics, however, of a more limited nature that can be analyzed first—the behavior of single lines and single terminations, for example—the more extensive systems can be studied in a qualitative way. First we consider switching operations involving a single transmission line.

13.2 The Short Line or Kilometric Fault

Some years ago a number of incidents occurred which indicated that certain circuit breakers had less difficulty interrupting current to a fault located close to their terminals, than the current to a similar fault some distance out along a transmission line, where, because of the added impedance of the line, the current should have been appreciably less and easier to interrupt. That this is not the case appeared to indicate some interaction between the circuit breaker and the transmission line. Skeats et al. (3) described and explained the phenomenon, showing that faults of this kind present the circuit breaker with a severe recovery condition. Since that time many papers have appeared on the subject; see, for example, Bolton et al. (4).

The significant difference between the fault close to the breaker and the fault some distance out along the line is that the line impedance not only limits the current in some measure but, as a consequence, supports some of the system voltage. The generated voltage is divided on either side of the breaker in proportion to the impedance of the source and the line. This is shown in Fig. 13.1. The further out the fault is, the greater fraction of the voltage the line sustains. At current zero, when the circuit breaker interrupts, the generated voltage will be near its peak. Therefore the line, now severed from the system, momentarily has on it a charge distribution which is a maximum at the circuit breaker and declines more or less linearly to zero at the fault. It was shown in Section 9.2 that an unbound distribution of charge on a line will not remain static but will travel, or attempt to do so, in both directions along the line. In this particular instance the charge is somewhat

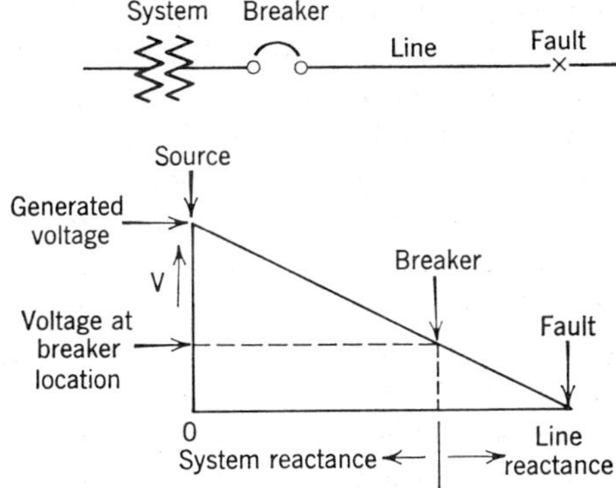

Fig. 13.1. Distribution of voltage on a faulted system.

confined, in that a traveling wave cannot proceed into the source since it is blocked by the open circuit breaker, nor can the charge propagate as a wave beyond the fault, assuming that the fault is a short circuit. On the other hand, we are quite familiar with waves encountering open circuits and short circuits. In Section 9.3 it was shown that when a wave impinges on any line discontinuity, a new reflected wave is initiated. In the situation presently under review, waves of this kind are initiated immediately, since the initial waves immediately encounter the open circuit and short circuit terminal conditions. The new waves proceed to interfere with the existing traveling charge system.

This problem can be analyzed in a very formal manner by setting down the wave equation and proceeding to solve it for the particular initial conditions of voltage distribution and the constraints imposed by the short circuit and open circuit terminations. But this is rather heavy going for a relatively simple problem. A good appreciation of events can be obtained without recourse to mathematics. Figure 13.2*a* shows the initial voltage distribution on the line; L is here the source inductance and L_1 the line inductance to the fault; E is the instantaneous value of the emf driving the fault current. If this is to disperse as a pair of traveling waves, each wave must be wedge shaped and have a peak amplitude of

$$\frac{1}{2}\frac{L_1}{L + L_1}E$$

The wave traveling to the left will immediately encounter the open circuit and will reflect with the same amplitude and sign. The identical wave starting to the right will at once encounter the short circuit and reflect with sign reversed. The voltage distribution a short time later is given in Fig. 13.2*b*. In this figure the initial wedge-shaped waves are shaded; their reflections are unshaded and identified by R. The two reflected waves will themselves suffer reflection as they reach the opposite terminations, and so the phenomenon will proceed, subjected to the effects of the losses of the line. The waves behave somewhat like a pair of tigers marching to and fro inside their cage. From Fig. 13.2 it is not difficult to compute the voltage variations at any point on the line as a function of time, given the velocity of the waves and the distance to the fault. Such voltage variations are plotted in Fig. 13.3 for the breaker end of the line, the point midway to the fault, and the point three-quarters of the distance out to the fault.

The voltage at the breaker (Fig. 13.3*a*) is of special interest in that this is part of what the circuit breaker experiences as its transient recovery voltage after it has interrupted the fault current. It will be seen that the line-side terminal of the circuit breaker can make a considerable swing in potential, $+V$ to $-V$, in a comparatively short time, $2T$. Evidently V is the initial peak voltage on the line, and T the time for a wave to travel from the circuit breaker to the fault. It is this fast rate of rise of the transient recovery voltage that prejudices the successful operation of the circuit breaker.

This explanation of the short-line fault is accurate and satisfactory but now that the physical picture has been established a different analysis will be

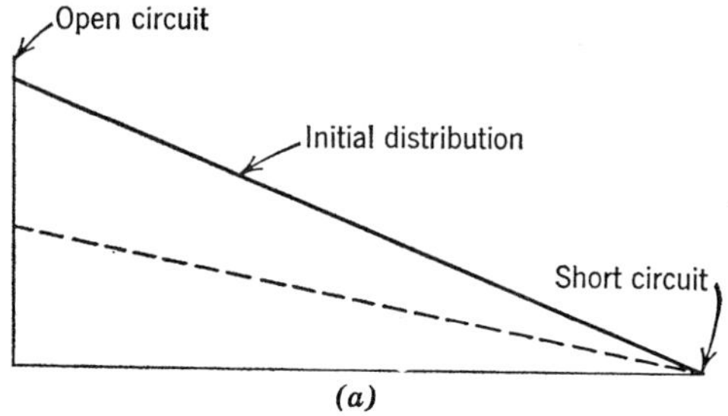

(*a*)

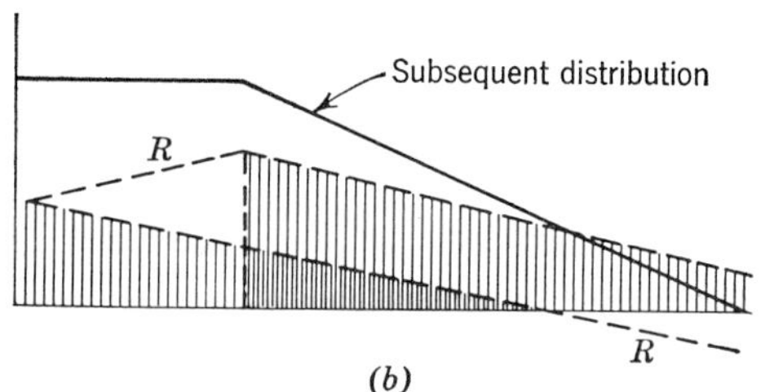

(*b*)

Fig. 13.2. Voltage distribution on a short faulted line. (*a*) Initial. (*b*) After the waves have begun to separate.

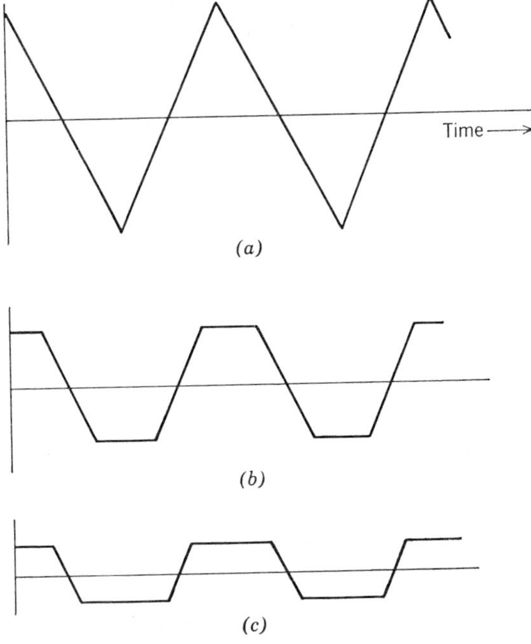

Fig. 13.3. Voltage variations following the clearing of a short line fault. (*a*) At the breaker. (*b*) Halfway to the fault. (*c*) Three-quarters of the way to the fault.

given in order to illustrate a different technique which has applications beyond the immediate problem. It can be shown (5) that the current and voltage at the sending end of a transmission line, V_S and I_S, are related to the current and voltage at the receiving end, V_R and I_R, through the equations

$$V_S = V_R \cosh j\omega(LC)^{1/2} + I_R Z_0 \sinh j\omega(LC)^{1/2} \tag{13.2.1}$$

$$I_S = I_R \cosh j\omega(LC)^{1/2} + \frac{V_R}{Z_0} \sinh j\omega(LC)^{1/2} \tag{13.2.2}$$

Here L and C are the total inductance and capacitance of the line, and Z_0 is the surge impedance $(L/C)^{1/2}$; all losses are neglected. The ratio of these two expressions gives the apparent impedance of the line, Z_S, looking into it from the sending end:

$$\begin{aligned} Z_S &= \frac{V_R \cosh j\omega(LC)^{1/2} + I_R Z_0 \sinh j\omega(LC)^{1/2}}{I_R \cosh j\omega(LC)^{1/2} + (V_R/Z_0) \sinh j\omega(LC)^{1/2}} \\ &= \frac{V_R/I_R + Z_0 \tanh j\omega(LC)^{1/2}}{1 + (V_R/I_R) Z_0 \tanh j\omega(LC)^{1/2}} \end{aligned} \tag{13.2.3}$$

If we now introduce $Z_R = V_R/I_R$, being the impedance of the circuit at the receiving end, Eq. 13.2.3 can be rewritten as

$$Z_S = Z_0\left[\frac{Z_R + Z_0 \tanh j\omega(LC)^{1/2}}{Z_0 + Z_R \tanh j\omega(LC)^{1/2}}\right] \tag{13.2.4}$$

If the line is a short circuit at the receiving end, as is the case with the short-line fault, $Z_R = 0$, so that Eq. 13.2.4 reduces to

$$|Z_S|_{sc} = Z_0 \tanh j\omega(LC)^{1/2} \tag{13.2.5}$$

We now make use of the concept of operational impedance, introduced in Section 2.5, and substitute s for $j\omega$. For simplicity the operational impedance of Eq. 13.2.5 will be called $z(s)$. Then

$$z(s) = Z_0 \tanh (LC)^{1/2}s \tag{13.2.6}$$

Now, $(LC)^{1/2} = T$, the time for a wave to travel the length of the line, and

$$\tanh Ts = \frac{\sinh Ts}{\cosh Ts} = \frac{\epsilon^{Ts} - \epsilon^{-Ts}}{\epsilon^{Ts} + \epsilon^{-Ts}}$$

$$= \frac{1 - \epsilon^{-2Ts}}{1 + \epsilon^{-2Ts}}$$

Let $\epsilon^{-2Ts} = \alpha$, then

$$\tanh Ts = (1 - \alpha)(1 + \alpha)^{-1}$$

$$= (1 - \alpha)[1 - \alpha + \alpha^2 - \alpha^3 + \cdots]$$

$$= (1 - 2\alpha + 2\alpha^2 - 2\alpha^3 + \cdots)$$

Thus, from Eq. 13.2.6,

$$z(s) = Z_0[1 - 2\epsilon^{-2Ts} + 2\epsilon^{-4Ts} - 2\epsilon^{-6Ts} \cdots] \tag{13.2.7}$$

In order to determine the recovery voltage when interrupting a short-line fault, we apply the well-tried method of injecting into the circuit at the circuit breaker terminals, minus the fault current. For the time of interest this can be represented by a ramp of current with a slope $-E/(L + L_1)$:

$$I = \frac{-E}{L + L_1} t \tag{13.2.8}$$

(if we wish to be more accurate we can use a sine wave with whatever degree of asymmetry we choose). The transform of Eq. 13.2.8 is

$$i(s) = \frac{-E}{L + L_1} \frac{1}{s^2} \tag{13.2.9}$$

The response of the shorted transmission line is given by the product $i(s)z(s)$ obtained from Eqs. 13.2.9 and 13.2.7:

$$i(s)z(s) = \frac{-E}{L + L_1} Z_0 \frac{1}{s^2} [1 - 2\epsilon^{-2Ts} + 2\epsilon^{-4Ts} - 2\epsilon^{-6Ts} + \cdots] \quad (13.2.10)$$

To obtain the actual voltage, we take the inverse transform of this expression term by term. The first term is a negative ramp with a slope Z_0 times that of the current. Successive terms can be evaluated by Heaviside's shifting theorem, which was introduced in Chapter 7. These terms are also ramps of voltage, but of twice the slope of the first term and delayed respectively $2T$, $4T$, $6T$, ... We also note that they alternate in sign. The summation of these ramp functions, as they are superimposed, produces the saw-tooth waveform arrived at earlier by the graphical approach. The period of this saw tooth wave is $4T$, as before, and its amplitude, peak to peak, is given by

$$\frac{E}{L + L_1} Z_0 2T$$

Since

$$Z_0 T = \left[\frac{L_1}{C_1}\right]^{1/2} (L_1 C_1)^{1/2} = L_1$$

this expression reduces to

$$2E \frac{L_1}{L + L_1}$$

as before.

What we have so far calculated is the response of the shorted line. The source-side response can be calculated in the same fashion. If the source can be represented by a parallel L and C, it will give rise to the familiar (1-cosine) recovery wave with a period $2\pi(LC)^{1/2}$ and an amplitude of

$$\left[E - \frac{EL_1}{L + L_1}\right] \quad \text{or} \quad \frac{EL}{L + L_1}$$

that is,

$$E \frac{L}{L + L_1}\left(1 - \cos \frac{t}{(LC)^{1/2}}\right) \quad (13.2.11)$$

The breaker sees the difference of expressions (13.2.10) and (13.2.11). This is shown in Fig. 13.4.

13.3 Line Dropping and Load Rejection

The subject of capacitance switching has been dealt with in some detail for single-phase circuits in Section 5.3, and for three-phase capacitor banks

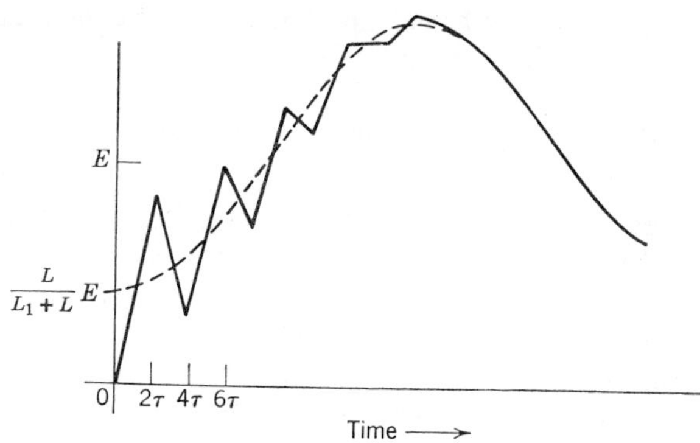

Fig. 13.4. Circuit breaker recovery voltage following a short line fault.

in Section 6.4. It was pointed out that in determining the voltage trapped on the capacitance following switching, an important consideration is the ratio of the positive sequence capacitance C_1 (total capacitance per phase) to the zero sequence capacitance C_0 (capacitance from line-to-ground). Unless the capacitor bank has a grounded neutral, C_0 is very small, and consequently C_1/C_0 is very large. For this condition Fig. 6.7 shows that the voltage across the first phase to clear approaches three times the line-to-neutral voltage. For transmission lines and cables the ratio C_1/C_0 is typically in the range 1.6 to 2.0 (6). This means, according to Fig. 6.7, that the maximum voltage appearing across the first pole to clear, when a transmission line is switched, is somewhere between 2.2 and 2.4 times the line-to-neutral system voltage. This is important because it reduces the chance of a restrike and also results in a significant reduction in the transient voltage should a restrike occur.

If a restrike should occur, the voltage across the switch at the time will divide between the surge impedance of the line, on the one hand, and the surge impedance of the source and such parallel lines and circuits as may be connected, on the other. Voltage waves proportional to these line and source side impedances will travel from the breaker into their respective circuits, the wave traveling up and down the line corresponding to the oscillatory swing of the bank in the case of a lumped capacitance.

In Section 4.5 a method of reducing switching transients was described whereby a resistor is used to shunt the contacts of the circuit breaker. In the situation described, which relates to the clearing of a fault close to the circuit breaker terminals, the resistor is virtually in parallel with the source. If a low enough value of resistance is used, it can very effectively damp the transient oscillations that the switching operation would otherwise evoke in the source. It is necessary, of course, to interrupt the resistor current subsequently.

Resistors can be of considerable value in capacitance switching or line dropping (7), (8). The disconnection of the capacitor bank or transmission line is actually executed by the resistor switch; thus the load at that time is a series combination of R and C. The system voltage is therefore no longer at a peak at current zero when interruption is effected, and consequently less voltage is trapped on the line, which reduces the chance of a restrike. The importance of this factor will depend upon the relative values of the circuit breaker resistor and the capacitive reactance of the load. This is illustrated in Fig. 13.5, adapted from reference (8). If in spite of this effect, a restrike should still occur. The resistor provides a further useful function in damping the restrike current and limiting the voltage swing.

For a long energized line, the line voltage is not univalued but varies with position on the line at any instant. Thus when the line is cut loose by a switching operation, a nonuniform charge will be left trapped on the line. The condition is somewhat like that described in Section 13.2, but the nonuniformity is not nearly as pronounced. It will result in waves with amplitudes depending upon the degree of nonuniformity, sloshing up and down the line, like water in a dish, until losses damp them out.

Sometimes a line is switched with the transformer at the remote end still connected. This leads to an interesting situation. The line has a path through the magnetizing impedance of the transformer, by which it can discharge. However, as long as the transformer core is unsaturated, this represents a high impedance and the discharge is very slow. The transformer appears as though it were connected to a battery: it can continue to support the line voltage as long as the flux in its core is changing, but it runs into saturation fairly quickly, whereupon its impedance drops abruptly. This is accompanied by a sharp increase in current and a rapid discharge of the line. Unless the transformer is overdesigned this will occur after about half a cycle of the power frequency. At voltage zero, the current is near its maximum and it continues flowing so that the polarity of the line voltage reverses. Once again, as the core comes out of saturation, the voltage will level out. This process keeps repeating itself; the result is that the line voltage oscillates with a relatively square waveform. With each oscillation a certain amount of the original trapped energy is dissipated; the voltage therefore diminishes with succeeding half cycles. Since the flux linkage to saturate the core, or the volt seconds that this represents, is constant, successive cycles of this oscillation have progressively longer periods. It has been reported (2) that the incidence of circuit breaker restrikes with a transformer connected is less than for unloaded lines. Field tests of this kind have been described by McElroy et al. (9).

The condition just described usually arises after prior switching operations have shed the load from the secondary side of the transformer. This secondary

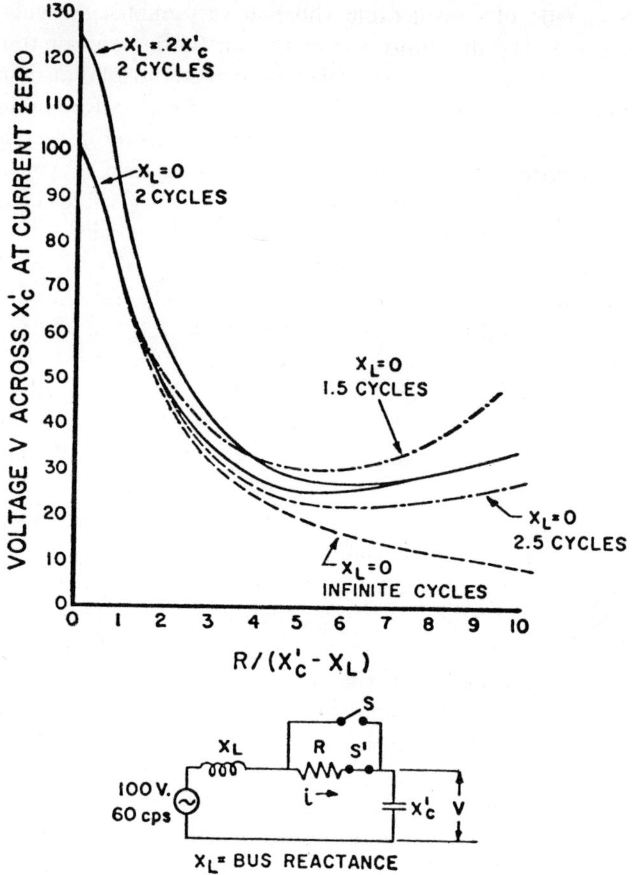

Fig. 13.5. Circuit and curves of trapped charge on X_c' at instant of opening switch S'. Solid curves are for S' opening approximately two cycles after opening of switch S.

side switching of transformers connected to high voltage lines can itself create overvoltages (14), (15) even without restriking. This comes about again as a consequence of the nonlinear properties of the transformer core. It is well known that magnetizing current of a transformer contains harmonics, notably the third. It is possible for the line feeding the transformer to form a relatively high impedance to the flow of such harmonics, so that following the load switching considerable harmonic voltages are generated. In the same way, where the lines concerned constitute the major load on a generating station, sudden load rejection will result in overspeeding of the machines and a rise in voltage until such time as this can be checked by the governors and exciters. These are longer term phenomena than we are concerned with in this book, but they can be very important if further switching operations occur during this period, for it is against this existing overvoltage background that the new transients are created.

13.4 Voltage Transients on Closing and Reclosing Lines

Thus far discussion has been directed toward the transient voltages arising from de-energizing systems. Quite severe overvoltages can also result from energizing a line, and yet higher voltages as a consequence of re-energizing after a momentary interruption (10), (11), (12). This problem has attracted a great deal of attention with the advent of EHV systems, for, as explained, switching surges are becoming a limiting factor in the choice of insulation levels and arrester characteristics for these very high-voltage circuits.

The basic considerations involved are relatively simple, but the number of variables makes study difficult without the help of a computer. As the switch is closing and immediately prior to the circuit being completed, a certain voltage exists across the switch contacts. If the line is de-energized this could be anywhere between zero and the peak system voltage. At the moment the contacts make or are joined by a prestriking discharge, this voltage disappears, only to reappear distributed around the circuit. The distribution will be in accordance with the impedance of the parts of the circuit. Looking back into the system from the circuit breaker, we see the source impedances parallelled by the surge impedance of however many lines are connected at the time. This combined impedance we can call Z_1. Looking toward the line we see its surge impedance, which momentarily will appear resistive, i.e. $Z_0 = R$. If the instantaneous voltage is V, a fraction,

$$V_L = \frac{R}{Z_1 + R} V \tag{13.4.1}$$

will momentarily appear across the line, and across the source a fraction,

$$V_S = \frac{Z_1}{R + Z_1} V \tag{13.4.2}$$

will appear. Since Z_1 is in general complex, Eqs. 13.4.1 and 13.4.2 are most readily evaluated by operational methods:

$$v_L(s) = \frac{z_1(s)}{R + z_1(s)} \frac{V}{s} \tag{13.4.3}$$

Evidently, if $R \gg Z_1$, most of the voltage will appear across the line. This is the condition that will prevail on a stiff system.

This voltage will travel down the line as a wave and be reflected from its remote end, returning to the source in due course. If the far end of the line is open-circuited the reflected wave will add to the incident wave, so that a voltage approaching 2 per unit will be impressed on the line.

A common practice on utility systems is to reclose a breaker as rapidly as possible after it has interrupted a fault. This allows service to be restored quickly if the fault is of a transient nature, as many of them are. It also improves the overall stability of the system. In the event that the transient fault was a single line-to-ground, a common type, it is possible to leave a considerable voltage trapped on the unfaulted lines at the time of disconnecting, if they are open-circuited. When the switch recloses it may do so on the opposite polarity to the trapped charge, leading to the theoretical doubling of the traveling wave voltages described above.

In an attempt to reduce these closing voltage transients on EHV systems, the practice of resistance preinsertion has been adopted. This follows the normal resistance switching operation but is in the reverse order, that is, first the circuit is completed through a resistor, and then, as a second step, the resistor is shorted out. On 500-kV circuit breakers, 400 Ω is typical (10). It has been pointed out how the voltage across the switch, before completing the circuit, divides across the impedances of the circuit when closing is completed.

The preinserted resistor provides another element to share the voltage, so that Eqs. 13.4.1 and 13.4.2 must be modified. The result is that less voltage is impressed across the line. When the resistors are later shorted out, a new transient is initiated, but with careful design the overall peak voltage can be significantly reduced. The effectiveness of this approach is evident from Fig. 13.6, which is taken from reference (11).

We have treated the subject of closing transients as if it were a single-phase problem. It was approached in this manner so we would not be burdened with consideration of multiple conductors until the basic factors involved were well understood. Hopefully, that has now been done and we can move on to consideration of more complicated, although more common, transmission lines.

The groundwork for this was really done in Chapter 9, specifically in Section 9.7. It was shown there that multiconductor circuits can support more than one mode of traveling wave and that different modes have different velocities. When one pole of a three-phase breaker closes, it initiates a ground mode wave on its conductor. It also induces line and ground mode waves on the other conductor by induction. When the second and third poles subsequently close, they will set off new systems of waves. As these waves are dispatched up and down the line, they and their reflections quickly give rise to a very complex voltage patterns on the line conductors.

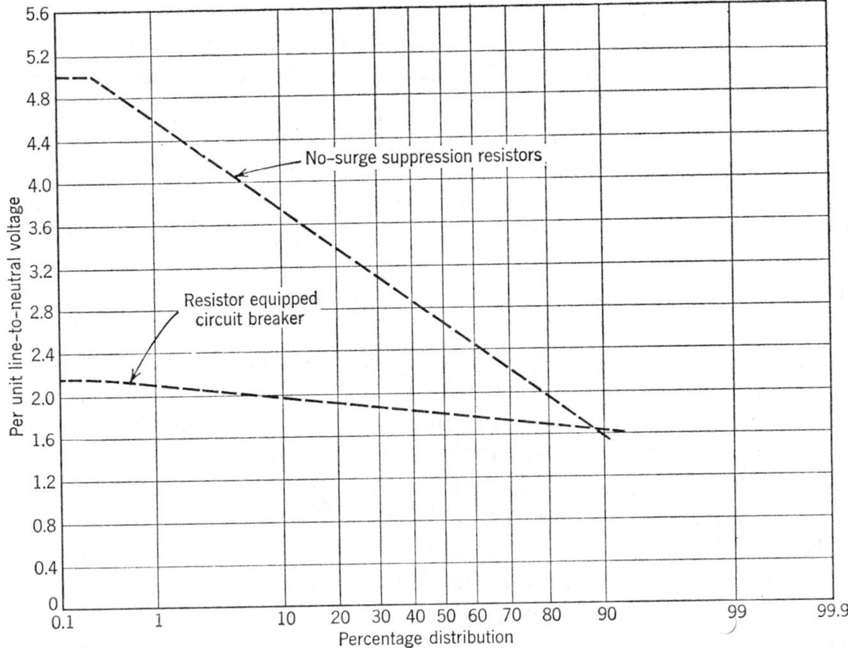

Fig. 13.6. Frequency distribution of reclosing transient voltage for 200-mile line.

The designer of the overall system wants to know whether anything is to be gained by closing at a particular point on the voltage wave, perhaps by individually closing the different poles. He is also curious as to how the resistor value affects the voltages generated, and how long he should wait before shorting out the resistors in order to minimize the transient voltages. To study all of these variables requires either a vast amount of testing or a great deal of analytical work with the Transient Network Analyzer (TNA) or a digital computer. This will be described in Chapter 14.

13.5 Overvoltage Induced by Faults

In Section 9.7, where we discussed the theory of multivelocity waves, an example was given of the waves that would be developed by lightning striking one conductor of a two-conductor line. It was shown that two pairs of waves would be initiated in either direction on both lines. On the struck line these dispersed the current injected into the line by the lightning stroke. The surges on the adjacent conductor were induced by electromagnetic coupling. A somewhat analogous situation arises when a ground fault occurs on one of the conductors. Here, instead of a current being injecting, a voltage is suddenly applied. The voltage is equal and opposite to that existing on the conductor at that time. The effect is very similar. Systems of waves are sent out on the faulted conductor and on adjacent conductors. It is the latter that concerns us here.

The induced waves will travel to the extremities of the line and be reflected, to return again, pass through each other, and so on. They will, of course, be superimposed on the existing power frequency voltages of the unfaulted phases. It is not difficult to see that with an unhappy coincidence of events, brought about by the line length, location of the fault, and time at which the fault occurs, a reinforcing of waves and power frequency peak could occur, which could produce a significant overvoltage.

This subject has been investigated by Kimbark and Legate (13). They point out that it is possible by careful design of circuit breaker closing and opening resistors, instant of closing and reclosing, and so on, to reduce the overvoltages described in Section 13.5 to modest proportions. On the other hand, faults of the kind just described are completely beyond the control of system designer. He may succeed in locking the front door of the barn only to have the horse bolt out the back.

Kimbark and Legate (13) conclude that a line-to-ground fault can produce an overvoltage on an unfaulted phase as high as 2.1 times the normal line-to-neutral voltage on a 3-phase line. Figure 13.7, from their paper, is an example of such an overvoltage, obtained from a TNA study. They determined that the worst fault location from this standpoint is at the midpoint of the line, and the maximum overvoltage is at the midpoint. The overvoltage decreases

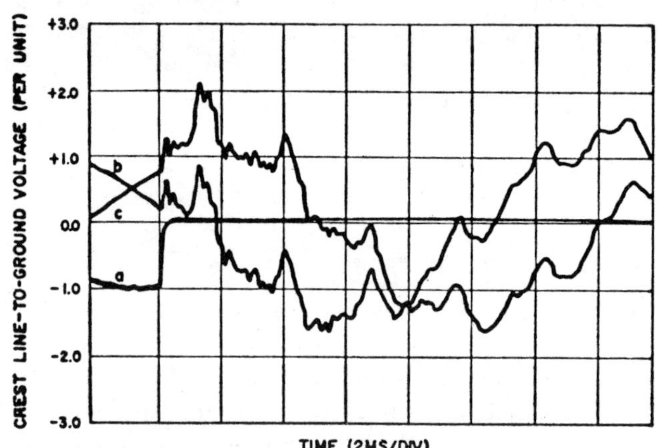

Fig. 13.7. Oscillogram of line-to-ground voltages obtained on TNA at midpoint of 180-mile, three-phase line for single line-to-ground fault at that point. Zero-impedance terminations.

slowly as the fault location changes, but the maximum remains above 1.75 per unit for locations over about 67% of the line length. The worst termination, as one might suppose, is zero impedance, which is approximated by a bus having several other lines.

We might also suppose that the phenomenon just described would have considerable relevance to HVDC transmission lines, which are frequently arranged with a pair of conductors on the same towers at $\pm V$ to ground. A fault on one circuit will cause a disturbance on the other. It is particularly desirable that this should not cause lightning arrester operation on the unfaulted phase.

13.6 Switching High-Voltage Direct-Current Lines

At the time of writing there are no HVDC circuit breakers in active system operation, although a number have been described in the literature. There are two good reasons for the absence of these breakers: they are very difficult to design and the HVDC systems presently in operation are strictly for point-to-point transmission of electrical energy. In the event of a line fault or a commutation failure at the inverter, the line can be removed from service by control of the valves (by blocking their firing) and re-energizing later when the fault has been removed. It seems likely that HVDC networks will be developed in due course; that is, lines will be tapped and have spurs. In these circumstances it will be necessary to have circuit breakers if the whole system is not to be shut down as a consequence of a fault on one line. Also, in situations where the power flow is in one direction only, there would appear to be economic advantages in operating with simple diodes at the sending end rather than controlled rectifiers. This too would require a breaker to protect against faults.

There are two likely ways that HVDC breakers will operate. The conventional approach follows the lines described in Section 7.2, that is, the device develops an opposing arc voltage, in excess of the system voltage, which drives the current to zero. The second approach would use the commutating technique of the static current limiting breaker, which is discussed in Section 7.5, although the interrupters themselves and the commutating switch may not be static devices. The second method presents some interesting problems which allow us to demonstrate a quite powerful analytical technique. It also serves as an introduction to the next section.

We saw in Section 7.5 that this mode of interruption is a form of current suppression and that it can lead to very high transient overvoltages if the circuit has significant inductance. In an a-c circuit the current is zero at the time of interruption, and therefore there is no magnetic energy stored. This is not the case with the commutated d-c interrupter. The energy stored in the circuit inductance will be released and transfer to the system capacitance, driving up the voltage, perhaps to a dangerous level. Long transmission lines have considerable inductance, of the order of 1 mH per mile. The HVDC lines are usually terminated in a large series reactor. It is therefore possible for total circuit inductance to be 1 H or even more. If the fault current reaches 5000 A, the stored inductive energy is 12.5 megajoules, which is very considerable. It is not difficult to dissipate the energy of the reactor since it is localized and can be spanned by some form of surge absorber. This course is not open for the line, the surge absorber must be connected line-to-ground and in some locations must therefore be able to clear a follow current.

To put the problem into perspective, consider a 500-kV line with an inductance of 0.4 H. Let us suppose that its surge impedance is 400 Ω. If a fault current of 5000 A is suddenly interrupted, the line voltage would rise to

$$IZ_0 = 5000 \times 400 = 2 \text{ MV}$$

if no steps are taken to control it. A 4 per unit rise of voltage would scarcely be acceptable. The values for Z_0 and L indicate that the line capacitance is 2.5 μF. This condition could be corrected by increasing the capacitance. To limit the voltage to 2 per unit, it would require a total of 10 μF since the energy stored in a capacitor is proportional to the voltage squared. This would require the addition of 7.5 μF of capacitance. Since this must be capable of being charged to 800 kV, it represents a prodigious capacitor bank. This could not be considered a satisfactory solution economically. A practical HVDC circuit breaker must have the inherent capability of dissipating a vast amount of energy either through arc voltage, as in the conventional approach, or by an associated high-voltage direct-current surge suppressor that will perform the energy dissipating function for it, once the circuit breaker has interrupted the current. This is the commutation approach.

What we wish to explore here is how to marry the lumpy circuit of the breaker with the distributed circuit of the line, under the transient conditions prevailing when a switching operation is performed. Consider Fig. 13.8, which represents in the most elementary way a d-c line, with a circuit breaker of the capacitor-commutated type and a series reactor. Here C is switched

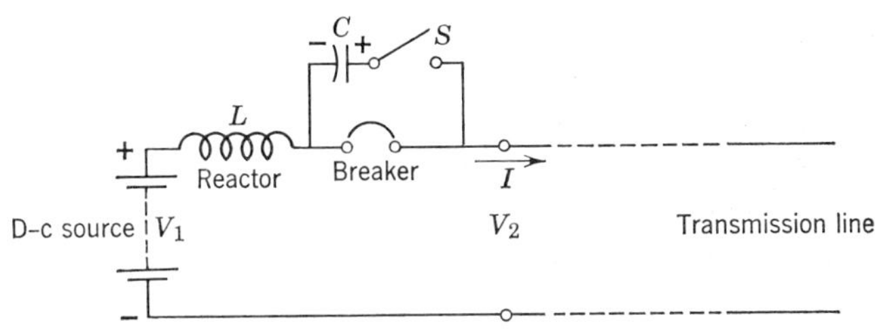

Fig. 13.8. Switching an HVDC transmission line.

into the circuit by closing the auxiliary switch S to execute the commutating operation. Current from the capacitor drives the breaker current to zero. The inductance L represents the reactor, behind which is the source or converter station. For the purpose of our analysis here, we assume that this is an infinite bus. This is obviously an approximation for the circuit, but it simplifies the analysis. A more rigorous representation could be substituted later.

When a short circuit occurs on the line a wave will propagate from the fault toward the source. This wave will remove voltage from the line in accordance with conditions at the short circuit. The wave will in fact be a voltage wave of $-V$, where $+V$ is the system voltage. The current wave associated with it will be $+V/Z_0$, so that the first intimation at the source that a fault exists will be the arrival of these waves. This is the beginning of a train of waves that will sweep up and down the line reflecting at the source and fault according to the conditions prevailing at these locations. As long as the circuit breaker remains closed, the source will appear as an inductive termination. When the circuit breaker opens and switch S is closed, whatever current is flowing in the breaker at that time is quickly commutated into the capacitor. These events we can assume are confined to the local loop comprising the breaker, the switch S, the capacitor, and the small loop inductance (not indicated in the figure), and are therefore decoupled from the main circuit. Once the breaker current has been removed, however, C becomes a part of the circuit proper. Let us suppose hereafter we are interested in computing the terminal voltage of the line, which we will call V_2. We must look at the problem as it affects (1) the terminal equipment and (2) the line.

The transients in both the terminal system and the line must conform to their appropriate circuit equations, including the initial conditions, but, in addition, the solutions must match where the circuits join. Our approach will be to use superposition. The relevant equation for the terminal circuit is

$$V_1 = L\frac{dI}{dt} + V_C(0) + \frac{1}{C}\int I\,dt + V_2 \tag{13.6.1}$$

$V_C(0)$ is the residual voltage left on the capacitor following commutation. The initial value of V_2 will depend upon the previous traveling wave activity on the line, which was described above. The initial current, $I(0)$, will be the fault current at the instant of commutation.

Viewed from the terminal equipment, the line behaves like a resistor equal to the characteristic impedance of the line. Thus a change ΔI in the current will be accompanied by a change in V_2 given by

$$\Delta V_2 = \Delta I Z_0$$

At any time, therefore, we may write for V_2

$$V_2 = V_2(0) + [I - I(0)]Z_0 \tag{13.6.2}$$

Eliminating the current from Eqs. 13.6.1 and 13.6.2 gives

$$\frac{d^2V_2}{dt^2} + \frac{Z_0}{L}\frac{dV_2}{dt} + \frac{V_2}{LC} = \frac{V_2(0) - Z_0I(0)}{LC} \tag{13.6.3}$$

which is the standard type of equation for a series *RLC* circuit, *R* being Z_0 in this instance. Similar equations have been treated in Chapter 4. The operational form of Eq. 13.6.3 is as follows:

$$v_2(s)\left[s^2 + \frac{Z_0}{L}s + \frac{1}{LC}\right] = \frac{V_2(0) - Z_0I(0)}{LC_s} + V_2(0)\left[s + \frac{Z_0}{L}\right] + V_2'(0) \tag{13.6.4}$$

From Eqs. 13.6.2 and 13.6.1,

$$V_2'(0) = Z_0I'(0) = \frac{Z_0[V_1 - V_C(0) - V_2(0)]}{L} \tag{13.6.5}$$

Substituting Eq. 13.6.5 in 13.6.4 and rearranging,

$$\left.\begin{aligned} v_2(s) = {} & \frac{V_2(0) - Z_0I(0)}{LC}\frac{1}{s(s^2 + (Z_0/L)s + 1/LC)} \\ & + \frac{Z_0}{L}[V_1 - V_C(0)]\frac{1}{s^2 + (Z_0/L)s + 1/LC} \\ & + V_2(0)\frac{s}{s^2 + (Z_0/L)s + 1/LC} \end{aligned}\right\} \tag{13.6.6}$$

Equation 13.6.6 contains one of each of the three standard transforms for the series *RLC* circuit, which were evaluated in Chapter 4, and which were the subject of the generalized damping curves developed in that chapter. It is therefore possible to determine $V_2(t)$ quite readily for any set of values for the circuit constants and the initial conditions.

We now direct attention to the line. As the terminal voltage V_2 changes, following the dictates of the solution of Eq. 13.6.6, it spreads as a traveling wave down the transmission line, adding to the existing voltage on the line. It reflects from the short circuit at the fault with its sign reversed and returns to the breaker. At this point it encounters the *C* and *L* of the terminal equipment and undergoes a further change. More precisely, a new reflected wave is generated, which can be obtained by multiplying the impinging wave by the reflection coefficient (Eq. 9.3.8) appropriate to the termination.

At this point we invoke the principle of superposition and assert that following the return of the reflection from the fault, the voltage at the line terminal is the sum of these terms:

1. The initial transient $V_2(t)$ from Eq. 13.6.6.
2. The reflection from the fault of this wave, which we can call $-V_2(t - 2T)$, where T is the time for a wave to travel from the breaker to the fault.
3. The second wave resulting from $-V_2(t - 2T)$ being reflected from the breaker; this can be designated $-aV_2(t - 2T)$, where a is the breaker end reflection coefficient.

As the disturbance continues to travel up and down the line, new terms or waves will be added at intervals of $2T$.

13.7 Switching Surges on an Integrated System

In Sections 13.2 to 13.6 we have discussed switching problems as they relate to individual lines. A new dimension is added when other parts of the interconnected system are introduced. The disturbance produced by the switching operation as modified by the system spreads through the system, setting up waves that travel out along the connected lines and reflect to and fro as discontinuities are encountered.

Consider the circuit of Fig. 13.9*a*, which, although a great simplification, typifies many stations. There are buses at 138 kV and 345 kV interconnected by autotransformer(s). The 138-kV bus is fed through the generator transformer(s). There are lines and/or ties connected to both buses. The circuit is redrawn in Fig. 13.9*b*, using a single-phase representation, and the significant capacitances are included. We will concern ourselves in this section with what will happen if one of the 345-kV circuit breakers is opened to clear a fault on its line. We might suppose that the fault is a short distance down the line. The switching operation will evoke a response from both the line and the system. The former has been adequately described in Section 13.2; our concern here, therefore, is with the latter.

If the impedance are reduced to a common voltage base, we see the equivalent circuit of Fig. 13.9*c* when we look back into the system. The components have the various significances indicated in the legend. Note that the lines are represented by the resistors R_1 and R_2. As pointed out in Chapter 9, this is an accurate representation of a line under transient conditions until it is modified by reflection from points down the line. The diagram, therefore, is only valid within this limitation.

It is possible to determine the response of the circuit in Fig. 13.9*c* to a ramp of current by hand calculations, but in practice we would use some of the

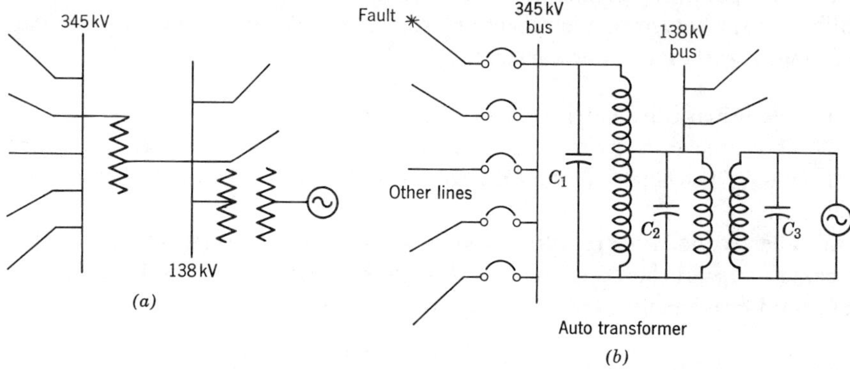

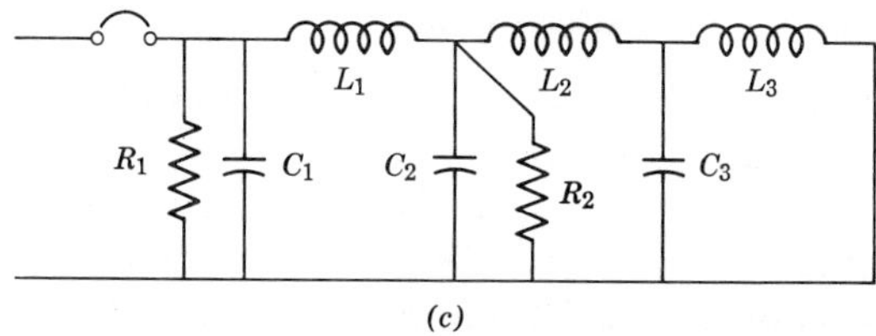

Fig. 13.9. Various representation of a generating and switching station.
C_1 = capacitance of 345 kV bus
C_2 = capacitance of 138 kV bus
C_3 = capacitance of generator bus
L_1 = inductance of auto trans.
L_2 = inductance of generator trans.
L_3 = generator inductance
R_1 and R_2 = parallel surge impedances of the lines

computer techniques described in Chapter 14, especially since we would be carrying out a three-phase analysis. Although we might suppose that the circuit would yield three frequencies, parameter values are such that the resistance R_1 effectively suppresses all oscillations. The following values, which are referred to 345 kV, are typical:

$C_1 = 0.025\,\mu\text{F}$ $\quad L_1 = 0.11\text{ H}$ $\quad R_1 = 80\ \Omega$ (4 parallel lines with $Z_0 = 320\ \Omega$)

$C_2 = 0.032\,\mu\text{F}$ $\quad L_2 = 0.07\text{ H}$

$C_3 = 0.006\,\mu\text{F}$ $\quad L_3 = 0.04\text{ H}$ $\quad R_2 = 670\ \Omega$ $\left(3 \text{ parallel lines } 320 \times \left[\dfrac{345}{138}\right]^2\right)$

A significant surge impedance in this circuit is

$$\left(\frac{L_1}{C_1}\right)^{1/2} = \left[\frac{0.11}{2.5 \times 10^{-8}}\right]^{1/2} = 2100\ \Omega$$

The 80 Ω of the 345-kV lines is so much lower than this value that it will overdamp the circuit, as explained in Chapter 4. In fact, for the parameter values given above, the source circuit can be replaced with very little error by a parallel RL circuit in which $L = L_1 + L_2 + L_3$ and $R = R_1$. We will so treat it, and consider later the small modification introduced by the capacitance.

The response of a parallel RL circuit to a ramp of current $I_0 t$ is readily shown to be

$$v(s) = \frac{RI_0}{s(s + R/L)}$$

in operational form or

$$V(t) = LI_0[1 - \epsilon^{-Rt/L}] \tag{13.7.1}$$

as a time function. But if the impedance limiting the fault being switched resides almost exclusively in the source, rather than the line, the slope I_0 of the current ramp, simulating the fault current, will be

$$I_0 = \frac{V_P}{L} \tag{13.7.2}$$

where V_P is the peak system voltage. Substituting Eq. 13.7.2 in 13.7.1 gives

$$V(t) = V_P(1 - \epsilon^{-Rt/L}) = V_P(1 - \epsilon^{-\alpha t}) \tag{13.7.3}$$

This is illustrated in Fig. 13.10*a*.

This analysis asserts that following interruption the bus voltage which was pulled down to near zero by the fault will rise in a manner defined by Eq.

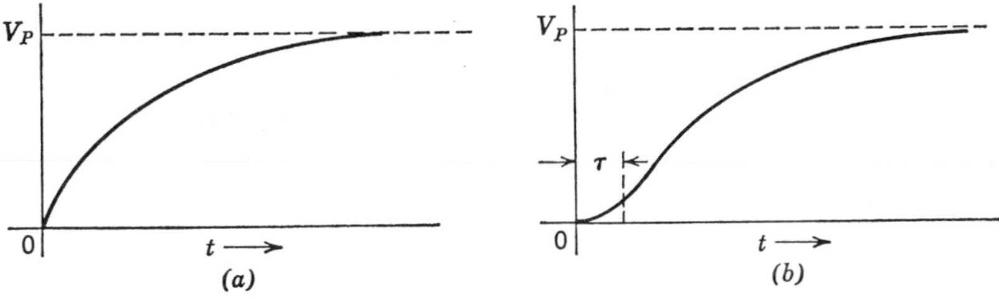

Fig. 13.10. Approximate response of Fig. 13.9*c* to a ramp of current. (*a*) Capacitance neglected. (*b*) Capacitance included.

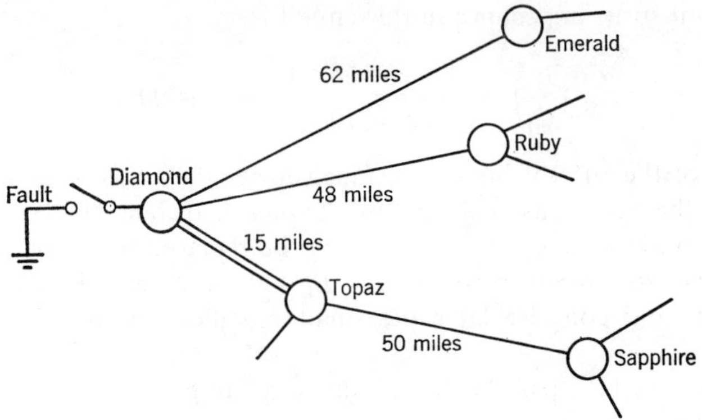

Fig. 13.11. Section of a hypothetical power system.

13.7.3 with a time constant, in this instance, of

$$\frac{L}{R} = \frac{0.22}{80} = 2.75 \times 10^{-3}\ \text{sec}$$

whence

$$V(t) = V_P(1 - \epsilon^{-364t}) \tag{13.7.4}$$

Furthermore, a voltage of this form will travel down each of the connected lines.

We have chosen to neglect the effect of the capacitance at the stations and will continue to do so. But let us consider what in fact the capacitance will do. Its principal effect is to introduce what amounts to a delay, as indicated in Fig. 13.10*b*, which increases as C increases. In most cases T will be quite short, typically 3–5 μsec.

The hypothetical system sketched in Fig. 13.11 provides the additional details needed for further analysis. Each of the circles named for some precious stone represents a station. The lines between are the tie circuits. The lengths of these transmission lines are given and we note that between Diamond and Topaz there is a double circuit line. We assume that the events described above have been occurring at the Diamond station and that the voltage described by Eq. 13.7.4 is the bus voltage at that station following the breaker operation. Waves of this form, generated by this switching, will travel to Emerald, Ruby, and Topaz, suffer reflection and refraction at these points, and the reflected waves, returning to Diamond, will set off new waves when they arrive. The times of arrival will depend upon the travel times to the distant stations and therefore on the circuit miles covered. In order to follow events, it is important to tabulate data of this kind. This is done in Table 13.1 on the assumption of a travel time of 10.8 μsec per round trip mile.

Table 13.1

From	To and Back	Time (μsecs)
Diamond	Emerald	670
Diamond	Ruby	518
Diamond	Topaz	162
Topaz	Sapphire	540

It is essential to have a code to identify the different waves. The following designation is quite useful. The stations are represented by letters arranged in sequence in the order in which the stations have been encountered by the wave concerned. For example,

EDTD (832)

describes a wave that travels first to Emerald, is reflected and returns to Diamond. Thence it has continued to Topaz from which it has subsequently traveled back to Diamond. The time for the whole procedure is indicated by the number in parenthesis following the letters, in this case 832 μsec.

To circumscribe the problem, attention will be confined to events at the Diamond station bus during the first 1000 μsec after current interruption. Waves arriving after that time will be disregarded. Stations not shown in Fig. 13.11 are assumed to be so remote as to preclude their returning waves during the period of interest. This notwithstanding, it will be found that numerous waves are involved. They are summarized in Table 13.2. The notation $(TD)^2$, $(TD)^3$, and so on, indicates two and three sequential round trips, respectively, on this circuit.

The effect of the many waves shown in Table 13.2 on the bus voltage at Diamond will depend upon the impedance of that station and connected lines, and also the impedance of the other stations that the wave has encountered. For example, the wave TSTD (702) is a remnant of the initial wave described by Eq. 13.7.4, which traveled to Topaz. There it was reflected and refracted. The part that traveled on to Sapphire to be modified by the reflection coefficient there, returned to Topaz to be modified again by the refraction coefficient there, and finally returned to Diamond, affects the Diamond bus voltage after 720 μsec.

As a first approximation, each station can be represented by a parallel *RL* circuit in the same way that the Diamond station has been represented. The inductance will depend upon the source impedance of the station and the resistance upon the transmission lines connected to its bus. Figure 13.12 represents the first generation of encounters to be evaluated. This requires the application of the reflection and refraction coefficients designated in Eqs.

Table 13.2 Waves Arriving at Diamond Station Bus in the First Millisecond after Interrupting a Fault

TD	(162)
$(TD)^2$	(324)
$(TD)^3$	(486)
RD	(518)
$(TD)^4$	(648)
ED	(670)
TDRD	(680)
RDTD	(680)
TSTD	(702)
$(TD)^5$	(810)
EDTD	(832)
TDED	(832)
$(TD)^2RD$	(842)
$(RD)(TD)^2$	(842)
TDRDTD	(842)
$TS(TD)^2$	(864)
TDTSTD	(864)
$(TD)^6$	(972)
$ED(TD)^2$	(994)
$(TD)^2ED$	(994)
TDEDTD	(994)

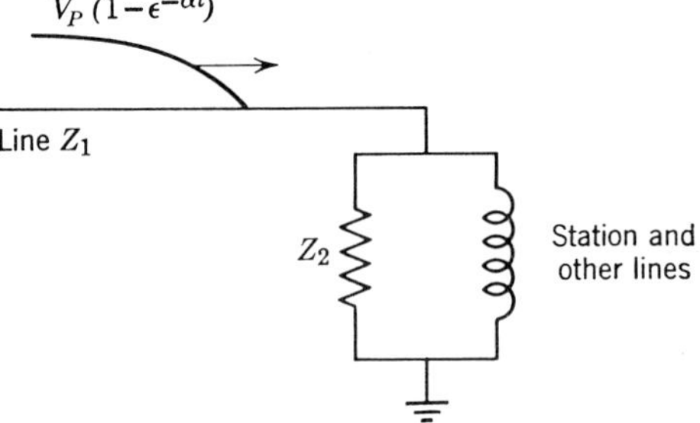

Fig. 13.12. First-generation encounter between wave and station.

9.3.7 and 9.3.8. The reflected wave will be

$$V_r = aV_i = V_i\left[\frac{Z_2 - Z_1}{Z_2 + Z_1}\right] \tag{13.7.5}$$

where V_i is the incident wave, Z_2 is the impedance of the station being encountered, and Z_1 the impedance of the line along which the incident wave is advancing. In operational form,

$$v_i(s) = \frac{V_P}{R}\frac{L}{s[s + R/L]} \tag{13.7.6}$$

$$Z_2(s) = \frac{RLs}{R + Ls} \tag{13.7.7}$$

Thus the reflection coefficient is $a(s)$:

$$\frac{Z(_2s) - Z_1(s)}{Z_2(s) + Z_1(s)} = \frac{Z_2(s)}{Z_2(s) + Z_1(s)} - \frac{Z_1(s)}{Z_2(s) + Z_1(s)}$$

Substituting from Eq. 13.7.7,

$$a(s) = \frac{R}{R + Z_1}\frac{s}{\{s + Z_1R/L(R + Z_1)\}} - \frac{Z_1}{L(R + Z_1)}\frac{(R + Ls)}{\{s + Z_1R/L(R + Z_1)\}} \tag{13.7.8}$$

Let

$$\beta = \frac{Z_1R}{L(R + Z_1)}$$

then reflection coefficient is

$$a(s) = \left[\frac{R - Z_1}{R + Z_1}\right]\frac{s}{s + \beta} - \frac{\beta}{s + \beta} \tag{13.7.9}$$

The incident wave is given in Eq. 13.7.3, whose transform is

$$v_i(s) = V_P\frac{1}{s(s + \alpha)} \tag{13.7.10}$$

The transform of the reflected wave is therefore

$$v_r(s) = V_P\left\{\left[\frac{R - Z_1}{R + Z_1}\right]\frac{1}{(s + \alpha)(s + \beta)} - \frac{\beta}{s(s + \alpha)(s + \beta)}\right\} \tag{13.7.11}$$

The transmitted or refracted wave at such an encounter can be determined in the same manner:

$$V_t = bV_i = V_i\left[\frac{2Z_2}{Z_1 + Z_2}\right] \tag{13.7.12}$$

The operational form of the transmission coefficient is given by

$$b(s) = \frac{2R}{R + Z_1} \frac{s}{s + \beta} \tag{13.7.13}$$

From 13.7.10 and 13.7.13,

$$v_t(s) = b(s)v_i(s) = V_P\left\{\frac{2R}{R + Z_1} \frac{1}{(s + \alpha)(s + \beta)}\right\} \tag{13.7.14}$$

The inverse transform of Eq. 13.7.14 gives the bus voltage at the station after the incident wave arrives. It also describes the voltage waves that travel down the several lines that R represents.

The terms of Eqs. 13.7.11 and 13.7.14 have the following inverse transforms:

$$\left.\begin{aligned} \mathcal{L}^{-1}\frac{1}{(s + \alpha)(s + \beta)} &= \frac{1}{\beta - \alpha}[\epsilon^{-\alpha t} - \epsilon^{-\beta t}] \\ \mathcal{L}^{-1}\frac{1}{s(s + \alpha)(s + \beta)} &= \frac{-\beta\epsilon^{-\alpha t} + \alpha\epsilon^{-\beta t} + (\beta - \alpha)}{\alpha\beta(\beta - \alpha)} \end{aligned}\right\} \tag{13.7.15}$$

by which the equations can be interpreted.

After encounters of this type the reflected and refracted waves return whence they came or proceed to more distant stations; in either event they experience what we might term second-generation encounters. The waves described by Eqs. 13.7.11 and 13.7.14 and evaluated by Eqs. 13.7.15 are modified by the reflection and refraction coefficients appropriate to the next station or discontinuity. These coefficients will have the same general form as Eqs. 13.7.8 and 13.7.13, but the R, Z_1, and β will differ numerically because of the different station and line characteristics. Evidently the products of the second-generation encounters will lead to transforms of the following kind:

$$\frac{1}{(s + \alpha)(s + \beta)(s + \gamma)}, \qquad \frac{1}{s(s + \alpha)(s + \beta)(s + \gamma)}, \qquad \cdots$$

This description should convey the nature of the problem; the bookkeeping is best left to the computer.

The modification of switching surges wrought by the presence of transmission lines is now sufficiently well understood for efforts to have been made (14) to incorporate the details into specifications for circuit breaker performance.

13.8 Transients in the Industrial Power Network

The behavior of industrial power systems somewhat parallels that of the utility networks described in the last few sections. However, the relatively short distances involved and the generally lumpy character of many of the components create what is better described as interfering oscillations rather than traveling waves. We describe here a particularly striking example. The system involved is the rectifier power supply for a mill motor; a unit is shown in Fig. 13.13. The individual rectifier tanks are fed through short lengths of bus from the secondaries of the rectifier transformers, which in turn are connected by interphase transformers.

Occasionally one of the mercury arc rectifier tanks will suppress its current (15). The effect, analogous to current chopping described in Section 5.2, is like a sudden injection of a negative step of current at the offending rectifier. This current permeates the entire system, dividing according to the impedances of the paths it encounters, and evoking some kind of response from

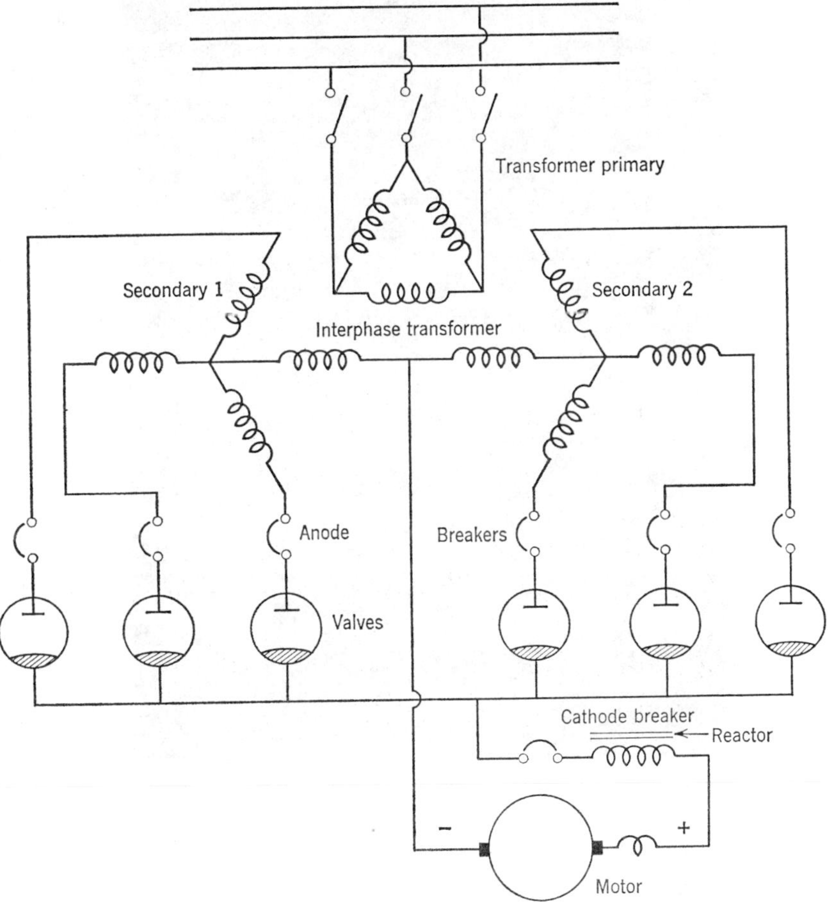

Fig. 13.13. Circuit of a mill motor drive using mercury arc rectifiers.

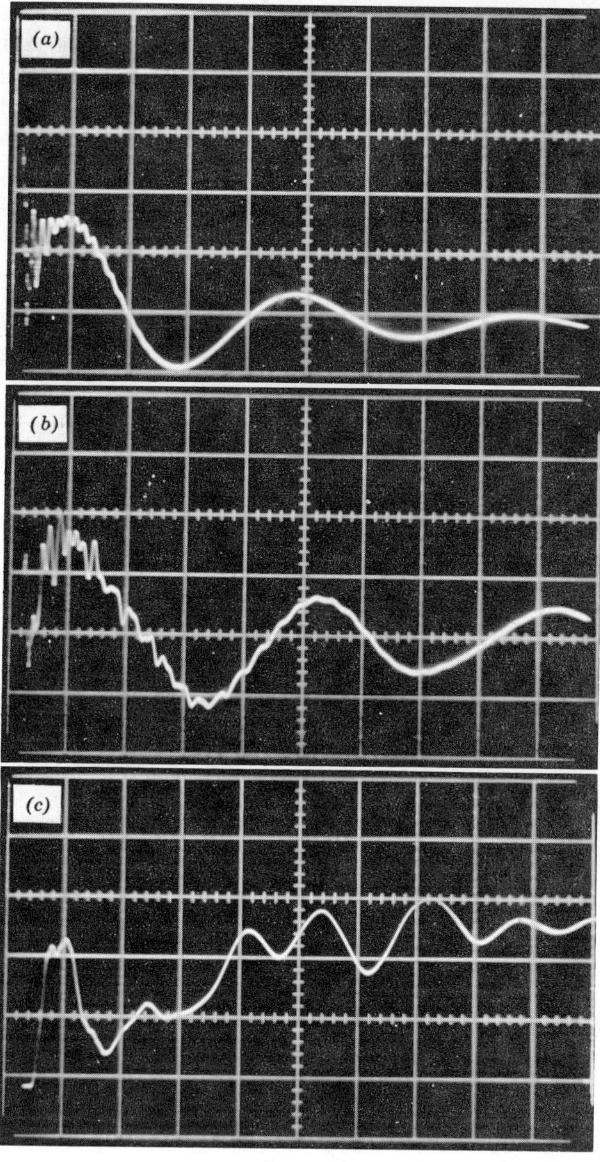

Fig. 13.14. Response of the circuit shown in Fig. 13.13 to a sudden suppression of current in a rectifier tank. (*a*) One major division = 50 μsec. (*b*) One major division = 2 μsec. (*c*) One major division = 0.2 μsec.

each branch. It was explained in Chapter 5 that the response is proportional to the current and the surge impedance of the branch or component involved. At the source of the disturbance, in this case the rectifier tank where the current is suppressed, the aggregate of the many individual responses will occur, whereas across individual elements of the circuit the individual responses will be experienced.

This is illustrated in Fig. 13.14 by measurements on an actual system. Details of the actual experimental method will be found in Chapter 17. The oscillograms were all taken at the rectifier itself and therefore represent the response of the whole system. However, they have different time scales, so only in Fig. 13.14*a* can the complete story be seen. The other two figures are more detailed resolutions of events at the beginning of the disturbance. There are three distinct frequencies involved, which it was possible to isolate by measurements at other locations. The lowest frequency, approximately 5 kHz, only observable in Fig. 13.14*a*, is contributed by the interphase transformer. The intermediate frequency, approximately 125 kHz, is associated with the transformer. The very high frequency, approximately 3 MHz, is a product of the buswork between the rectifier tank and the transformer. Only this last has any real semblance of a wave phenomenon. The first pulse in Fig. 13.14*c* is a consequence of the initial wave being wiped out by a reflection from the first significant discontinuity. This happens to be the transformer, whose capacitance momentarily appears as a short circuit. The round-trip travel time is about 0.12 μsec, or about 120 ft for open bus. This corresponds to the physical locations of the rectifier and the transformers. The wave subsequently is modified by further reflections and by current penetration into the transformer. It was actually possible to associate the dimple in the top of the initial voltage pulse with a minor physical discontinuity in the bus structure. A more extensive investigation of this type has been described by Greenwood, Kotheimer, and Langlois (16) as it affects the power supply for an aluminum smelter.

REFERENCES

1. A. R. Hileman et al., "Insulation Coordination in APS 500 kV Stations," *Trans. IEEE*, PAS Vol. 86 (1967), p. 655.
2. AIEE Committee Report, "Switching Surges, I—Phase-to-Ground Voltages," *Trans. AIEE*, Vol. 80 (1961), p. 240.
3. W. F. Skeats, C. H. Titus and W. R. Wilson, "Severe Rate of Rise of Recovery Voltage Associated with Transmission Line Short Circuits, *Trans. AIEE*, Vol. 76 (1957), p. 1256.
4. E. Bolton, A. C. Ehrenberg, F. L. Hamilton, A. G. Hawkins, F. P. Matravers, and J. A. Thomas, "British Investigations of Short Line Fault Phenomena," CIGRE Report No. 109 (1964).

5. E. W. Kimbark, *Electrical Transmission of Power and Signals*, John Wiley & Sons, New York, (1949).
6. I. B. Johnson, A. J. Schultz, N. R. Schultz, and R. B. Shores, "Some Fundamentals on Capacitance Switching," *Trans. AIEE*, Vol. 74, Part III (1955), p. 727.
7. N. E. Dillow, I. B. Johnson, N. R. Schultz and A. E. Were, "Switching Capacitor Kilowatt-Amperes with Power Circuit Breakers," *Trans. AIEE*, Vol. 11 (1952), p. 188.
8. A. G. McElroy, Discussion of paper "Switching of Extra-High Voltage Circuits II—Surge Reduction with Circuit Breaker Resistors" by D. E. Hedman, I. B. Johnson, C. H. Titus, and D. D. Wilson, *Trans. IEEE*, Vol. 83 (1964), p. 1196.
9. A. G. McElroy, W. C. Price, H. M. Smith and D. F. Shankle, "Field Measurements of Switching Surges as Modified by Unloaded 345KV Transformers," *Trans. IEEE*, Vol. 82, PAS (1963), p. 500.
10. D. E. Hedman et al., "Switching of Extra-High Voltage Circuits II Surge Reduction with Circuit Breaker Resistors," *Trans. IEEE*, Vol. 83 (1964), p. 1196.
11. I. B. Johnson, V. E. Phillips and H. O. Simmons, "Switching of Extra-High-Voltage Circuits I—System Requirements for Circuit Breakers," *Trans. IEEE*, Vol. 83 (1964), p. 1187.
12. I. B. Johnson, R. F. Silva and D. D. Wilson, "Switching Surges due to Energization or Reclosing," *Proc. American Power Conference*, Vol. 23 (1961), p. 229.
13. E. W. Kimbark and A. C. Legate, "Fault Surge Versus Switching Surge, A Study of Transient Voltages Caused by Line to Ground Faults," *Trans. IEEE*, Vol. 87 (1968), p. 1762.
14. O. Naef, C. P. Zimmerman and J. E. Beehler, "Proposed Transient Recovery Voltage Ratings for Power Circuit Breakers," *Trans. IEEE*, PAS Vol. 86 (1965), p. 580.,
15. A. W. Hull and F. R. Elder, "The Cause of High Voltage Surges in Rectifier Circuits," *Jour. Appl. Phys.*, Vol. 13 (1943), p. 372.
16. A. N. Greenwood, W. C. Kotheimer and C. A. Langlois, "An Investigation of the Transient Behavior of an Aluminum Potline Installation," *Trans. AIEE.*, Vol. 80, Part II (1961), p. 1.

14 Computer Aids to the Calculation of Electrical Transients

14.1 Introduction

Many computational aids can help the student with a firm understanding of electrical transients in solving electrical transient problems. The "apprenticeship" of working analytically and experimentally with transients is a prerequisite if the computer techniques are to be exercised with the necessary judgment. It is evident from the material of Chapter 13, if not from analysis of much simpler problems such as that of Section 3.4, that short cuts must be used if solutions of acceptable engineering accuracy are to be found within a reasonable time. This is where a computer can help and, indeed, where without such assistance, some problems would be quite intractable. On the other hand, it should be stressed that the computer is an important adjunct to analysis and experiment, not a substitute for it.

Computer aids span a spectrum of techniques and devices. It is our purpose in this chapter to examine the more important of these and show how they can be used to advantage in power system transient problems. It should be recognized, however, that this is a highly specialized field which is developing rapidly in some areas. This book presents an introduction to the subject only.

We use the word computer here in a fairly broad sense. It is not confined to hardware, but includes nomograms and precalculated curves such as the damping curves included in Chapter 4. Section 14.2 will be concerned with these aspects of the subject. Beyond this, in hardware, we recognize two types of computers, the analog computer and the digital computer. Both of these are capable of further subdivision. Analog computers, especially, come in several different forms. They attempt to simulate components, circuits, or systems so that when one applies a stimulus to the computer, it responds with an answer that the system would give to a corresponding stimulus. This is essentially an experimental approach. Model systems, transient network analyzers, electrolytic tanks, and so forth, fall into this category. The operational electronic analog computer is somewhat different in that it sets out to solve the equations of the system by building an analog to them.

The digital computer also recognizes that transient events can be expressed in mathematical terms and set down as equations. Given a statement of these relationships, it can grind out solutions. It is, of course, the extreme rapidity with which it executes these operations that make the digital computer so useful. In fact, as we shall see, the relationships can be quite complex, so that their translation into forms that the computer can handle demands considerable ingenuity on the part of the engineer. The computer is a tool; the artistry comes in maximizing its potential and economizing in its time. These are the considerations that govern the way it is used.

14.2 Precalculated Curves for Computing Switching Transient Recovery Voltages

In Section 3.4 we calculated the transient recovery voltage following a switching operation in a two-mesh circuit. For the seeming triviality of the circuit, it was a protracted calculation. Many practical power system transient switching problems can be safely reduced to a circuit of this kind. There are other two frequency circuits, which, although less prevalent, are of considerable practical importance (Fig. 14.1). All these circuits have two common features, across the terminal, on the right, which represent the contacts of the circuit breaker, there is a continuous path for current through capacitance alone, and there is similar path through inductance alone. The second path is the course of the power current when the switch is closed. This group of circuits covers all two-frequency circuits likely to be encountered in power systems. The operation of a circuit breaker to clear a fault in any one of these circuits can be simulated by injecting minus the fault current into the circuit at the right-hand terminals.

It was pointed out by K. S. Johnson (1), who worked with similar circuits in the communications field, that an equivalence could be established between any pair of these circuits, that is, any one could be transformed into any other by manipulating the values of the *L*s and *C*s. Boehne (2) recognized the usefulness of this for the calculation of transient recovery voltages in power circuit switching. Specifically, the response of the *RS* circuit (Fig. 14.1*f*) is particularly easy to compute, since the two frequencies are completely decoupled. They are given by

$$f_R = \frac{1}{2\pi(L_R C_R)^{1/2}} \quad \text{and} \quad f_S = \frac{1}{2\pi(L_S C_S)^{1/2}} \tag{14.2.1}$$

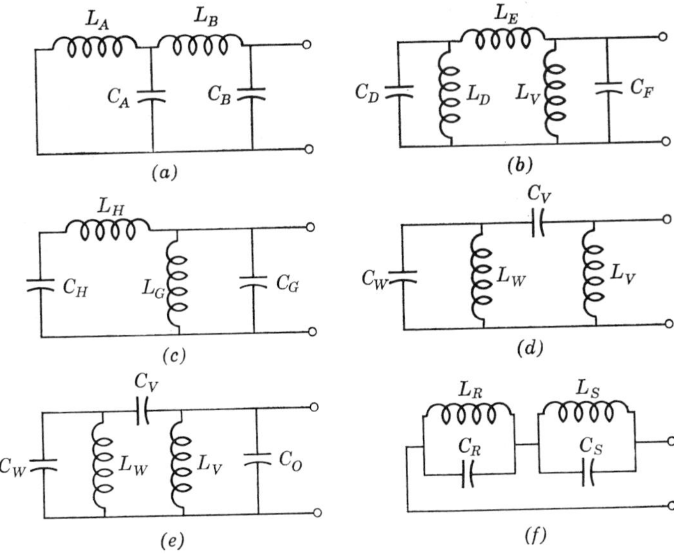

Fig. 14.1. Two frequency circuits encountered in switching transients. (*a*) *AB* circuit. (*b*) *DEF* circuit. (*c*) *HG* circuit. (*d*) *WV* circuit. (*e*) *WVO* circuit. (*f*) *RS* circuit.

The magnitudes of the voltage components at these frequencies are also readily calculable, inasmuch as they are proportional to the inductances L_R and L_S. The response of the *RS* circuit to a ramp of current $V/(L_S + L_S)$ amperes per second is given by

$$V(t) = \frac{V}{L_R + L_S}[L_R(1 - \cos 2\pi f_R t) + L_S(1 - \cos 2\pi f_S t)]$$

$$= V[a_R(1 - \cos \omega_R t) + a_S(1 - \cos \omega_S t)] \qquad (14.2.2)$$

where

$$a_R = \frac{L_R}{L_R + L_S}, \qquad a_S = \frac{L_S}{L_R + L_S}$$

Now, if relationships can be established between the *L*s and *C*s of the circuits in Fig. 14.1*a* to 14.1*e* and the components $L_R C_R L_S C_S$ of circuit Fig. 14.1*f*, the transient response of the other five circuits can be expressed in the same form as Eq. 14.2.2. This was what Boehne (2) did. The same approach was used by Hammarlund (3), who went on to extend the method to three-loop circuits. Since the *AB* circuit is by far the most common, it was chosen as the prototype or reference circuit, and formulas were derived for transposing the other circuits into this circuit. The results of an evaluation of the *AB* circuit given by Hammarlund shows that the circuit can be written quite simply as follows:

$$\left.\begin{aligned} \omega_R &= \left[\frac{a-b}{a+b}\right]^{1/4}(\omega_A\omega_B)^{1/2}, & \omega_S &= \left[\frac{a-b}{a-b}\right]^{1/4}(\omega_A\omega_B)^{1/2} \\ a_R &= \frac{(a+b-\alpha^2)(a+b)}{2(\beta+1)b}, & a_S &= 1 - a_R \end{aligned}\right\} \tag{14.2.3}$$

where

$$a = \tfrac{1}{2}(1+\beta+\alpha^2), \qquad b = (a^2-\alpha^2)^{1/2}$$

$$\alpha = \left[\frac{L_A C_A}{L_B C_B}\right]^{1/2} = \frac{\omega_B}{\omega_A}, \qquad \beta = \frac{L_A}{L_B}$$

$$\omega_A = \frac{1}{(L_A C_A)^{1/2}}, \qquad \omega_B = \frac{1}{(L_B C_B)^{1/2}}$$

If we introduce

$$k = \left[\frac{a+b}{a-b}\right]^{1/4}$$

and

$$\omega_M = (\omega_A\omega_B)^{1/2}$$

we can write

$$\left.\begin{aligned} \omega_S &= k\omega_M \\ \omega_R &= \frac{1}{k}\omega \end{aligned}\right\} \tag{14.2.4}$$

Hammarlund proceeded to calculate and plot k as a function of α and β and a_R as a function of α and β. The results of these calculations are presented in Figs. 14.2 and 14.3. The curves allow us to calculate the response of any *AB* circuit (or any of the other circuits, as we shall see shortly) by the following simple procedure:

1. Compute α, β, and ω_M.
2. Enter the frequency curves with α and β and find k, then compute ω_S and ω_R from Eq. 14.2.4.
3. Enter the amplitude curves with α and β and find a_R, then compute a_S from Eq. 14.2.3. The total response is given by Eq. 14.2.2.

As previously pointed out, there is an equivalence between any pair of circuits in Fig. 14.1; thus it is possible to transpose any one into any other. Specifically, having established the curves of Figs. 14.2 and 14.3 for the *AB* circuit, it is natural to transpose the circuits of Fig. 14.1*b* to 14.1*c* into this

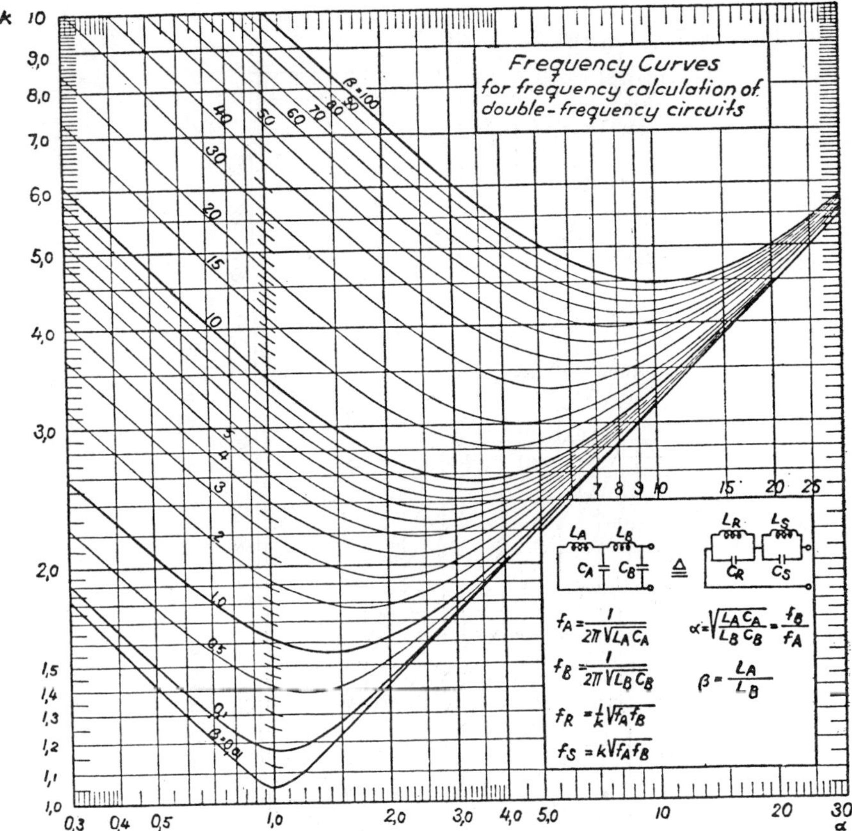

Fig. 14.2. Curves for frequency computation of two-frequency circuits.

prototype circuit so that these curves can be used for these circuits also. The formulas for transposing these circuits into the prototype circuit are given in Table 14.1.

We can now use the curves to obtain a numerical solution to a specific problem. This is represented in Fig. 14.4*a*. We are interested in determining the form of the recovery transient appearing after the circuit breaker operates to clear a short circuit near its terminals. The reactor has been added to limit the short circuit current to 5000 A. Figure 14.4*b* shows the equivalent circuit. The components have the following significance and values:

L_A = source inductance (transformer, etc.) = 2.66 mH (1 Ω at 60 Hz)
L_B = reactor inductance = 5.31 mH (2 Ω at 60 Hz)
C_A = capacitance of equipment on the bus = 0.008 μF
C_B = reactor and breaker capacitance = 450 pF

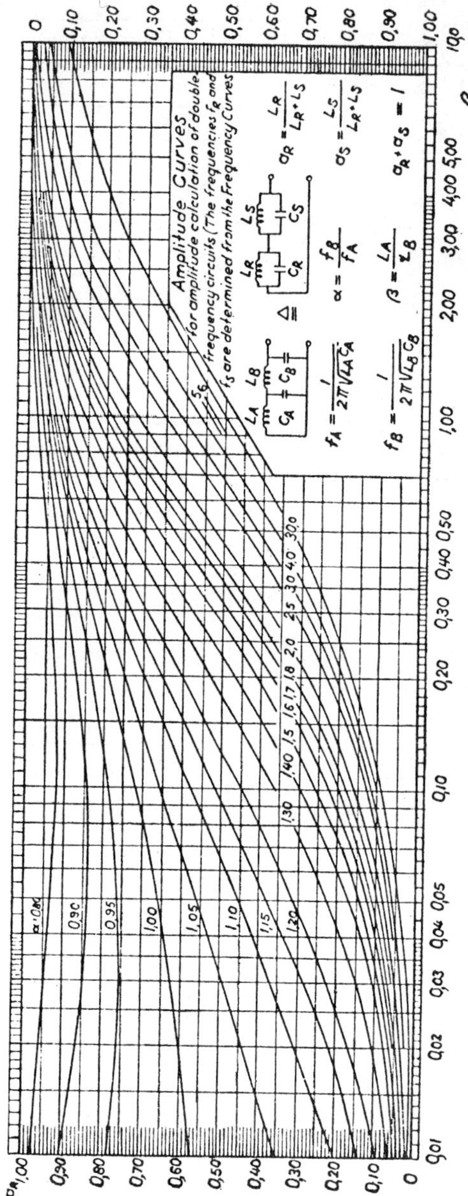

Fig. 14.3. Curves for amplitude computation of two-frequency circuits.

Table 14.1 Formulas for the Transfer of Different Circuits to the Prototype Circuit and for Direct Computation of the Parameters To Be Used in Figs. 14.2 and 14.3

AB	DEF	HG	WV	WVO
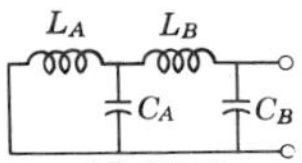 *AB* Circuit (L_A, L_B, C_A, C_B)	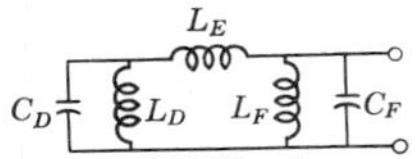 *DEF* Circuit (L_E, C_D, L_D, L_F, C_F)	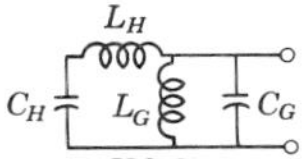 *HG* Circuit (L_H, C_H, L_G, C_G)	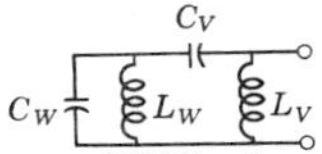 *WV* Circuit (C_V, C_W, L_W, L_V)	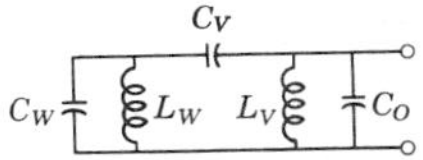 *WVO* Circuit (C_V, C_W, L_W, L_V, C_O)
L_A	$L_E \dfrac{L_F}{L_E + L_F} \beta$	$L_G \dfrac{L_G}{L_G + L_H}$	$L_V \dfrac{\beta}{1 + \beta}$	$L_V \dfrac{\beta}{1 + \beta}$
L_B	$L_E \dfrac{L_F}{L_E + L_F}$	$L_H \dfrac{L_G}{L_G + L_H}$	$L_V \dfrac{1}{1 + \beta}$	$L_V \dfrac{1}{1 + \beta}$
C_A	$C_D \left(\dfrac{L_E + L_F}{L_F}\right)^2$	$C_H \left(\dfrac{L_H + L_G}{L_G}\right)^2$	$\dfrac{C_V{}^2}{C_V + C_W} \left(\dfrac{1 + \beta}{\beta}\right)^2$	$\dfrac{C_V{}^2}{C_V + C_W} \left(\dfrac{1 + \beta}{\beta}\right)^2$
C_B	C_F	C_G	$\dfrac{C_V C_W}{C_V + C_W}$	$\dfrac{C_V C_W + C_0 C_V + C_0 C_W}{C_V + C_W}$
$\alpha = \left(\dfrac{L_A C_A}{L_B C_B}\right)^{1/2}$	$\dfrac{L_E + L_F}{(L_E L_F)^{1/2}} \left(\dfrac{L_D}{L_D + L_E + L_E} \cdot \dfrac{C_D}{C_F}\right)^{1/2}$	$\dfrac{L_G + L_H}{(L_G L_H)^{1/2}} \left(\dfrac{C_H}{C_G}\right)^{1/2}$	$(\beta + 1) \dfrac{C_V + C_W}{(C_V C_W)^{1/2}} \left(\dfrac{L_W}{L_V}\right)^{1/2}$	$(\beta + 1) \dfrac{C_V + C_W}{(C_V C_W)^{1/2}} \left(\dfrac{L_W}{L_V}\right)^{1/2} \dfrac{1}{(1 + C_0/C_W + C_0/C_V)^{1/2}}$
$\beta = \dfrac{L_A}{L_B}$	$\dfrac{L_D L_F}{L_E(L_D + L_E + L_F)}$	$\dfrac{L_G}{L_H}$	$\dfrac{L_V}{L_W} \left(\dfrac{C_V}{C_V + C_W}\right)^2$	$\dfrac{L_V}{L_W} \left(\dfrac{C_V}{C_V + C_W}\right)^2$
$\omega_A = \dfrac{1}{(L_A C_A)^{1/2}}$	$\dfrac{1}{(L_D C_D)^{1/2}} \left(\dfrac{L_D + L_E + L_F}{L_E + L_F}\right)^{1/2}$	$\dfrac{1}{[(L_G + L_H) C_H]^{1/2}}$	$\dfrac{1}{(L_V C_V)^{1/2}} \times \left(\dfrac{C_V + C_W}{C_V} \cdot \dfrac{\beta}{1 + \beta}\right)^{1/2}$	$\dfrac{1}{(L_V C_V)^{1/2}} \left(\dfrac{C_V + C_W}{C_V} \dfrac{\beta}{1 + \beta}\right)^{1/2}$
$\omega_B = \dfrac{1}{(L_B C_B)^{1/2}}$	$\dfrac{1}{(L_F C_F)^{1/2}} \left(\dfrac{L_E + L_F}{L_E}\right)^{1/2}$	$\dfrac{1}{(L_G C_G)^{1/2}} \times \left(\dfrac{L_G + L_H}{L_H}\right)^{1/2}$	$\dfrac{1}{(L_V C_V)^{1/2}} \times \left[\dfrac{C_V + C_W}{C_W} (1 + \beta)\right]^{1/2}$	$\dfrac{1}{(L_V C_V)^{1/2}} \left[\dfrac{(C_V + C_W) C_V}{C_V C_W + C_0 C_V + C_0 C_W} (1 + \beta)\right]^{1/2}$
$\omega_M = (\omega_A \omega_B)^{1/2}$	$(\omega_D \omega_F)^{1/2} \left(1 + \dfrac{L_D}{L_E} + \dfrac{L_F}{L_E}\right)^{1/4}$	$(\omega_G \omega_H)^{1/2}$	$(\omega_V \omega_W)^{1/2}$	$(\omega_V \omega_W)^{1/2} \dfrac{1}{(1 + C_0/C_W + C_0/C_V)^{1/4}}$

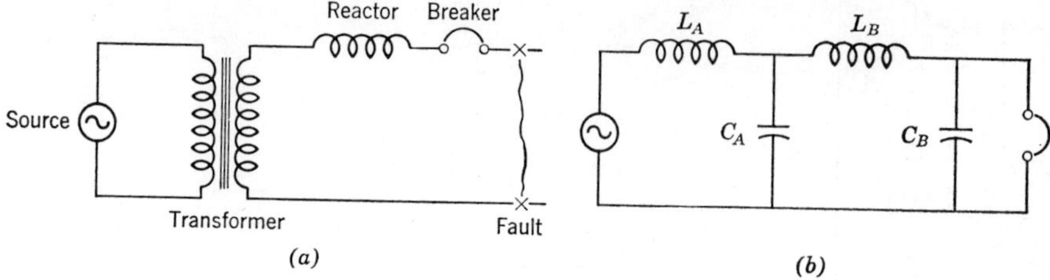

Fig. 14.4. Clearing a fault. (*a*) Actual circuit. (*b*) Equivalent circuit.

Using this data we calculate that

$$\alpha = \left[\frac{L_A C_A}{L_B C_B}\right]^{1/2} = \left[\frac{2.66 \times 8 \times 10^{-12}}{5.31 \times 4.5 \times 10^{-13}}\right]^{1/2} = 2.98, \qquad \beta = 0.5$$

$$\omega_A = \frac{10^6}{(2.66 \times 8)^{1/2}} = 2.17 \times 10^5, \qquad \omega_B = \frac{10^7}{(5.31 \times 45)^{1/2}} = 4.45 \times 10^5$$

$$\omega_M = (2.17 \times 4.45)^{1/2} \times 10^5 = 3.13 \times 10^5$$

From Fig. 14.2, $k = 1.77$, whence the two natural frequencies are

$$\omega_S = k\omega_M = 5.32 \times 10^5 \text{ rad/sec}$$

$$\omega_R = \frac{1}{k}\,\omega_M = 1.76 \times 10^5 \text{ rad/sec}$$

From Fig. 14.3, $a_S = 0.58$, $a_R = 0.42$, which gives for the recovery voltage transient

$$V = 15\sqrt{2}[0.58(1 - \cos 5.32 \times 10^5 t) + 0.42(1 - \cos 1.76 \times 10^5 t)]$$

No damping is included in this calculation.

14.3 The Electronic Analog Computer

This computer is well suited for solving electrical transient problems in lumpy circuits and it is especially attractive for investigations of the effect of varying one, or more of the circuit elements over a range of values. It is thus very valuable in design where we may be attempting to optimize a desired condition or minimize a particular transient. This computer basically "solves" ordinary differential equations and displays the results in a very informative manner.

The computer comprises a variety of units, each being designed to perform a specific function. Three of the more important of these units are shown in Fig. 14.5. The first, designated A, is capable of adding a number of electrical

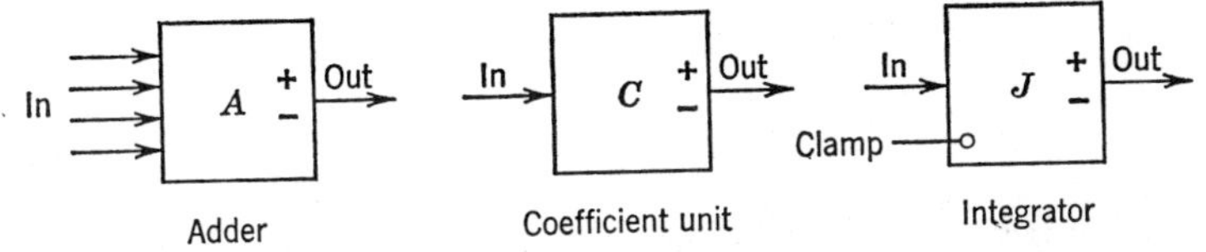

Fig. 14.5. Three important units in the electronic analog computer.

signals. It takes these signals as input and gives their sum as output. The coefficient unit, C, is used to multiply a voltage by a constant. The third unit, J, integrates whatever signal is put into it and produces the integral at its output. There are several other units with different functions, but these three illustrate the technique here. Each of these units has a negative as well as a positive output. It is therefore possible to obtain minus the integral of the input to a J unit, or minus the sum of the signals being fed into an A unit. In addition to these devices the computer has a display unit, which is a multichannel cathode ray oscilloscope (CRO), and a number of signal generators.

The best way to understand how to use a computer of this kind is to consider a simple problem. Let us take the very simple circuit of Fig. 14.6*a* and ask what happens when we close the switch. We start by setting down the first-order differential equation for the circuit,

$$V = RI + L\frac{dI}{dt} \tag{14.3.1}$$

But when using the analog computer, the equations always should be arranged as if to solve for the term containing the highest order derivative,

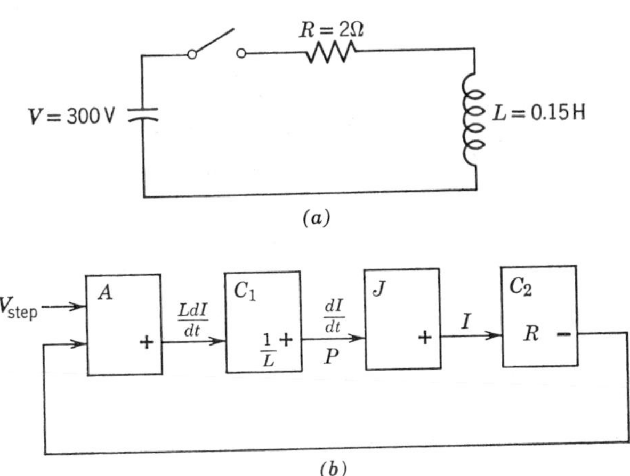

Fig. 14.6. Interconnections for the electronic analog computer to solve the simple *RL* circuit transient.

so that Eq. 14.3.1 is rearranged to read

$$L\frac{dI}{dt} = V - IR \tag{14.3.1a}$$

This equation is now solved by connecting together units from the computer in a particular way. The connections for this problem are shown in Fig. 14.6*b*. We start at point P by assuming that we know dI/dt. We use this as an input to a J unit and thereby obtain the current I from the positive output of the J unit. This signal is then fed into the coefficient unit C_2, which multiplies it by R. The negative output then gives $-RI$, which is added to a step of voltage V by the adder unit. According to Eq. 14.3.1a, this sum is equal to $L(dI/dt)$, so we can take the adder output, multiply it by $1/L$ in coefficient unit C_1, and obtain the dI/dt we assumed we had in the first place. The block diagram forms a closed loop.

The input to the loop is the step voltage V, which usually, although not always, is applied at some steady repetition rate; 60 or 100 times per second is typical. On each occasion it stimulates the circuit into its transient response. After every such operation a signal is sent to each integrator unit to clear it, or reset it to its initial conditions, ready for the next jolt. The display unit is usually a cathode ray oscilloscope which can be plugged to see the output of any unit or with a multichannel CRO, to see several outputs simultaneously. Thus, if we wish to observe the current in the simple problem just described, we would connect the CRO to the output of the J unit; if our interest is the voltage across the reactor, we would look at the output of the adder. The time base on the oscilloscope is triggered synchronously with the initiation of the transient, so that repeated identical sweeps are superimposed, giving the appearance of a steady trace on the CRO screen.

The preceding description provides a general idea of the principle of the electronic analog computer, but more must be said before we can solve a specific problem numerically. We must tackle the problem of scaling. Regardless of the symbols denoting the inputs and outputs of the A, C, and J units in Fig. 14.6*b*, we are really dealing with voltages in the computer. The input to a unit is a voltage, the output is another voltage. Scaling is the process whereby we relate these voltages to each other, to the quantities they represent, and to the dial settings on the units. Scaling also determines the relationship between the computer time scale and the actual time scale of the problem. Computer time and real time do not have to be, and are usually not the same.

Let us consider a coefficient unit first, and let us suppose its purpose in the equation is to multiply a current I by a resistance R to obtain a voltage V. This is shown in Fig. 14.7*a*. This diagram indicates that $I \times R = V$. Figure 14.7*b* indicates how the scaling is to be designated; the current scale

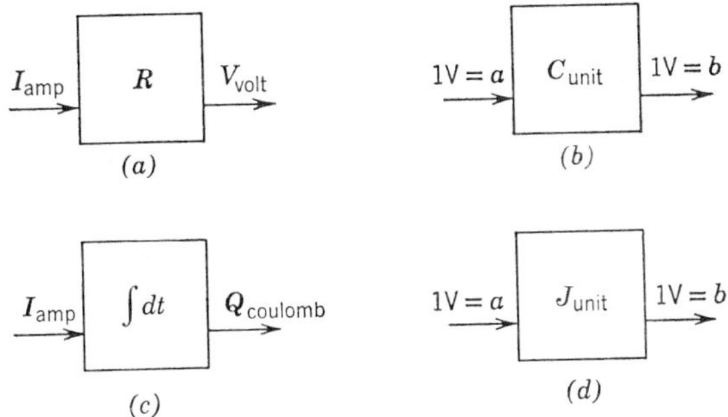

Fig. 14.7. Scaling for coefficient and integrator units.

is 1 V = a A, and the voltage scale is 1 V = b V. Thus, to put in a current I, we actually apply a voltage I/a at the input. If the problem voltage out is V volt, the output from the C unit would be V/b V. But the dial setting on the coefficient unit, which we will designate C, also relates the input and output voltages:

$$e_o = Ce_i \tag{14.3.2}$$

where e_o and e_i are the output and input signals of the unit. Substituting for these quantities gives

$$\frac{V}{b} = \frac{CI}{a} \tag{14.3.3}$$

If we now solve for C and let $V/I = R$, we obtain

$$C = \frac{a}{b} R \qquad \text{or} \qquad R = \frac{b}{a} C \tag{14.3.4}$$

This equation relates the setting C, the input and output scales a and b, and the parameter R that the C unit represents. Once the input and output scales are established, the coefficient unit is calibrated; but in setting the values for a and b we must look at the block diagram as a whole, not just one component.

Moving next to the integrator, we again set up two blocks Figs. 14.7c and 14.7d, one for the actual integration on the left, the other, on the right, for the computer integration. Suppose we wish to integrate a current to obtain charge. This can be written

$$Q = \int_0^\tau I\,dt \tag{14.3.5}$$

which relates input and output of Fig. 14.7c. The upper limit τ is the period of interest. The input and output of a J unit are related by the following equations:

$$e_0 = k\int_0^{t_c} e_i\, dt \tag{14.3.6}$$

where t_c is the time required for the computer to perform the integration, which will be the same as the sweep duration on the CRO, and $1/k$ is the time constant of the integrator unit. These units have three position settings or J numbers, which determine k. For units manufactured by the Philbrick Co. these are

$$Jk = 2500$$

We now substitute for the computer input and output voltages their values in terms of the scales a and b:

$$e_0 = \frac{Q}{b}$$

$$e_i = \frac{I}{a} \tag{14.3.7}$$

Substituting from Eq. 14.3.7 in Eq. 14.3.6,

$$\frac{Q}{b} = k\int_0^{t_c} \frac{I}{a}\, dt \tag{14.3.8}$$

Combining Eqs. 14.3.8 and 14.3.5,

$$Q = \frac{bk}{a}\int_0^{t_c} I\, dt = \int_0^{\tau} I\, dt \tag{14.3.9}$$

Let $I = 1$ A; this equation then gives a simple algebraic relationship:

$$\frac{bk}{a} t_c = \tau = \frac{2500}{J} \times \frac{b}{a} \times t_c \tag{14.3.10}$$

Suppose that $t_c = 40$ m sec; Eq. 14.3.10 then becomes

$$T = \frac{100}{J}\frac{b}{a}$$

or

$$\frac{100}{JT} = \frac{a}{b} \tag{14.3.11}$$

This is the integrator scaling unit.

The only other kind of unit we have used so far in our sample problem is the adder, or A unit. For this, the input and output scales are the same, $a/b = 1$.

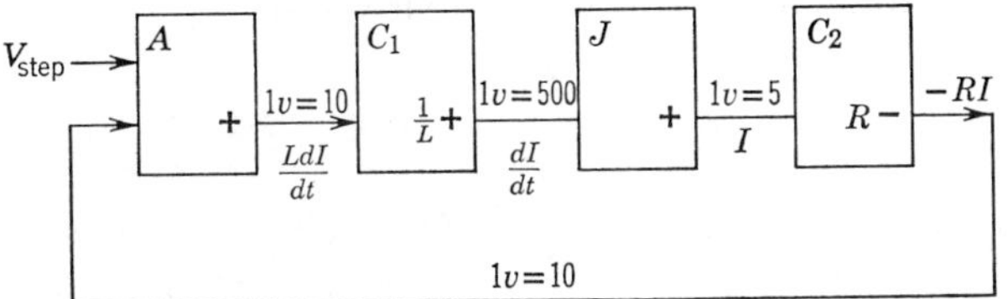

Fig. 14.8. Block diagram for the problem of Fig. 14.6 with appropriate scaling.

We demonstrate scaling by going through the procedure for the problem of Fig. 14.6 with some specific numerical values. To this end, this figure has been redrawn with its scaling factor in Fig. 14.8. There is no single solution to scaling a problem; we look for a and b values that will cover the period of our interest and give satisfactory magnitudes for the display unit. In this regard there is always a limit to the voltage that the CRO can display without going off scale, and also there is a limit to the voltage that the J units can handle without saturating. This is typically 50 V. We assume that we have to work within this constraint.

We start by making an estimate of the likely maximum values for the current and voltage and deciding on the total time we are interested in. This is fairly straightforward here; 300 V is a good number for the maximum voltage, 150 A for the maximum current, and 0.1 sec serves the time satisfactorily. In more complicated problems these values will not be so easily selected, but if it should turn out that our first choice is not a good one, the situation is readily correctable by a second choice. With these data and the scaling equations (14.3.4) and (14.3.11), we first calculate the setting of C_2. If we expect a maximum current of 150 A, and must not exceed 50 V, an input scale $a = 5$ will suffice, that is, 1 V equals 5 A. Since $R = 2\ \Omega$, we can write, from Eq. 14.3.4,

$$C_2 = \tfrac{5}{10} \times 2 = 1$$

The setting of the C_2 coefficient unit should be unity.

Turning now to the integrator, we tentatively select $J = 10$, which is the middle value. Since b is already established at 5 A/V, it is apparent from Eq. 14.3.4 that

$$\frac{100}{10 \times 0.1} = \frac{a}{5}$$

or

$$a = 500$$

The setting of coefficient unit C_1 follows the same lines as C_2:

$$C_1 = \frac{1}{0.15} \times \frac{10}{500} = 0.133$$

The step voltage to represent the 300-V input will be

$$V_{\text{step}} = \frac{300}{10} = 30 \text{ V}$$

This completes the scaling for the sample problem. Since it is a simple problem for which estimates for the range of variables could be readily made, the C values reduced to reasonable values on the first try. On the computer being considered, any value between 0.1 and 5 is an acceptable setting for a C unit. The integrator unit was set at 10 only because it is the middle setting. Notice that in the scaling equation the product JT occurs. Therefore, to observe ten times the period, we would change J from 10 to 1; to observe one-tenth of the time, we would change J from 10 to 100.

This example indicates how a problem is set up on the electronic analog computer, but it conveys little of the versatility of the device. Accordingly, a second example will be taken of more practical nature.

In Section 7.5 a static current-limiting breaker is described in which thyristors are used for switches. A circuit diagram is shown in Fig. 7.10 and the transients evoked by "opening" the circuit breaker are illustrated in Fig. 7.11. We use a related problem as our next example.

It is pointed out in Section 7.5 that a considerable overvoltage can develop when the fault current is commutated from thyristor T_1 in Fig. 7.10 into the capacitor C. This is primarily a consequence of the magnetic energy of the current stored in the inductance L being transferred to the capacitor C. Our analysis shows that this contribution to the voltage transient is $I(0)(L/C)^{1/2}$, where $I(0)$ is the current at the time of commutation. This is given in Eq. 7.5.2. Apparently, the voltage would be reduced if we used a larger capacitor for C. But this would be an expensive solution. Our example explores an alternative that augments C by connecting a polarized electrolytic capacitor C_x in parallel with C in the manner shown in Fig. 14.9. This type of capacitor

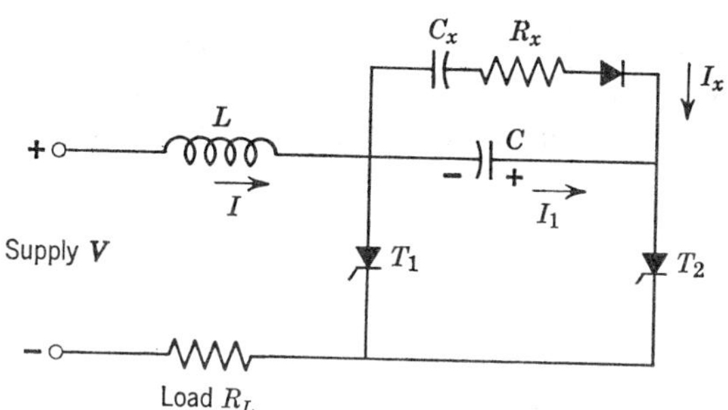

Fig. 14.9. Modified current-limiting static circuit breaker, subject of computer study.

has the virtue of being comparatively cheap, and it allows a considerable capacitance to be packaged in a small space. However, it can be charged with only one polarity, and it is for this reason that the diode is placed in series with it. The diode precludes C_x's being charged by the precharging circuit which is used to charge C. The plus and minus symbols on C_x play no part in the initial stages of interrupting the current in T_1. This quenching current is supplied entirely from C by gating T_2. In this respect the circuit behaves exactly as the circuit of Fig. 7.10. The capacitor C_x comes into the act after current-interruption in T_1 has been effected; specifically, when the voltage across that thyristor changes sign in the ensuing transient, for then the diode becomes forward biased. It will be seen from Fig. 14.9 that a resistor R_x has been added. Its purpose is to damp the transient voltage to best advantage. The object of the computer investigation is to determine what values of C_x and R_x provide the best combination to keep the transient overvoltage within any specified limit. We therefore set up a block diagram, which later can be interconnected physically on the electronic analog computer to represent this circuit; in this diagram R_x and C_x can be varied freely by simply turning dials on coefficient units. The effects of these variations will be observable at once, on the display oscillograph screen. It will be similarly possible, indeed quite easy, to vary the load being switched by changing the value of R_L, making it zero, to represent a short circuit, if need be. It should be understood that we are studying the transient following the suppression of current in T_1 from the time at which the voltage across this device reverses and the C_xR_x branch comes into the act. We must know the initial conditions at that time, $I(0)$, for example. This can be computed from our analysis in Section 7.5 or from a separate analog computer study.

The equations describing the behavior of the circuit in Fig. 14.9 are

$$V_C = V - L\frac{dI}{dt} - R_LI \tag{14.3.12}$$

$$V_C = \frac{1}{C}\int I_1\,dt \tag{14.3.13}$$

$$V_C = R_xI_x + \frac{1}{C_x}\int I_x\,dt \tag{14.3.14}$$

and

$$I = I_1 + I_x \tag{14.3.15}$$

The block diagram which interprets these equations into computer units is shown in Fig. 14.10. The various equations are identified by the components within the boxes outlined by broken lines.

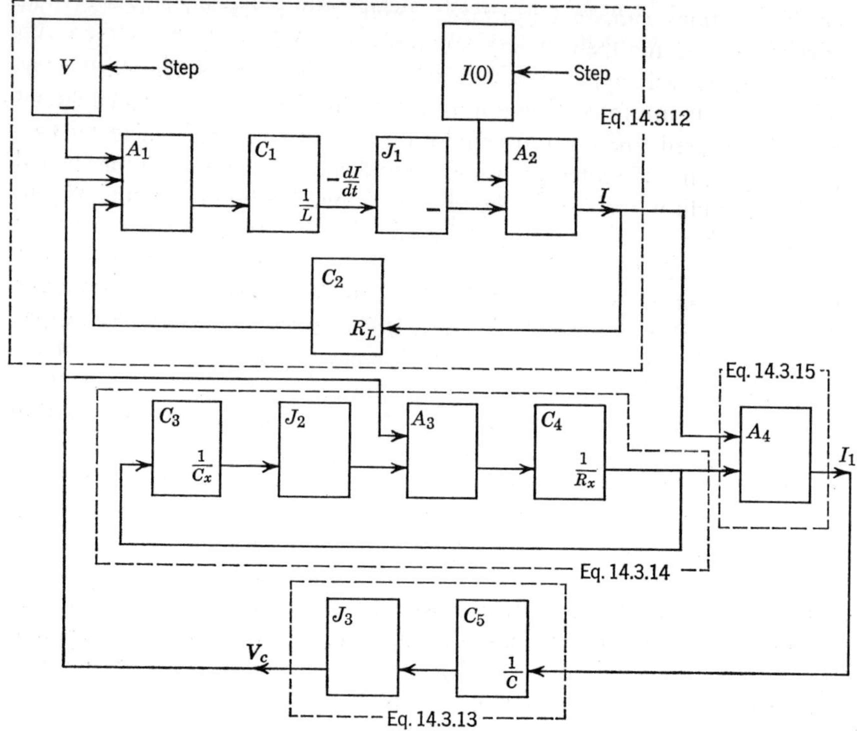

Fig. 14.10. Computer block diagram corresponding to the circuit of Fig. 14.9.

To solve Eqs. 14.3.12 to 14.3.15 by hand, for one set of parameter values and initial conditions, would take several hours. To do it for a range of values for C, C_x, R_1, and R_x, and for different initial values of $I(0)$, would be an appalling chore. But once the block diagram of Fig. 14.10 has been embodied on the computer, solutions can be generated as quickly as one can turn the dials on the units.

These solutions are valid until either I or I_x attempts to change sign. This cannot occur in practice because of the rectifying devices. We must terminate the problem at this time. However, if we wish to find out what happens when the diode blocks or the current in T_2 comes to zero, we could set up new equations appropriate to these conditions, construct a new block diagram, and interconnect the computer components. It would be necessary to include the stray capacitance across the semiconductor whose current is zero. The initial conditions for this new problem would be obtained as the final conditions of the old one.

We have restricted ourselves to problems involving A, C, and J units only.

Other functions can be performed; Table 14.2 gives some of the more important ones and their scaling equations.

The reader will appreciate that the electronic analog computer is not confined in its use to the solution of electrical transients in lumped circuits but can be used for any transient study that can be specified in terms of ordinary linear differential equations. It finds favor, therefore, in mechanical transient applications also. Although best suited for lumpy circuits, be they electrical, thermal, or mechanical, the use of this computer for circuits with distributed constants is also possible. A number of interesting techniques have been developed. Dupont and Robert (4) have used it to study the discharge of 735-kV lines. The simulation was rather limited but it yielded meaningful information concerning the ability of potential transformers and shunt reactors to remove trapped charge from lines after switching operations. Thomas and Hedin (5) have used the electronic analog computer to solve switching surge problems involving transmission lines. The simulation was

Table 14.2 Scaling Equations for Analog Computer Units

Unit	Scaling Equation
$1V = a \to$ (three inputs) $\boxed{A} \to 1V - b$ Adder	Same scale in and out
$1V = a \to \boxed{CK} \to 1V = b$ Coefficient	$C = \frac{a}{b} K$
$1V = a \to \boxed{J} \to 1V = b$ Integrator	$\frac{100}{JT} = \frac{a}{b}$
$1V = a \to$, $1V = b \to \boxed{MU} \to 1V = c$ Multiplier	$\frac{C}{25ab} = 1$
$1V = a \to \boxed{S} \to 1V = b$ Square	$\frac{b}{25a^2} = 1$
$1V = a \to \boxed{\sqrt{\ }} \to 1V = b$ Square root	$\frac{5b}{\sqrt{a}} = 1$
$1V = a \to$, $1V = b \to \boxed{\div \frac{a}{b}} \to 1V = c$ Divider	$\frac{25bc}{a} = 1$

achieved by constructing a multichannel pure transport delay time unit. This is capable of storing surge waveshapes of arbitrary form, operating on them (to invert them, for example), and delivering them back to the computer after a preselected time interval. In this way travel times of waves on lines and the effects of reflections can be simulated. For three-phase circuits, with multivelocity waves of the kind described in Section 9.7, the amount of equipment required is considerable. This approach is therefore limited to relatively simple circuit arrangements.

14.4 The Electrolytic Tank

If we took a system of conductors and applied charges to them, thereby establishing potential differences between them, we would create an electric field in their vicinity. Electric flux can be considered as emanating from positive charge on some parts of the conductor surfaces and converging on equal negative charges on other surface locations. If we now took the same system of conductors and placed them in a conducting medium whose conductivity is much less than that of the conductors themselves, and establish potential differences in proportion to those in the former model, currents would flow between the conductors, following paths exactly analogous to the flux lines in the dielectric model. The underlying mathematical justification for this assertion is the fact that the electric field in the dielectric model which is responsible for the flux, and the electric field in the conduction model which is responsible for the current, are both prescribed by Laplace's equation,

$$\nabla^2\varphi = 0$$

Interpreted in terms of the dielectric field, this says the electric flux, or the flux of **D**, out of any closed surface within a field containing no net charge, is zero. This property of having zero flux out of any closed surface is characteristic of the current density vector **J**, in a homogeneous conductor. There is a direct analogy between **D** and **J** in these instances, and between their fluxes Φ and I.

This analogy is exploited in the electrolytic tank. If, for example, we wish to study the electric field around a high-voltage conductor, or perhaps between a bushing of complex shape and neighboring ground conductors, one makes measurements on model electrodes immersed in a conducting fluid or electrolyte. If the field is two dimensional or possesses an axis of symmetry, the procedure is fairly straightforward. A simple example is shown in Fig. 14.11. There we are concerned with plotting the electric field within a cable which has rather odd shaped conductors. This type of problem is well suited to the technique since the field is confined within a well-defined boundary (the sheath of the cable). Potentials are applied to the conductors

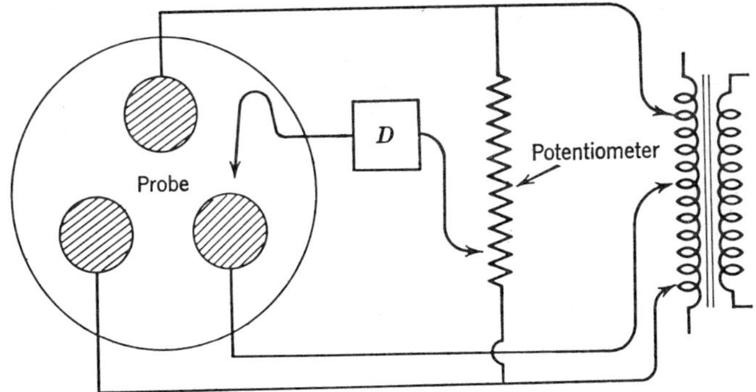

Fig. 14.11. Exploring the electric field in a three-phase cable by means of an electrolytic tank.

appropriate to the instant in the cycle being studied. The space between the electrodes is filled to a uniform depth with electrolyte and the potential field is explored with a probe connected through a detector, usually a galvanometer or a CRO, to a potential divider. When the detector indicates zero, the probe is known to be at the potential of the tap point P on the potential divider. Moving the probe around, it is possible to trace out this particular equipotential by maintaining zero on the galvanometer. Alternating current is used, sometimes at high frequency, to avoid polarizing the liquid.

Carter (6) gives the following simple analysis of the analogy. The conductance between the electrodes in the model measures the capacitance in the full-scale system. To calculate one from the other, suppose linear dimensions in the system are to those in the model in the ratio $\lambda : 1$. Let V, $\mathbf{E}$ be potential and electric force in the system, V', $\mathbf{E}'$, in the model. Then

$$\frac{\mathbf{E}'}{\mathbf{E}} = \lambda \frac{V'}{V} \tag{14.4.1}$$

But $\mathbf{D} = \epsilon \mathbf{E}$ in the system, and $\mathbf{E}' = \rho \mathbf{J}'$ in the model, where ρ is the resistivity of the electrolyte; therefore

$$\mathbf{D} = \frac{\epsilon \rho V}{\lambda V'} \mathbf{J}' \tag{14.4.2}$$

Areas in the system are λ^2 times as great as areas in the model; therefore

$$\text{flux } \mathbf{D} = \frac{\lambda^2 \epsilon \rho V}{\lambda V'} \text{ flux } \mathbf{J}' \tag{14.4.3}$$

the fluxes being taken across corresponding areas. But the total flux of $\mathbf{D}$ from one electrode is the charge Q on that electrode; the total flux of $\mathbf{J}'$ from an electrode in the model is the current I' from that electrode. Therefore

$$Q = \frac{\lambda \epsilon \rho V}{V'} I'$$

or

$$\frac{Q}{V} = \lambda \epsilon \rho \left(\frac{I'}{V'} \right) \tag{14.4.4}$$

or

$$\frac{C}{G'} = \lambda \epsilon \rho$$

Here C is the capacitance in the system and G' the conductance in the model. If the constants on the right-hand side are known, C may be deduced from a measurement of G'. For a two-dimensional field measured with electrodes immersed to depth d, the capacitance will be that of a length λd of the electrodes in the full-scale system.

The analogy can be extended to complex dielectrics, when materials of several different permittivities, oil, porcelain, air, may be involved. Evidently this can be accommodated in the model, according to Eq. 14.4.4, by varying the depth of the liquid or its resistivity in inverse proportion to the permittivity. Both methods are used in practice.

The technique we have described is widely used for plotting steady-state Laplacian fields be they electric, thermal, or magnetic. Its applicability to transient studies is less widely appreciated. Under transient conditions capacitance considerations dominate voltage distributions, at least initially. Transient voltage distributions can therefore be studied by this technique, as long as this condition prevails. We take an example from Chapter 11, where it was shown that the initial voltage distribution in a transformer winding, following the impact of a steep-fronted surge, is dictated by the complicated capacitance network formed by the winding, core, tank, and so on. We found that this distribution would be very nonuniform and could concentrate high stress in certain locations. Some simplifying assumptions had to be made, such as the uniformity of the capacitance distribution, in order to make the analysis tractable. The electrolytic tank offers more freedom in this regard. The relatively simple model shown in Fig. 14.12*a* demonstrates the method. This diagram shows part of a shell-type, single-phase transformer with primary and secondary windings. The construction is assumed to have rotational symmetry about the center line on the left. It is as if someone sawed through the transformer in a plane containing the center line, leaving a cross section of the core and all the conductors exposed. The conductors would be represented by isolated blocks of metal and the

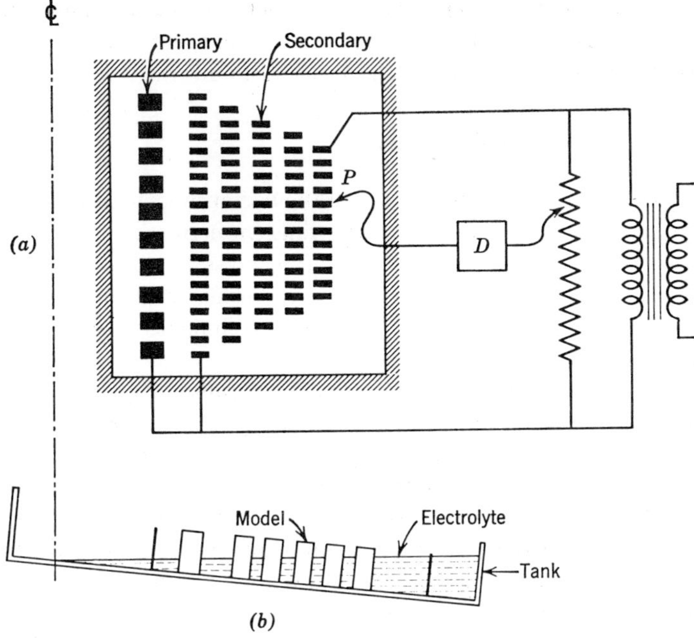

Fig. 14.12. Use of electrolytic tank for determining voltage distribution in a transformer under surge conditions.

inner boundary of the core by a metal strip, but to conform with the profile of the core section. These elements would then be arranged in the tank in the correct manner shown in Fig. 14.12*a*.

To comply with the rotational symmetry, the tank would be tilted as indicated in Fig. 14.12*b* with the "shore-line" along the axis of the core. If we wish to allow for a difference between the permittivity of the liquid and solid insulation in the transformer, this can be done by placing thin barriers at the boundaries between the different dielectrics. Electrolytes of different conductivity can then be put in the different enclosures, or they can be filled to different depths with the same electrolyte. Communication from one side to the other of such a barrier is provided by applying connecting vertical strips of conducting paint on both sides of the barrier.

To investigate the initial consequence of a steep-fronted voltage wave striking such a winding, a voltage is applied between the high-voltage line conductor and ground, represented by the core. If one end of the low voltage winding is ground, that particular conductor would also be connected to the core. The potential assumed by all the other conductors is then explored with a probe in the manner already described.

As an indication of the versatility of the method, Johnson and Schultz (7) used the electrolytic tank to derive capacitance data for a study of lightning phenomena involving towers, insulator strings, and transmission lines.

14.5 The Transient Network Analyzer

In this section we are concerned with the use of models to study electrical transients, and most especially with the transient network analyzer, or TNA, whose name very aptly describes its function as a simulator of power systems. Individual power system components, such as transformers and transmission towers, have also been studied by modelling techniques; a brief description is included at the end of the section.

The transient network analyzer extends to transient conditions the idea of the steady-state analyzer or a-c calculating board, which, for many years prior to the advent of the digital computer was used to investigate load flow, determine fault currents, and for stability studies on a-c systems. Such analyzers comprised circuit elements, most notably inductors and resistors, connected to represent the various components of the power system being studied. The energy sources were endowed with inertial attributes and regulator and prime mover governor characteristics, so as to respond to system changes, by phase swing and output voltage, in the way the actual machines would respond to similar disturbances in the system proper. However, the time scale for events in such a system, if time was involved at all, was measured in cycles or even seconds, whereas to investigate the transients that concern us in this book, we must consider much shorter times, for which the relative amplitude and phase of source emfs are to all intents and purposes constant. In one respect this simplifies the representation of such sources. But this is much more than offset by the requirement that *all* components possess the high-frequency transient characteristics of the power system equipment they represent. Thus one important addition is the capacitance of various simulated components. Such installations have been in use now for many years (8); they have been progressively refined over that period.

The model approach finds its virtue in the relative ease with which individual components can duplicate their actual counterparts as compared with the difficulty of representing combinations of interconnected elements in an analytical solution. The inductive, capacitive, and reactive attributes of the various power system components are simulated by reactors, capacitors and resistors in the analyzer. Usually, although not always, these have the same ohmic value as the actual units at system frequency. Where this is the case, voltage provides the scale. Thus, if 1 V in the model represents 1 kV in the system, 1 A in the model will be equivalent to 1000 A in the system.

Fairly precise simulation of transformers is difficult but possible. It requires that the magnetic characteristics (saturation curve and losses) match the corresponding real characteristics as closely as possible. In addition, the leakage inductance and capacitance must correspond. Requirements in different areas often conflict, so that sometimes a compromise must be made. The most difficult quality to model is the resistance of the transformer, which tends to be higher in the model than in practice. It is always possible to add resistance, inductance, or capacitance, but it is a challenge to the designer to take it out. Most modern TNAs have managed to reduce these problems to acceptable proportions and transformers of different construction, single-phase, core- and shell-type, can be adequately represented. Transmission lines are modelled by lumped constant networks. This is clearly an approximation, but one which we shall show can be minimized by selecting enough sections. Traveling wave effects obviously cannot be reproduced satisfactorily on a line simulated by only one or two T or π sections.

The simplest representation of a transmission line, which considers it only from the terminal point of view and only at one frequency, would be a T or π circuit like those indicated in Fig. 14.13. Starting with the equations relating the terminal conditions of a line under steady-state conditions, which were given as Eqs. 13.2.1 and 13.2.2 and which for convenience are repeated below, it is possible to determine the values of the shunt and series elements in the T and π circuits, for exact equivalence between these circuits and the true line:

$$V_S = V_R \cosh j\omega(LC)^{1/2} + I_R Z_0 \sinh j\omega(LC)^{1/2} \qquad (13.2.1$$

$$I_S = I_R \cosh j\omega(LC)^{1/2} + \frac{V_R}{Z_0} \sinh j\omega(LC)^{1/2} \qquad (13.2.2)$$

In the equations the subscripts S and R refer to the sending and receiving end currents and voltages, respectively, Z_0 is the characteristic impedance of the line, and L and C are the total inductance and capacitance of the line. Line resistance and leakage are neglected. For equivalence between the true line and a T or π circuit to exist, the terminal conditions of the T or π circuit must be coupled by an identical relationship. This situation prevails when the two circuits being compared have the same open-circuit and short-circuit

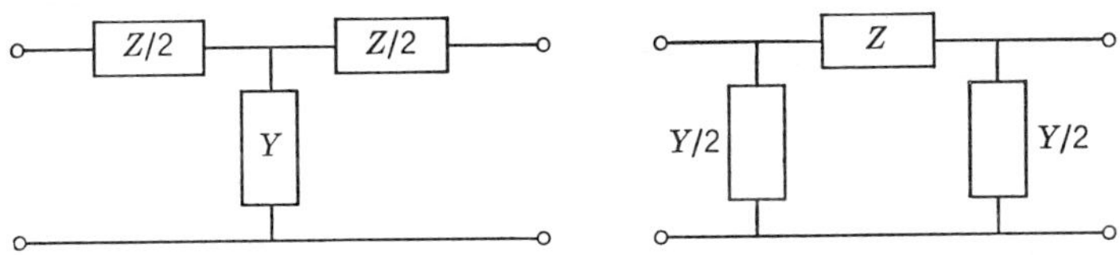

Fig. 14.13. T and π circuits.

impedances (9), (10). For the true line, the open-circuit impedance at frequency ω can be determined by setting $I_R = 0$ in Eqs. 13.3.2 and 13.2.2. Then

$$Z_{oc} = \frac{V_R \cosh j\omega(LC)^{1/2}}{(V_R/Z_0)\sinh j\omega(LC)^{1/2}} = Z_0 \coth j\omega(LC)^{1/2} \tag{14.5.1}$$

The short-circuit impedance is found by setting $V_R = 0$. This gives

$$Z_{sc} = \frac{I_R Z_0 \sinh j\omega(LC)^{1/2}}{I_R \cosh j\omega(LC)^{1/2}} = Z_0 \tanh j\omega(LC)^{1/2} \tag{14.5.2}$$

It is a simple matter to find the open-circuit and short-circuit impedances of the T and π circuits in terms of the Ys and Zs of their elements. For the T circuit,

$$Z_{oc} = \frac{ZY + 2}{2Y} \tag{14.5.3}$$

$$Z_{sc} = \frac{Z}{2}\frac{ZY + 4}{ZY + 2} \tag{14.5.4}$$

For the π circuit,

$$Z_{oc} = \frac{2}{Y}\frac{ZY + 2}{ZY + 4} \tag{14.5.5}$$

$$Z_{sc} = \frac{2ZY}{ZY + 2} \tag{14.5.6}$$

By equating (14.5.1), (14.5.3), and so on, it is readily shown that for the T circuit

$$\frac{Z}{2} = \frac{\omega L}{2}\left[\frac{\tanh \frac{1}{2}j\omega(LC)^{1/2}}{\frac{1}{2}j\omega(LC)^{1/2}}\right] \tag{14.5.7}$$

$$Y = \omega c\left[\frac{\sinh j\omega(LC)^{1/2}}{j\omega(LC)^{1/2}}\right] \tag{14.5.8}$$

and for the π circuit

$$Z = \omega L\left[\frac{\sinh j\omega(LC)^{1/2}}{j\omega(LC)^{1/2}}\right] \tag{14.5.9}$$

$$\frac{Y}{2} = \omega C\left[\frac{\tanh \frac{1}{2}j\omega(LC)^{1/2}}{\frac{1}{2}j\omega(LC)^{1/2}}\right] \tag{14.5.10}$$

Equations 15.4.7 to 14.5.10 indicate that a transmission line can be represented exactly by a T or π circuit *at one frequency*. The error for other frequencies will depend upon those parts of the preceding expressions contained within the brackets. Thus, if

$$\frac{\sinh j\omega(LC)^{1/2}}{j\omega(LC)^{1/2}} \approx 1 \approx \frac{\tanh \frac{1}{2}j\omega(LC)^{1/2}}{\frac{1}{2}j\omega(LC)^{1/2}} \tag{14.5.11}$$

the error will be small. This situation prevails when $\omega(LC)^{1/2}$ is small. Now, L and C are the total inductance and total capacitance of the line, so that if the line length is x, we may write

$$(LC)^{1/2} = (lc)^{1/2}x \tag{14.5.12}$$

where l and c are the inductance and capacitance per unit length of the line. This quantity, of course, diminishes as x diminishes so that the frequency error is decreased with the line length.

If we let $Z = \omega l$ and $Y = \omega c$ in Fig. 14.13, so that the circuit may be constituted from a series inductance equal to the total line inductance, with half the total line capacitance symmetrically disposed on either side, we would have the so-called nominal T and π circuits. These would be coarser approximations than the equivalent circuits we have already described and, in fact, would not be correct at any frequency. The error would still depend upon the terms given in Eq. 14.5.11 and can be evaluated for any frequency or line length by expressing the error term in series form:

$$\frac{\sinh j\omega(lc)^{1/2}x}{j\omega(lc)^{1/2}x} = 1 + \frac{j\omega(lc)^{1/2}x}{3!} + \frac{[j\omega(lc)^{1/2}x]^2}{5!} + \cdots \tag{14.5.13}$$

The lower the frequency and the shorter the line, the less will the error be. It is found by substitution that the error is small provided the line length is short compared with a quarter wavelength of the highest frequency being considered. Figure 14.14 indicates how the function $\sinh \theta/\theta$ increases with increasing θ. But for a 1-km line at 10 kHz, θ is typically 0.2, so a nominal T or π would be quite accurate.

The obvious way to improve the line representation is to divide the line into relatively small sections, with an appropriate fraction of the total inductance and capacitance assigned to each. The accuracy of such a ladder network improves progressively as the number of rungs is increased.

It must be remembered that all the foregoing analysis on artificial lines relates to steady-state conditions. This can be somewhat misleading from a transient point of view, since the frequency spectrum of a transient may be both broad and complicated. It is more proper to consider the response of an equivalent circuit, T or π, or ladder network composed therefrom, to the application of a step function. This type of stimulus is approached by some surges, such as line flashovers. Moreover, from the response to a step, the response to other functions can be obtained by Duhamel's integral as

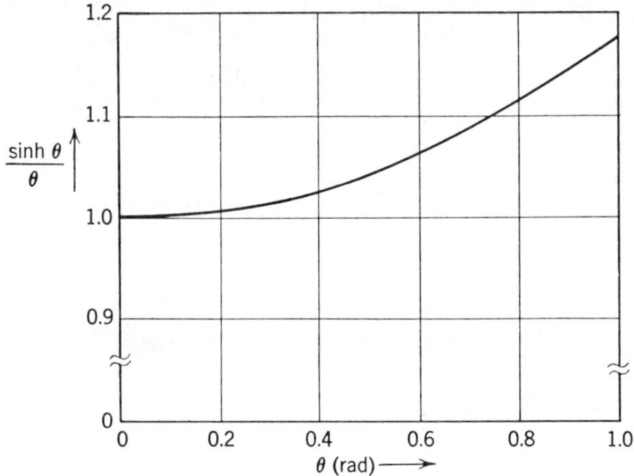

Fig. 14.14. Function $\sinh \theta/\theta$ plotted against θ to indicate error when lines are replaced by T or π circuits.

explained in Chapter 2. We know that for a true line, in the absence of losses, a step travels along the line undistorted. In Section 9.1 we pointed out in connection with Fig. 9.1*b* that when a ladder network is energized, some influence, no matter how slight, is felt at the remote end immediately after energization. The question is, how big is this effect? Carslaw and Jaeger (11) have addressed themselves to this problem.

They consider a step of voltage **E** applied to a line with n sections per unit length. A line of length x will therefore have nx sections. If the inductance and capacitance per unit length are l and c, respectively, Carslaw and Jaeger show that the current emerging from the nxth element is

$$I_{nx} = E\left(\frac{c}{l}\right)^{1/2} \int_0^{2nvt} J_{2nx}(\tau)\, d\tau \tag{14.5.14}$$

where $v = (lc)^{-1/2}$ and $J_{2nx}(\tau)$ is the Bessel function of the first kind, of order $2nx$. The current at the point x of the uniform loss-free transmission line would be

$$I_x = E\left(\frac{c}{l}\right)^{1/2}\left(t - \frac{x}{e}\right) \tag{14.5.15}$$

that is, it would be square, like the applied voltage wave, being related to it by the surge admittance $(c/l)^{1/2}$, but it would be delayed by a time x/v, the time for the wave to travel the distance x. In Fig. 14.15, which is taken from Carslaw and Jaeger (11), I_{nx} is plotted against vt/x for the values 2, 4, 6, 8 of nx; the curve marked ∞ is the value of I_x given by Eq. 14.5.15. It appears for these small values of n the approximation is not good and improves rather slowly. The oscillation appears to be equivalent to the Gibbs oscillation associated with Fourier series. Also, the "foot" which appears in the approximation prior to the arrival of the true wave, confirms a point made earlier, that in the artificial line, some influence, no matter how slight, is felt at remote points immediately after the stimulus is applied. In the true line this must await the arrival of the wave.

For a fair representation on the TNA, lines should comprise 10–20 elements, so that a typical element might represent a 10-mile section. The bandwidth of the line is then roughly given by the natural frequency of the elementary π circuit. It is important to be able to recognize the minor disturbances introduced by imperfections in the representation and differentiate between them and the true effects of the system. This sometimes requires a certain amount of judgment derived from experience.

It is necessary, of course, in three-phase circuits to represent zero sequence as well as positive sequence impedances of components. Thus a section for a transmission line might be represented by the connection of elements shown in Fig. 14.16, where the subscripts 1 and 0 identify positive and zero sequence inductances and capacitances, respectively.

The TNA is energized by three-phase generators that appear as an infinite bus. The characteristics of any source can be simulated by adding appropriate source inductance and capacitance.

We have reiterated many times that electrical transients are initiated by sudden changes of circuit conditions such as switching operations and the

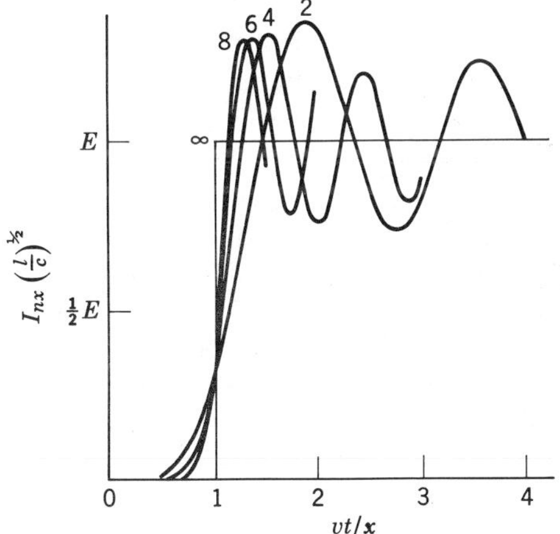

Fig. 14.15. Current leaving the nxth section of an artificial line as a consequence of a step stimulus.

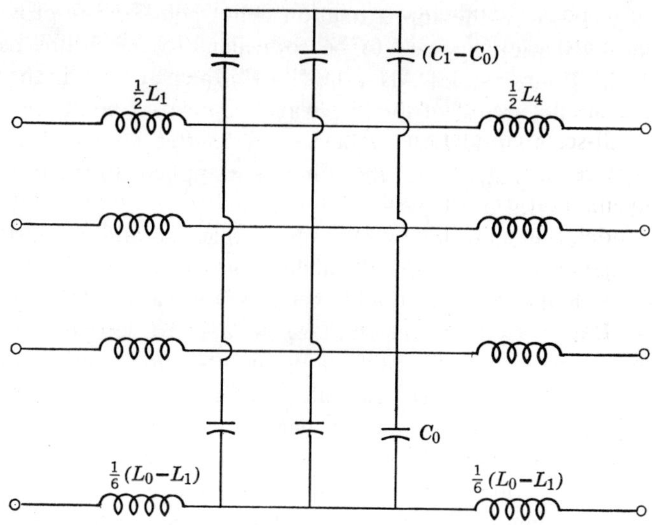

Fig. 14.16. *T* section for transmission line, three-phase representation.

occurrence of faults. It is necessary to be able to perform such operations on the TNA, and to have the event occur at any desired instant in the cycle. This may be done mechanically with a rotary type of switch, which is synchronized with the supply or, alternatively, by using a solid-state switch such as a silicon controlled rectifier, which offers more precision. The same repetitive scheme used for the electronic analog computer (described in Section 14.3) is applied here: the switching is performed many times per second, and the response is observed on CRO(s), synchronously triggered with the switching event. The repetition rate is limited by the duration of the transient, which is usually a fraction of a power frequency cycle, and the ability of the analyzer to be reset between switching events. By this means, it is possible to obtain a sharp, steady image on the oscilloscope screen, even for very short duration transients. It is possible to program a sequence of events, such as the occurrence of a fault and the subsequent opening of a circuit breaker to clear it, or the closing of a switch, wherein the different poles do not close in unison. In accordance with good measuring practice, the measuring elements should disturb the analyzer network as little as possible, and together with the oscilloscope and its amplifiers and attenuators, have a wide enough frequency response to faithfully reproduce the analyzer output.

In some studies it is necessary to represent lightning arresters. These can be modelled quite accurately as to their sparkover voltage and nonlinear resistance characteristic while conducting. For the gap element (refer to Section 12.2) a solid-state switching device such as a thyristor is used. It is triggered in response to voltage level. The nonlinear resistor uses a small wafer of the material used in the actual device. To simulate lightning itself or the surges produced thereby, model impulse generators are used, comprising a simple arrangement of capacitors and resistors much like a singlestage full-scale counterpart. Again the switching device is usually a solid-state device.

The problems that can be studied with the TNA are limited only by the size of the model, that is, by the number and variety of components available, and by the ingenuity of the operator. Concordia (12), in an extensive bibliography, cites some 80 references relating to transient network analyzer applications. Barthold and Silva (13) give the following partial list of topics examined on the TNA:

1. Lightning surges in complex station arrangements and optimum arrester location.
2. Recovery voltages on distribution and transmission systems.
3. Switching and restrike phenomena in dropping lines, cable, and capacitor banks, including modification by nonlinear transformer and arrester elements.
4. Neutral instability and reversed phase rotation.
5. Ferroresonance.
6. Magnification of switching surges.
7. Sudden loss of load.
8. Switching of light inductive current.
9. Surge transfer through transformers.
10. Duty requirements of protective devices.
11. Lightning and switching surge response of motors, generators, transformers, transmission towers, cables, and so on.
12. Rectifier transients.
13. Voltage flicker.

It will be recognized that many of these represent problems which it would be virtually impossible to solve without computational aid of some kind. In some instances the complication arises as a consequence of the large number of components involved; others owe their complexity to the nonlinear properties of one or more components. Figure 14.17, from reference (14) is an example of this latter kind. It shows the record obtained on interrupting the charging kVA of a simulated high-voltage transmission line, shunted by a model of a power transformer. In this instance the line was made up of 10-mile, 3-phase sections.

The TNA has become a useful tool in the design of power systems and system components. Croft et al. (15) describe an application of this kind. It was pointed out in Section 4.5 how switching transients of the kind

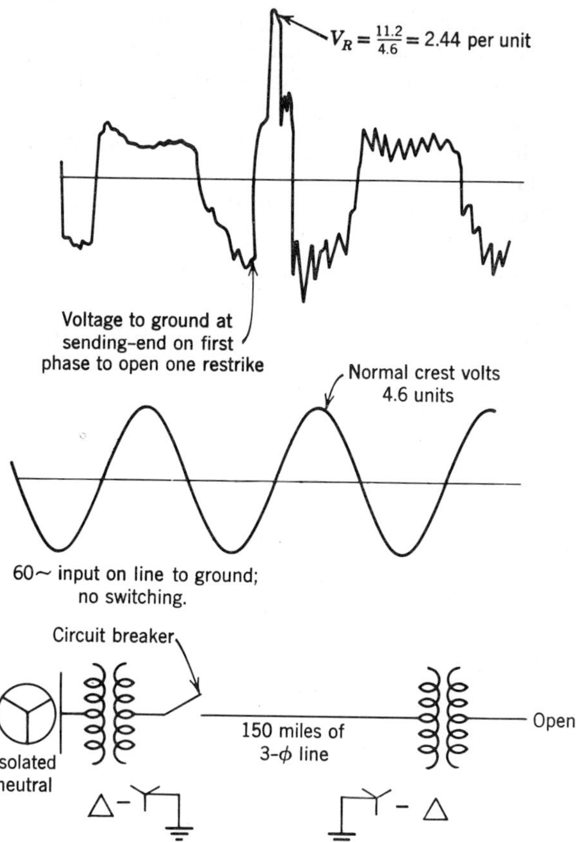

Fig. 14.17. Typical voltage obtained on TNA. Charging current of long transmission line interrupted with shunt magnetizing impedance of connected transformer.

described in the last paragraph can be reduced by shunting the circuit breaker with a resistor. In Section 13.4 it was shown how overvoltage on closing could be reduced by preinserting resistors. Resistor values can be optimized on the model. In a similar way, the selection and location of arresters can be optimized. This is described by Wilson and Hedman (16).

Perhaps the greatest benefit that the TNA provides arises when one is collecting statistical data. It is recognized that there is a considerable spread in the transient voltages developed by a specific switching operation, because of the many factors that influence the transient. These factors include the instant in the cycle at which the switching event occurs and the variations between the closing or opening of the different phases. It is relatively easy to perform the switching operation on the TNA repeatedly, in random fashion, and thereby collect quite quickly and at relatively low costs, sufficient data to insure its statistical significance. To do the same thing by field tests on an actual system would be costly and inconvenient, and in the design stage the line would not be available for tests. An example of this kind is reported by Glavitsch and Ruoss (17).

How accurate are the results obtained on the TNA? Most system parameters are functions of frequency; examples of this frequency dependence will be found in Chapter 15. The challenge is to build the same frequency dependence into the model. The fidelity of modelling has improved considerably over the years so that even complicated items such as transformers can now be represented accurately. Transmission lines still appear to present a problem and are probably the cause of the errors that arise. Generally the predicted overvoltages determined from switching surges err on the high side. The shortcomings of artificial line representations were pointed out above. Undoubtedly, some of the discrepancy can be explained by the phenomena depicted in Fig. 14.15. Another source of error is probably associated with transient resistance which is not readily scaled. It was pointed out in Chapter 8, notably in Figs. 8.11 and 8.17, that the resistance of a circular conductor to a transient current is a function of time; it takes time for the current to penetrate from the outer skin. The rate of penetration depends upon the material but also on the scale, as shown by Eq. 8.5.6. This indicates that the transient resistance will be less effective in the thin wire of the model than in the heavy conductor of the line. This is also very important as far as ground resistance is concerned. As a consequence of these effects, the representation tends to deteriorate above around 4 kHz.

Another factor that is not accounted for that can be very important is corona. No satisfactory solution to this problem has yet been devized.

In spite of these limitations, in many instances results obtained in TNA investigations have shown remarkable agreement with the measurements made during field tests on actual systems. Figure 14.18 of Wilson's (18) substantiates this contention. This shows records taken to determine the effectiveness of preinserting resistors when closing a circuit breaker in the location shown. The actual tests were made on the System of the Bonneville Power Administration whose courtesy in permitting reproduction is acknowledged. In particular, resistors were inserted for approximately $1\frac{1}{2}$ cycles on a 230-kV breaker when energizing a 230/345-kV autotransformer, 175 miles of 345-kV line and terminating stepdown autotransformer. Figure 14.18 illustrates the field oscillogram of the 3-phase voltage on the receiving end autotransformer. The accompanying TNA oscillogram indicates the excellent duplication. The initial voltages which appear nearly sinusoidal occur due

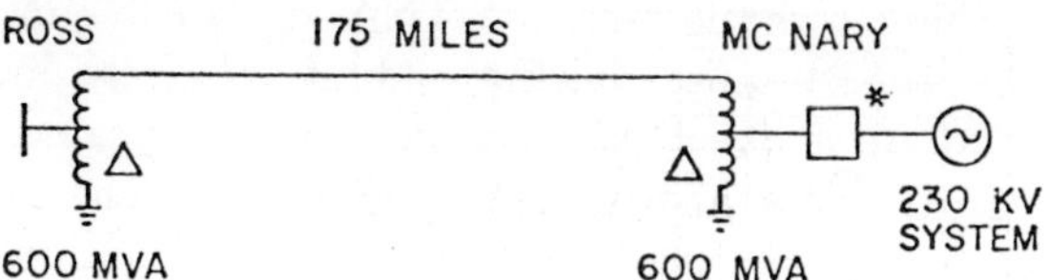

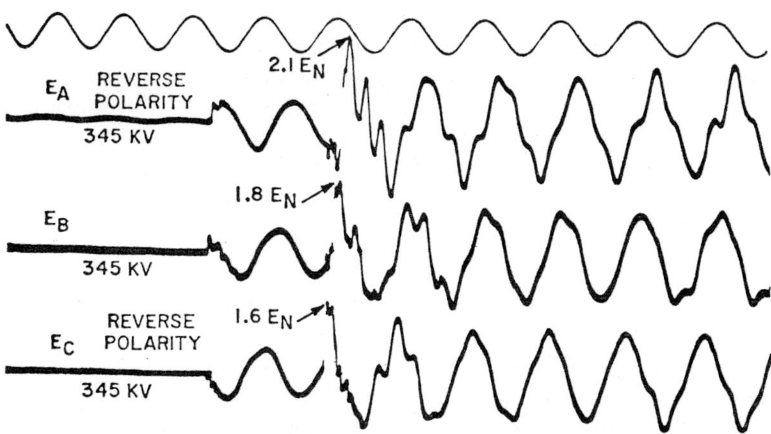

Fig. 14.18. Comparison of field text oscillograms and TNA records of a transmission line switching operation. (A) Ross and voltage—field oscillogram. (B) TNA oscillogram; 2 div = 1.0 per unit.

to the resistor in the circuit and the following, more oscillatory voltages occur due to shorting of the resistor. Other examples of the use of the TNA in power system transient analysis are included in references (19) to (23), from which it will be seen that it is a well established device for this purpose. Undoubtedly efforts will continue to improve its accuracy.

At the beginning of this section we mentioned that the modelling approach had been applied to the study of transients in components as well as circuits or systems. One such investigation on transformers is reported by Abetti (24). In Chapter 11 the responses of transformer windings to voltage and current surges were discussed. Relatively simple representations of the windings yielded analytical solutions of considerable complexity. The much more complicated winding arrangements of practical power transformers, which all but defy analytical study, succumbed to the model approach. Abetti (24) points out some of the difficulties of a purely geometrical model and shows how these can be circumvented by an electromagnetic model.

Mention has already been made of transmission tower models (7) in connection with the electrolytic tank. To study the effect of tower impedance

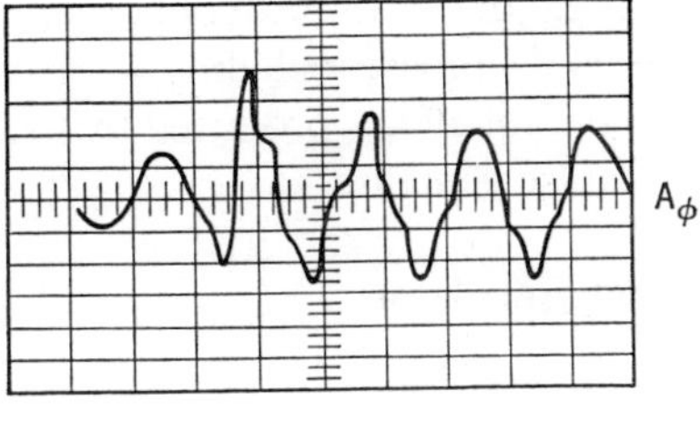

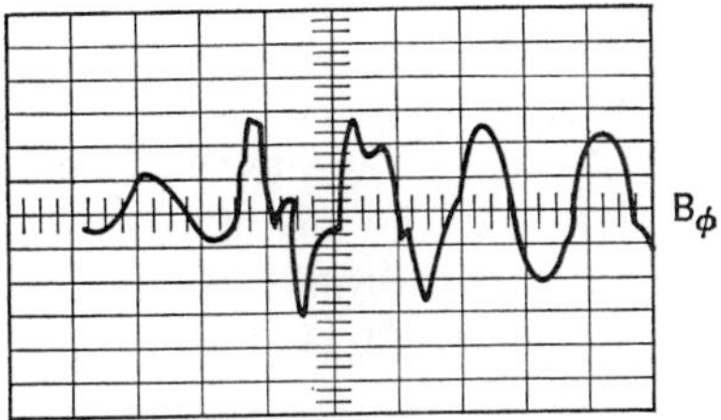

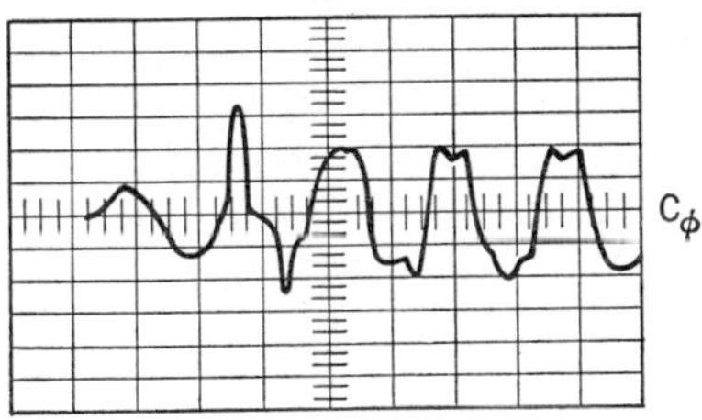

Fig. 18. (*Continued*).

and the nature of the stroke current in producing flashover voltages across an insulator string, a model of the tower and its conductors was designed and built. By this means certain circuit characteristics such as the following, could be more easily included:

1. Representation of the tower as a distributed constant circuit of series inductance and shunt capacitance having a finite time of wave travel and a surge impedance which is not constant along its height.
2. Representation of coupling to the phase conductors to include the effect of the tower and not just for the condition of the conductors isolated in space above ground.

The model was built to 1/28 scale. In order to not have the time scale reduced in the same ratio, a time-base change was introduced. This can be done by changing the scale of inductance and/or capacitance. Johnson and Schultz (7) did this by artificially increasing the capacitance of the model.

In a quite different area the author's colleague, Dr. T. W. Liao, has modelled entire laboratory facilities, including their equipment and ground mat, in order to study transient effects produced by such violent disturbances as the discharging of impulse generators.

14.6 The Direct-Current Simulator

With the advent of high-voltage direct-current transmission there was a need for a direct current simulator with the capability of representing the d-c system in the way that the a-c network analyzer and the TNA represent the a-c system. Such a simulator has been described by Hudson et al. (25). It was designed with unit components so that specific as well as generalized systems may be studied. A total of 8 converter units, 2000 miles of quasi-distributed d-c transmission line, and 850 miles of a-c transmission line can be represented. The model transformers (both converter and auto) are designed to duplicate the magnetic characteristics of the actual power systems while maintaining low loss. Harmonic filters, shunt capacitor banks for reactive kilovar supply, and shunt and series capacitor compensation on the a-c transmission lines are included.

The valve groups (six working valves and one bypass valve) are composed of silicon controlled rectifiers, which are suited to the application as they exhibit a low voltage drop when in the conducting state and have very rapid switching time and small physical size. The current and voltage bases for the model have been selected to make it compatible with a TNA, with which it can be operated to study a d-c line and its terminal equipment as part of an a-c system.

Much of the work done on this d-c simulator falls into the category of dynamic control studies, but it can and has been used to investigate electrical transients due to lightning, faults and d-c breaker switching surges. The simulator approach for combined a-c/d-c studies has been described by Weber et al. (26).

14.7 The Digital Computer

The application of the digital computer to power system transient studies has been and remains a burgeoning field of endeavor. The type of problems for which these machines are most often used are those described in Chapter 13, that is, transient investigations on integrated systems. But, as we shall see, they can be equally well applied to smaller systems or components.

The appeal of the digital computer is its ability to process a vast amount of data in a systematic way, and do so in an extremely short time. The procedure of numerical integration is a good example. Standard routines are also available for various matrix operations. The computer is very adept at storing, retrieving, operating on, and restoring volumes of data. System transient studies can be stated in these terms, for they are concerned with describing events in space and time at many different locations, which may be set down as a large number of differential equations.

The field is so extensive that in this section, we can provide little more than a superficial survey. Our object will be to point up different methods that have been applied and direct the reader to a few selected references wherein he can delve more deeply. A few ingenious techniques that allow certain difficulties to be circumvented or assist in reducing the amount of computation will be described. Although computers are extremely fast, the cost of supporting a large machine is considerable, so that every effort should be made to utilize it to the best advantage. In an interesting comparative study of transmission line closing transients Bickford and Doepel (27) discuss this point, stressing the fact that one should be careful not to seek accuracy greater than the validity of the input data will allow.

For many jobs the digital computer is in direct competition with the TNA, and one method of digital solutions is to virtually set up a mathematical model of the TNA.

Lumped Parameter Digital Solutions. We saw in Section 14.5 that the TNA is a lumped parameter device; even transmission lines and cables, which are distributed components, are represented by lumped T, π, or L sections. A system of this kind is therefore represented by a very large number of meshes, generally containing Rs, Ls and Cs. Each mesh or nodal point can be described by a second-order differential equation. For simplicity, we consider a circuit that reduces to a small number of meshes. This problem has merit in its own right, for it illustrates a class of problem too complicated to be tackled readily by hand calculations, yet seemingly trivial for the computer. But in fact this is not so, for the advent of time-sharing and easy access to large computers from desk side terminals has shown that the digital approach is well suited to such investigations. It will be apparent that the method can be extended to much more complicated circuits of the type we began discussing earlier.

In the example chosen, we compute the recovery of the first phase to clear in the three-phase switching operation depicted by Fig. 14.19*a*. Many practical problems can be reduced to an equivalent circuit of about this complexity. The switch is operating to clear a three-phase fault on the right, and it is assumed that phase B interrupts first. Using techniques described in Chapter 6, we simplify the circuit and then inject into the circuit at the contacts of the circuit breaker a ramp of current equal and opposite to the fault current. The circuit is simplified by recognizing the symmetry of phases A and C and proceeding to parallel them (Fig. 14.19*b*). This is a

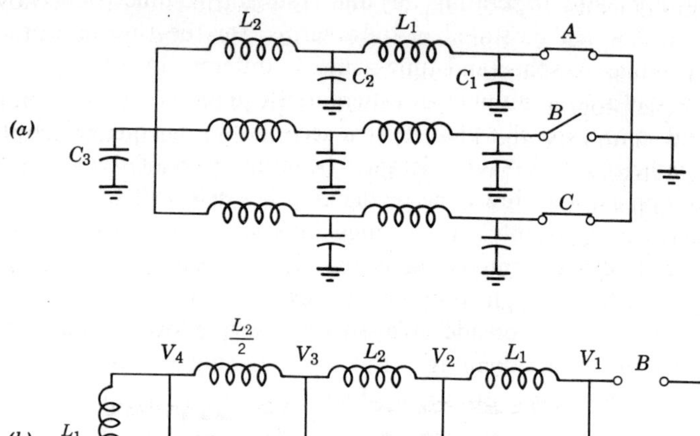

Fig. 14.19. Three-phase fault—equivalent circuit for first phase to clear.

$C_1 = 0.02\ \mu F$ $\quad L_1 = 800\ \mu H$
$C_2 = 0.66\ \mu F$ $\quad L_2 = 390\ \mu H$
$C_3 = 0.46\ \mu F$ $\quad \omega = 377$

four-mesh circuit, for which a hand calculation, with just one set of parameter values, would be lengthy and tedious. A starting point for such an analysis could be the setting down of four simultaneous second-order differential equations with the nodal voltages V_1 to V_4 as variables:

$$\frac{1}{L_2}\int (V_3 - V_2)\,dt - \frac{1}{L_1}\int (V_2 - V_1)\,dt = C_2\frac{dV_2}{dt}$$

or

$$\frac{V_3}{L_2} - V_2\left[\frac{1}{L_2} - \frac{1}{L_1}\right] + \frac{V_1}{L_1} = C_2\frac{d^2V_2}{dt^2} \tag{14.7.1}$$

There will be three other similar equations. The computer is adept at solving first-order differential equations by the so-called Runge–Cutta routine. Equation 15.7.1 is changed into a first order equation by the substitution

$$y_2 = \frac{dV_2}{dt} \tag{14.7.2}$$

The other three equations can be treated similarly by corresponding substitutions. The equations so modified, together with the equations of the substitutions themselves (Eq. 14.7.2 and other necessary substitutions), give eight simultaneous first-order differential equations with unknowns V_1, $V_2, \ldots$, which the computer proceeds to solve and print out. We are particularly interested in the solution for V_1.

The simple program* for this problem is given below. It has been included to show its simplicity and brevity. It is written in a particular adaptation of Fortran. Actually, it is two programs: lines 0 to 10 identify the variables, put in the parameter values for the Ls and Cs, call for the equation solving subroutine, and specify the data to be printed out; lines 11 to 23 take the subroutine and solve the equations, which themselves are set down in lines 13 to 20.

A plot of the recovery voltage for the following parameter values:

$$\begin{aligned} C_1 &= 0.02\ \mu\text{F} & L_1 &= 800\ \mu\text{H} \\ C_2 &= 0.66\ \mu\text{F} & L_2 &= 245\ \mu\text{H} \\ C_3 &= 0.46\ \mu\text{F} & \omega &= 377\ \text{rads/sec} \end{aligned}$$

when a 60-Hz current wave of peak amplitude 1 A is injected at the circuit breaker, is presented in Fig. 14.20. It is interesting to note that although this is a four-mesh circuit there are only two frequencies of any consequence. No damping was included in this computation, but it could have been added readily with little complication to the program and only a modest increase in the running time.

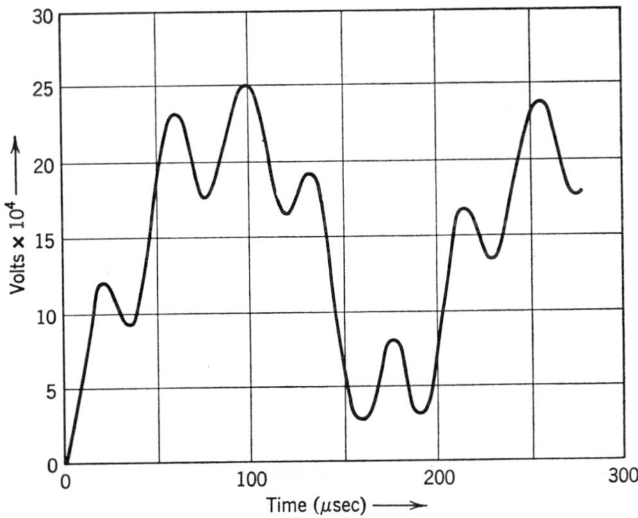

Fig. 14.20. Response of the circuit shown in Fig. 14.19 to an injected current of 1 A.

* The author wishes to thank his colleague, Mr. E. J. Touhey, for providing this program.

The reader will recognize that the problem just described would be a good candidate for the electronic analog computer described in Section 14.3. The choice between the two is often a matter of personal predilection, although in general the digital approach leads to greater accuracy, whereas the analog computer offers greater flexibility while the program is running. As was pointed out, one can readily observe the effect of varying a parameter over a range, or one can adjust one or more parameters to achieve a specified end.

For the large digital computer it is a matter of small consequence whether there are 8, 80, or even 800 first-order differential equations to solve. Thus the method we have just outlined can be applied to much more complicated circuits, such as those that evolve with the TNA. In a study of this kind a line might typically be represented by 16 sections. The terminal conditions could be related, if one had the time and patience, by a single differential equation of the thirty-second degree! But the computer handles quantity, and would simply be programmed for 32 simultaneous equations of the first order. For three-phase lines even this would be inadequate, since each line section requires a representation such as Fig. 14.16.

Digital program libraries have been set up in which digital programs are stored, much as books are stored in a regular library. Those adjacent to a time-sharing system (for example, The General Electric Co. Program Library, Information Service Dept.) are accessible from the time-sharing terminals. One such program, designated TRANS $, performs Laplace transform inversions. We are now familiar with the fact that linear circuits in their transient state can be described in terms of differential equations, and we have shown that these equations can be transformed out of the time domain into the s domain by the Laplace transform. The system response to a particular stimulus is then typically expressed as the quotient of two polynomials. This is what we have called the operational solution. The particular digital program being described computes from this operational solution the solution in the time domain. The option is available for plotting the output of the program versus time for any interval specified by the user.

The digital computer is also well suited to lumped circuits where nonlinear elements are involved. Nonlinear resistance elements are used in lightning arresters and surge suppressors (see Chapter 14); iron cored reactors and transformers are examples of nonlinear inductors if their cores are pushed into saturation. As an example, we will treat the subject of ferroresonance described in Section 5.6.

It was shown in Section 5.6, that under certain circumstances the capacitance of power circuits can come into series resonance with the circuit inductance to produce excessive voltages. This phenomenon is referred to as ferroresonance if an iron-cored inductance is present. Typically, the components involved might be the capacitance of a cable and the inductance of a transformer.

Because of the shape of the magnetizing curve, I versus Φ, of the transformer, its inductance is a function of current, being high when the current is low and vice versa. True resonance therefore can occur only at one point on the I/Φ curve. Nevertheless, serious overvoltage can result.

To program such a problem we could specify the magnetic circuit by a curve or table which the computer would draw from, or we could approximate it by an analytical expression; for example, the characteristic might be described by the expression

$$L = \frac{d\Phi}{dI} = A\epsilon^{-I/B} \tag{14.7.3}$$

which indicates that when $I = 0$, the inductance is A henrys and diminishes as I increases in a manner dependent on the constant B, diminishing to A/ϵ when $I = B$.

To illustrate this method, suppose a 13.8-kV stepdown transformer comes in resonance with cables on its primary side. Typically, A might be 8 H and, if Eq. 14.7.3 is to fit a magnetization curve, B might be 1.76. Then

$$L = \frac{d\Phi}{dI} = 8\epsilon^{-I/1.76} \tag{14.7.4}$$

Now let us suppose that 60-Hz resonance occurs when $I = 4$ H, so that

$$2\pi(4 \times 10^{-6} \times C)^{1/2} = \tfrac{1}{60}$$

$$C = 1.75\ \mu\text{F}$$

Neglecting resistance, the equation describing the circuit is

$$L\frac{dI}{dt} + \frac{1}{C}\int I\,dt = V_P \sin(\omega t + \phi)$$

or, substituting for L from Eq. 14.7.4,

$$8\epsilon^{I/1.76}\frac{dI}{dt} + \frac{10^6}{1.75}\int I\,dt = \frac{13.8\sqrt{2}}{\sqrt{3}}\sin(377t + \phi) \tag{14.7.5}$$

This is the nonlinear differential equation that the computer solves for I. It is also possible to compute the voltage across the transformer (the first term in Eq. 14.7.5) or across the cable (the second term in Eq. 14.7.5).

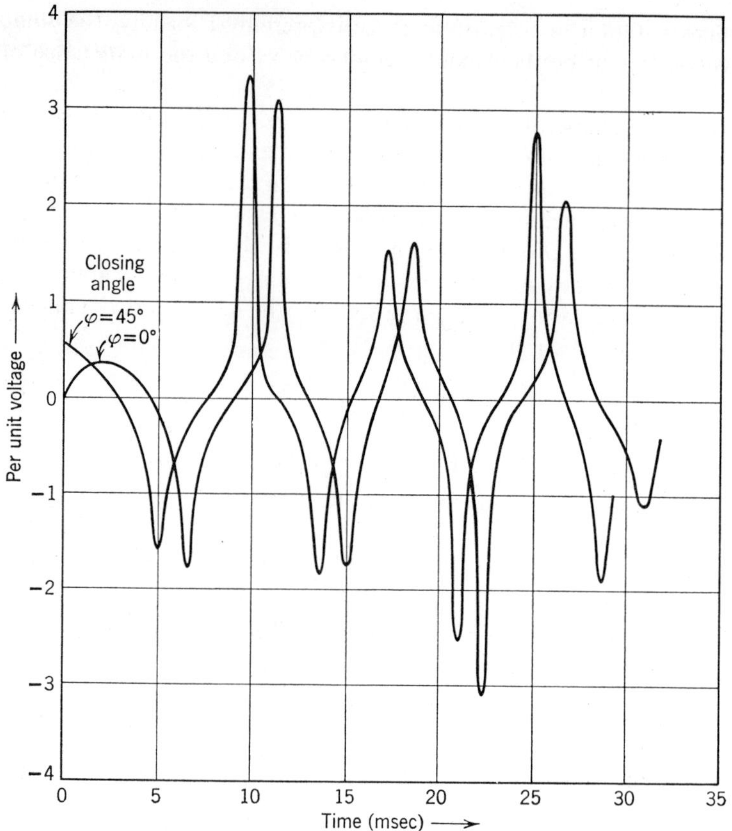

Fig. 14.21. Computer plot of ferroresonance voltage obtained from Eq. 14.7.5.

The first two cycles of the transformer voltage, for closing angles $\varphi = 0°$ and $\varphi = 45°$, are plotted in Fig. 14.21.

Lattice Diagram Solutions. A very different digital approach for extensive power systems is based on Bewley's (28) lattice diagrams, which were discussed in Section 9.5. The application of this method to single-phase representations has been described by Barthold and Carter (29). The method is capable of accommodating any specified input waveshape, real or complex line terminations, any system configuration, and operation of one or more arresters. It could therefore be applied, for example, to switching surge studies in systems like that described in Section 13.7 and illustrated in Fig. 13.11. In principle it is a healthy application of superposition combined with an ingenious system of bookkeeping.

The behavior of a traveling wave when it encounters a line discontinuity has been described in terms of reflection and refraction coefficients (Section 9.3):

$$\text{reflection coefficient} = a = \frac{Z_B - Z_A}{Z_B + Z_A}$$

$$\text{refraction coefficient} = b = (1 - a) = \frac{2Z_B}{Z_B + Z_A}$$

This is shown in Fig. 9.7 and repeated in Fig. 14.22*a*. Barthold and Carter (29) point out that the phenomenon can be represented equally well by the combination of waves in Fig. 14.22*b*, where the incoming unit wave proceeds through the discontinuity undiminished, but generates a new wave equal to *a* at the instant it reaches the discontinuity. This new wave emanates from the discontinuity in both directions, and the sum of the original wave and the newly generated wave represents the total response of the discontinuity to the impinging wave.

In much the same manner, the response of a complex network can be represented as the superposition of the undiminished transmission of the original input plus similar transmission of secondary wave components generated as the original wave arrives at each bus in the system. The secondary waves produce a third generation of waves; and the process continues *ad infinitum*. In the ensuing discussion, the term "response waves" is used to describe the entire chain of wave components produced by the input wave.

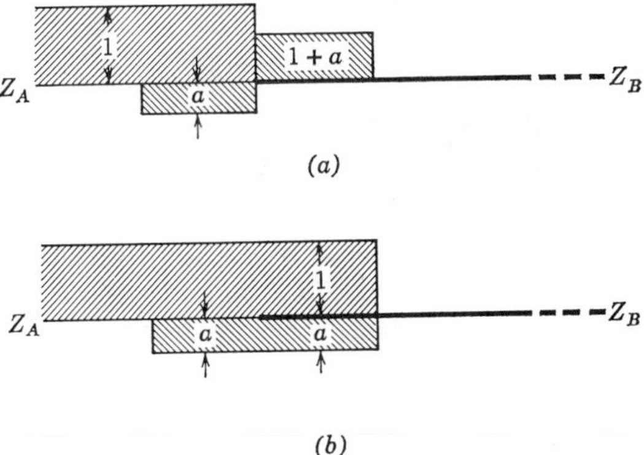

Fig. 14.22. Space diagram of a unit wave passing through a discontinuity. (*a*) Reflected and refracted components. (*b*) Undiminished and response components.

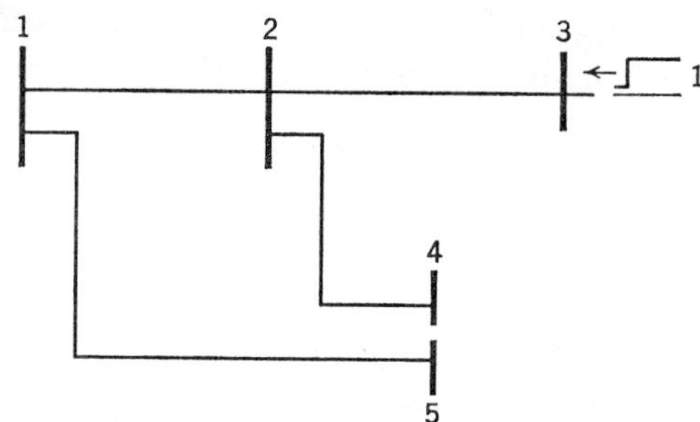

Fig. 14.23. One-line diagram of a five-bus system.

Barthold and Carter (29) illustrate the method by considering the system shown in Fig. 14.23. The voltage applied at bus 3 is considered to consist of a unit step function supplied from a source having zero internal impedance. In accordance with the concept described above, the unit wave generated on bus 3 will arrive undiminished at buses 1, 2, 4, and 5 at times equal to the travel times from bus 3 to each of these buses. Since a great many such analyses must be made before a complete solution is reached, it is convenient to describe a travel time array for the system under study. Such an array is shown in Fig. 14.24*a*. Entries in the third row of this figure, for example, give the correct time of arrival at buses 1, 2, 4, and 5 for the unit wave originating at bus 3 at time zero.

The travel time array is also used to determine the time of arrival of all response waves. The array gives the times that must be added to the time of origin of a particular response wave to establish its time of arrival, on an absolute scale, at each other bus in the system.

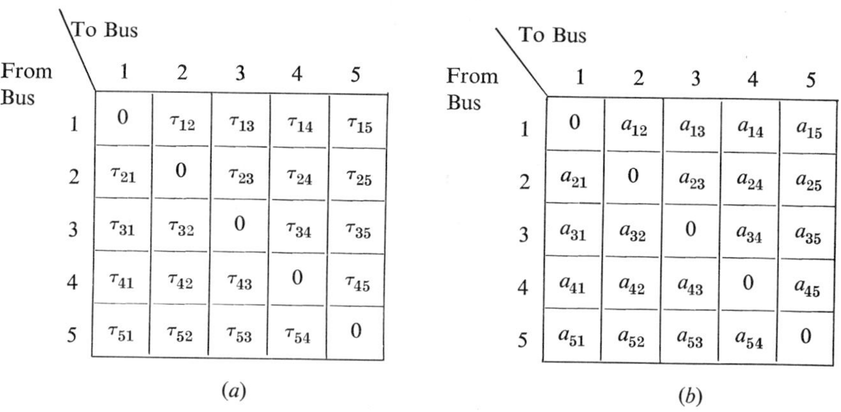

From Bus \ To Bus	1	2	3	4	5
1	0	τ_{12}	τ_{13}	τ_{14}	τ_{15}
2	τ_{21}	0	τ_{23}	τ_{24}	τ_{25}
3	τ_{31}	τ_{32}	0	τ_{34}	τ_{35}
4	τ_{41}	τ_{42}	τ_{43}	0	τ_{45}
5	τ_{51}	τ_{52}	τ_{53}	τ_{54}	0

(*a*)

From Bus \ To Bus	1	2	3	4	5
1	0	a_{12}	a_{13}	a_{14}	a_{15}
2	a_{21}	0	a_{23}	a_{24}	a_{25}
3	a_{31}	a_{32}	0	a_{34}	a_{35}
4	a_{41}	a_{42}	a_{43}	0	a_{45}
5	a_{51}	a_{52}	a_{53}	a_{54}	0

(*b*)

Fig. 14.24. System constants. (*a*) Travel time (τ) array. (*b*) Reflection coefficient (a) array.

A similar array for reflection coefficients can be defined to establish the magnitude of response waves. This array for the system of Fig. 14.23 is shown in Fig. 14.24*b*. At any location the magnitudes of response waves to the original wave, or to any other response wave, are determined by multiplying the incident wave by the appropriate entry in that row of the *a* array corresponding to the bus being considered.

The results of all of these events are recorded in a voltage register which specifies voltage at the various buses at sequential time intervals. Once a wave is entered at a particular location, it retains its value in that location for all subsequent times during the life of the problem. It is convenient to follow a procedure which periodically updates each bus voltage by accumulating all prior waves recorded at the bus in question. This is done through a signal table, similar in form to the voltage register. An impression of how this operates can be reached with the aid of Fig. 14.25, which shows a flow diagram of the digital traveling wave program.

The method just described is obviously well suited to distributed circuits, but what of the many lumped components such as transformers, shunt reactors and capacitor banks that are found in power systems? As it is possible to represent distributed circuits by lumpy circuits, so is it possible to approximate lumpy circuits by distributed circuits or stub lines. Capacitors are represented by open-circuited stub lines, inductors by short-circuited stubs. The characteristic impedance and travel time of a line are given by

$$Z_0 = \left(\frac{L}{C}\right)^{1/2} \tag{14.7.6}$$

$$\tau = (LC)^{1/2} \tag{14.7.7}$$

from which

$$Z_0 = \frac{\tau}{C} \tag{14.7.8}$$

and the associated line inductance is

$$L = \frac{\tau^2}{C} = \tau Z_0 \tag{14.7.9}$$

For a stub line to be a good representation of a capacitor, the line capacitance should equal the capacitor value; τ, and therefore L, should be small, and preferably Z_0 should be less than one-tenth of all other surge impedances connected to the same busbar.

For inductors the inductance of the stub should be the same as that of the stub line, and

$$C = \frac{\tau^2}{L} = \frac{\tau}{Z_0} \tag{14.7.10}$$

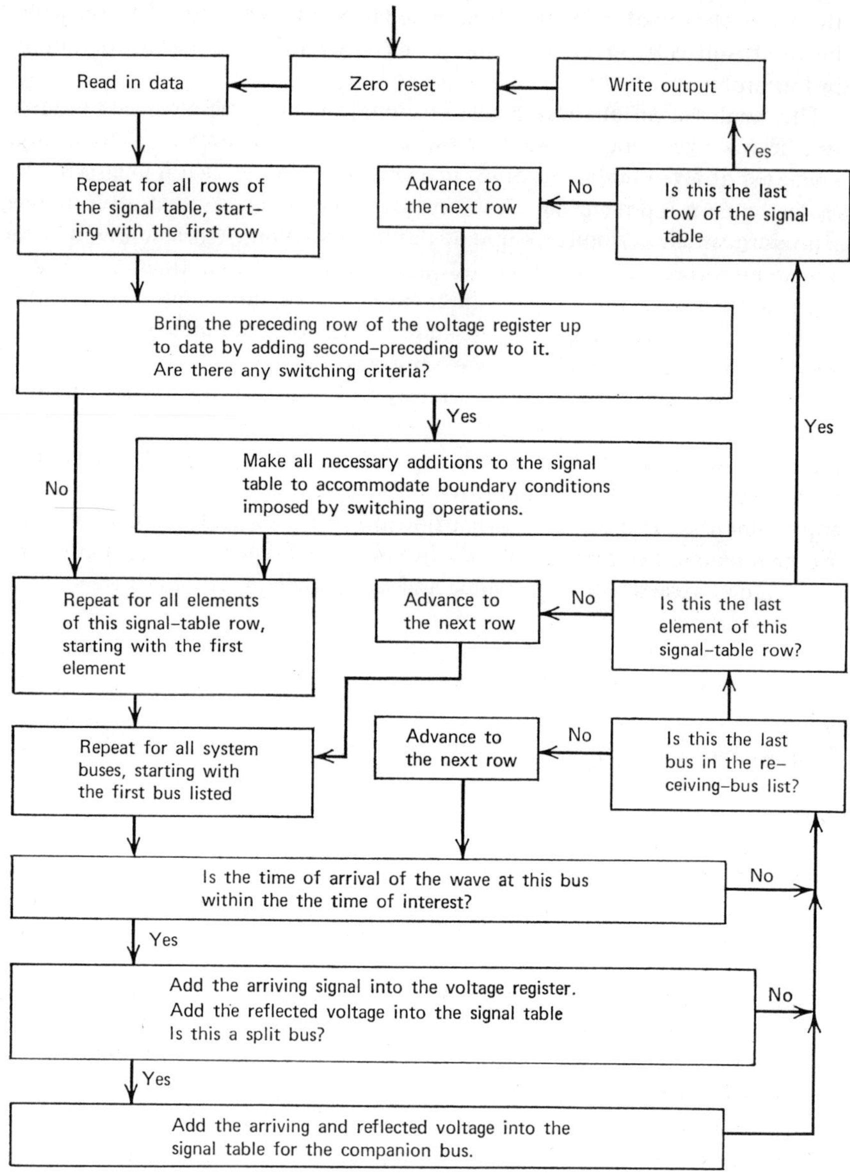

Fig. 14.25. Flow diagram for digital lattice diagram traveling-wave programs.

should be small, so that τ must be small. Preferably Z_0 should be at least ten times the surge impedance of all circuits connected to the same bus.

A lumped resistor can always be represented by an infinitely long line with the same ohmic value for its surge impedance. In fact, a finite length would be used, but one whose travel time is longer than the period of interest in the problem.

The analysis for the traveling wave digital method was described in terms of a unit step function wave, but waveshape is no restriction. A step input could be achieved by initially storing a value of unity in the input signal table in the column representing the input bus. By a similar procedure a stepped equivalent of sinusoidal, ramp, or arbitrary functions can be entered in the signal table to give the system response to the desired function. This is equivalent to combining a large number of positive and negative step functions, displaced from one another in time of origin. Any degree of accuracy can be achieved by making the time intervals ever smaller.

Although we have chosen to illustrate this method with voltage stimuli, it is clear that since currents and voltages are related by surge impedances, current stimuli can be just as readily studied. For example, the technique of determining transient recovery voltages after opening circuit breakers, by injecting a current into the circuit, can still be used by multiplying the current by the circuit surge impedance, thereby converting it into a voltage stimulus.

Bickford and Doepal (27) show how nonlinear shunt resistors, such as those used in lightning arresters, can be represented by what amounts to an application of Thevenin's theorem. The voltage at the busbar to which the resistor is connected is calculated for each time interval, ignoring the presence of the nonlinear resistor. This busbar is then represented by a generator having an internal voltage equal to that just calculated, neglecting the nonlinear resistor, and an internal impedance looking into the busbar with the nonlinear resistor removed. The circuit consisting of this generator and the nonlinear resistor is then solved for the resistor current. The difference between this current and that similarly determined for the previous time interval represents the current element to be injected into the busbar, in order to give the busbar voltage when the nonlinear resistor is in circuit.

As previously stated, it is necessary to convert the injected currents into applied voltages, and this is done here by multiplying the current by the effective surge impedance seen looking into the busbar to which the resistor is connected. The characteristic of the nonlinear resistor is approximated in the program by a number of straight-line sections.

This technique may be used to represent a lightning arrester and gives good agreement with test results. In this application the lightning arrester is represented by a nonlinear resistor, which is switched in circuit when the voltage of the busbar to which it is connected exceeds the sparkover voltage, and is switched out of circuit when the resistor current falls below the extinguishing current of the arrester.

When the subject of circuit reduction was reviewed in Section 6.6, it was pointed out that those parts of the network remote from the point of disturbance need not be represented as rigorously as those parts close to the disturbance. Digital computer studies can be handled in the same way. Part of a complex system is represented by a reduced equivalent circuit, either because the available computational facilities cannot accommodate a full representation or because it is not economical or necessary to represent all of the system in full detail (27). Since in the lattice diagram method, all wave forms are synthesized by means of small step functions, an alternate method is to represent the more remote parts of the system by their step responses. In this way, savings can often be made both in computer storage and in computation time, particularly when a number of switching studies are required at one point in a large system. The step response representation is not an approximation but gives exactly the same result as a full representation.

The lattice diagram program written by Bickford and Doepal (27) provides step response generators representing parts of a system in this way, a condition being that the part of the system to be represented by a step response should be linear or a reasonable approximation thereto. A program limitation is that is should be connected to the remainder of the system at one point only.

In effect, the calculation must be done in two parts, since before the main calculation is carried out, the step response of part of the system must be determined. If a system is considered to consist of two parts, A and B, and if B is to be represented by step response, this may be calculated using the lattice diagram program by applying a unit step to system B through a resistor R at the junction of A and B. Now R is given by the effective surge impedance of A as seen by B. Alternatively, if the measured step response of part of the circuit is available, this could equally well be used.

Once the step response is determined, it is specified by a series of time and voltage coordinates and used in this form as data for the main calculation. In this calculation, voltage waves incident at the busbar to which a step response generator is connected are reflected with a waveform given by a step response and an amplitude determined by the per unit voltage of the incident wave.

Fourier Transform Solutions. Another method for using digital computers to solve power system switching transients has been described by Battisson et al. (30). It is based on the numerical inversion of the modified Fourier transform. They prefer this approach to the traveling wave method, where lumped parameter and/or extensive networks are to be investigated. They also state that with the traveling wave method it is difficult to include the frequency dependence of the line parameters owing to skin effects and earth penetration. Further, to deduce the state of the system at any time, the computation must be carried through stage by stage from the instant of switching; in the process, errors inevitably accumulate. The Fourier method overcomes these difficulties by deriving the time response from a simple transformation of the frequency spectrum of the network, and thus permits the direct computation over any time interval and for only the points of interest on the network. Computer time and space spent on what are, effectively, numerical bookkeeping arrangements for other parts of the network and from zero time are eliminated. However, the Fourier method itself requires considerable computation and much data from the system, which frequently are not available.

Fundamentally, the method requires the calculation of the response of the system over a range of frequencies and the use of the inverse Fourier transform to transform the response from the frequency domain into the time domain. In earlier chapters of this book we became familiar with the Laplace transform, which is here repeated:

$$\mathfrak{L}F(t) = \int_0^\infty F(t)\epsilon^{-st}\,dt = f(s) \tag{2.2.1}$$

Here we are transforming a function $F(t)$ in the time domain into another function $f(s)$ in the s domain. We found it convenient to do this in order to solve transient circuit problems. Specifically, we were able to transform differential equations into algebraic equations, which could be solved in terms of s and then transformed back into the time domain to give the time solution. In Section 2.2 it was explained that s in general is complex, and can be written

$$s = \sigma + j\omega$$

In the problems that we have dealt with in this book it has been possible to perform the inverse transformation, that is, the finding of the t function corresponding to the s function, by reference to standard tables of transforms. More generally the inverse transform is given by the inverse transform integral,

$$F(t) = \frac{1}{2\pi j}\int_{\sigma-j\omega}^{\sigma+j\omega} f(s)\epsilon^{st}\,ds = \mathfrak{L}^{-1}f(s) \tag{14.7.11}$$

The Laplace transform pair given by Eqs. 2.2.1 and 14.7.11 are more general forms of the Fourier transform pair, which are defined as

$$\mathfrak{F}F(t) = \int_0^\infty f(t)\epsilon^{-(\sigma+j\omega)t}\,dt = f(\omega) \tag{14.7.12}$$

and

$$F(t) = \frac{1}{2\pi}\int_{-\infty}^{\infty} \mathfrak{F}(\sigma + j\omega)\epsilon^{(\sigma+j\omega)t}\,d\omega = \mathfrak{F}^{-1}f(\omega) \tag{14.7.13}$$

It is seen that Eqs. 14.7.12 and 14.7.13 convert to Eqs. 2.2.1 and 14.7.11 if we substitute $s = \sigma + j\omega$. The Fourier transform is preferred for the present purpose because the transform parameter can be interpreted physically as the angular frequency ω.

The analytical evaluation of inverse Fourier transforms is frequently very difficult or impossible. However, the integral can be evaluated numerically on the computer by techniques described in reference (30). This reference includes several representative problems solved by this method.

Consider a single-phase transmission line as an example. Its frequency response is given by

$$f(\omega) = \epsilon^{-\gamma(\omega)x} \tag{14.7.14}$$

where x is the length of the line and γ is the propagation coefficient:

$$\gamma(\omega) = [Z_L(\omega)Y_C(\omega)]^{1/2} \tag{14.7.15}$$

Here Z_1 and Y_C are, respectively, the series impedance and the shunt admittance of the line. The propagation coefficient γ may be calculated at various frequencies from Carson's formulas (31).

The frequency response of the line to a step input is given by

$$f'(\omega) = -\frac{\gamma(\omega)x}{j\omega} \tag{14.7.16}$$

The response in the time domain is calculated by applying the inverse Fourier transform to Eq. 14.7.16:

$$F(t) = \frac{1}{2\pi}\int_{-\infty}^{\infty} \frac{\epsilon^{-\gamma(\omega)x}\epsilon^{j\omega t}}{j\omega}\, d\omega \tag{14.7.17}$$

Traveling Wave Solutions. Another form of digital solution that has been applied to power system transient studies is called the traveling wave method. As we have seen, methods previously described can be used for studying traveling waves, but this name perhaps best suits this method since it does in fact solve the wave equations. The ability of the digital computer to manipulate matrices is the important factor in the method.

The method has been described by Uram, Miller, and Ferro (32), (33). It best can be followed by referring to the derivation of the simple wave equation in Section 9.2 and the somewhat more complicated treatment given in Section 9.7, where a ground plane is introduced.

We started by considering a differential element of a single-phase line in Fig. 9.3 and wrote down two partial differential equations for this element, one for current (Eq. 9.2.1) and one for voltage (Eq. 9.2.2), derived by applying Kirchhoff's laws to the small network. In Section 9.7 the same treatment yielded four equations (Eqs. 9.7.1 to 9.7.4) because of the extra circuit introduced by the ground plane. These are partial differential equations since they describe voltage and current as functions of time and position on the transmission line. Uram et al. consider a three-phase line and ground, which could be a ground plane, ground wire(s), or both. Consequently, their elementary circuit for a short length Δx of the line is more complicated. It is shown in Fig. 14.26. It will be seen that the overhead conductors are described by their positive-sequence parameters, while the effects of the ground return are accounted for with zero-sequence parameters. The equations derived from this representation are 14.7.18 and 14.7.19. They are more numerous because of the additional circuits and the coefficients are more complicated because of the introduction of resistance. These equations should be compared with the general case for an n-conductor line given by McElroy and Smith (34) and described briefly at the end of Section 9.7.

$$\left.\begin{aligned}
-\frac{\partial E_1(x,t)}{\partial x} &= \frac{1}{3}\left\{\left[\left(R_0 + L_0\frac{\partial}{\partial t}\right) + 2\left(R_1 + L_1\frac{\partial}{\partial t}\right)\right]I_1(x,t)\right.\\
&\quad + \left[\left(R_0 + L_0\frac{\partial}{\partial t}\right) - \left(R_1 + L_1\frac{\partial}{\partial t}\right)\right]I_2(x,t)\\
&\quad \left. + \left[\left(R_0 + L_0\frac{\partial}{\partial t}\right) - \left(R_1 + L_1\frac{\partial}{\partial t}\right)\right]I_3(x,t)\right\}\\
-\frac{\partial E_2(x,t)}{\partial x} &= \frac{1}{3}\left\{\left[\left(R_0 + L_0\frac{\partial}{\partial t}\right) - \left(R_1 + L_1\frac{\partial}{\partial t}\right)\right]I_1(x,t)\right.\\
&\quad + \left[\left(R_0 + L_0\frac{\partial}{\partial t}\right) + 2\left(R_1 + L_1\frac{\partial}{\partial t}\right)\right]I_2(x,t)\\
&\quad \left. + \left[\left(R_0 + L_0\frac{\partial}{\partial t}\right) - \left(R_1 + L_1\frac{\partial}{\partial t}\right)\right]I_3(x,t)\right\}\\
-\frac{\partial E_3(x,t)}{\partial x} &= \frac{1}{3}\left\{\left[\left(R_0 + L_0\frac{\partial}{\partial t}\right) - \left(R_1 + L_1\frac{\partial}{\partial t}\right)\right]I_1(x,t)\right.\\
&\quad + \left[\left(R_0 + L_0\frac{\partial}{\partial t}\right) - \left(R_1 + L_1\frac{\partial}{\partial t}\right)\right]I_2(x,t)\\
&\quad \left. + \left[\left(R_0 + L_0\frac{\partial}{\partial t}\right) + 2\left(R_1 + L_1\frac{\partial}{\partial t}\right)\right]I_3(x,t)\right\}
\end{aligned}\right\} \qquad (14.7.18)$$

$$
\left.
\begin{aligned}
-\frac{\partial E_1(x,t)}{\partial t} &= \frac{1}{3}\left[\left(\frac{1}{C_0}+\frac{2}{C_1}\right)\frac{\partial I_1(x,t)}{\partial x}+\left(\frac{1}{C_0}-\frac{1}{C_1}\right)\frac{\partial I_2(x,t)}{\partial x}\right.\\
&\qquad\left.+\left(\frac{1}{C_0}-\frac{1}{C_1}\right)\frac{\partial I_3(x,t)}{\partial x}\right]\\
-\frac{\partial E_2(x,t)}{\partial t} &= \frac{1}{3}\left[\left(\frac{1}{C_0}-\frac{1}{C_1}\right)\frac{\partial I_1(x,t)}{\partial x}+\left(\frac{1}{C_0}+\frac{2}{C_1}\right)\frac{\partial I_2(x,t)}{\partial x}\right.\\
&\qquad\left.+\left(\frac{1}{C_0}-\frac{1}{C_1}\right)\frac{\partial I_3(x,t)}{\partial x}\right]\\
-\frac{\partial E_3(x,t)}{\partial t} &= \frac{1}{3}\left[\left(\frac{1}{C_0}-\frac{1}{C_1}\right)\frac{\partial I_1(x,t)}{\partial x}+\left(\frac{1}{C_0}-\frac{1}{C_1}\right)\frac{\partial I_2(x,t)}{\partial x}\right.\\
&\qquad\left.+\left(\frac{1}{C_0}+\frac{2}{C_1}\right)\frac{\partial I_3(x,t)}{\partial x}\right]
\end{aligned}
\right\}
\tag{14.7.19}
$$

The partial differentials appearing in Eqs. 14.7.18 and 14.7.19 can be converted to ordinary differentials by taking the Laplace transform with respect to time. This procedure, as described by Churchill (35), leads to ordinary differential equations in x and the Laplace operator s. The systematic implementation of this transformation and the further development of the analysis requires that the proliferating mass of algebra be expressed in compact form. Thus, transforming Eqs. 14.7.18 and 14.7.19, Uram and Miller

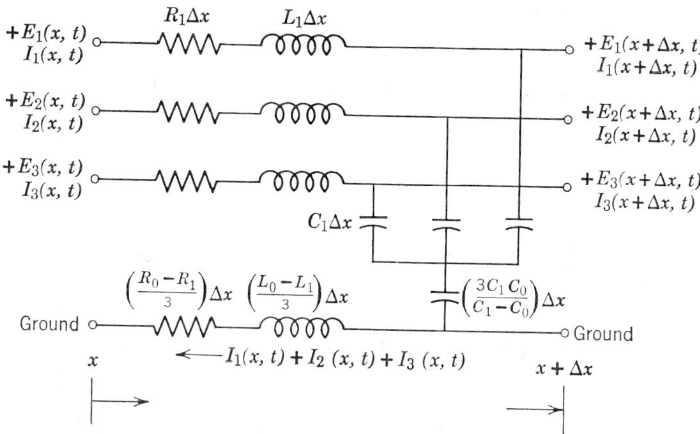

Fig. 14.26. Equivalent circuit for differential element of three-phase transmission line.

(31) express the results as follows:

$$-\begin{bmatrix} \dfrac{dE_1(x,s)}{dx} \\ \dfrac{dE_2(x,s)}{dx} \\ \dfrac{dE_3(x,s)}{dx} \end{bmatrix} = \frac{1}{3}\begin{bmatrix} (Z_0 + 2Z_1)(Z_0 - Z_1)(Z_0 - Z_1) \\ (Z_0 - Z_1)(Z_0 + 2Z_1)(Z_0 - Z_1) \\ (Z_0 - Z_1)(Z_0 - Z_1)(Z_0 + 2Z_1) \end{bmatrix}\begin{bmatrix} I_1(x,s) \\ I_2(x,s) \\ I_3(x,s) \end{bmatrix} \tag{14.7.20}$$

$$-\begin{bmatrix} E_1(x,s) \\ E_2(x,s) \\ E_3(x,s) \end{bmatrix} = \frac{1}{3}\begin{bmatrix} \left(\frac{1}{Y_0}+\frac{2}{Y_1}\right)\left(\frac{1}{Y_0}-\frac{1}{Y_1}\right)\left(\frac{1}{Y_0}-\frac{1}{Y_1}\right) \\ \left(\frac{1}{Y_0}-\frac{1}{Y_1}\right)\left(\frac{1}{Y_0}+\frac{2}{Y_1}\right)\left(\frac{1}{Y_0}-\frac{1}{Y_1}\right) \\ \left(\frac{1}{Y_0}-\frac{1}{Y_1}\right)\left(\frac{1}{Y_0}-\frac{1}{Y_1}\right)\left(\frac{1}{Y_0}+\frac{2}{Y_1}\right) \end{bmatrix}\begin{bmatrix} \dfrac{dI_1(x,s)}{dx} \\ \dfrac{dI_2(x,s)}{dx} \\ \dfrac{dI_3(x,s)}{dx} \end{bmatrix} \tag{14.7.21}$$

where

$$Z_0 = R_0 + sL_0$$
$$Z_1 = R_1 + sL_1$$
$$Y_0 = sC_0$$
$$Y_1 = sC_1$$

or, more concisely,

$$\left.\begin{aligned} -\frac{d}{dx}[E] &= \tfrac{1}{3}[Z_A][I] \\ -[E] &= \tfrac{1}{3}[Z_B]\frac{d}{dx}[I] \end{aligned}\right\} \tag{14.7.22}$$

The details of the steps of changes of variable and further transformation required to reach solutions for Eqs. 14.7 20 and 14 7.21 are described in references (32) and (33). Rather than repeating them, we will attempt to relate some of the steps and results to theory and physical understanding developed earlier in this book and point out the part played by the computer.

The procedures we applied to the simple single-phase line in Section 9.2 are followed for Eqs. 14.7.20 and 14.7.21. To determine voltages, currents are eliminated to produce second-order differential equations in E. The currents and voltages are shown to be related through surge impedance matrices which describe the ratios of voltages to currents in line and ground modes of propagation. Terms emerge in the s domain of the form $F(s) \exp [-(LC)^{1/2}s]$. We know from Section 7.3, and more specifically from Eq. 7.3.1, that this represents a time delayed function. Thus, if

$$\mathcal{L}F(s) = f(t)$$

then

$$\mathcal{L}F(s)\epsilon^{-\tau s} = f(t - \tau) \qquad (t > \tau)$$

Here $f(t - \tau)$ is zero until $t = \tau$; thereafter it has the same value $f(t)$ had τ sec before. If $\tau = (LC)^{-1/2}x$, this has a special significance since $(LC)^{-1/2}$ is the velocity of propagation of a wave on a line. Thus τ becomes the time it takes for the wave to reach a point x down the line. The traveling wave effect is implicit in the solution. We observed the same thing when analyzing the short line fault in Section 13.2. In the present instance the different modes of propagation are manifest by the different velocities $(L_1C_1)^{-1/2}$ and $(L_0C_0)^{-1/2}$.

Equations 14.7.18 and 14.7.19 are general even when all the Rs, Ls and Cs have been specified. For solutions to any particular problem the initial conditions must be known—that is how the circuit is stimulated—and the boundary conditions must be stated—that is how the line is terminated.

The basic procedures we have described involve matrix manipulations, the table storage of data, their subsequent retrieval after appropriate delay times, incremental changes, followed by iterations, and so on. The digital computer is readily programmed to perform these functions at high speed. In an example, in which the effects of nonsimultaneous closing of the three poles of a transmission circuit breaker onto a three-phase line were being studied, a solution for any given set of conditions could be found in nine seconds (32).

This method of solving power system transient problems as yet has been applied only to relatively simple systems involving one or two transmission lines, although the authors (33) conclude:

> The digital computer solution of transmission line transients for a completely general system with many lines and many buses, appears to be feasible, but will be a most complex problem. A difficulty already being encountered is the computer memory required for the large amount of data. This is especially true for tabular storage of the delay functions which occur in the transmission line mathematical model. However, improved programming techniques, should help in this area, so that general solutions should be possible in the near future.

14.8 The Statistical Treatment of Transients—The Monte Carlo Method

It is well known that many transient phenomena are statistical in nature. For example, the magnitude and wave shape of overvoltages occurring on an industrial power system cover a considerable range. The magnitude and wave shape of lightning currents on a utility system are also statistical. The ability of a system to sustain transient overvoltages without getting into difficulties depends upon many things, some, though not all, within the designer's control. In an industrial power system, some of these factors are the thickness and age of insulation, the characteristics and disposition of protective devices, and the amount of contamination. As far as lightning on transmission lines is concerned, the consequence will depend upon conductor spacings, tower and insulator configurations, tower footing resistances, and so forth.

It is possible to attain a high degree of integrity on a power system by paying attention to all these matters and providing accordingly. However, this can become very costly, and there is always a point beyond which it is uneconomical to proceed because the gains to be achieved cannot be justified by the costs involved.

Sometimes a judgment is required on a different basis. For example, when flashover occurs on a line and the resulting fault is removed by opening circuit breakers, an ionized cloud of gas is left behind. It takes a significant time for the dielectric strength of such a postarc path to be restored by the natural processes of cooling and convection. If one attempts to re-energize the line too soon, the path may well be unable to support the closing transient that attends the reclosing of the breakers, leading to a refault. On the other hand, undue delay in reclosing the circuit breakers brings the risk of causing an instability in the system. A judgment must then be made on the risks of refault versus the risks of instability. Before such a judgment can be made we must have information on the recovery rates or deionization times of postarc paths and on the magnitude and waveshape of switching surges. The waveshape can be determined in several ways, perhaps most readily by making many tests on a transient network analyzer. The data on recovery strengths are more difficult to obtain, but tests have been conducted by Johnson et al. (36) to obtain such data. The point we wish to make here is that both these sets of data, the amplitude and waveshape of closing transients and the deionization times, are statistical. Some method must be found for predicting the outcome of the interplay of such probabilistic data. The Monte Carlo method is such an approach. Its application to the calculation of transmission line lightning performance has been described by Anderson (37). We will use this example to describe the method.

Consider what happens when lightning strikes a line. It may strike the phase conductors or the ground wires, it may be near or at a tower, or it may be in midspan. Its current magnitude and rate of rise can vary over a wide range. Whether or not it causes a flashover depends upon all of these things, as well as on the volt/time characteristics of the insulators, the footing resistance of the tower, the point on the system voltage wave where the incident occurs, and several other factors. There are obviously myriad combinations of all of these possibilities; the method must select the conditions in a meaningful way and predict the outcome. As Anderson points out, it is like playing a game with nature; you can deal the cards and see who wins. However, in order to predict ahead of time how a particular line design is going to fare in a particular location, it is necessary that the cards be distributed as they occur in nature.

The best distribution of tower footing resistance for the location must be known; similarly, data must be provided on the frequency distribution of lightning currents of different magnitudes and wavefronts. Anderson (37) gives histograms for these latter distributions; information of this kind was presented in Fig. 10.5. Similar data, as well as they are known, must be also available for the many other variables mentioned above. The computer must then select from these inputs *in random manner* and operate on them using the knowledge of how a system responds to a lightning stroke, which has been described in Section 10.3, and predict the outcome.

It is necessary to perform many such computations, or play many such games, if you will, before a reasonably reliable assessment of the ultimate behavior of the line can be predicted. This is where the computer is invaluable. The random selection of data is akin to the way the cards will be dealt following a thorough shuffle. The computer does not shuffle cards, because this would be too slow. Instead it generates random numbers (or refers to random number tables) and uses these numbers to determine which values of currents, times, and so on, to choose for a particular play of the game.

The Monte Carlo technique has been used by Suzuki (38) for computing the incidence of switching surge flashovers.

REFERENCES

1. K. S. Johnson, *Transmission Circuits for Telephone Communication—Method of Analysis and Design*, Van Nostrand, New York (1927).
2. E. W. Boehne, "The Determination of Circuit Recovery Rates," *Trans. AIEE*, Vol. 54 (1935), p. 530.
3. P. Hammarlund, "Transient Recovery Voltage Subsequent to Short Circuit Interruption with Special Reference to Swedish Power Systems," *Proc. Ingeniers Vetenskaps Akademien*, No. 189 (1946).
4. A. Dupont and J. Robert, "Discharge Study of 735 kV Lines by Analog Computer and Field Tests on 230- and 345-kV Lines," *Trans. IEEE*, Vol. 87 (1968), p. 1010.
5. C. H. Thomas and R. A. Hedin, "Switching Surges on Transmission Lines Studied by Differential Analyzer Simulation," *Trans. IEEE.*, Vol. 87 (1968).
6. G. W. Carter, *The Electromagnetic Field in its Engineering Aspects*, Longmans, Green and Company, London, New York and Toronto (1954).

7. I. B. Johnson and A. J. Schultz, "Analytical Studies on Lightning Phenomena, Involving Towers, Insulator Strings, and Transmission Lines," *Trans. AIEE*, Vol. 77, Part III (1958), p. 1310.
8. H. A. Peterson, "An Electric Circuit Transient Analyzer," *General Electric Review* (September 1939), p. 394.
9. H. H. Skilling, *Electric Transmission Lines*, McGraw-Hill Book Co., New York (1951).
10. E. W. Kimbark, *The Electrical Transmission of Power and Signals*, John Wiley & Sons, New York (1949).
11. H. S. Carslaw and J. C. Jaeger, *Operational Methods in Applied Mathematics*, 2nd ed., Oxford University Press (1947).
12. C. Concordia, "The Transient Network Analyzer for Electric Power System Problems," Supplement to CIGRE Committee No. 13 Report for 1956.
13. L. O. Barthold and R. F. Silva, "Model Techniques in the Transient Network Analyzer Solutions of Power Systems Problems," *Proc. American Power Conference*, Vol. 22 (1960), p. 722.
14. AIEE Committee Report, "Switching Surges I—Phase-to-Ground Voltages," *Trans. AIEE*, Vol. 80, Part III (1961), p. 240.
15. W. H. Croft, R. D. Hartly, R. L. Linden and D. D. Wilson, "Switching Surge and Dynamic Voltage Study of Arizona Public Service Company's Proposed 345 kV Transmission System Utilizing Miniature Analyzer Techniques," *Trans. AIEE*, Vol. 81, Part III (1962), p. 302.
16. D. D. Wilson and D. E. Hedman, "Predict and Control Switching-Surge Levels on EHV Systems," *Electrical World*, Vol. 165 (1963).
17. H. Glavitsch and E. Ruoss, "Transient Voltages Caused by the Operation of Circuit Breakers in High Voltage Systems" CIGRE (1968) Report 13-05.
18. D. D. Wilson, Unpublished data.
19. D. E. Hedman et al., "Switching of Extra High Voltage Circuits II—Surge Reduction with Circuit-Breaker Resistors," *Trans. IEEE*, Vol. 83 (1964), p. 1196.
20. E. W. Kimbark and A. C. Legate, "Fault Surge Versus Switching Surge: A Study of Transient Overvoltages Caused by Line-to-Ground Faults," *Trans. IEEE*, Vol. 87 (1968).
21. I. B. Johnson, R. F. Silva and D. D. Wilson, "Switching Surges due to Energization or Reclosing," *Proc. American Power Conference*, Vol. 23 (1961), p. 229.
22. J. Sabath, H. M. Smith and R. C. Johnson, "Analog Computer Study of Switching Surge Transients for a 500 kV System," *Trans. IEEE*, Vol. 85 (1966), p. 1.
23. R. Joetten and R. Foerst, "Model Plant Tests on the Behavior of a HVDC Transmission During AC and DC Disturbances," CIGRE Report No. 43-08 (1968).
24. P. A. Abetti, "Transformer Models for the Determination of Transient Voltages," *Trans. AIEE*, Vol. 72 (1953), p. 468.
25. J. E. Hudson, E. M. Hunter and D. D. Wilson, "EHV-DC Simulator," *Trans. IEEE*, Vol. 83 (1966), p. 1101.

26. L. C. Weber, D. K. Reitan and A. E. Kilgour, "Design Considerations for a Simulator of an EHV AC/DC Transmission System," *Proc. Am. Power Conf.*, Vol. 27 (1965), p. 929.
27. J. P. Bickford and P. S. Doepel, "Calculation of Switching Transients with Particular Reference to Line Energization," *Proc. IEE*, Vol. 114 (1967), p. 465.
28. L. V. Bewley, *Traveling Waves on Transmission Systems*, John Wiley & Sons, New York (1951).
29. L. O. Barthold and G. K. Carter, "Digital Traveling Wave Solutions," *Trans. AIEE*, Vol. 80, Part III (1961), p. 812.
30. M. J. Battisson, S. J. Day, N. Mullineux, K. C. Parton and J. R. Reed, "Calculation of Switching Phenomena in Power Systems," *Proc. IEE*, Vol. 114 (1967), p. 478.
31. J. R. Carson, "Wave Propagation in Overhead Wires with Ground Return," *Bell System Tech. Journ.*, Vol. 5 (1926), p. 539.
32. R. Uram and R. W. Miller, "Mathematical Analysis and Solution of Transmission-Line Transients I—Theory," *Trans IEEE*, Vol. 83 (1964), p. 1116.
33. R. Uram and W. E. Ferro, "Mathematical Analysis and Solution of Transmission-Line Transients II—Applications," *Trans. IEEE*, Vol. 83 (1964), p. 1123.
34. A. J. McElroy and H. M. Smith, "Propagation of Switching Surge Wavefronts on EHV Transmission Lines," *Trans AIEE*, Vol. 81, Part III (1962), p. 983.
35. R. V. Churchill, *Operational Mathematics*, 2nd ed., McGraw-Hill Book Co., New York (1958).
36. I. B. Johnson et al., "Fault Arc Deionization and Circuit Breaker Reclosing on EHV Lines—Field Laboratory and Analytical Results," CIGRE Report No. 307 (1962).
37. J. G. Anderson, "Monte Carlo Computer Calculation of Transmission-Line Performance," *Trans. AIEE*, Vol. 80, Part III (1961), p. 414.
38. T. Suzuki, "Computation of Switching Surge Flashover Probability by Monte Carlo Simulation," IEEE Summer Power Meeting (1968) CPGG7 PWR.

```
EXTERNAL DERIV.
      COMMON TEMP (40), U(8), F(8), N, C1, C2, C3, H1, H2, H
90    INPUT, C1, C2, C3, H1, H2, T, DT, TF
      INPUT, U, N
      H = 1/H1 + 1/H2
10    CALL RKPB1 (DERIV, TEMP, DT, U, F, N)
      IF(T-TF) 15, 90, 90
15    PRINT, T, U (1)
      CALL RKPB2 (DERIV, TEMP, T, DT, U, F, N)
      GO TO 10
      END
      SUBROUTINE DERIV
      COMMON TEMP (40), U(8), F(8), N, C1, C2, C3, H1, H2, H
      F(1) = U(5)
      F(2) = U(6)
      F(3) = U(7)
```

```
F(4) = U(8)
F(5) = 1 + (U(2) - U(1)/H1)/C1
F(6) = (U(1)/H1 + U(3)/H2-H*U(2))/C2
F(7) = (2*U(4) + U(2) - U(3)*3)/H2/C3
F(8) = (U(3)*2/H2-U(4)*2*H)/2./C2
RETURN
$USE RKPBX$***
END
```

15 System and Circuit Parameter Values for Use in Transient Calculations

15.1 Introduction

The theory of transients expounded in some of the earlier chapters of this book would be of little more than academic interest if it could not be translated into terms of practical everyday power system circuits The material of this chapter, mostly a compendium of parameter values—*L*s, *C*s, *R*s, and so on, for typical power system components—is intended to provide the wherewithal to make such a translation.

To predict the transient response of a circuit to a particular stimulus, it is first necessary to reduce the circuit to a network of *R*s, *L*s, and *C*s Then we derive and solve the equations which describe the transient behavior of this network. Finally, values must be assigned to the branch elements. Only then can we determine what the natural frequencies will be, what the peak voltage or current will be at particular locations, and how quickly the disturbance will be damped out.

Consider the simple circuit shown in Fig. 15.1*a*. Suppose we wish to determine the transient inrush current that will flow when the switch *S* is closed and the capacitor bank is energized from the power system through the transformer. If we consider the power system as an infinite bus (zero impedance), the simplest representation we could use for the circuit, which acknowledges the facts of the situation, would be that shown in Fig. 15.1*b*, where L_T is the leakage reactance of the transformer, C_T is its effective capacitance, and L is the inductance of the loop between the transformer and the capacitor bank. Developing an equation for the current in terms of L_T, L, C_T, and C and the initial conditions of the circuit when the switch is closed is a relatively straightforward procedure. But to solve it numerically,

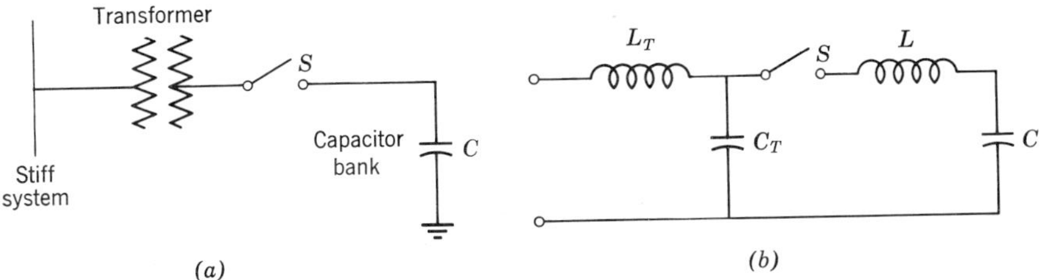

Fig. 15.1. Energizing a capacitor bank through a transformer.

we must assign values to the circuit parameters. What value should we assign to the transformer capacitance or the loop inductance out to the transformer bank? The purpose of this chapter is to furnish data of this kind.

15.2 Transient Parameter Values for Transformers and Reactors

Our principal interest in transformers is their inductance and capacitance, from which we can obtain the natural frequency at which they oscillate when disturbed. We are thinking here of the transformer as a whole. Multi-frequency oscillations within the windings of a transformer have been described in Chapter 12.

We are also interested in the effective resistance of transformers insofar as it affects the damping of oscillations and contributes to the X/R ratio of power circuits, from which can be determined the decrement of transient fault currents, for example.

In most instances, the inductance that matters when a transformer is party to a system oscillation is the leakage inductance, which is usually obtainable by a simple calculation from the per unit reactance, X_{pu}, or the percent reactance, $\%X$, which more often than not is stated on the nameplate. By definition,

$$X_{pu} = \frac{IX}{V}; \qquad \%X = \frac{IX}{V} \times 100 \tag{15.2.1}$$

where X is the leakage reactance, and V and I are the rated current and voltage. Since $X = \omega L$, the leakage inductance is

$$L = \frac{X_{pu}V}{\omega I} = \frac{\%XV}{100\omega I} \tag{15.2.2}$$

This can be put in a more useful form by multiplying numerator and denominator by V:

$$L = \frac{X_{pu}V^2}{\omega VI}$$

$$= \frac{X_{pu}(kV)^2}{\omega(MVA)} \tag{15.2.3}$$

which puts it in terms of the rated voltage in kV, and the transformer rating in MVA. If line-to-line voltage is used, then the MVA is the total MVA of the three-phase transformer. If line-to-neutral voltage is specified, then the MVA is the MVA/phase.

Consider, as an example, a 34.5-kV, 15-MVA transformer which might typically have a 7% reactance. According to Eq. 15.2.3,

$$L = \frac{0.07 \times (34.5)^2}{377 \times 15} = 14.7 \text{ mH}$$

This is the value of the leakage inductance referred to the 34.5-kV winding. If the other winding is 13.8 kV, the transformer's inductance referred to that winding would be reduced in proportion to the square of the turns ratio:

$$L = 14.7 \times \left[\frac{13.8}{34.5}\right]^2 = 2.36 \text{ mH}$$

The leakage inductance of a transformer depends upon its design or, more specifically, upon the disposition of its windings, for this determines the leakage flux. The range of percent impedance values for some typical transformers is given in Table 15.1. As we shall see, winding resistance is small, so these can be taken as percent reactance values for our purpose and can be used to calculate the leakage inductance in accordance with Eq. 15.2.3.

It will be noted that the percent impedance increases with increasing voltage, which is a natural consequence of the increased separation between the windings themselves and the windings and the core, demanded by the need for extra insulation. This same factor is responsible for the difference between the delta and ungrounded wye transformers as against those with grounded wye windings. As shown in the expansion of the table for 1175-kV BIL, the impedance increases as the voltage of the low-voltage winding increases. This is true at the other voltage levels. From the point of view of transient studies, these variations are of secondary importance, since in the computation of natural frequencies and surge impedances, we are interested in $L^{1/2}$.

It is important to be clear about the kVA base on which an impedance is calculated. It is possible to increase the kVA rating of a transformer by 50% or more by forced cooling. On the new base the percent impedance would be correspondingly increased, but we clearly have not changed the leakage inductance of the transformer.

Autotransformers are commonly used to interconnect transmission circuits at different voltage levels. Since primary and secondary share a common winding on such transformers, they tend to have lower leakage inductance than two-winding transformers of comparable rating. The high

Table 15.1 Typical Positive-Sequence, 60-Hz, Per Unit Impedance Values for Transformers Rated 15,000 kVA and Above

High-Voltage Winding BIL	Low-Voltage Winding BIL	Percent Reactance	
		Δ or Ungrounded Y	Grounded Y
150	110	5–7.5	—
200	150	5.5–8.25	—
250	150	5.75–8.5	—
350	200	6.25–9.25	—
450	250	7.0–10.5	—
550	350	7.75–11.5	—
650	350	8.25–12.5	—
750	450	9.25–14.0	—
825	450	9.5–14.25	—
900	550	10.25–15.5	—
1050	650	11.0–16.5	—
—	150	—	8.75–13.25
1175	350	—	9.0–13.5
—	650	—	10.0–15.0
—	900	—	11.5–16.5
1300	650	—	10.5–15.75
1425	650	—	11.5–16.5
1550	750	—	11.75–17.5
1675	750	—	12.00–18
1800	750	—	12.0–18
2050	750	—	12.75–19.25
2300	750	—	13.25–19.75

to low percent impedance of an autotransformer can be calculated from Table 15.1 by the relationship

$$Z_{\mathrm{H-L}} = \alpha^2 Z$$

where

$$\alpha = \text{co-ratio} = \frac{\text{high voltage} - \text{low voltage}}{\text{high voltage}}$$

and Z is the impedance for a similar two-winding transformer.

There are occasions when the magnetizing inductance, rather than the leakage inductance, of a transformer is of interest. This would be the case, for example, when an unloaded transformer is being disconnected. This is much higher than the leakage inductance. For large power transformers, the magnetizing current is typically less than 1% of the load current. If the same transformer has 12% leakage reactance, the ratio of magnetizing inductance to leakage inductance is

$$\frac{L_m}{L_s} = \frac{1}{0.01 \times 0.12} = 833$$

Some typical figures for magnetizing current are given in Table 15.2, from which it will be seen that, as a general rule, when stated as a percentage of full load current, the magnetizing current diminishes as the size of the transformer increases. There is a significant spread in values depending upon the amount of iron that the manufacturer chooses to put into his product. This is seen in Fig. 15.2, where typical ranges of magnetizing current are plotted for 15-kV class transformers of different kVA ratings.

Another important fact regarding the magnetizing inductance of a transformer is that it is not constant. The flux is carried by iron for the most part, and, as is well known, the relationship between the flux and current is nonlinear. Under steady-state conditions this may introduce harmonics into the current or voltage waveforms. Under transient conditions it can lead to very high transient currents when the transformer is energized; this phenomenon is described in Section 5.5. Sometimes in the course of a subsidence transient, when an unloaded transformer has been switched off, it can exhibit alternately high and low values of inductance as its core passes in and out of saturation. This is discussed briefly in Section 13.3.

We now turn to the matter of transformer capacitance. Let us start by considering a very simple device comprising two cylindrical windings, a core and a tank. This is represented schematically in Fig. 15.3. The inner winding, closest to the core, is the low voltage winding; it is surrounded by the high-voltage winding. It is not difficult to think of these windings as coaxial metal

Table 15.2 Typical Values of Magnetizing Current, as a Percentage of Normal Load Current, for Power Transformers

Transformer MVA	Magnetizing Current as Percent of Load Current			
	BIL = 350	650	900	1300
20	0.8	0.90	1.0	1.2
40	0.65	0.74	0.82	0.94
60	0.58	0.65	0.73	0.84
80	0.54	0.61	0.68	0.77
100	0.51	0.59	0.65	0.73
150	0.47	0.53	0.61	0.67
200	—	0.51	0.58	0.64
300	—	0.49	0.55	0.61
500	—	0.47	0.53	0.59

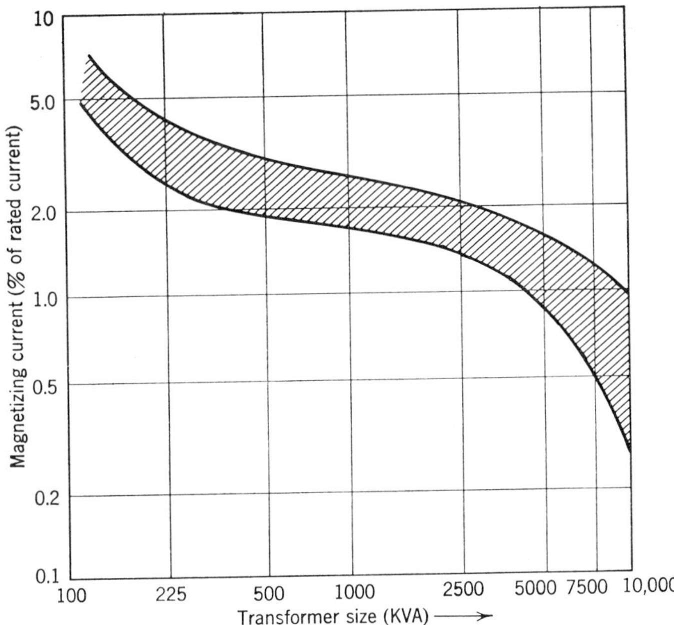

Fig. 15.2. Magnetizing current of 15-kV class three-phase transformers.

cylinders, which form the plates of a capacitor. Similarly, there will be a capacitor formed between the inner winding and the core. This is assuredly an oversimplification, but it is a good starting point. Like any other capacitors, their capacitance values will depend upon the area of the plates, the separation of the plates, and the permittivity of the material separating them. We would expect, therefore, that transformers of high kVA would have higher capacitance than those of low kVA, simply because they are physically bigger. We

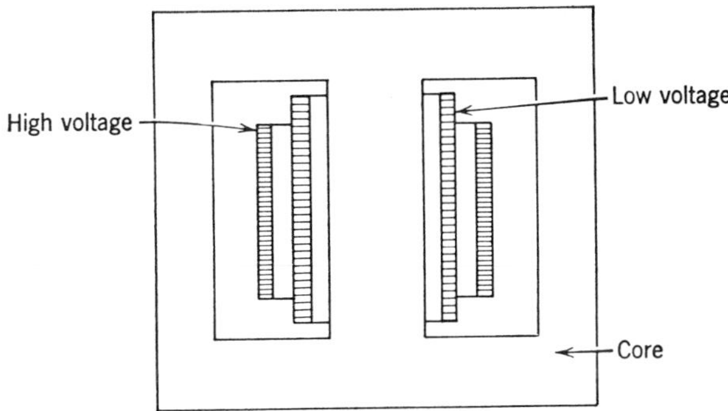

Fig. 15.3. Simplified cutaway model of a transformer.

might also suppose that the capacitance of a high-voltage transformer would tend to be less than a low-voltage transformer of comparable rating, since the fact of high voltage calls for more separation between the windings, and between the windings and core. It is also reasonable to expect that a so-called "dry type" transformer will have less capacitance than a corresponding transformer where the interstices of the windings are filled with oil or some similar fluid of higher permittivity. In practice it is found that these observations do correspond to the facts. We have been considering a very simple model here. Details of design, notably the type of winding used, can significantly affect capacitance values, as can electrostatic grading shields (Fig. 11.3). In Section 11.2 two styles of winding are described, the layer type and the disk type; they are illustrated in Fig. 11.3. From a capacitance point of view the disk type is more like a stack of phonograph records than like the coaxial cylinders of the layer type winding. There is capacitance between the disks and from the disks to the grounded structure surrounding them and to the other winding. It is found that the effective capacitance to ground for disk windings is often no more than 30–40% of the value of a layer type winding of the same kVA rating.

Sometimes one end of a winding is connected to ground; the star point of a Y winding is often connected so. The oversimplification of our model may make it difficult for the reader to see how we can have a capacitor when the plates are joined together. It should be remembered that the capacitance is really distributed and that when a voltage is applied across the winding an electric field is set up between the winding and adjacent conductors.

Different parts of the capacitance are charged to different voltages; only at the neutral is the voltage zero. When we look at a winding as a whole, we are really asking whether we can place at the terminal some capacitance that would effectively behave in the same way as this complex distributed capacitance. To answer this question, consider the simple winding shown in Fig. 15.4. Here the capacitance to ground is uniformly distributed, each length Δx of the winding has a capacitance $C_g \, \Delta x$, where C_g is the capacitance to ground

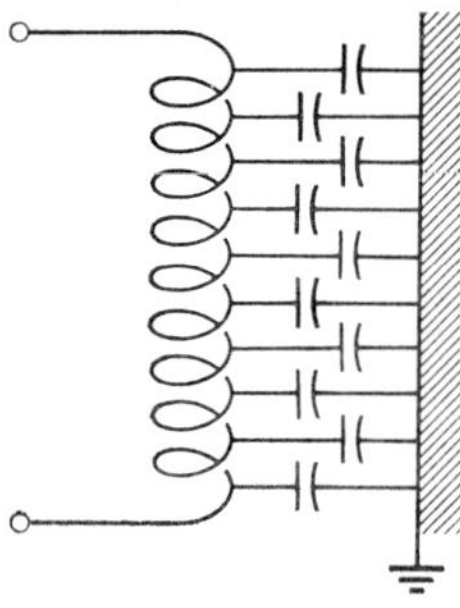

Fig. 15.4. Transformer winding with uniformly distributed capacitance.

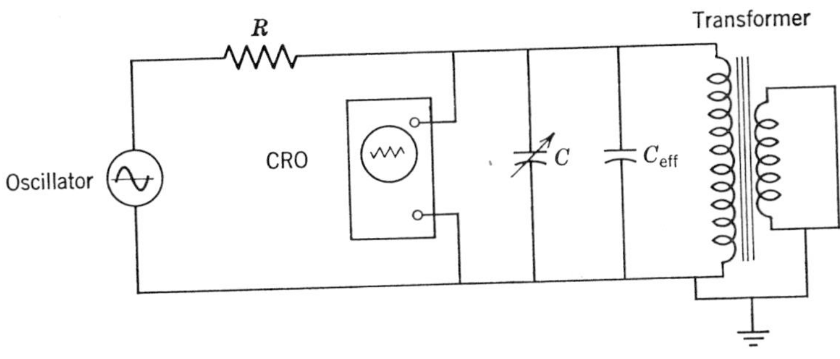

Fig. 15.5. Circuit for measuring the effective capacitance of a transformer.

per unit length of the winding. Any such element, distance x along the winding, can be effectively transferred to the terminal by looking upon it as a load on an autotransformer. The elementary capacitance is "seen" through the turns ratio x/l, and therefore has a value

$$C_g \Delta x \left(\frac{x}{l}\right)^2$$

If we integrate the effect of all such elementary lengths for the entire winding, we have for the effective capacitance at the terminal

$$\begin{aligned} C_{\text{eff}} &= C_g \int_0^l \frac{x^2}{l^2}\, dx \\ &= \tfrac{1}{3} C_g l \end{aligned} \tag{15.2.4}$$

which is one-third of the total capacitance to ground of the winding. Again, this is a simplified approach; it assumes the winding is perfectly uniform and it ignores series capacitance down the winding. It does give a physical interpretation of the capacitance we are talking about.

How can this effective capacitance be measured? A method suggested to the author by H. W. Lord has proved very useful. It is illustrated in Fig. 15.5. The winding whose capacitance is to be determined is connected through a resistor to a variable frequency oscillator. The resistor should be high, relative to the short-circuit reactance of the transformer over the frequency range of interest. A variable capacitor C is placed across the winding, and therefore in parallel with C_{eff}; the other winding is short-circuited. The transformer is now excited from the oscillator and a resonance is sought between the leakage inductance of the transformer L_s, and the parallel combination of C and C_{eff}. This point is established by a peaking of the voltage across the winding as seen on the CRO. The procedure is repeated for several different values of C. Now, the natural frequency is given by

$$f_0 = \frac{1}{2\pi[L_s(C + C_{\text{eff}})]^{1/2}}$$

or

$$(C + C_{\text{eff}}) \propto \frac{1}{f_0^2} \tag{15.2.5}$$

C_{eff} is determined by plotting $1/f_0^2$ against C as shown in Fig. 15.6. An extrapolation of this plot to the capacitance axis gives C_{eff}. The numbers shown in Fig. 15.6 were taken from actual measurements on several different 13.8 kV/480 V transformers that might typically be used in industrial substations.

Figures 15.7*a* to 15.7*h* have been generously supplied by the IEEE Working Group on Transient Recovery Voltages; they provide values of the total

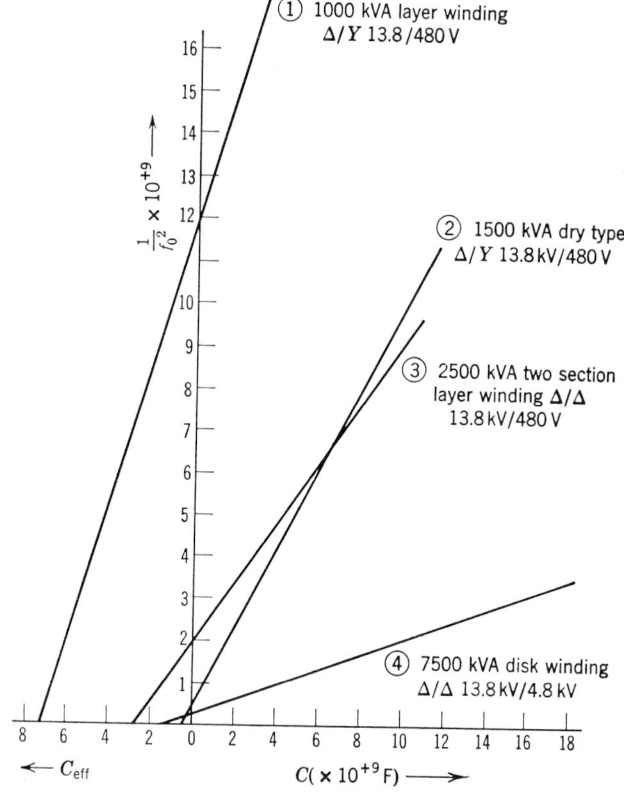

Fig. 15.6. Experimental data for measuring the effective capacitance of 15-kV class transformers.

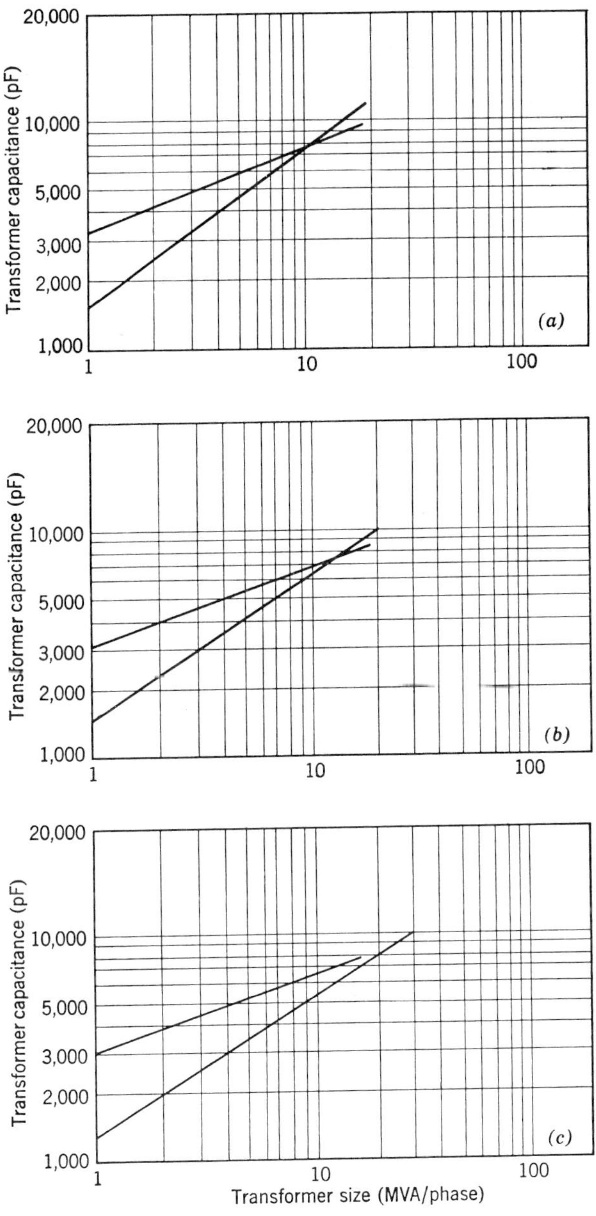

Fig. 15.7. Transformer winding capacitance to ground range of values for various BIL of highest voltage winding. (*a*) 110 kV BIL. (*b*) 150 kV BIL. (*c*) 200 kV BIL. (*d*) 250 kV BIL. (*e*) 350 kV BIL. (*f*) 450 kV BIL. (*g*) 550 kV BIL. (*h*) 650 kV BIL.

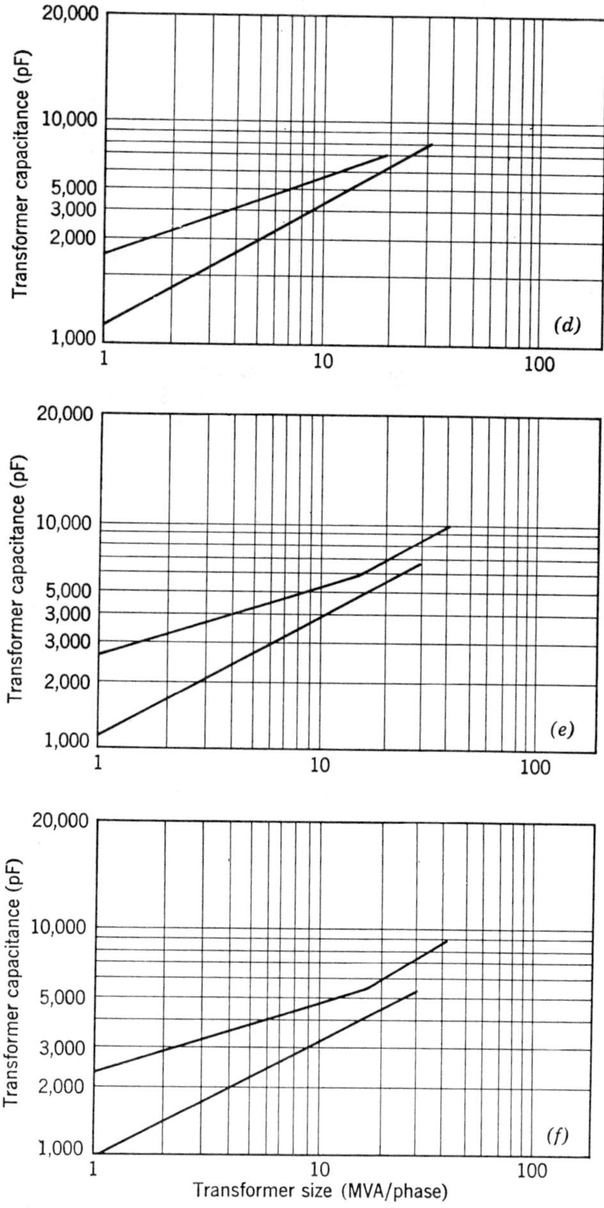

Fig. 15.7. (*continued*)

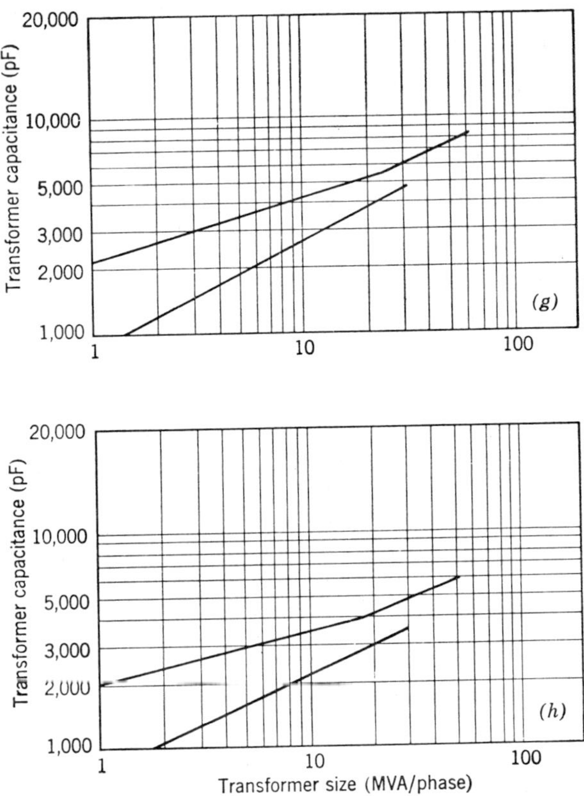

Fig. 15.7. *(continued)*

capacitance to ground of the highest voltage winding on a per phase basis. It is suggested that half of the winding capacitance of each phase, as given in the figures, be considered lumped at each terminal of the winding. For a Y-connected bank, 0.33 to 0.406 of the capacitance values taken from the appropriate figure is the effective value component for the winding. For a delta-connected bank, the full phase values from the figures is used because there are two windings providing lumped terminal capacitance. It will be seen that the values lie in a range which is wide at times. Values at the limit of the range may occur as frequently as those in the middle.

In addition to the capacitance of the winding, there is the capacitance of the bushing to which it is connected. This is included in Fig. 15.6 but not in Fig. 15.7. Table 15.3 gives typical capacitance values for outdoor bushings. In the case of a delta-connected bank of single-phase transformers, the capacitance of two bushings must be added.

Table 15.3 Capacitance to Ground of Outdoor Bushings

Kilovolts	Ampere Rating	Capacitance Range (pF)
15	600	160–180
	1200	190–220
23	400	200–450
	600	280
	1200	190–450
	2000	280–650
	3000	370–560
	4000	500–620
34.5	400	200–390
	600	150–220
	1200	170–390
	2000	240–360
	3000	350–620
46	400	180–330
	600	150–280
	1200	170–330
	2000	200–330
69	400	180–270
	600	250
	1200	160–290
	2000	210–320
115	800	250–450
	1200	250–420
	1600	250–430
138	800	250–450
	1200	250–420
	1600	250–460
161	800	260–440
	1200	260–440
	1600	260–440
196	800	350–550
	1200	350–550
	1600	350–550
330	1600	530
345	800–2000	
	BIL: 1050	550
	1175	500
	1300	450
500	800–2000	
	BIL: 1425	500
	1550	500
	1675	520

Interest in transformer inductance and capacitance centers around their influence on the natural frequencies at which transformers oscillate when disturbed. This has an important bearing on switching operations. In a paper by Dienne and Frisson (1), tests on many transformers are reported in which the natural frequency of the transformers was measured. Similar data is presented by Paillet and Dufour (2). Let us consider a 69-kV, 15-MVA transformer with a 10% reactance. According to Eq. 15.2.3 its leakage inductance/phase would be

$$\frac{0.1 \times 69^2}{377 \times 15} = 85\,\text{mH}$$

If the transformer is delta-wound, its effective winding capacitance from Fig. 15.7*d* (250 kV BIL) might be 5000 pF. To this we might add 100 pF for the bushing. When a switching operation is performed at the terminals of delta-wound transformers, the first pole to open disconnects two phases of the winding. The capacitance involved is therefore the parallel capacitance of two phases. However, the inductance involved is the parallel inductance of two phases, the product, therefore, is still $L_s C_{\text{eff}}$. For our example, the natural frequency is

$$f_0 = \frac{1}{2\pi(L_s C_{\text{eff}})^{1/2}} = \frac{1}{2\pi[8.5 \times 10^{-2} \times 5.1 \times 10^{-9}]^{1/2}}$$

$$= 7.700\ \text{Hz},$$

Dienne and Frisson attempt to summarize as follows the considerable data they present:

It is difficult to give a mean value of the transient recovery voltage frequencies for all measurements and for the whole range of transformers. But if we consider only three-phase to earth faults, the following figures can be given for the high voltage windings of the transformers:

150 kV/70 kV/MV	Transformers: 5 to 6 kHz
150/MV	Transformers: 6 to 8 kHz
70/MV	Transformers: 6 to 16 kHz

From the few measurements made on the MV windings of transformers or on MV/LT (medium voltage/low tension) transformers, we can only point out that the frequencies are very variable and high: 20-100 kHz.

Some data on three-winding transformers is reported by Tchernyshev et al. (3).

When an unloaded transformer is switched, its natural frequency is much lower since it is its magnetizing inductance rather than its leakage inductance that is involved in the oscillation, and, as pointed out earlier in this section, this is usually 100 to 1000 times greater than the leakage reactance. The natural frequency is therefore 3 to 10% of the short-circuit natural frequency. Under these conditions it is found that the oscillation is quite heavily damped. This is related to our next topic.

Transformer resistance, if we are talking about the true ohmic resistance of the windings, is very low. Typically it is in the range of 0.2–1.5%; some values are plotted in Fig. 15.8, adapted from reference (4). Iron losses in the transformer core may also be manifest as resistance; the criterion here is whether or not the important flux is carried by the core or whether it is principally airborn leakage flux. The losses can be represented as a resistor in parallel with the tank circuit formed by L_m and C_{eff}. It very effectively damps the oscillations (refer to Chapter 4).

Under fault conditions in a power system, much of the flux in transformers follows air paths; the effective resistance shunting L_s and C_{eff} is much higher. The true series resistance of the windings is then the principal damping force, and it is far less effective. This is evident from Fig. 15.9, adapted from reference (1). These factors are important to the extent that they affect the damping of switching transients.

In Section 3.2 we discussed the transient that occurs on closing a circuit and pointed out that the current is more or less asymmetrical depending upon

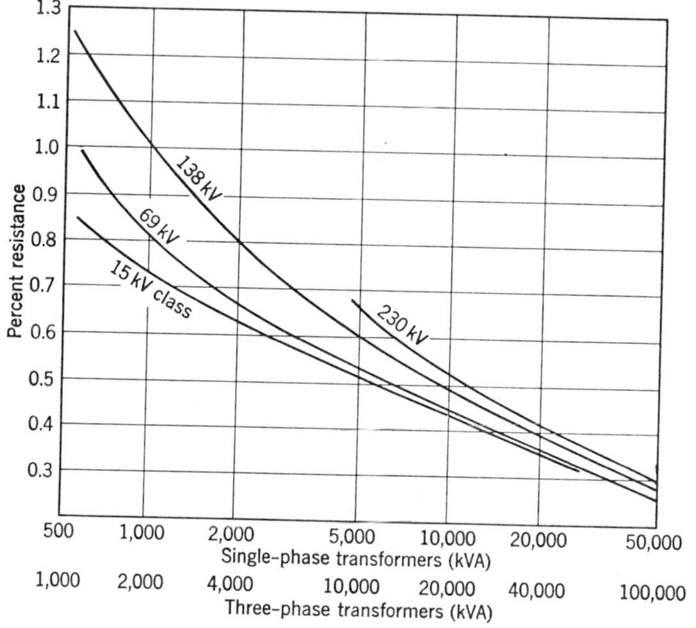

Fig. 15.8. Percent resistance of oil-immersed self-cooled transformers.

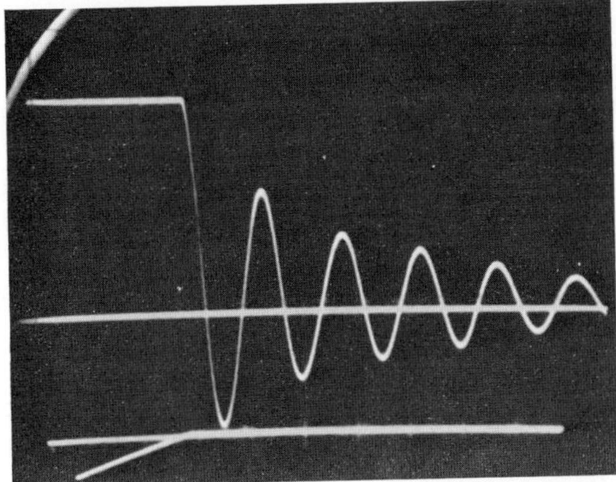

Fig. 15.9. Natural frequency oscillation of a short-circuited transformer (6.7 kHz) showing low damping.

the instant in the cycle when the switch is closed. The time the current takes to become symmetrical depends upon the L/R time constant or X/R ratio of the circuit. This also applies to fault currents when a new circuit is inadvertently closed by a fault. In many instances a major part of the circuit impedance is contributed by transformers, in which case it is their X/R ratios that are important. These can be derived from Eq. 15.2.3 and Fig. 15.8; they are typically in the range 10–30, being lowest for distribution transformers and highest for large, high-voltage transformers. We shall have occasion to refer to the X/R ratio of other system components; it is therefore helpful to show quantitatively what this means in terms of the decay of asymmetry in fault currents. This is shown in Fig. 15.10.

Power system reactors are of two kinds: those with a magnetic core, which are similar to open circuited transformers and can be treated as such; and the air-cored variety, which are more like a short-circuited transformer, except that they have considerably less capacitance. The iron-clad type are usually connected as a shunt component, whereas the cast-concrete air-cored type are series connected for fault current limiting.

The inductance of the latter kind can be determined from its nameplate rating. The effective capacitance, as mentioned above, is low, which we would expect with the removal of the close ground surface that the core represents. If the reactor is viewed as a π circuit, 75 to 150 pF at each terminal is a reasonable value for the capacitance. This means that the natural frequency of such a device is high, especially in low-voltage circuits. Figure 15.11 shows the response of such a reactor to a shock excitation. It will be seen that its natural frequency is approximately 33 kHz. Significantly, higher values are sometimes found. Moreover, damping is slight.

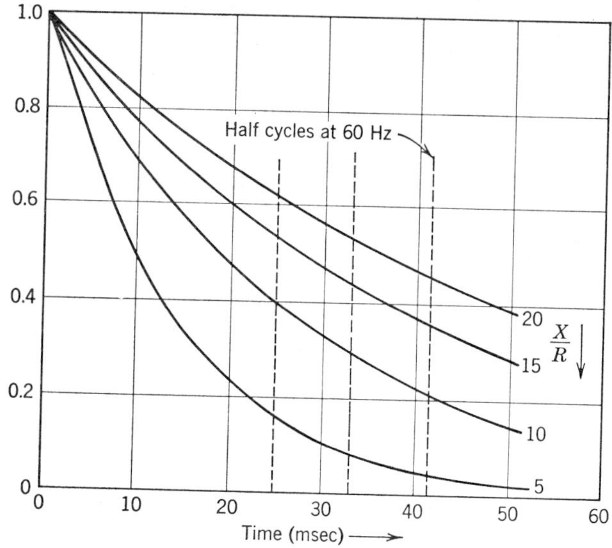

Fig. 15.10. Decay of asymmetry of different X/R ratios (60 Hz).

15.3 Transient Parameter Values for Generators

We have said relatively little in this book about generators. This is because most transient disturbances created by faults and switching operations occur at points fairly remote from the generator terminals. Consequently, the generator often plays only a minor role in the subsequent events. However, this is not always the case.

The transient behavior of generators is rather complex, because of the way the magnetic flux patterns change with time. Consequently no single number

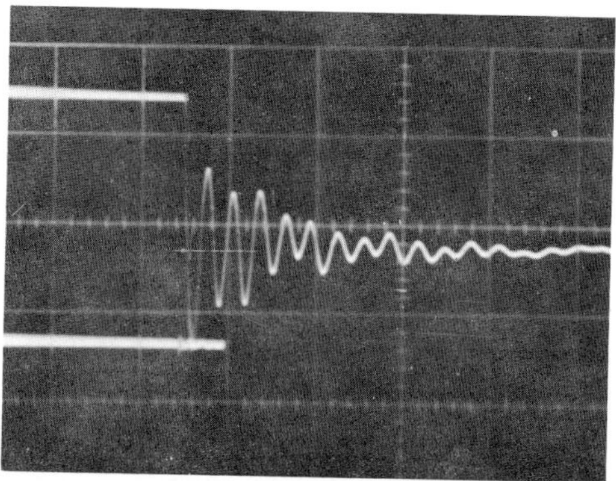

Fig 15.11. Transient response of a reactor.

Table 15.4 Sixty-Hertz Reactance Values of Synchronous Generators

Two-pole turbine generator	0.12–0.21	0.07–0.14
Four-pole turbine generator	0.20–0.28	0.12–0.17
Salient pole generator		
With dampers	0.20–0.50	0.13–0.32
Without dampers	0.20–0.50	0.20–0.50
Synchronous condenser	0.3–0.50	0.19–0.30

can be attached to the inductance of such a machine. It would not be appropriate to discuss all the details of this in this text; it is well covered in the literature, by Concordia (5), for example. It suffices to say here that the effective reactance of a synchronous generator starts at a relatively low value, the so-called subtransient reactance x_d'', on the occurrence of a fault it increases to the so-called transient reactance, x_d'; and then it increases more slowly to the so-called synchronous reactance value, which usually takes many cycles to attain. This often exceeds 1 pu, that is, the sustained short-circuit current may be less than the full-load current. Since most of the events we are interested in occur quickly, we will list only the subtransient and transient reactances (Table 15.4). The corresponding inductances can be obtained from Eq. 15.2.3.

The capacitance of generators is higher than that of transformers of comparable rating because the stator windings are embedded in the stator steel and therefore are in closer physical proximity to the ground plane which the steel represents. Generally speaking the capacitance of generators increases with size, but there can be a considerable spread in a given rating, because of design. This is apparent from Fig. 15.12, which shows values for 2-pole, 3600-rpm machines. This capacitance is the total capacitance-to-ground of the 3-phase stator winding. Using the π model again for each phase, each of the π capacitors would be $\frac{1}{6}$ of the value in the figure.

Forced cooled machines tend to have smaller capacitance because the conductors can be smaller; a 500-MW set generator may have only 0.5 μF total. The slower rpm machines, particularly hydromachines, tend to have higher capacitance.

Capacitance values for motors vary considerably with design. Some typical values, obtained from reference (4), are presented in Fig. 15.13.

15.4 Transient Parameter Values for Transmission Lines and Cables

The L, C, and R of transmission lines and cables are of interest in transient problems for several reasons. If the lines or cables are relatively short, these

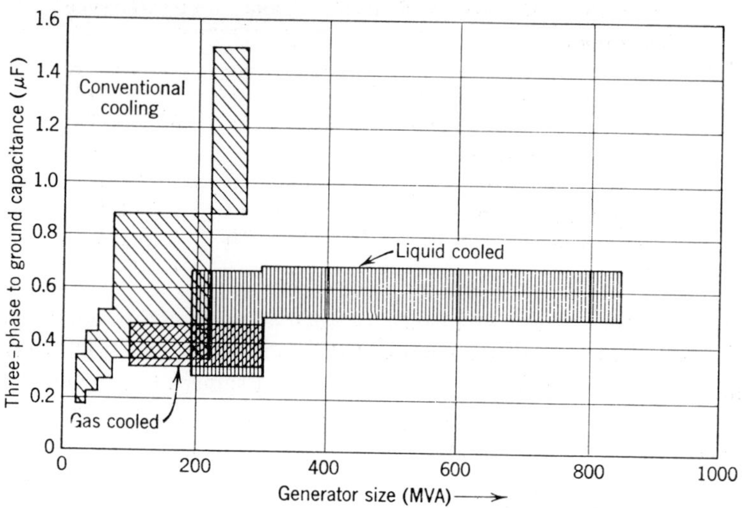

Fig. 15.12. Capacitance of 3600-rpm 2-pole generators.

parameters can be treated as lumpy elements and as such they can have an important influence on transient currents and voltage when system disturbances occur. When the circuits are long, the distributed nature of the inductance, capacitance, and resistance is apparent. These properties then determine the characteristic or surge impedance of the lines, the velocity of wave propagation, and the amount of attenuation experienced by traveling waves.

For a simple transmission line comprising a pair of parallel cylindrical conductors, it was shown in Section 8.3 that the inductance per unit length of such a circuit is

$$L = \frac{\mu_0}{\pi}\left[\ln \frac{d-r}{r} + \frac{1}{4}\right] \text{ henry/meter} \tag{8.3.15}$$

When the conductor spacing d is large compared with the conductor radius r, this simplifies, as pointed out in Section 9.1 to

$$L \approx \frac{\mu_0}{\pi} \ln \frac{d}{r} \text{ henry/meter} \tag{9.1.7}$$

With similar approximations, the capacitance between the conductors of such a line is

$$C = \frac{\pi\epsilon_0}{\ln d/r} \text{ farad/meter} \tag{9.1.8}$$

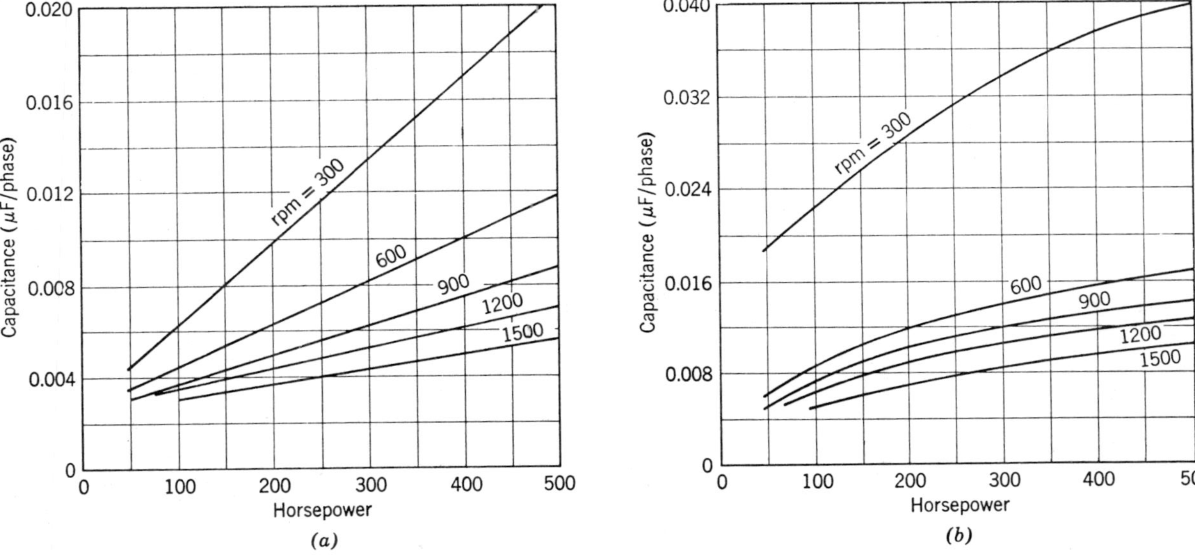

Fig. 15.13. Capacitance of 2300-V motors. Values for motors up to 6600 V comparable. (*a*) Synchronous motors. (*b*) Induction motors.

Whence the velocity of propagation is given by

$$v = \frac{1}{(LC)^{1/2}} = \frac{1}{(\mu_0\epsilon_0)^{1/2}} = 3 \times 10^8 \text{ m/sec} \tag{9.1.9}$$

or a little less than 5 miles/μsec. It is independent of the conductor size and spacing.

The surge impedance, on the other hand, is given by

$$\left.\begin{aligned} Z_0 = \left(\frac{L}{C}\right)^{1/2} &= \frac{1}{\pi}\left(\frac{\mu_0}{\epsilon_0}\right)^{1/2} \ln\frac{d}{r} \\ &= 120 \ln\frac{d}{r}\ \Omega \\ &= 276 \log\frac{d}{r}\ \Omega \end{aligned}\right\} \tag{15.4.1}$$

It is interesting that Z_0 is relatively insensitive to conductor spacing. Consider a pair of conductors 1 in. in diameter spaced 15 ft apart:

$$Z_0 \approx 276 \log 360 = 708\ \Omega$$

If we double the spacing, Z_0 increases to 786 Ω only. The same change would be wrought by reducing the conductor diameter to $\frac{1}{2}$ in. It is found that the surge impedance of most practical lines of this type lies in the range 600–900 Ω. The approximations made in the preceding computations are to neglect magnetic flux within the conductor and ignore r compared with d. These are justified in this type of calculation except where the conductors of the line are unusually close together; cables might be considered this way, for example. Refinements for different types of conductor, using the concept of "geometrical mean radius" (GMR) are presented by Wagner and Evans (6).

What we have just calculated and the formulas that apply should now be clearly understood. Equations 9.1.9 and 15.4.1 give approximations for the velocity of wave propagation and the surge impedance of an isolated transmission linc comprising two parallel cylindrical conductors. The L and C are for the circuit, not for the individual conductors. The presence of other conductors, notably ground, will modify these values somewhat, because of mutual effects between the original circuit and new circuits introduced by the other conductor(s). The presence of ground was examined in Section 9.7, where it was shown that for this particular arrangement of conductors there are two modes of wave propagation with their separate velocities and surge impedances. These may be summarized as follows:

$$\left.\begin{aligned} v_F &= \frac{1}{[(L - M)(C_1 + 2C_2)]^{1/2}} && \text{(fast or line mode)} \\ v_S &= \frac{1}{[C_2(L + M)]^{1/2}} && \text{(slow or ground mode)} \\ Z_F &= \left[\frac{L - M}{C_1 + 2C_2}\right]^{1/2} && \\ Z_S &= \left[\frac{L + M}{C_1}\right]^{1/2} && \end{aligned}\right\} \qquad (15.4.2)$$

The constants are defined by Fig. 9.19. It will be seen that, except for M, they can be calculated from the simple expression of Eqs. 9.1.7 and 9.1.8 for inductance and capacitance, which are repeated above. The M can be obtained from Eq. 8.3.17.

The same approach can be applied to more complicated arrays of conductors, such as three-phase lines with or without ground wires and double-circuit lines. This idea is introduced at the end of Section 9.7, starting with Eq. 9.7.26. However, in these circumstances, the precise circuits we are considering are less clearly defined and the practice is to describe the surge impedance of the individual conductors with respect to neutral. This could, of course, also be applied to our idealized single-phase circuit, far removed from the influence of other conductors, in which case we would choose the plane midway between the conductors as our neutral; C would then be twice as big and L half as big. This would have no influence on the velocity of propagation but would mean that the line-to-neutral surge impedance is half the surge impedance of the line-to-line circuit. If we inject 1 A of current into a particular conductor, this approach tells us what voltage will be associated with that current ($1 \cdot Z_0$) in that conductor. It recognizes that this current will have a return path in one or more other conductors and will give rise to voltage waves there appropriate to their share of the current and their surge impedances.

When the conductors of a transmission line are asymmetrically spaced, the inductance and capacitance of the three phases are different. It is, of course, possible to calculate their values by the method described above. As a matter of practicality the unequal inductive reactance and capacitive susceptance cause an undesirable asymmetry in the volt drop and charging currents of the three conductors under steady-state conditions. For this reason the conductors of the three phases are sometimes transposed at intervals along the line, that is, they interchange positions so that on the average they occupy each position for a third of the length of the line. From

Copper
NO. 7
NO. 5
NO. 3
NO. 1
2/0
4/0
300,000
500,000
750,000
1,000,000
Reactance at 60 Hz (ohm/mile)
0.4
0.5
0.6
0.7
0.8
0.9
1.0
1
2
3
4
5
7
10
20
30
40
Equivalent conductor spacing (ft)

(*a*)

ACSR
NO. 1
2/0
3/0
Single-layer conductors
Multiple-layer conductors
226,800
336,400
477,000
795,000
1,192,300
1,590,000
Reactance at 60 Hz (ohm/mile)
0.4
0.5
0.6
0.7
0.8
0.9
1.0
1
2
3
4
5
7
10
20
30
40
Equivalent conductor spacing (ft)

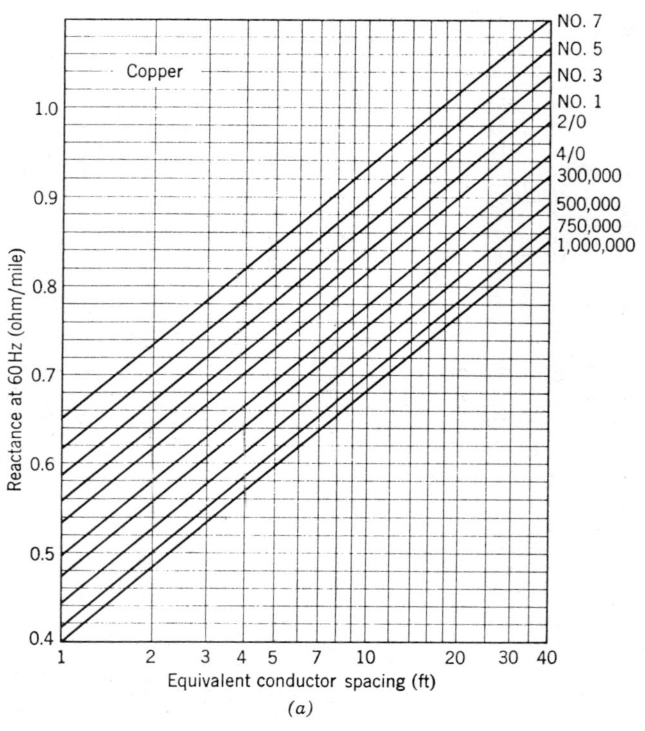

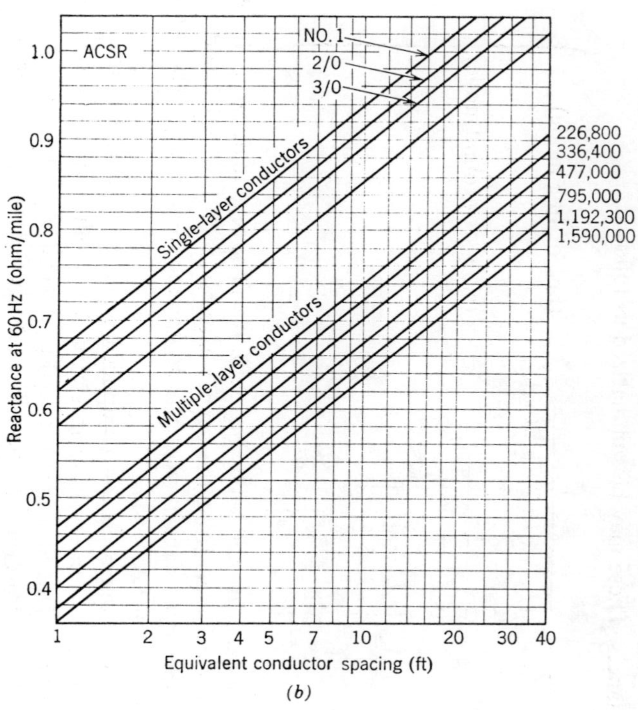

(*b*)

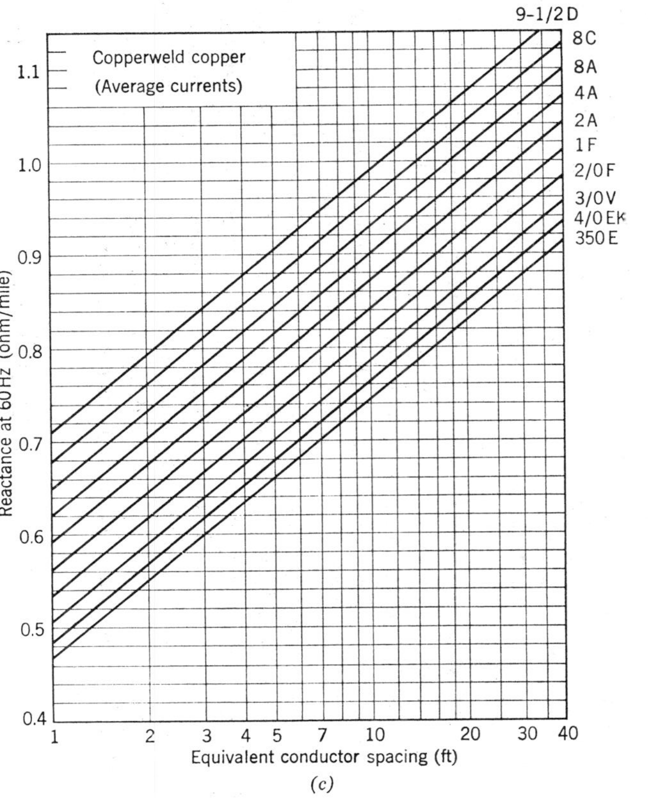

(c)

Fig. 15.14. Inductive reactance of three-phase transmission lines (per phase values) with different conductors and spacings.

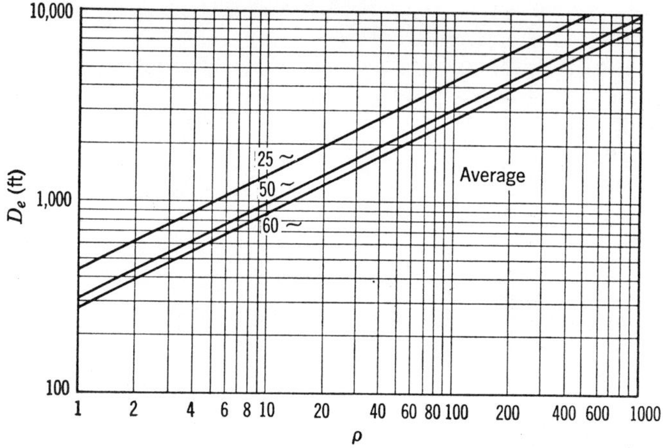

Fig. 15.15. Equivalent depth of ground return circuit as a function of the earth resistivity ρ in ohm meters and frequency.

a steady-state and a transient point of view, the circuit then has the attributes of a symmetrical line with conductors at the corners of an equilateral triangle. The equivalent spacing is given by the geometrical mean distance (GMD) of the asymmetrical spacings. If the conductor spacings are a, b, and c, then

$$\text{GMD} = [a \cdot b \cdot c]^{1/3} \tag{15.4.3}$$

This simplifies the calculation of L and C, for it is readily shown that the inductance and capacitance per conductor of such a line are the same as the per conductor values of a single-phase line with the same conductor size and spacing.

Typical per conductor values of the inductive reactance at 60 Hz for several different conductors (4) are given in Fig. 15.14. The inductance can be readily determined from these diagrams.

As mentioned earlier, the presence of ground influences the line parameter values. This is taken into account by substituting fictitious image conductors in place of the ground, as we have shown. The image conductor is only a true image of the actual conductor where the ground has zero resistivity; for soils of finite resistivity the image is deeper. This matter has been studied by Carson (7), who showed that the location of the image depends on frequency as well as ground resistivity. This is indicated in Fig. 15.15, which is taken from reference (6). Here D_e is the distance from the outgoing conductor to the fictitious image conductor and ρ is the resistivity. These plots are satisfied by the equation

$$D_e = 2160\left(\frac{\rho}{f}\right)^{1/2} \text{ feet} \tag{15.4.4}$$

Table 15.5 Earth Resistivities

Average of large number of determinations	100 ohm meter
Sea water	0.01–1.0 ohm meter
Swampy ground	10–100 ohm meter
Dry earth	1000 ohm meter
Pure slate	10^7 ohm meter
Sandstone	10^8 ohm meter

Some typical resistivity values are given in Table 15.5, which is also from reference (6).

We have an interesting situation here, in that for those modes of wave propagation which involve ground, the resistance of the ground return helps to determine the inductance and thereby the surge impedance and wave velocity. The fact that frequency is also involved means that in multifrequency surges distortion is introduced. The higher the frequency, the less do the currents penetrate into the ground. This means lower inductance for the path and is reflected in a somewhat higher propagation velocity for higher frequency components of traveling waves.

For a number of reasons the individual phases of EHV transmission lines often comprise several parallel conductors, usually four, which are arranged in what is termed a bundle. The conductors are held apart by separators. Such lines are somewhat easier to string than single conductors with comparable current-carrying capacity. They also tend to produce less corona since the high field intensity region is reduced. Bundled conductors have less inductance than single conductors because there is less flux adjacent to conductors. This is indicated by the flux pattern in Fig. 15.16. The capacitance of such bundles, on the other hand, is greater than the single conductor, because they appear to have a greater surface area. The combination of these two effects means that such lines have a lower surge impedance and higher

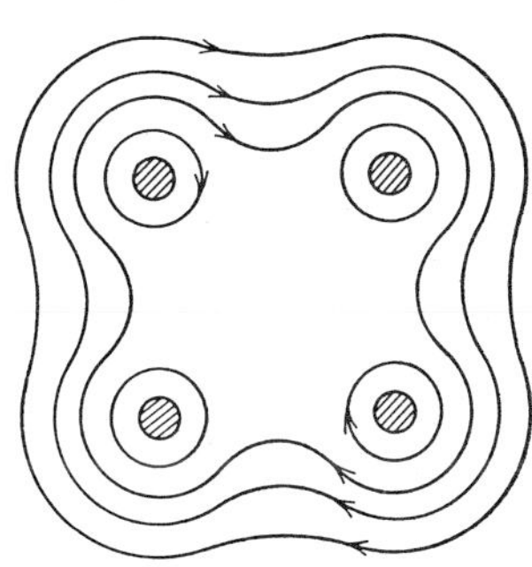

Fig. 15.16. Flux pattern around bundled conductors.

charging current. Some typical values are:

for a 500-kV line:

$$Z_0 = 275\ \Omega$$

charging current 2–2.3 A/mile at 60 Hz

For the calculation of transient fault currents, the line resistance is not unimportant, since the X/R ratio of lines is lower than most other system apparatus. Approximate values for the a-c resistance of copper conductors can be obtained from Fig. 15.17. The values vary somewhat with the style of conductor (number of strands, and so on) but these variation are not sufficient to materially affect the considerations under review. Taken in conjunction with Fig. 15.14, it is apparent that X/R ratios are in the range 1–5. It should be recognized that if ground is involved, this value will change; R will increase more than will X.

Because the conductors of a cable are almost invariably closer together than those of an aerial line, the inductance of cables is lower and the capacitance higher than for overhead lines. The capacitive effect is accentuated by the higher permittivity of cable insulating materials. These differences are reflected in the velocity of propagation of surges along cables and, more especially, in the surge impedance of cables. Typical values are:

$$v = 400\ \text{ft}/\mu\text{sec}$$

$$Z_0 = 50\ \Omega$$

Another difference with cables is that the electric field is confined, since almost invariably the conductors have a grounded shield around them, either

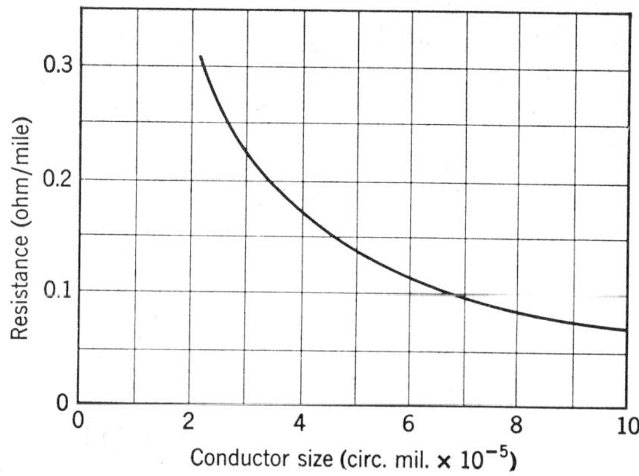

Fig. 15.17. A-c (60 Hz) resistance of copper conductors allowing for stranding and spiraling.

Table 15.6 Typical Capacitance Values for Power Cables

Conductor Size (Circ. mils or AWG)	Capacitance-to-Ground (μF/mile)					
	Shielded, Paper Insulated			Oil-Filled, Paper Insulated		
	15 kV	23 kV	35 kV	35 kV	46 kV	69 kV
4	0.33	—	—	—	—	—
2	0.40	0.32	—	—	—	—
1	0.44	0.35	—	—	—	—
0	0.49	0.39	0.27	—	—	—
00	0.51	0.40	0.29	0.44	0.40	0.32
000	0.55	0.44	0.31	0.48	0.435	0.35
0000	0.60	0.47	0.37	0.55	0.48	0.39
250,000	0.66	0.51	0.39	0.58	0.51	0.41
300,000	0.68	0.54	0.41	0.63	0.55	0.44
350,000	0.74	0.58	0.44	0.68	0.59	0.465
400,000	0.78	0.60	0.465	0.72	0.63	0.49
500,000	0.86	0.68	0.51	0.78	0.685	0.525
600,000	0.92	0.72	0.54	0.83	0.73	0.56
750,000	1.02	0.78	0.59	0.86	0.79	0.61

individually, or around the three phases (the so-called belted type). For the single conductor concentric cable, it is readily shown that the capacitance is given by

$$C = \frac{2\pi\epsilon_0\kappa}{\ln r_2/r_1} \tag{15.4.5}$$

where $\epsilon_0 = 8.854 \times 10^{-12}$ is the permittivity of free space, κ is the relative permittivity of the cable dielectric, and r_1 and r_2 are the conductor radius and the inside radius of the sheath, respectively. This is the capacitance to ground. A pair of such cables forming a single-phase circuit would have half of this capacitance since there would be two such capacitors virtually in series. Some typical values of cable capacitance are given in Table 15.6. These are for three-conductor cables with the conductors individually shielded.

Compared with overhead lines, power cables are relatively short, but in some instances there may be many parallel cables connected to a bus. Their capacitance markedly affects switching transients that may be generated adjacent to the bus. The cables behave much like a lumped capacitance, which can be calculated from the data in Table 15.6 if the total cable length is known.

The magnetic field too is confined if the current goes down the center conductor and returns through the concentric sheath. This is not the normal way of using the cable; in three-phase circuits the return path will be through the other phase conductors. However, most cable faults are single line-to-ground faults, and in such circumstances the sheath is involved. Neglecting the flux in the conductors themselves, the inductance of a concentric cable is then

$$\left.\begin{aligned} L &= \frac{\mu_0}{2\pi} \ln \frac{r_2}{r_1} \text{ henry/meter} \\ &= 7.43 \times 10^{-4} \log \frac{r_2}{r_1} \text{ henry/mile} \end{aligned}\right\} \tag{15.4.6}$$

If it is desired to include the contribution of the center conductor and sheath, the center conductor adds $\mu_0/4\pi$ (see Eq. 8.3.15) and the shield contributes approximately the same amount. In fact transient currents flowing in this mode will probably not penetrate deeply into either conductor, so that Eq. 15.4.6 usually provides a good approximation. Taken with Eq. 15.4.5 it provides the surge impedance of the concentric cable:

$$Z_0 = \frac{1}{2\pi}\left(\frac{\mu_0}{\epsilon_0 \kappa}\right) \ln \frac{r_2}{r_1}\ \Omega \tag{15.4.7}$$

In three-phase operation, the shields intercept the pulsating flux and sheath currents are induced. But these are insufficient to radically affect inductance values; the same procedures that we outlined above for aerial lines suffice for calculating the phase inductance. This leads to values in the range 400–700 μH/mile.

The X/R values for cables are lower even than those for aerial lines because the inductive reactance is lower. Where the shield is involved as a current path, the difference is more pronounced because of the ıelatively poor conductivity of lead sheaths. Sheath resistance ranges between about 0.4 and 1 Ω/mile, tending to be lowest for the higher voltage cables, since these have the greatest shield diameters.

15.5 Characteristics of Bus Work

There are many situations where conductors cannot be described as lines, and where they are certainly not cables, but yet where their impedance represents a significant factor in determining a transient current or voltage. Good examples are the conductors that connect apparatus—transformers, reactors, switchgear and capacitor banks—in a substation or switchyard. The phase conductors and the ground return are important. Similar buses are to be found in industrial plants running through and between pieces of equipment.

To establish the inductance of the paths which such conductors constitute, it is usually sufficient to estimate the running length of the circuit and calculate the inductance of the basis of 0.4 μH/ft. This corresponds approximately to the inductance of each conductor of a transmission line formed by 2-in. diameter conductors spaced 25 ft apart. In almost every instance, the capacitance of such conductors is negligible compared with the capacitance of the equipment it joins.

Bus runs comprising parallel bars, usually of rectangular cross section, are a commonplace in industrial power circuits. The spacing is usually from 2 to 12 in. Such conductors range from 0.1 to 0.2 μH/ft. The capacitance of such buses are typically in the range 1–3 pF/ft, depending on spacing.

Concentric bus structures used about power stations constitute a significant contribution to the total station capacitance. Typical values are given in Table 15.7.

The characteristics of buses are often affected by the proximity of ground planes, especially structural steel work. It is not unusual to find such bus runs taking a course parallel and close to the steel wall of cubicles or similar structures. This obviously affects the ground capacitance, but it also

Table 15.7 Bus Capacitances

Isolated Phase Bus Capacitance			Segregated Phase Bus Capacitance
Ampere Rating	15-kV Class 110-kv BIL (pF/ft)	23-kV Class 150-kv BIL (pF/ft)	15-kV Class 110-kv BIL (pF/ft)
1,200	8.9–14.3	8.0–12.4	10.0
2,000	10.2–14.3	9.0–12.4	10.0–10.2
2,500	10.2–14.3	9.0–12.4	
3,000	10.2–14.3	9.0–12.4	10.0–10.2
3,500	10.2–14.3	9.0–12.4	
4,000	14.0–14.3	12.4–13.5	10.0–12.6
4,500	14.0–14.3	12.7–13.5	
5,000	14.0–19.0	12.7–15.8	12.5–14.9
5,500	14.0–19.0	12.7–15.8	
6,000	14.0–19.0	13.5–15.8	15.0–17.1
6,500	14.0–19.0	13.5–15.8	
7,000	17.3–22.6	14.4–17.6	17.1
7,500	17.3–22.6	14.4–17.6	
8,000	21.7	17.6	—
9,000	21.7	18.1	
10,000	21.7	18.1	—
11,000	23.7	20.5	
12,000	23.7	20.5	—

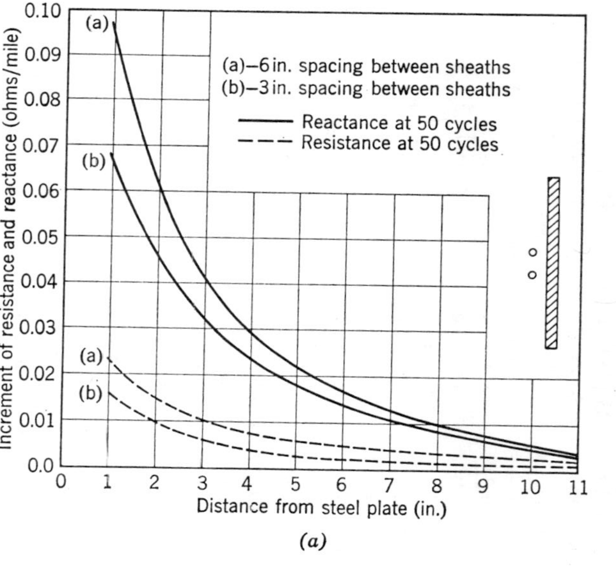

(*a*)

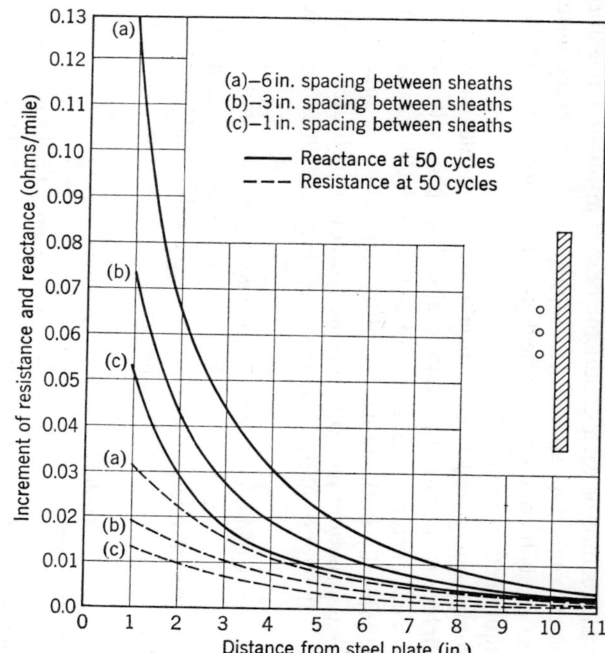

(*b*)

Fig. 15.18. Increase in cable resistance and reactance caused by the proximity to steel plate (cable sheaths are isolated). (*a*) For single-phase circuit. (*b*) For three-phase circuit.

influences the inductance and resistance of bus in that the magnetic field of the currents in the bus link the adjacent "circuits." Under steady-state conditions, the presence of such a steel plate can materially increase the inductance by providing a lower reluctance path than the air. Currents induced in the adjacent structure entail losses, which are reflected as an increase in the effective resistance of the bus. Figure 15.18, adapted from a paper by Booth et al. (8), illustrates these effects.

How important these factors are under transient conditions will depend on the time scale of the transient. In the calculation of power frequency transient fault currents, they are certainly influential. For high-frequency transients the inductive effects may be quite different. It was shown in Chapter 8 that it is not possible for flux changes to quickly penetrate a conducting surface. The adjacent conducting plane, be it magnetic or non-magnetic, will act as a barrier to flux, at least momentarily, and thereby tend to reduce the inductance. The section on shielding in Chapter 8 would be relevant.

15.6 Capacitance of Instrument Transformers

Potential transformers and current transformers used for metering and relaying introduce a certain amount of capacitance onto a bus. Individually, this does not amount to much, but when a number of such transformers are situated in reasonably close proximity their cumulative effect can be significant. The data given in Tables 15.8 and 15.9 (again supplied by the IEEE

Table 15.8 Capacitance of Potential Transformers, Primary Winding to Ground and to Secondary with Terminals Shorted and Grounded

Insulation Class-kV	Capacitance (pF)	
	Line-Line Type	Line-Neutral Type
15	260	
25	250–440	270–800
34.5	310–440	270–900
46	350–430	300–970
69	360–440	340–1300
115	470–520	480–610
138	490–550	530–660
161	510–580	510–700
196	—	580–820
230	600–680	600–810
345	—	920

Table 15.9 Capacitance of Current Transformers, Primary Winding to Ground and Secondary with Terminals Shorted and Grounded

Insulation Class-kV	Capacitance (pF)
25	180–260
34.5	160–250
46	170–220
69	170–260
115	210–320
161	310–380
196	330–390
230	350–420

Working Group on Transient Recovery Voltages) are typical of this class of equipment.

REFERENCES

1. G. Dienne and J. M. Frisson, "Electrical Problems of Power Stations and Networks—Contribution to the Experimental Study of Recovery Voltages when Switching-Off Short-Circuits on Transformer Secondaries," CIGRE Report No. 13-07 (1968).
2. G. Paillet and N. Dufour, "An Investigation of the Natural Frequencies and Surge Impedances of Transformer Windings," CIGRE Report No. 109 (1948).
3. N. M. Tchernyshev, J. P. Schelglov and G. A. Dorf, "Experimental Determination of Restriking Voltage on Circuit Breakers in Power Transformer Circuits. Some Data on Restriking Voltage at Interruption of Short Line Fault," CIGRE Report No. 138 (1966).
4. *Electrical Transmission and Distribution Reference Book*, 4th ed., Westinghouse Electric Corp. (1950).
5. C. Concordia, *Synchronous Machines Theory and Performance*, John Wiley & Sons, New York (1951).
6. C. F. Wagner and R. D. Evans, *Symmetrical Components*, McGraw-Hill Book Co., New York and London (1933).
7. J. R. Carson, "Wave Propagation in Overhead Wires with Ground Return," *Bell System Tech. Jour.*, Vol. 5 (1926), p. 539.
8. H. C. Booth, E. E. Hutchings and S. Whitehead, "Current Rating and Impedance of Cables in Buildings and Ships," *Jour. IEE*, Vol. 83 (1938), p. 497.

16 Equipment for Measuring Transients

16.1 Some General Observations on the Measurement of Transients

In this chapter we are concerned with equipment for measuring transients. In Chapter 17 the subject of transient measurements will be extended to measuring techniques and surge testing. This material could be expanded into a treatise by itself. For this reason the treatment here can serve only as an introduction to the subject, it cannot be comprehensive. Before proceeding to discuss actual devices, there are a few fundamental concepts that require consideration.

To measure is to reduce the matter at hand to numbers, and we do not truly know the subject until this has been done. Transient measurements have much in common with measurements of steady-state quantities, but the fleeting nature of the phenomenon we are attempting to record introduces some special problems of its own. The ideal result of a transient measurement would be that we would have a perfect replica of the transient, the voltage, or whatever, as a function of time, with the scale of the replica precisely known. The more sophisticated measurements, perhaps the majority of measurements, aim at this ideal. There are occasions, however, when something much less will be acceptable, and we give attention to these, also.

The ideal can only be approached, it cannot be attained, if only for the reason that we cannot take any measurement in a system without disturbing the system to some extent. The reader will recognize this applies equally to steady-state measurements. For example, if we measure a current, the shunt will introduce an impedance, or if we measure a voltage, the voltmeter will take some current. Such errors can usually be made insignificant. But in the transient environment this is often more difficult to do. This is a point that we will frequently consider as we proceed.

Any time-varying quantity can be expressed as a series of terms. Alternating waveforms comprise a fundamental and a number of harmonics. The reader is probably familiar with the method whereby these harmonics are extracted and the waveform expressed as the summation of a Fourier series. In many practical cases such waveforms can be represented quite closely by a relatively small number of terms. Another representation would be to make a frequency plot and, at each harmonic, draw a vertical line proportional to the amplitude of that harmonic. Transients are aperiodic time functions. We close a switch, the transient occurs and is gone. However, if we think of such a disturbance as a periodic time function whose period is infinitely long, it is still possible to apply Fourier techniques. The frequency plot, instead of being a number of discrete lines, becomes a continuous spectrum, and its summation becomes an integral [see, for example, Tropper (1)]. What is the relevance of this to transient measurements? When a transient current or voltage is applied to a circuit, the response of the circuit can be found by the principle of superposition—we add the responses of the constituent parts of the stimulus. These are the harmonics in the case of the periodic stimulus. A measuring instrument is such a circuit, and if it is to give a true record of the stimulus, it must respond equally to the various component frequencies. Often it cannot do this; the frequency response of the instrument or measuring circuit is limited. An example will make this clear.

A galvanometer coil experiences a torque when a current passes through it, and can be deflected, or cause a small magnet to be deflected, in proportion to the current. If the current is changing, the torque changes, and the deflection changes. Suppose now that we pass an alternating current through the coil. The coil will be set into mechanical oscillation at the same frequency as the current. However, it is fairly obvious that as the frequency is increased, the ability of the coil to follow the alternations will diminish because of the inertia of the coil structure. Thus, although the current magnitude may be held fixed, the deflection or excursions of the galvanometer will diminish with increasing frequency until eventually its movement will become imperceptible. Such performance can be plotted as a frequency response curve. This has been done in Fig. 16.1 in a way which allows the effect of damping to be illustrated. To that extent these curves are somewhat analogous to the generalized damping curves for series and parallel *RLC* circuits given in Chapter 4. Frequency is expressed in terms of the natural frequency of the galvanometer. Thus, if a galvanometer has a natural frequency of 10 kHz and damping factor of 0.75, its response will be down about 10% at 6.2 kHz.

The frequency response of a measuring device is clearly of great importance. We consider it constantly as we review different measuring equipments, for it sets definite bounds for the application of such equipments. The frequency response of the galvanometer in Fig. 16.1 is limited by mechanical

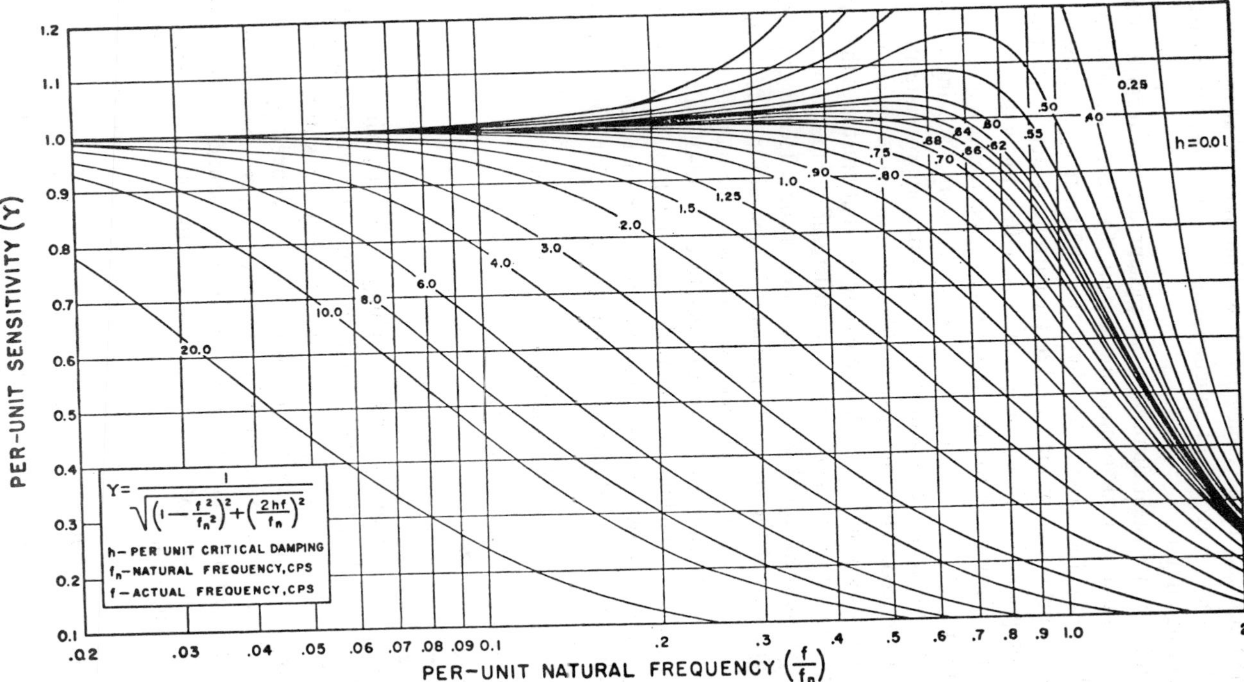

Fig. 16.1. Typical galvanometer frequency response curves showing effect of damping. (Courtesy Consolidated Electrodynamics.)

considerations. It is possible to reduce the inertia of the movement and thereby extend the frequency range somewhat. It will be shown that there are many other factors that control the frequency response of measuring instruments and measuring circuits.

Frequently the transient quantity to be measured is not recorded directly. Whenever a shunt is used to measure current, for example, what is really measured is the voltage across the shunt that the current gives rise to. This voltage is assumed to be proportional to the current, a fact which should not be taken for granted with transient currents. Often the voltage appearing on the shunt is insufficient to drive the measuring device; it requires amplification. Conversely, on many occasions when a voltage is being measured, the voltage is too great to be handled. At these times, appropriate attenuation must be introduced. This suggests the idea of a *measuring system*, rather than a single measuring device. A prime example that illustrates these points is the measurement of transient stress in a structural member subjected to an impact. A common method is to use a strain gauge. Hence the stress is measured in terms of the strain it produces. This strain in fact changes the resistance of the strain gauge element, which upsets the balance of a bridge. The ultimate indication of the current is therefore an out-of-balance voltage, which, with suitable amplification, provides the final display. Intermediaries in these types of physical transformations are referred to as transducers. The characteristics of such devices have an important bearing on the accuracy of the measurement.

There are basically three types of transients to consider:

1. Single transients under our control.
2. Recurrent transients.
3. Random transients.

Those in the first group are typically switching transients, where we are in a position to open or close the switch at our discretion and are therefore able to anticipate the consequences. The second category constitutes transients occurring regularly as commutation transients in converting equipment. Finally there are those generated by extraneous operations beyond our control, which appear in an unpredictable random manner on our system. In this chapter and the next, we discuss the instrumentation and techniques for measuring all of these.

16.2 Magnetic Oscillographs

Oscillographs and oscilloscopes are capable, in theory at least, of producing a replica of the waveform of a transient and are therefore popular transient measuring devices. Basically, they comprise a moving system which deflects in response to an applied signal. The moving system is coupled to an optical system of some kind, by which the motion of the moving system can be recorded. The deflection of the moving system must be proportional, at any instant, to the current or voltage being measured. The inertia of the moving system must be extremely low if it is to respond quickly to changes in the applied signal. In particular, its own natural period must be much shorter, or its natural frequency much higher, than that of the waveform being examined. The movement must also be critically damped.

In the magnetic oscillograph the moving system is mechanical; there is therefore a relatively low limit to the frequency response. In the cathode ray oscilloscope, which is the subject of the next section, the moving system is a beam of electrons, with an extremely low inertia. The frequency response of oscilloscopes is limited by other considerations, in most instances.

Magnetic oscillographs are essentially vibration galvanometers specially designed for a low natural period. Current designs use the same basic principle developed originally by Duddell (2). A crude representation of the movement is shown in Fig. 16.2*a*. It comprises a loop of fine phosphor bronze wire, similar to a hairpin in form, a magnetic field, usually provided by a permanent magnet, and a small mirror cemented to the wires. The wires are braced across two bridges much like strings on a guitar; tension is maintained by an adjustable spring. When current is passed through the wire loop, the loop is caused to deflect by the magnetic field and thereby impart a movement to the mirror. Restoring torque is provided by the wire itself. A more modern adaptation is shown in Fig. 16.2*b*, which indicates, in schematic form, the recording system. A beam of light focused on the mirror and reflected to a screen has its image on the screen deflected by the passage of current. In a well designed system the deflection will be proportional to the current. The beam of light acts like a long, low inertia pointer. Damping of the movement is obtained by having the entire mechanism submersed in oil. The degree of damping affects the frequency response curve, as indicated in Fig. 16.1. Damping also introduces phase shift between the signal and the response as manifest by the deflection. For this reason a condition somewhat less than critically damped is often used, for then the phase shift is almost linear with frequency. In this way, harmonics, although attenuated with respect to the fundamental, at least retain the correct time relationship.

It is possible with modern galvanometer construction to obtain a natural frequency as high as 15–20 kHz, although this characteristic in most oscillograph galvanometers is not as high as that. Magnetic oscillographs are therefore restricted to recording in the frequency range of 0–7 kHz. These instruments typically require a few milliamperes to drive them, although high sensitivities of the order of 1 in./μA are possible. If they are used for measuring voltage, the element has a high resistance connected in series with

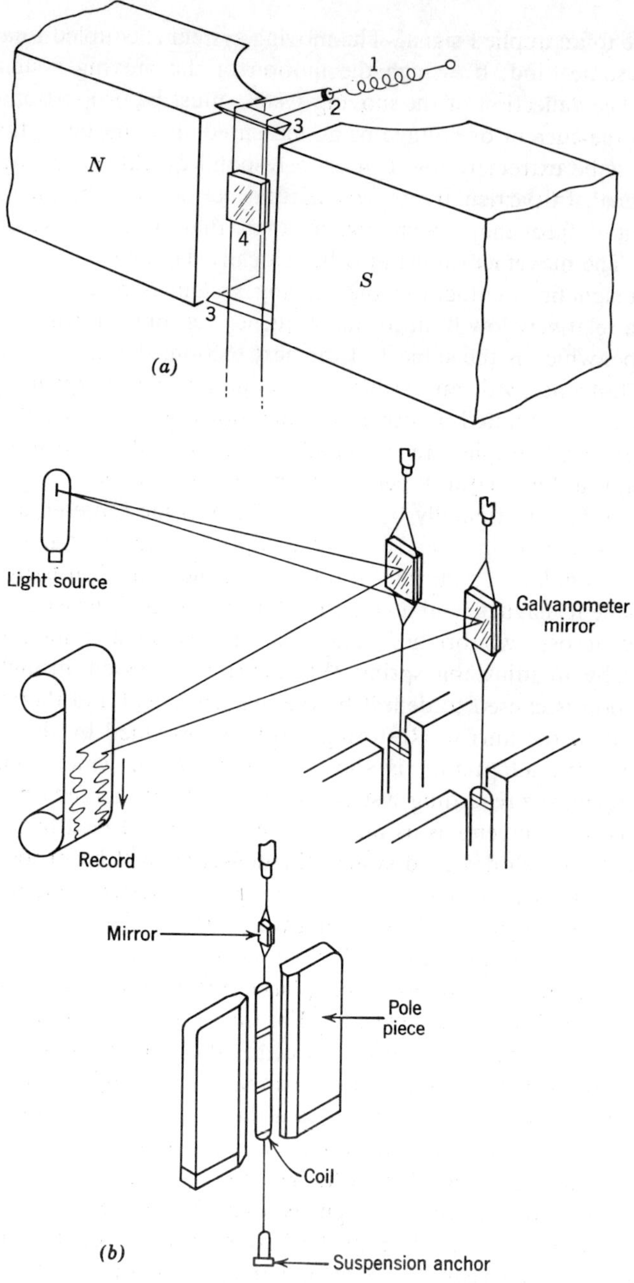

Fig. 16.2. Types of galvanometers for oscillographs. (*a*) 1. Spring. 2. Pulley. 3. Bridge. 4. Mirror. (Courtesy of Consolidated Electrodynamics.)

it. For current measurements they would be in parallel with a shunt and take a small fraction of the total current.

One of the advantages of magnetic oscillographs is that they can have many channels, that is, one instrument can be used to measure a number of transient quantities simultaneously. This is done by parallel-mounting a number of galvanometers of the type described. They are usually arranged in the arc of a circle, but so adjusted as to produce their traces at different points on the film or recording paper being used. Obviously, with a large number of such traces, the excursion of any individual trace must be limited, if they are not to overlap and become impossible to read. The positions of the traces are individually adjustable to permit different excursions on different traces.

Measurements of this kind can be made reasonably well if the transient differences of potential of the points in the circuit where the measurements are being made is not very great. It must be recognized that these potentials are brought into the oscillograph. Frequently the insulation between the different galvanometer circuits will support only a few hundred volts. Some manufacturers have recognized this problem and, by a sacrifice in the number of channels, allow up to 10 kV between circuits. Where more is required, separate instruments must be used. The insulation to local ground, including the power supply to lamps, must not be overlooked. Safety of personnel is another important consideration, for high transient potentials may be brought into the measuring area.

The mechanics of what we have described so far causes a spot or spots of light to move in response to a current or currents in one or more galvanometers. To obtain a record as a function of time, the spot(s) must impinge on a light-sensitive surface, which is itself moving perpendicular to the spot(s). The light-sensitive material is either film or paper, film being less convenient but in general more sensitive and more easily reproduced. The film or paper may be rapidly transferred from a storage spool to a take-up spool during the interval of the measurement, or it may be wrapped around a drum, which is quickly brought up to speed before exposure. Still other instruments disgorge the paper at a preset speed as long as their motors are running. Such instruments can operate at paper speeds up to 10 ft/sec for resolving relatively fast disturbances, but they are perhaps better suited for recording slower transients or random disturbances where the paper may run through at a few inches per minute only, but perhaps for hours at a stretch. In the case of high-speed operation, the speed may not be uniform. It is therefore desirable to reserve one channel for a timing signal to calibrate the others.

With film it is necessary to have photographic facilities at hand for developing, fixing, and washing the film. The exposed film is transferred from the instrument to the darkroom in a light-tight cassette. In general, it is not necessary to load the film in a dark room. The trace on the paper appears after a brief exposure to light following the actual exposure in the instrument.

A relatively versatile instrument made by the Siemens Company is shown in Fig. 16.3. Light from the source (24) is taken through a lens system (19, 22) and by mirrors (25, 26) to the galvanometer (29), whence after deflection it passes through lenses (29, 31) to the recording paper, which is moving vertically. As many as 36 such systems can be set up in the instrument. Part of the light from the galvanometer is intercepted by an auxiliary mirror (34) and bounced to a prism (35), whence it is reflected onto a many-sided rotating mirror. This causes an image similar to that on the film to traverse a ground-glass screen (27), allowing the transient to be observed by the operator. The number of light spots imaged in this way depends upon the number of mechanisms employed. Their displacements relative to each other on the ground glass screen correspond to the distance between the light spots on the recording paper. The phenomena to be recorded can therefore be positioned on the ground glass screen in the way they are required on the oscillogram.

Time ordinate recording is obtained in this instrument by taking yet another beam of light through a lens (42) and mirrors (43, 45). But this light is controlled by a shutter formed by a slotted disk (44), which is driven synchronously by a motor (47). It is claimed that this instrument has a maximum writing speed of 10,000 meters/sec.

An interesting type of galvanometer/recording system has been described by Kaiser (3) and Sima (4). In this system the light beam and mirror are replaced by a fine jet of ink which, coupled to a galvanometer, writes directly on a moving paper. Galvanometers of either the moving loop or the moving magnet type are available. These are illustrated in Fig. 16.4*a*.

The mechanism consists of a fine glass capillary (outer diameter approximately 80 μ, inner diameter approximately 40 μ) which is bent at right angles to the direction of rotation and which tapers to an opening of 10 μ. The capillary is molded at its lower end into a filter (porosity approximately 2 μ). The recording ink is pumped through this filter and through the capillary into the nozzle, which directs the fine jet onto the paper. A multichannel apparatus utilizing galvanometers of this kind is shown in Fig. 16.4*b*.

An alternative to film or paper is magnetic tape. Here, the signal is recorded by a high-fidelity magnetic recording head. It can be transcribed subsequently by a reading head, which may drive one of a variety of recorders. It is not necessary to transcribe at the same speed as the recording is made. It is often convenient to do so at a much slower speed, perhaps with a pen recorder, in order to obtain an expanded time scale and greater resolution.

The fact that magnetic tape can be used and re-used makes it suitable for continuous monitoring of circuits. The tape in this case would form a belt or

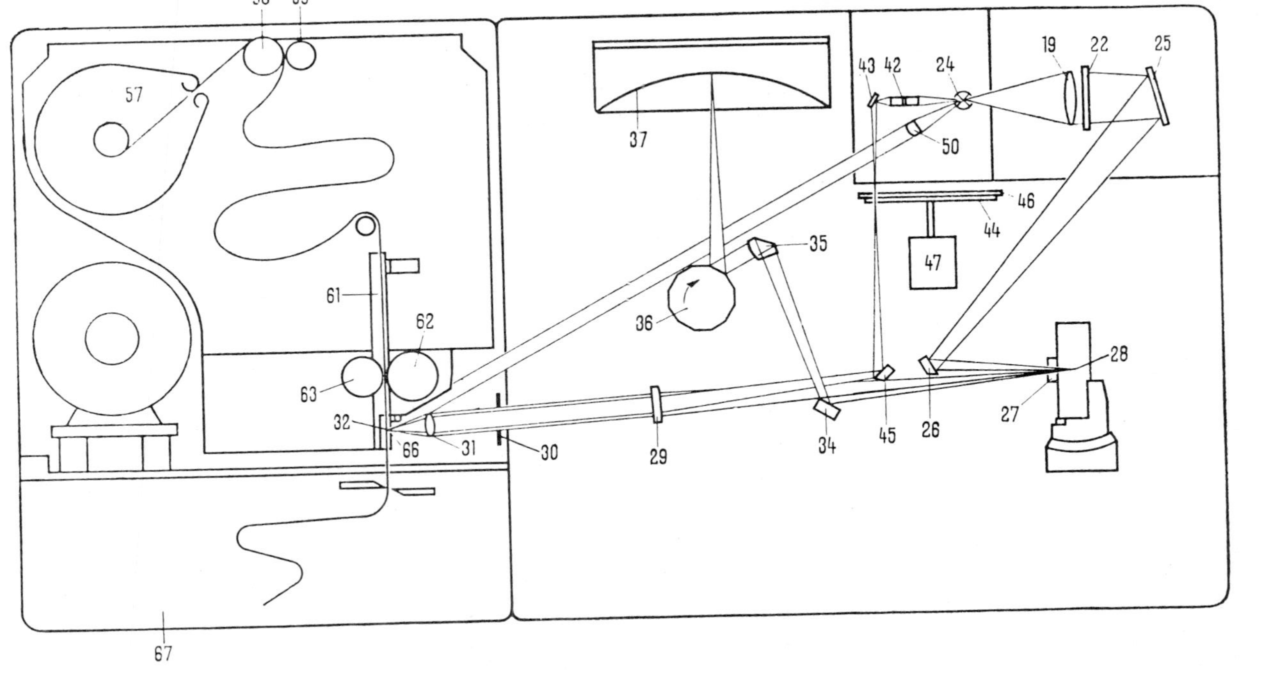

Fig. 16.3. Optical system of a modern oscillograph. (Courtesy of the Siemens Co.)

19. Condenser lens
22. Slit diaphragm
24. Light source
25. Deflection mirror, adjustable
26. Deflection mirror, fixed
27. Mechanism lens
28. Mechanism mirror
29. Intermediate lens
30. Adjusting diaphragm
31. Cylindrical objective
32. Light spot
34. Reflection mirror
35. Lens prism
36. Polygon rotary mirror
37. Ground glass observation screen
42. Condenser lens
43. Deflection mirror
44. Slotted disk
45. Intermediate lens
46. Diaphragm disk
47. Synchronous motor
50. Condenser lens
57. Supply magazine
58. Prefeeding roller
59. Coupling roller
61. Paper guide
62. Paper transport roller
63. Pressure roller
66. Exposure slit
67. Take-up magazine

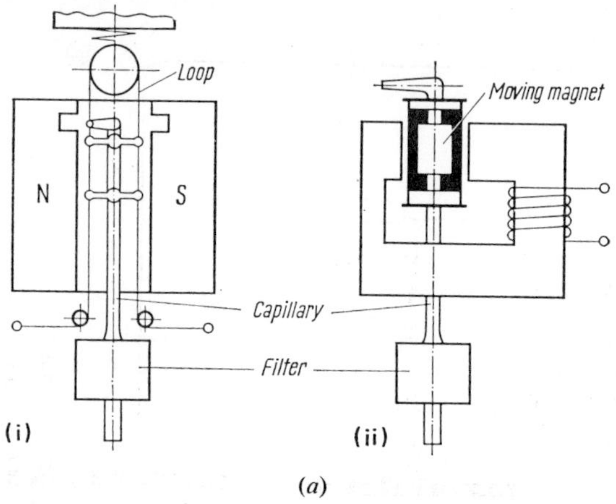

(*a*)

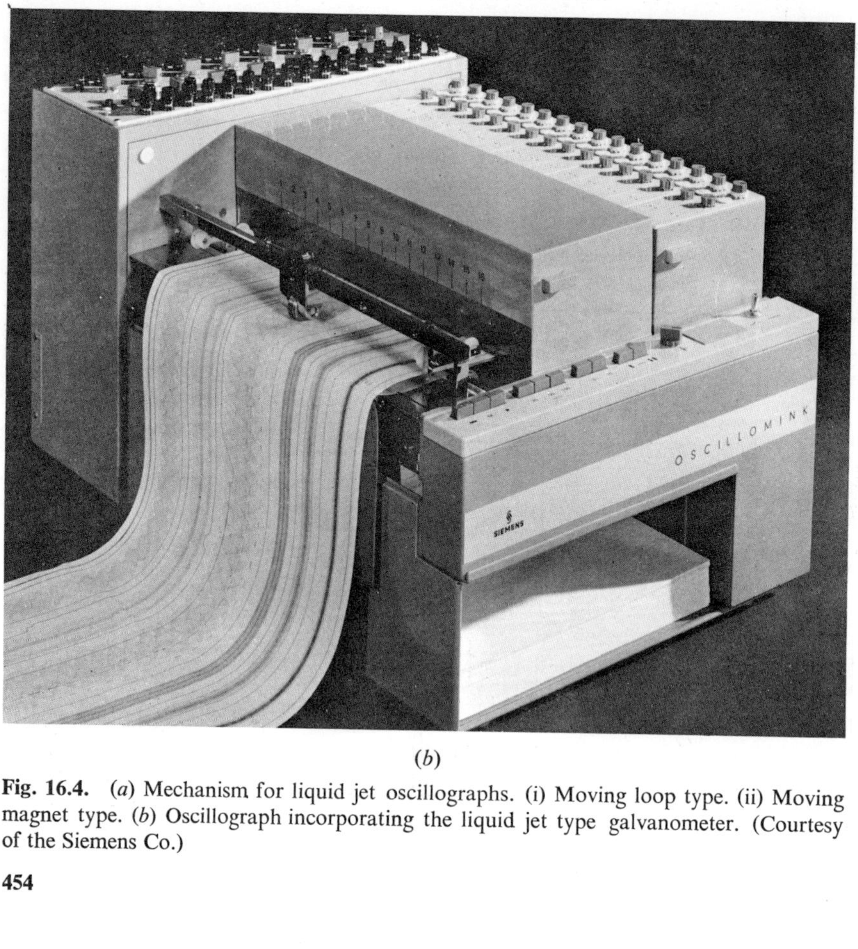

(*b*)

Fig. 16.4. (*a*) Mechanism for liquid jet oscillographs. (i) Moving loop type. (ii) Moving magnet type. (*b*) Oscillograph incorporating the liquid jet type galvanometer. (Courtesy of the Siemens Co.)

closed loop, which could be arrested following any transient disturbance. This tape would capture a record of the disturbance and, often of equal importance, would record events immediately preceding the disturbance. During tranquil periods the loop simply turns, wiping off past signals as it records new ones.

16.3 Oscilloscopes

The cathode ray oscilloscope has become the workhorse for electrical transient measurements. The principle of the device is well known, so we treat it only briefly, reserving detailed discussions for the finer points more relevant to our theme.

A cathode ray, that is, a beam of electrons, is produced by focusing and accelerating the emission from a hot filament. The focused beam is directed to a phosphor-coated screen where it causes a bright scintillation at the small spot on which it impinges. The beam, and therefore the spot, can be deflected by an electric field, applied transversely to the direction of the beam. If the beam is so deflected in a vertical direction by a field proportional to the quantity being measured, the spot deflection will be similarly proportional. If, simultaneously, a second electric field is applied, perpendicular to the first, which increases uniformly with time, the spot will be suffered to sweep horizontally across the screen at a uniform rate, at the same time that it is being deflected vertically in proportion to the signal. The result is a plot of the transient, in cartesian coordinates, on the oscilloscope screen, showing the precise variations of the signal quantity as a function of time. This is the basic principle of the cathode ray oscilloscope. It has many attributes which commend it for transient measurement work, and it has therefore been developed with many special features to better accomplish different requirements in this area.

The sensitivity S of an oscilloscope, the proportionality factor between the deflection and the applied signal, can be determined from the geometry of the cathode ray tube and the accelerating voltage. Millman and Seely (5) develop the expression

$$S = \frac{lL}{2dE_a} \text{ meter/volt} \tag{16.3.1}$$

The significance of the symbols is apparent from Fig. 16.5. For

$$l = 4\text{ cm}; \qquad L = 30\text{ cm}; \qquad d = 1\text{ cm}; \qquad E_a = 3\text{ kV}$$

$$S = 2 \times 10^{-4}\text{ m/V} = 0.2\text{ mm/V}$$

In the overall equipment this will be modified by the amplification or attenuation that may be introduced between the signal input and the deflection plates.

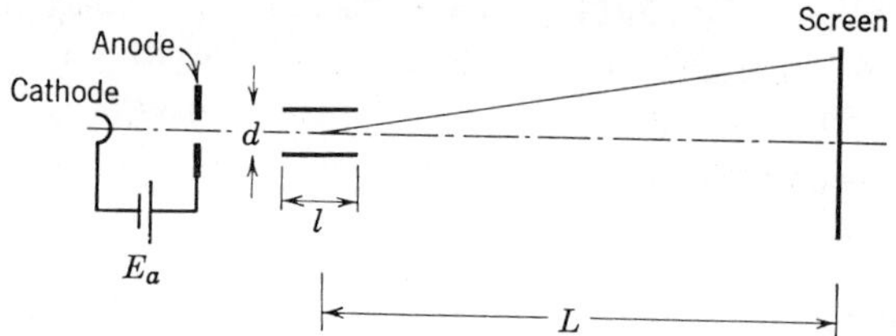

Fig. 16.5. Factors determining the sensitivity of a cathode ray tube.

Equation 16.3.1 indicates that to achieve high sensitivity we should use a low accelerating voltage. But for transient observations this may cause difficulties in another area, that of the so-called *writing speed.* In order to see or, better still, photograph a transient trace on an oscilloscope screen, it must have a minimal brilliance. The brilliance of the trace depends upon the brilliance of the spot itself and how fast it is caused to sweep across the screen. The brighter the spot, the faster can it traverse or write while still maintaining a satisfactory image. In turn, the brilliance of the spot depends upon the energy density of the electrons striking the screen, and it is here that the accelerating voltage is important since this is a determining factor in the energy density:

$$\text{electron energy input to the spot} = \frac{E_a \times I}{A} \qquad (16.3.2)$$

where I is the beam current and A the spot area. This indicates that a high accelerating voltage, a copious emission, and a sharply focused spot permit the highest writing speeds.

A sharp focus is in any case a desirable attribute since it contributes to better resolution and therefore to the accuracy of the record. In this regard, it is important to carry out the focusing under transient conditions, that is, with the sweep speed to be used in the measurement. Focusing under recurrent sweep conditions, where the brilliance is necessarily turned down, is usually unsatisfactory. The greater brilliance required for the transient sweep almost invariably affects the focus.

The normal way of controlling the brilliance is by adjustment of the beam current. This is carried out by regulating the potential of an electrode, called the Wehnelt cylinder, located close to the filament and biased somewhat negatively with respect to the filament. By increasing this bias the emission is reduced; it is readily carried to the point of complete suppression of the beam current. This is important for transient measurement work. It allows the screen to be blank and the shutter of a camera to be open before the transient is recorded. Just before the spot is swept, a positive pulse, a so-called brightening or unblanking pulse, is applied to the Wehnelt cylinder, which brings up the brilliance of the spot to a very high level for the short duration of the sweep. This level is such that it would burn the phosphor if allowed to persist for any significant time. The sharp increase of beam current causes defocusing; that is why it is important to focus under these conditions.

The contrary effects of accelerating voltage on the sensitivity and luminosity are reconciled to some extent in postdeflection acceleration tubes. The name suggests how they function. Acceleration is accomplished in two stages. Sensitivity is increased by using a relatively low accelerating voltage in the first stage. Luminosity and writing speeds are increased by a further stage of acceleration after deflection. This is achieved by a second anode or intensifier usually in the form of a conducting ring around the tube close to the screen indicated as Q in Fig. 16.6. The electrode P, connected to the anode, assures that the deflection region is free of longitudinal accelerating electric fields.

The time it takes for the electrons in the beam to pass between the deflecting plates is extremely short because the velocity of electrons is very high. For example, an electron accelerated through 5 kV will acquire a directed velocity given by

$$v = \left(\frac{2eV}{m}\right)^{1/2} = \left[\frac{2 \times 1.6 \times 10^{-19} \times 5 \times 10^{3}}{9.11 \times 10^{-31}}\right]^{1/2} = 4.2 \times 10^{7} \text{ m/sec}$$

(in this expression e and m are the electronic charge and mass and V is the accelerating voltage). Thus, if the deflecting plates are 3 cm long, the electrons will spend only 1.26×10^{-8} sec under their influence. However, if we are attempting to observe an extremely fast transient, or short pulse, for example, this time may be important. What we observe may be an averaging effect produced by the plates. For very fast transients of this kind, the *traveling wave tube* has been developed. In these devices the deflecting plates are segmented, each segment being joined to its neighbors by the equivalent of a

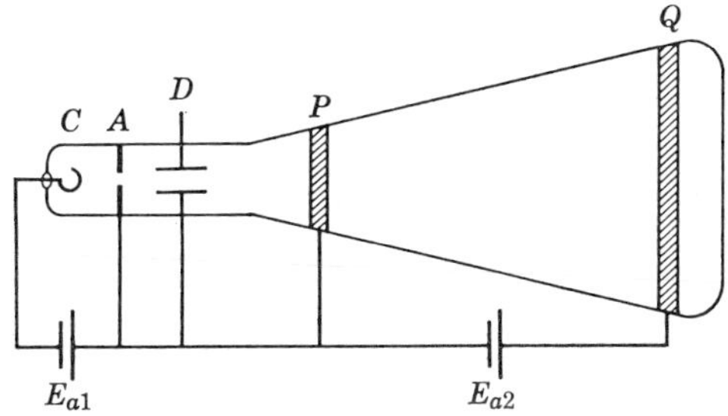

Fig. 16.6. Electrode arrangements for postdeflection acceleration.

short transmission line. The deflecting signal travels down this "line," influencing each successive segment, at the same speed as the beam electrons. The variable of plate transversing time is thereby removed. This subject is discussed further in Section 17.5.

One of the advantages mentioned for the magnetic oscillograph was its ability to display many separate traces. This is much more difficult with the CRO, although double-beam oscilloscopes are common. The two beams may be provided by separate electron guns, or in less expensive models a beam-splitting plate may be employed. The electrons then derive from one source but are split into two streams by a horizontal plate. With both types of oscilloscope the two beams then pursue their separate courses through independent deflecting systems. In this way it is possible to observe two independent signals simultaneously.

The beams may or may not be driven by the same time base. Sometimes it is convenient to use different sweep speeds for the two traces. For instance, we might wish to obtain greater resolution of the beginning of a transient. This can be achieved by displaying the same signal on both traces but at a higher sweep speed on one than the other. This has been done in Fig. 16.7*a*. The lower trace, the slower one, shows this transient to be a sequence of events. The upper trace, at five times the sweep speed, examines the first event in greater detail.

Some oscilloscopes have the facility to expand any portion of the trace, that is, display any portion at a higher speed (see Fig. 16.7*b*). This can be very useful when observing repetitive transients, such as the commutation transients of rectifiers or inverters, where we might wish to scrutinize a particular part of the wave, such as the overshoot at the instant of commutation.

Another way of obtaining two traces is to use a "chopper." This is a fast switching means whereby the beam is caused to flip back and forth rapidly between two locations in the tube, spending on the average half the time in one location and half in the other. The result is two series of spots rather than two continuous traces on the screen. However, unless the sweep is very fast, these spots merge to give the appearance of continuous traces. The switching rates usually operative are either 10^5 or 5×10^5 times per second. With these arrangements observations are clearly restricted to signals well below 100 kHz or 0.5 MHz, respectively. If choppers are used on both beams of a double-beam oscilloscope, it is possible to obtain four beams, but to do more than this requires specialized equipment.

Frequently it is necessary to amplify the transient signal to be observed before displaying it on the screen. Oscilloscopes are usually equipped with amplifiers for this purpose, either integral with the oscilloscope or as plug-in units, allowing interchangeable amplifiers with different characteristics to be

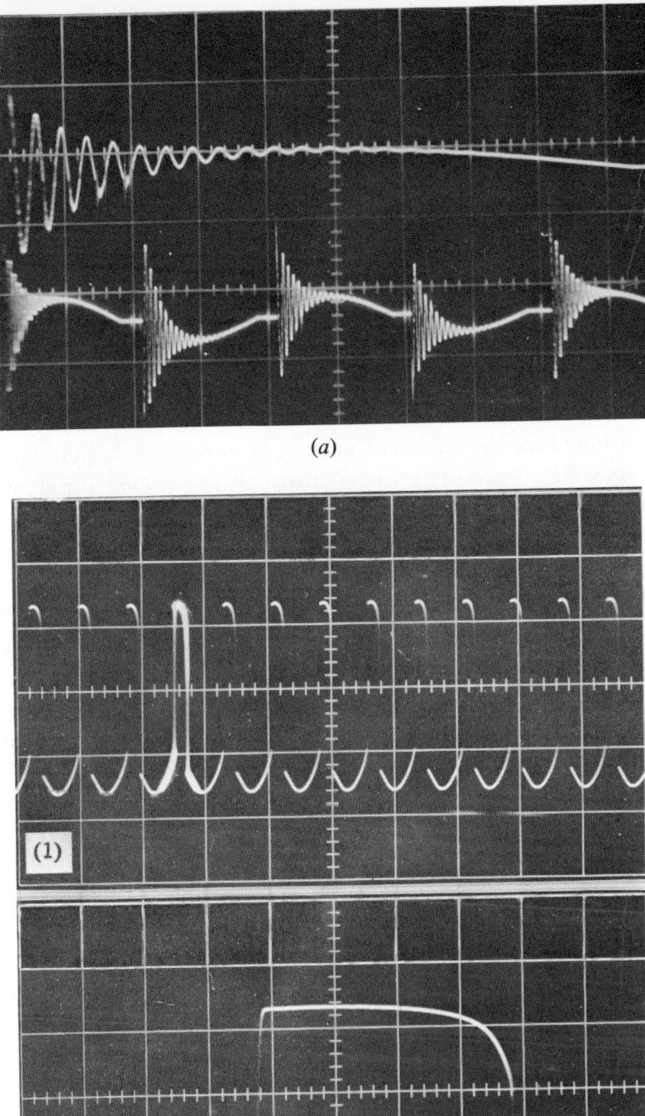

(*a*)

(*b*)

Fig. 16.7. (*a*) Transient recorded at two different writing speeds on double beam cathode ray oscilloscope. Top beam, 200 μsec/div. Bottom beam, 1 μsec/div. (*b*) Example of trace expansion. Trace (2) is a $\times 20$ expansion of the intensified portion of trace (1). (Courtesy of Tektronix Inc.)

used as the needs require. A wide selection of such amplifiers is available with a range of gain and frequency response. Sometimes a situation may demand a special requirement, for example, the ability to sustain a considerable overload momentarily and yet respond faithfully when the overload is removed. Such amplifiers can be obtained.

The input impedance to oscilloscopes is usually 1 megohm paralleled by a capacitance of 10 pF. Where this is unacceptable because of its effect on the circuit being measured, various probes and attenuators are available to be placed at the input to change the impedance (see Section 16.6).

Synchronization, the capability of having the time-base sweep the spot at the same instant that the transient occurs, is important in making transient measurements. The sweep is initiated through the trigger input, which, on versatile oscilloscopes, offers several modes of operation. These are polarity, level or magnitude, and slope of the trigger signal. Several different selections are shown in Fig. 16.8. The trigger signal may be derived from the transient itself, or it could come from some independent source. For example, if we wished to observe the transient occurring on a circuit when a circuit breaker opens, the sweep could be initiated by an auxiliary switch on the breaker mechanism, which opens slightly ahead of the circuit breaker contacts. A simple circuit to illustrate this procedure is shown in Fig. 16.9*a*. When the auxiliary contacts open, the trigger is exposed to the voltage V_2 that develops across R_2. The trigger mode in this instance would be set for a positive-going slope and a positive level somewhat below the voltage V_2. Such an independent arrangement of triggering leaves one free of direct coupling with the power circuit, and allows a trigger level to be selected which is above the local "noise" level, thereby effectively ruling out spurious trips.

Figure 16.9*b* shows a slightly more elaborate arrangement which permits some adjustment of the instant of trigger because of the time involved in

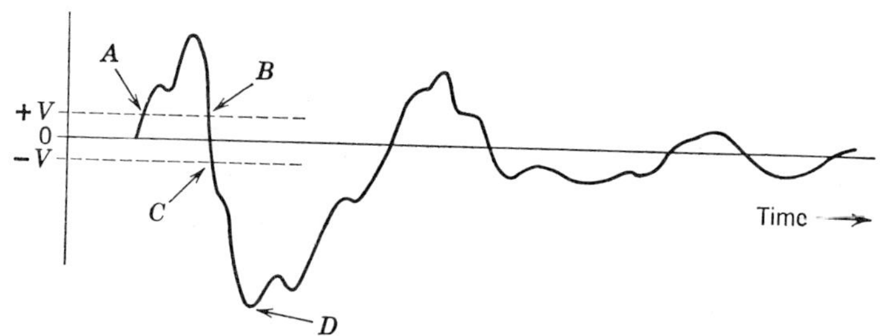

Fig. 16.8. Illustrating where, on a transient, the oscilloscope trace would be initiated for different trigger modes. (*A*) Positive signal or level, positive slope. (*B*) Positive level, negative slope. (*C*) Negative level, negative slope. (*D*) Negative level, positive slope. Trigger level is set $+V$ or $-V$ as the case may be.

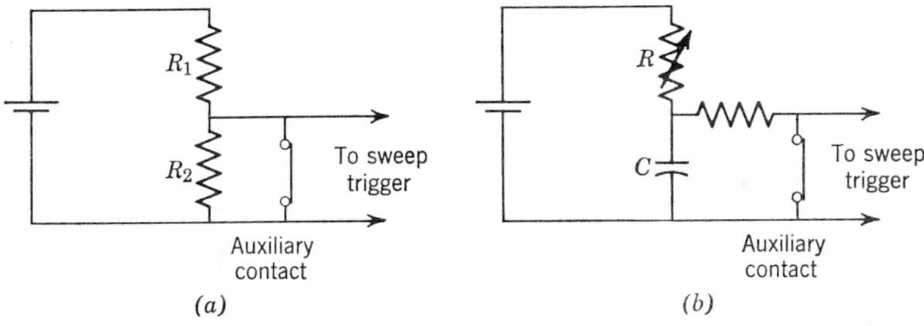

Fig. 16.9. Circuits containing an auxiliary contact for sweep initiation. (*a*) Fixed time. (*b*) Adjustable time delay.

charging *C* through *R*. Sometimes it is necessary to trigger separately a whole sequence of events (more than one CRO, the shutter opening of a camera, and so on), which calls for a multichannel timer. Such a device is described in Chapter 17.

There is sometimes a need to initiate an event, such as the triggering of an oscilloscope, immediately when a transient occurs, or perhaps the instant a circuit is energized. Frequently, perhaps because the circuit is at high voltage or because it is part of a power circuit, it may not be desirable to make electrical contact with the circuit. On such occasions signals can be obtained by magnetic coupling in the manner indicated in Fig. 16.10. A transient current of either polarity flowing in the main circuit will induce a voltage in the adjacent sensing circuit and produce a voltage of positive polarity across the resistor *R*. The two parts of the sensing circuit can be simple loops or they can be coils, depending upon the strength of the signal required and the magnitude of the inducing current.

The single sweep function for oscilloscopes is almost essential for studying single, as distinct from recurrent transients. This feature allows the time base to sweep only once on receiving a trigger signal, after which it locks out until it is reset. This avoids multiple sweeps, which may occur one on top of the other and obscure the photographic record.

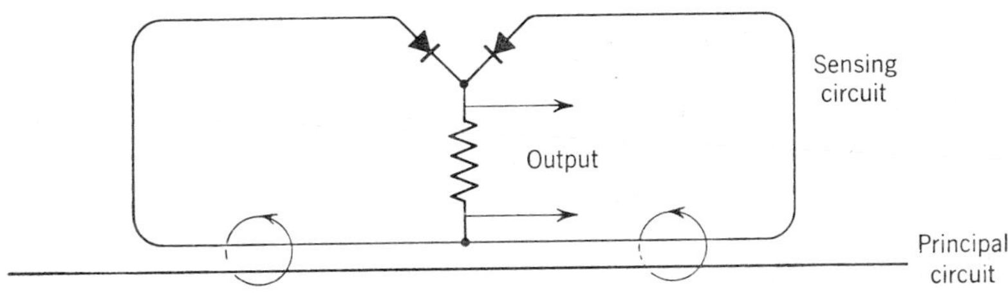

Fig. 16.10. Magnetically coupled sensing circuit.

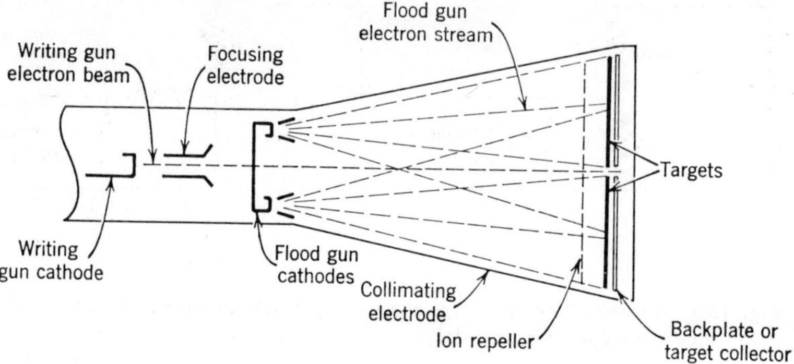

Fig. 16.11. Pictorial diagram of storage-type cathode ray tube. (Courtesy of Tekronix Inc.)

Another method of recording a transient has been developed in the *storage oscilloscope*. This instrument, as its name suggests, stores information on the transient and can retain an image of the phenomenon on its screen for hours if need be. In addition to the electron gun and the focusing and accelerating electrodes of a conventional oscilloscope, the storage oscilloscope has one or more "flood guns" and special storage elements. These latter elements store electric charges on an insulated surface by using the secondary emission properties of the surface. The stored charge is then used to control the flow of electrons to a phosphor screen to give a visual output corresponding to the location of the stored charge. By this means it is possible for the target, when bombarded by the flood gun electrons, to emit light at those locations whereon it had been previously written by the electrons of the writing beam.

Some oscilloscopes of this type have so-called split screen viewing areas, that is, the screen is divided into two parts, which can be individually operated in the conventional or the stored mode. A schematic of such a tube is shown in Fig. 16.11. A good storage oscilloscope can record single-shot transients up to a sweep speed of 10 μsec/cm without difficulty.

16.4 Cameras for Cathode Ray Oscilloscopes

There are now cameras that have been specifically designed for use with most oscilloscopes. They attach securely to the front of the oscilloscope and can swing away to view the screen, or there is a viewing port with a light-tight trapdoor through which the screen can be viewed. Most of these cameras use Polaroid film, which is sensitive and convenient, since the results can be obtained seconds after recording the event.

Cameras using regular film are sometimes preferred when a very large number of exposures are to be taken, or in special laboratory cameras which are arranged to observe several CRO screens simultaneously. Some such cameras are of the rotating drum type. The film is loaded on the inside of a drum, which is rotated at high speed. This then functions in lieu of a time base; the CROs need only provide a signal in a direction perpendicular to the direction of motion of film. The drum is brought up to a high constant speed before the transient event occurs, so that it gives traces with a linear time scale. This type of camera, although much more laborious to use, is capable of photographing several traces on a film perhaps six inches wide and four feet long, which give excellent resolution.

Light exclusion is very important when photographing single transients because the amount of light in the signal itself is very small and therefore a sensitive film with a fairly wide lens aperture may be necessary to record it. Under these circumstances any small light leak between the oscilloscope and the camera or through the oscilloscope itself may be devastating. This is particularly troublesome where the shutter must be left open for prolonged periods on the chance of a random transient occurring. Special cameras have been designed for this application, which move the film along at regular intervals whether or not the transient has occurred, so as to keep relatively fresh film always in readiness. When some event does cause the oscilloscope to sweep, the camera takes a picture and again moves the film, this time in response to the occurrence. As far as light leaks are concerned, it should be remembered that after an oscilloscope screen has been exposed to ambient light for a time, it will continue to glow slightly in the dark. This can affect a film. Time should be allowed for this emission to decay before the shutter is opened.

In those cases where one has control over the initiation of the transient to be recorded, advantage can be taken of a pair of contacts possessed by some cameras, which are closed in response to the opening of the shutter. These contacts can be used to start the transient event, with the shutter set for a fairly short exposure. In this way the shutter remains open no longer than necessary.

Some common errors which often mar photographic records are:

1. Failure to focus the camera under dynamic conditions.
2. Having the graticule lines too dim or too bright.
3. Failure to put a zero line in.
4. Failure to change spot brightness or camera aperture from shot to shot, as the circumstances may require, and, we should add, failure to put a film in!

Where a complicated setup is involved, it is a good idea to make a checklist which is conscientiously gone through on each test to avoid embarrassments.

16.5 Equipment for Measuring Transient Currents

Shunts. When a shunt can be used to measure transient currents, it is the best device for the job. However, it must be adequately designed. In principle, it functions like a shunt in a d-c circuit, that is, the voltage drop developed across it is a measure of the current flowing through it.

A shunt must have the necessary thermal capacity to take the normal and transient currents without heating to the point where a change of resistance would significantly affect the drop being measured. The shunt must have the mechanical strength to support the forces produced by the transient current. Above all, it must be as free as possible of inductive effects, and this is usually found to be the limitation.

Figure 16.12*a* schematically represents a single-element shunt of the type used with a millivoltmeter for measuring current in d-c circuits. Such a shunt

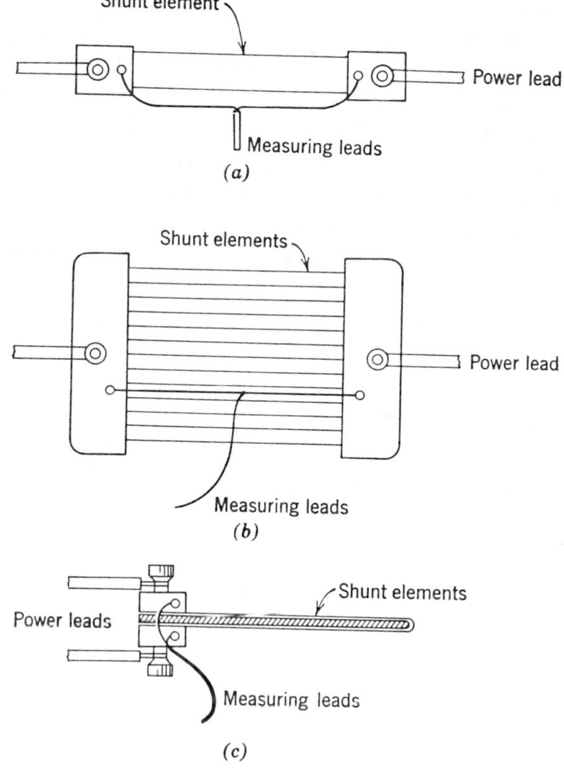

Fig. 16.12. Types of shunt. (*a*) Single element. (*b*) Multiple element for heavy current. (*c*) "Hairpin" shunt.

is of very limited value for transient measurements because of its inductance or, more accurately, because of the inductance of the measuring circuit of which the shunt is a part. This effect is best considered by thinking in terms of the magnetic flux that links the loop formed by the shunt and its measuring leads and, more especially, of the rate of change of this flux when a high-frequency transient current passes through the shunt. The voltage induced thereby will appear at the measuring instrument terminals, CRO or what have you, in addition to the true *IR* drop in the shunt. The voltage measured is no longer simply proportional to the current; an error is introduced both in the magnitude and phase. This error will increase with frequency and at very high frequencies may actually swamp out the true resistance effect. The situation can be improved by minimizing the area of the loop in the drop leads, but this is frequently not enough.

For high currents a multielement shunt is needed because of thermal requirements; such a shunt is sketched in Fig. 16.12*b*. Here the problem is manifest in a slightly different way. As a consequence of inductive effects, the current will tend to flow in greater concentration in the outer elements of the shunt, thus giving rise to a greater *IR* drop in those elements than the elements near the center. Where, then, should the drop leads be connected? In this type of shunt they are usually located off-center but, wherever they are, they are not giving an unequivocal indication of the total circuit current at all frequencies.

In Chapter 8 we considered the transient resistance of a conductor. This too is basically an inductive effect, which shows up as an increase in the apparent resistance with frequency. We can therefore expect it to be manifest in the simple type of shunt.

In an attempt to overcome, or reduce, inductive effects, shunts have been constructed from a thin strip of material in the form of a hairpin (Fig. 16.12*c*). The strip form tends to minimize high-frequency resistance effects and the geometrical shape tends to reduct the inductive loop. Some separation must exist, of course, between the two parts; moreover, mechanical forces are unfortunately maximized by this type of design.

The most widely accepted arrangement of shunt for transient measurements is the coaxial type. When high-quality measurements are to be made this should be used. The theory of this shunt has been described by Park (6); a schematic of such a shunt is given in Fig. 16.13.

The calibrated element is a metallic cylinder of a resistance material such as nichrome. The current is conducted down this cylinder and returns through a heavy coaxial sheath in the manner shown. One drop lead connects to a disk, or spider, which closes one end of the cylinder. The other drop lead passes down the axis of the shunt to the center of a similar disk at the other end. These leads are taken off in a coaxial cable, usually from the

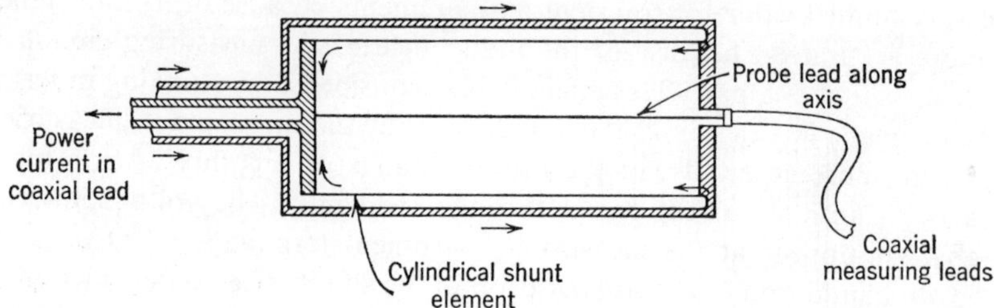

Fig. 16.13. Cross section through a coaxial shunt.

opposite end to which the power connections are made. The volume within the inner cylinder is essentially field-free. This is evident if one considers any closed path within this region; it embraces no current, that is, $\int \mathbf{H}\, dl = 0$.

Shunts of this kind are capable of high accuracy but are not perfect if only because of the finite thickness of the resistor cylinder, which allows eddy currents to flow. It can be shown, using the types of analyses developed in Chapter 8, that the voltage measured across the shunt is given by

$$V = I \frac{\rho l}{2\pi R} \frac{(1+j)}{\sqrt{2}\, d} \cdot \frac{1}{\sinh [(1+j)/\sqrt{2}](t/d)} \tag{16.5.1}$$

where R and l are the radius and length of the cylinder and t is its thickness; ρ is the resistivity of the material and d is defined in Eq. 8.4.8 as

$$d = \left[\frac{\rho}{\mu\mu_0\omega}\right]^{1/2} \tag{8.4.8}$$

Equation 16.5.1 indicates the magnitude and phase error introduced by the finite thickness t, but if t is small, these errors are also small and the expression reduces to

$$V = I\left[\frac{\rho l}{2\pi R t}\right]$$

Other errors arise from "end effects," that is, the distortion of the current paths in the cylinders by the magnetic fields of the power leads. This is minimized by making the power connections coaxial. Inaccuracies in construction, most notably lack of concentricity in the cylinders, can also cause errors.

Figure 16.14 records a current approaching and reaching current zero. The two oscillograms, although they are supposedly displaying the same current, are different in appearance. The first is the true current measured by a high-quality shunt; the second exhibits significant distortion introduced by the inductance of the shunt.

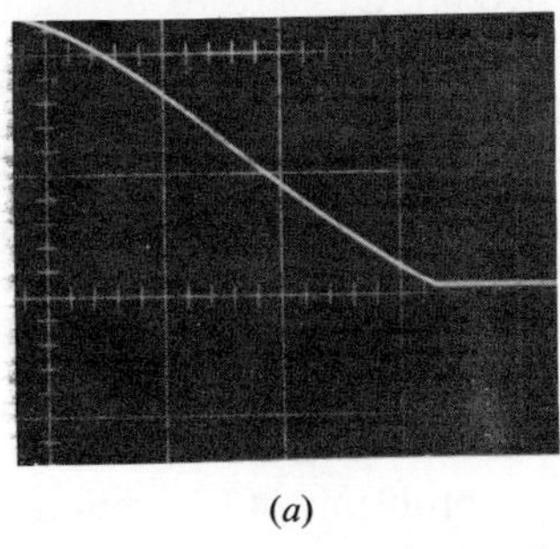

(a)

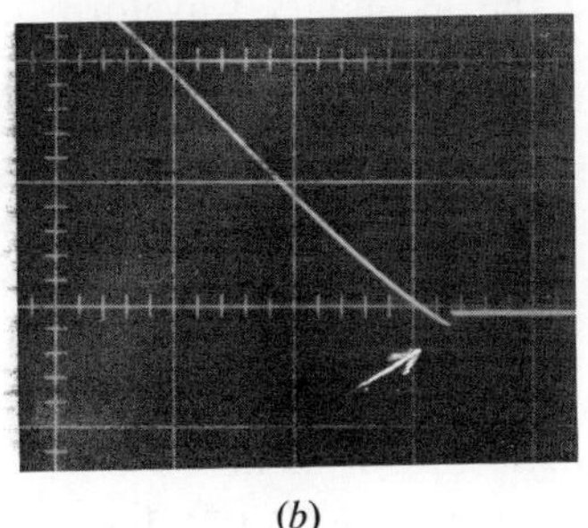

(b)

Fig. 16.14. Oscilloscopic record of a current measurement with: (a) a good coaxial shunt, (b) an indifferent shunt showing distortion at current xero.

Current Transformers and Linear Couplers. Shunts are not always convenient to use; for one thing they present problems when the point of measurement is at high potential, or momentarily assumes a high potential when the transient occurs. Under these conditions it is helpful to have some isolation between the measuring circuit and the circuit proper. This can be obtained by using a current transformer (CT).

The theory of current transformers under steady-state conditions is well documented in the literature; see, for example, Hague (7). Unlike a potential or power transformer, the primary current is not determined by the secondary load but by the impedance of other elements in the primary circuit. The secondary circuit is virtually shorted so that the secondary ampere-turns just about balance the primary ampere-turns and the core flux density is low. The primary and secondary currents are approximately in inverse proportion to the turns on the windings. Our concern here is to find the extent to which these relationships apply under transient conditions.

This question really addresses itself to the frequency response of the CT. Experience shows that errors creep in at both ends of the spectrum. At the upper end, errors are introduced by core losses, notably eddy current losses, which increase with frequency. A point eventually is reached where the capacitance of the winding becomes important because it offers an alternative path for current. Resonances may be set up between the transformer inductance and these stray capacitances or the lead capacitances. For some CT's, such problems are discernible at 5 kHz and are very pronounced at 10 kHz. The effects can be reduced by careful design.

At the bottom end of the frequency scale errors arise because of core saturation. Transient currents, such as inrush currents into capacitor banks or transformers, are frequently high in magnitude and often contain a significant d-c component. The d-c component biases the core to such an extent that when a-c peaks of the same polarity are added, saturation occurs and the secondary waveform is badly distorted. If CTs are to be used to measure transient asymmetrical currents of this kind, the core must be generously designed to avoid these problems.

In summary, the run-of-the-mill CT has serious shortcomings as a transient measuring device, but the problems can be designed out of it, at least to some extent.

Air-Cored Devices. An alternative to the regular CT is the family of devices known as air-cored CTs, linear couplers, or current probes. Since they have no steel core, they cannot suffer from problems of saturation.

In essence these devices are mutual inductors. They comprise a coil which links with a part of the flux surrounding the current-carrying conductor. The voltage induced in the coil is proportional to the rate of change of the current. This voltage is integrated with respect to time by an electronic integrating current, so that the output of the integrator, which can be displayed on an oscilloscope, is proportional to the current.

The current probe has one distinct advantage over other forms of current measuring devices in that it does not noticeably disturb the circuit being measured as do shunts and iron-cored current transformers. This is of particular importance in the study of current distribution in low-impedance circuits such as paralleled rectifiers where insertion of a shunt or transformer would profoundly affect the current distribution in the circuit.

The coil is typically toroidal in form, although it does not have to be. Other shapes may be more desirable for special applications. They can be large enough to embrace a considerable bus structure, or small enough to insert into a relatively inaccessible part of a circuit, as the need dictates. They can even be flexible, to conveniently fit a tortuous path.

It is very important that the progression of the winding not form a single turn in the plane of the toroid. This can be prevented by returning the wire back through the center of the winding to the beginning. In fact, small-diameter windings can be wound directly on a heavier wire and the end of the winding merely soldered to the larger wire. Care should be taken to see that the winding is uniform and that there are no gaps in the flux path.

The integrator is usually a high-gain direct-coupled transistor amplifier using an RC feedback loop to perform voltage time integration. This is the familiar operational amplifier commonly used in analog computers. Since the voltage induced in the probe coil is proportional to dI/dt and therefore to frequency, the amplifier gain must diminish linearly with increasing frequency. It is not difficult with this combination to secure a reasonably flat response from 10 Hz to 10 kHz. Such a device cannot, of course, measure a sustained direct current because it produces no changing flux. Direct-current biases are therefore not reproduced. Under transient conditions, as when current is suddenly applied, the d-c component is measured and the current waveform will have its zero level properly recorded. This condition will deteriorate with time if the bias is maintained. At the upper end of the frequency spectrum performance deteriorates because of capacitance effects, but it is not difficult to produce a device for frequencies up to 20 kHz.

A clip-on type of design for such a current probe is shown in Fig. 16.15. It consists of two half toroids cemented into a pivoted holder made of two pieces of insulating material. A steel spring holds the toroid halves closed. The sensitivity of such a device depends upon its mutual inductance with the measured circuit and the amplification of the integrator. For a simple toroidal coil linked to a conductor,

$$M = \frac{\mu_0 NA}{l} \tag{16.5.2}$$

where

M = mutual inductance (henries)
N = number of turns of wire
A = cross-sectional area of the winding (square meters)
l = mean length of winding (meters)
μ_0 = permeability of free space $= 4\pi \times 10^{-7}$

the sensitivity is given by

$$S = 2\pi f MJ \tag{16.5.3}$$

where J is the amplification of the integrator at the frequency f.

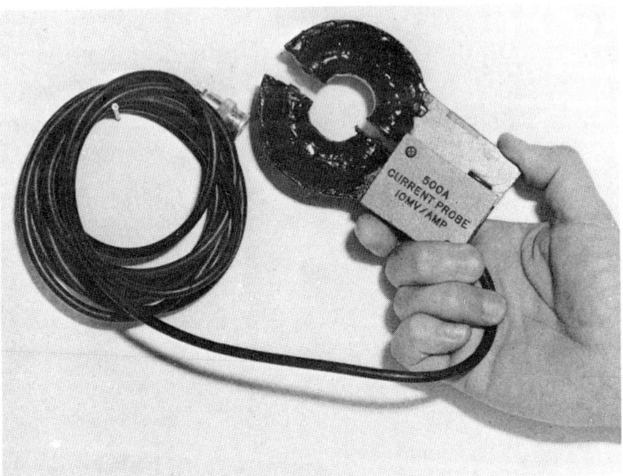

Fig. 16.15. Clip-on type of linear coupler or current probe. (Courtesy General Electric Co.)

16.6 Transient Voltage Measuring Equipment

Potential Dividers. When the voltage to be examined is in excess of that which the measuring instrument can handle, some form of potential divider must be placed between the circuit and the measuring equipment. Inasmuch as most oscilloscopes and oscillographs are designed to handle 50 V or less, a potential divider is a very common requirement when measurements are being made on power systems.

In its simplest form, the potential divider comprises two impedances in series, as shown in Fig. 16.16. The ratio of the divider is given by

$$\frac{V_2}{V_1} = \frac{Z_2}{Z_1 + Z_2} \tag{16.6.1}$$

In an ideal divider we would obtain, at the output, a perfect replica of the imput voltage, reduced in this proportion.

Since in transient work we are interested in a wide frequency spectrum, the impedances Z_1 and Z_2 must be of the same kind, that is, they must both be resistors, both be capacitors, or both be inductors; otherwise the ratio would change with frequency. As a matter of practicality inductors are very rarely used for this purpose—the principal components are either capacitors or resistors—but in theory inductors could be used.

Another requirement for potential dividers is that they have a minimal loading effect on the circuit so as not to influence the transient being measured. This rules out high values of capacitors or low resistors. In stiff power systems often this is not a problem, although power dissipation with resistance dividers becomes a limitation.

When we consider the design of a potential divider, the first problems arise because of the impurity of components. Unfortunately, we cannot say a capacitor is a capacitor is a capacitor, or a resistor is a resistor is a resistor!

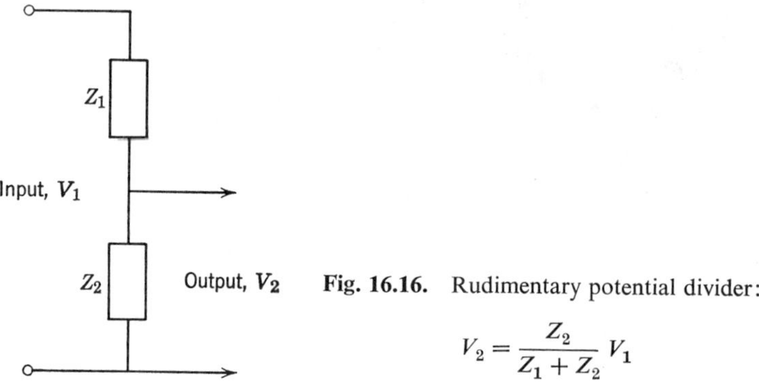

Fig. 16.16. Rudimentary potential divider:

$$V_2 = \frac{Z_2}{Z_1 + Z_2} V_1$$

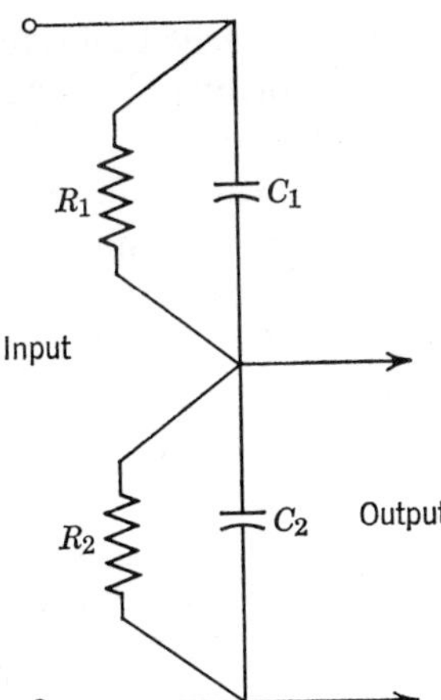

Fig. 16.17. Resistively-compensated capacitance divider:

$$R_1C_1 = R_2C_2$$

Each possesses attributes of the other in some measure, and both possess some inductance. This must be recognized and compensated for. Thus a capacitance divider may have to become resistance-compensated, as indicated in Fig. 16.17. The leakage R_1 of the high-voltage capacitor C_1 is compensated for by the resistor R_2 across C_2. Since the leakage of C_1 may be somewhat variable, depending upon atmospheric environment, R_1 may be paralleled by a constant swamping resistance. In any event, the overall objective is to preserve the divider ratio by maintaining

$$R_1C_1 = R_2C_2$$

It must also be recognized that other strays can influence a divider ratio, and that these may be influenced by purely geometrical environmental conditions. The stray capacitance of a resistance divider, for example, depends upon the proximity of other grounded structures. Some more elaborate dividers for laboratory use are shielded for this purpose; they take their environment with them, as it were. Just as we avoid a loop in the leads to a shunt because of the voltage that can be magnetically induced in the loop by stray linking flux, so we couple the measuring instrument as tightly as possible to the lower end of a divider. This is typically done with a short length of coaxial cable secured directly to the lower capacitor or resistor terminals.

As soon as one ceases to look upon the potential divider in isolation, other factors demand attention. The instrument connected across the lower end is in effect part of the divider. If it is a CRO, it will have a high impedance, but it should still be considered. In many instances the measurement must be made remotely, requiring a cable connection between the divider and the instrument. One begins to get the concept of a measuring system, rather than just one or two instruments. We will examine some of the consequences of this idea and try to arrive at some guidelines.

It is helpful to do this by studying a specific example, which can be taken in steps of increasing complexity. Consider a 500:1 capacitance divider and suppose in the first instance that a CRO is directly connected across the lower end. To minimize the loading of the circuit by the divider, C_1 might be typically 20 pF, whence C_2 would be 499 times as great, or approximately 0.01 μF. Compared with this, the input capacitance of the oscilloscope is quite negligible, but the shunting effect of a 1-megohm resistor at the input of the oscilloscope cannot be so lightly disregarded; it will materially affect the frequency response at the lower end. At 16 Hz the impedance of C_2 is also 1 megohm, at 160 Hz it is still one-tenth of this value. Thus power frequency voltages do not come through with the same ratio as higher transient frequencies. If the arrangement is calibrated at 60 Hz, as it may well be, it would be in error at the higher frequencies. On the other hand, should it happen, by chance, that the leakage across the capacitor C_1 is exactly 500 megohm, the divider would record faithfully down to zero frequency.

If a cable is placed between the divider and the CRO, the consequences will depend upon the length of the cable and the frequency range to be measured. If the cable is short, it simply adds to the capacitance of C_2 and can be taken account of in the ratio. As the cable becomes longer, we can no longer think of it as a lumped capacitor; it must be considered as a transmission line with a surge impedance and a travel time. A signal impressed at the divider end will travel to the CRO, arriving there some time later. This wave will suffer reflection at the CRO and in fact what will be observed there will be the refracted wave produced in the total CRO termination. Further reflections and refractions will follow. The net effect is that the cable will introduce into the record a high-frequency oscillation not present in the original signal. If we are observing relatively low frequencies, this is of little consequence, but not so if high frequency transients are being studied.

Figure 16.18*a* shows a first attempt to eliminate or reduce the undesirable effects of the cable. A resistor, numerically equal to the surge impedance of the cable, is interposed between the divider and the cable. Consider now a step of voltage, V, impressed across the divider; a fraction

$$\frac{C_1}{C_1 + C_2} V$$

will appear across C_2. Because of R, only one-half of this will be impressed on the cable and travel down the cable as a wave. At the oscilloscope the wave encounters a virtual open circuit; the reflected wave is $V/2$ also, and the

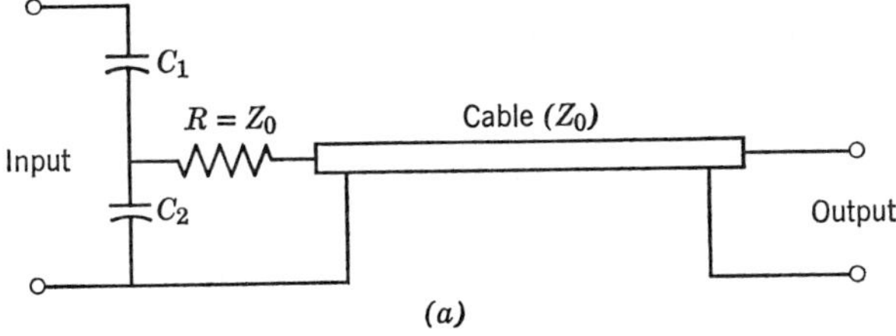

(a)

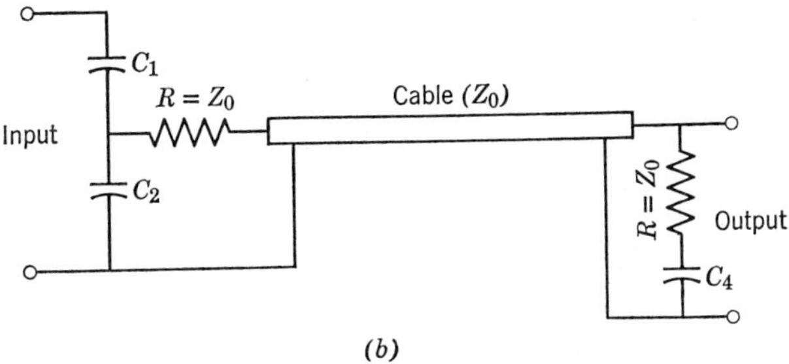

(b)

Fig. 16.18. Potential divider circuits containing cable.

voltage seen by the oscilloscope is

$$\frac{V}{2} + \frac{V}{2} = V$$

When the reflected wave returns to the divider end, it is absorbed without further reflection in R since R is matched with the line.

This analysis is accurate only for short times or high frequencies. An alternative circuit has been proposed by Burch (8); it is given in Fig. 16.18*b*. In this circuit the fact that the cable will tend to discharge C_2 and cause its voltage to sag is compensated for by the charging of C_4.

Burch (8) addresses himself to the more general case, which is illustrated in Fig. 16.19. He gives the following operational expression for the voltage, v, at the oscilloscope, in terms of the step voltage V applied across the divider:

$$v(s) = \frac{\rho\epsilon^{-s\tau}}{1 + \mu_1\mu_2\epsilon^{-s\tau}} \frac{V}{s} \tag{16.6.2}$$

wherein

$$\rho = \frac{Z_5}{Z_4} \frac{Z_2}{Z_1 + Z_2} \frac{(1 + \mu_1)(1 - \mu_2)}{2}$$

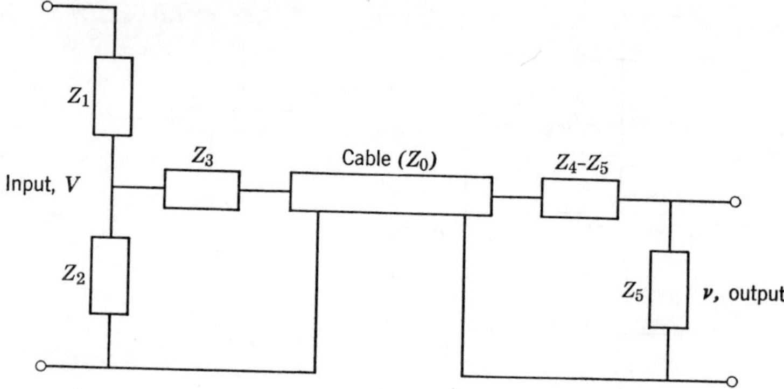

Fig. 16.19. General case of the potential divider with a cable.

and μ_1 and μ_2 are the reflection coefficients at the ends of the cable:

$$\mu_1 = \frac{Z_0(Z_1 + Z_2) - (Z_1Z_2 + Z_2Z_3 + Z_3Z_1)}{Z_0(Z_1 + Z_2) + (Z_1Z_2 + Z_2Z_3 + Z_3Z_1)}$$

$$\mu_2 = \frac{Z_0 - Z_4}{Z_0 + Z_4}$$

Here τ is the travel time for a wave making one journey along the length of the cable. The various Zs, of course, are expressed in their operational form.

Resistors for potential dividers should be selected with low inductance in mind. A simple bifilar type of construction gives this but presents a problem from an insulation point of view, for the resistors, at least in the top part of the divider, must support a considerable voltage. Resistors are specially wound for the purpose on flat cards. Many of them are then connected in series for high-voltage applications.

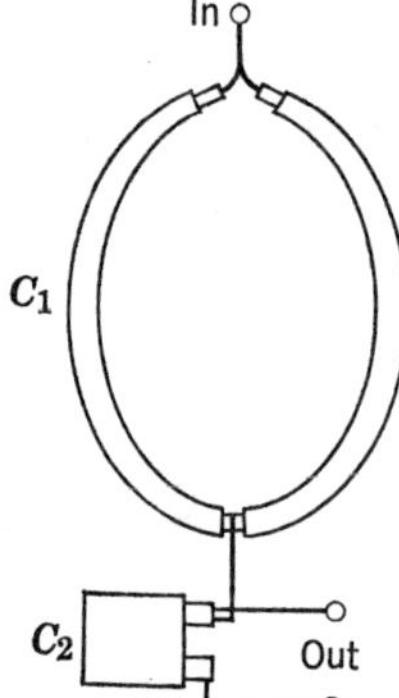

Fig. 16.20. Makeshift capacitance potential divider made from a short length of high-voltage cable.

Capacitors for potential dividers may take many forms; the paper/oil capacitors used for coupling carrier signals onto transmission lines may be used to advantage for the high-voltage, low-capacitance units in high-voltage testing. Another type for this purpose is the vacuum capacitor, which has an extremely low power factor. Compressed gas is also used as a dielectric. Paper/oil or mica capacitors are satisfactory for the lower end.

It is often necessary to improvise capacitance potential dividers. This can be done crudely, with a pin insulator or a bushing for the high voltage end. The author has had good success using a short length of cable for this purpose (Fig. 16.20). The sheath is stripped back at both ends and exposed in the middle to form the connection to the low-voltage capacitor. Such devices, though lacking the accuracy of a good commercial product, are quite adequate for many field investigations.

Voltage probes, which are a form of light, portable, potential divider are often supplied with modern CROs. They usually have a ratio of 10 or 100, although higher ratios can be obtained. They are compensated and typically insulated for 600 V or 2 kV. Their input impedance is of the order of 10 megohm shunted by a few picofarads. Higher voltage portable probes are available; one such probe, suitable for measurements up to 40 kV peak, is illustrated in Fig. 16.21. Such devices are very satisfactory within their rating, but a word of caution is injected in Section 17.5 where very fast transients are concerned. A good general reference on potential dividers is Craggs and Meek's book *High Voltage Laboratory Technique* (9).

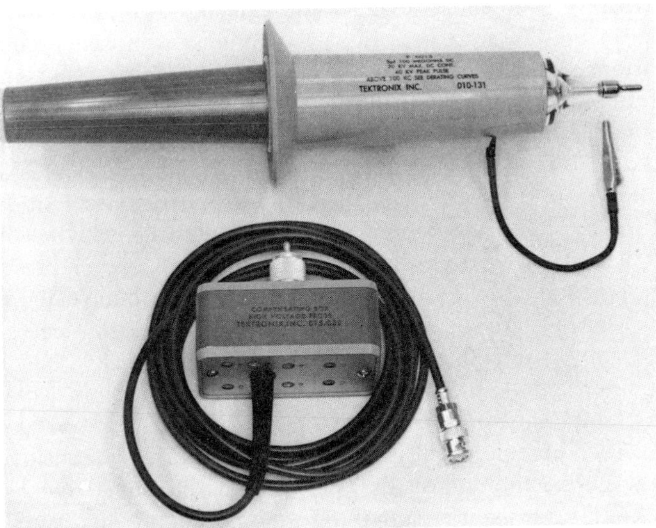

Fig. 16.21. 40-kV probe for voltage measurements. (Courtesy of Tektronix Inc.)

Potential Transformers. Potential transformers designed for 60-Hz operation are somewhat limited in transient capability. Most are adequate, however, for driving a magnetic oscillograph; that is, their frequency response is comparable with that of an oscillograph galvanometer. A potential transformer cannot, of course, record any direct component in the voltage. Indeed, any sustained direct component will quickly drive the core into saturation. It must also be remembered that for frequencies below 60 Hz, the peak value that will cause saturation diminishes as the frequency diminishes, which means that low-frequency response is limited. It is possible to improve this by special design through putting more iron into the core.

High-frequency response can also be extended by design. This calls for thinner laminations or a ferrite core and attention to minimizing capacitance effects. These are the same factors that control the design of good pulse transformers (10), which are transient PTs *par excellence.*

16.7 Equipment for Measuring Random Disturbances

The occurrence of many electrical transients is quite unpredictable; lightning is a good example. To detect and/or record such transients, it is necessary to monitor the circuit continuously. There are many devices available for this purpose, ranging in sophistication as the amount of information required and the amount of attention available dictates.

At one end of the scale it is sometimes necessary to set up one or more oscilloscopes with potential dividers and cameras to record transients that may appear on the screen(s). In such circumstances the oscilloscope is usually arranged to operate in the triggered mode. A sweep occurs in response to any transient above a particular level. Following such a sweep, the film in the camera is automatically racked along in readiness for the next event and the sweep is reset. Even in the best of such equipment there is some light leak, if only from the filament of the cathode ray tube itself. In due course this will fog the film. Accordingly, at regular intervals, whether or not any transient has occurred, the film is caused to move on, to preserve good recording potential. Obviously, such an installation is quite elaborate and is used only when a detailed record of the waveform of the transient is required. In many instances something much less elegant will suffice.

Sometimes the only question is how often do voltage transients exceed a particular limit. This calls for some simple "go" or "no go" device. The circuit for such an equipment is shown in Fig. 16.22. The sensitive element here is the unijunction transistor T_1. When the incoming signal exceeds a preset level, this transistor is caused to turn on. In doing so, it gates the small thyristor T_2, which behaves as a power amplifier, releasing sufficient energy from a charged capacitor to operate an electromagnetic counter. It then

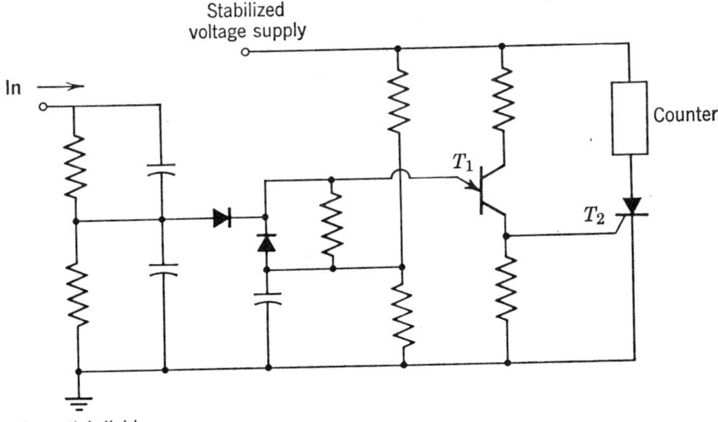

Fig. 16.22. Simple recorder to count random surges over a preset level.

takes some time for the equipment to reset, so that its ability to record repetitive transients is limited.

It is obvious, of course, that several devices of this kind can be operated in parallel at increasing sensing levels, V_1, V_2, V_3, and so on. It is then possible to establish, in a given period, how many transient voltages have exceeded V_1; what fraction of them have exceeded V_2, or V_3, and so on.

An ingenious device for measuring the incidence and amplitude of random transient voltages, utilizing magnetic tape, has been described by Rudolfsson and Skogstedt (11); it utilizes a scheme proposed by Wurz (12). It is based on the fact that the magnitization in a tape can be erased by the application of a rather well defined magnetic field. The measuring and recording principle is as follows. A sinusoidal oscillation (tone) of constant amplitude is recorded with a tape recorder on a tape. The prerecorded tape is placed in an elongated coil as shown in Fig. 16.23. If a current is passed through the coil, its magnetic field will tend to erase the tone. By winding the coil in the manner shown in Fig. 16.23, it is possible to profile the magnetic field as indicated. Only when the field exceeds a certain critical level will the tone be removed, so that a gap will be left in the signal on the tape whenever this occurs.

The coil has a resistor in series with it, which swamps the impedance of the coil; the current is then proportional to the voltage applied across the combination. The tape is slowly drawn through the coil from one spool to another, with the device connected across the circuit being investigated. The incidence of gaps in the tone will disclose the number of surges in excess of the critical level occurring on the circuit. The length of an individual gap will give a correct measure of the magnitude of the transient. Multichannel devices of this type have been made (10) and the readout has been automated.

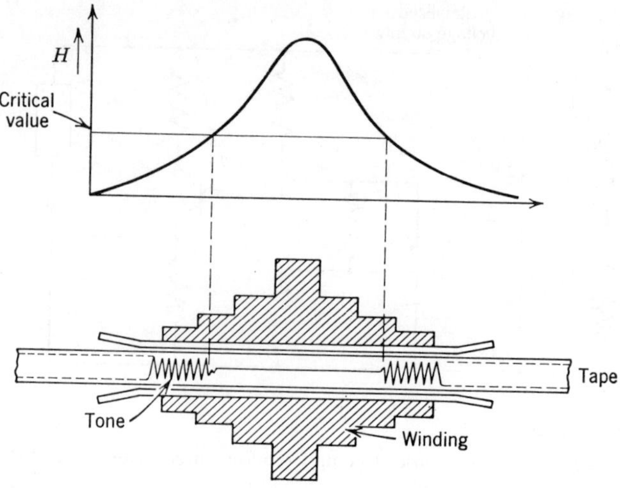

Fig. 16.23. Surge detector using magnetic tape. (Courtesy the ASEA Co.)

Another device for recording transient voltages is the klydonograph (13). This is a relatively inexpensive piece of equipment, which is used occasionally in profusion on power circuits. It has found considerable acceptance as a device for measuring voltage transients due to lightning. Klydonographs rely for their operation on peculiar patterns, known as Lichtenberg figures, that appear on a photographic film when it is subjected to a surface corona discharge. The film is placed, with a sheet of dielectric material, between two electrodes, one being a flat plate and the other a circular rod. This is shown schematically in Fig. 16.24. Its useful range is approximately 2–20 kV. In high-voltage systems the klydonograph is connected across a potential divider. This device can also be used to measure surge currents by connecting it across a high impedance shunt through which the current flows. It has been

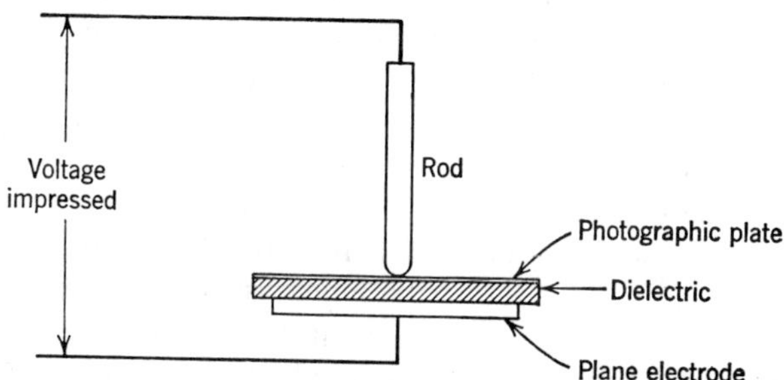

Fig. 16.24. Schematic diagram of the Klydonograph.

used in this manner for the study of lightning. A typical pattern from a klydonograph is shown in Fig. 16.25. To an experienced eye it reveals a considerable number of data about the polarity, magnitude, and duration of the surge, because the pattern varies with polarity, magnitude, and steepness of front. To help in this regard, instruments are often connected back-to-back in respect of their electrode arrangements. In this way both positive and negative records are obtained on each surge. As a peak voltage measuring instrument the klydonograph has an accuracy probably no better than ±25%.

Anderson and Giocomoni (14) describe an extension of the klydonograph idea, which comprises a gang of such devices operating in parallel with certain auxiliary equipment. The device is called a teinograph, but perhaps a better name would be a surge mousetrap, for it virtually captures the surge and then records it. The teinograph can operate in one of two modes, the time-sequential mode or the space-sequential mode; the names describe the functions. In the time sequential-mode, it takes "snapshots" of the surge at time intervals and thereby obtains time resolution. In the second mode it takes a number of "snapshots" of different parts of the transient simultaneously and thereby obtains space-resolution of the wave.

Klydonographs have been mechanized with slowly moving films to record random transients.

16.8 Sphere Gaps

The sphere gap has been used for a long time as a means of measuring peak voltages, at least in the laboratory environment. Many pieces of power equipment such as transformers, switchgear, and insulation structures are required by specification to support certain impulse and switching surge peak voltages and are tested in the laboratory for compliance (see Chapter 17). Sphere gaps are the standard measuring devices for much of this work.

The voltage at which a sphere gap flashes over depends primarily on the size of the spheres and their separation. However, the flashover voltage also depends upon the waveshape of the surge, more especially on the rate of rise of the voltage. Polarity is also a factor, as is the question of which sphere, if any, is grounded. For a given spacing, the negative flashover voltage (positive sphere grounded) is higher than the positive. Other factors that influence the flashover voltage are air density (temperature and pressure) and to some extent the humidity. A good review of these effects is to be found in reference (8).

The physical processes involved in the breakdown of a sphere gap under electrical stress are statistical by nature, consequently the breakdown voltage level is not unique. If a number of tests are made on a given gap, each time

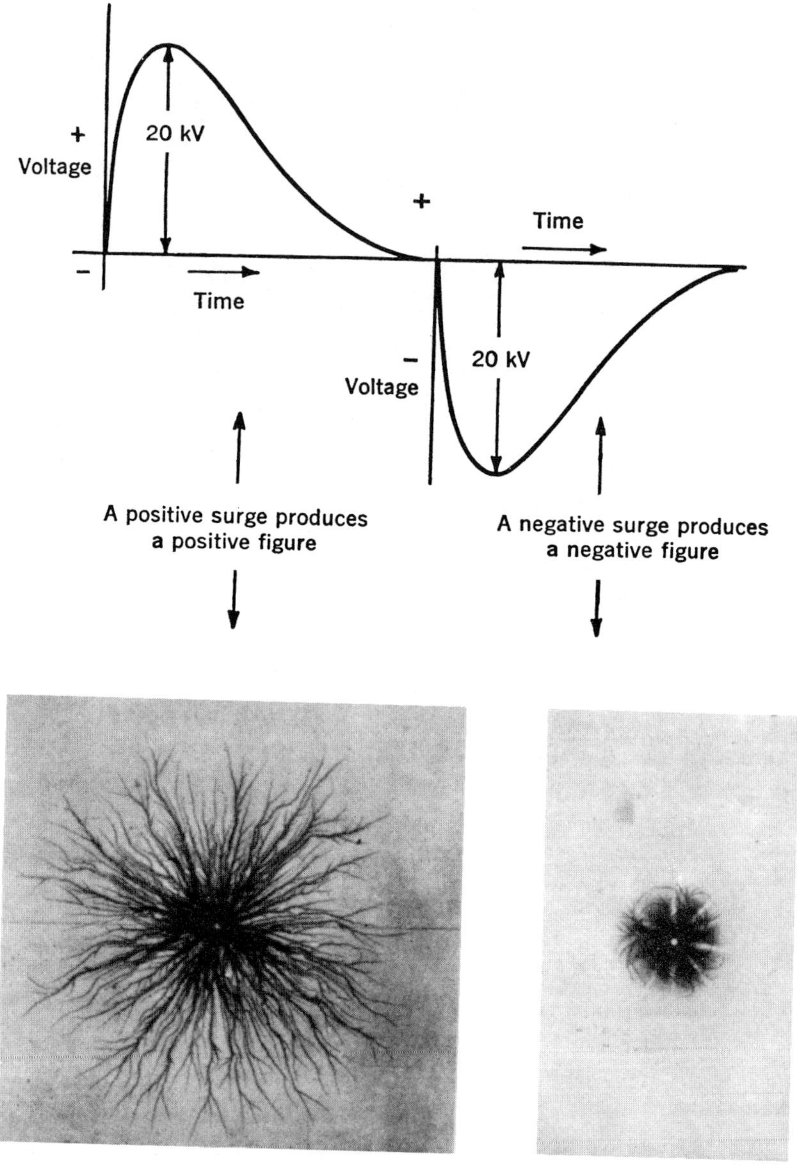

Fig. 16.25. Positive and negative photographic Lichtenberg figures produced by positive and negative surge voltages of the same magnitude and wave shape.

Table 16.1A Sphere Gap Sparkover Crest Voltages (One Sphere Grounded) 760 mm, 25°C

	Sphere Diameter (cm)			
	6.25		12.5	
Sphere Gap Spacing (cm)	Sixty Cycle and Negative Impulse kV Crest	Positive Impulse kV Crest	Sixty Cycle and Negative Impulse kV Crest	Postive Impulse kV Crest
0.5	17.0	17.0		
1.0	31.3	31.3	31.7	31.7
1.5	44.5	44.8	44.9	44.9
2.0	57.0	57.4	58.0	58.0
2.5	68.8	69.3	70.8	70.8
3.0	78.8	79.4	83.5	83.5
3.5	86.6	88.0	95.0	95.3
4.0	93.6		106.0	103.0
4.5	99.8		117.0	120.0
5.0	105.5		127.0	132.3
5.5	110.0*		135.3	142.5
6.0	114.0*		143.5	153.8
6.25	117.0*		147.5	158.0
7.0			157.7	171.0
8.0			170.5	
9.0			182.0	
10.0			192.0*	
11.0			200.0*	
12.0			208.0*	
12.5			211.0*	

	Sphere Diameter (cm)					
	25		50		75	
Sphere Gap Spacing (cm)	Sixty Cycle and Negative Impulse kV Crest	Positive Impulse kV Crest	Sixty Cycle and Negative Impulse kV Crest	Positive Impulse kV Crest	Sixty Cycle and Negative Impulse kV Crest	Positive Impulse kV Crest
2.5	72	62				
5.0	136	136	136	136	136	136
7.5	192	196	197	197	200	200
10.0	241	252	260	260	261	261
12.5	278	296	317	319	324	324
15.0	309	334	367	374	380	380
17.5	338	364	411	426	433	443
20.0	362	390	451	474	484	499
22.5	379	409	486	511	528	548
25.0	393	426	519	547	573	597
30.0			573	605	653	687
35.0			615	655	721	755
40.0			651	698	777	816
45.0			681	732	827	870
50.0			707	758	870	917
55.0					910	960
60.0					945	999
65.0					977	1031
70.0					1003	1058
75.0					1025	1081

	Sphere Diameter (cm)					
	100		150		200	
Sphere Gap Spacing (cm)	Sixty Cycle and Negative Impulse kV Crest	Positive Impulse kV Crest	Sixty Cycle and Negative Impulse kV Crest	Positive Impulse kV Crest	Sixty Cycle and Negative Impulse kV Crest	Positive Impulse kV Crest
10	261	261	261	261	261	261
20	504	504	505	505	506	506
30	700	715	736	736	746	746
40	862	888	947	955	973	973
50	985	1024	1120	1140	1172	1178
60	1084	1124	1254	1293	1346	1364
70	1163	1209	1360	1400	1505	1533
80	1234	1284	1458	1502	1635	1671
90	1295	1344	1552	1597	1752	1788
100	1338	1390	1628	1678	1857	1896
110			1695	1755		
120			1760	1824		
130			1815	1880		
140			1865	1920		
150			1900	1944		

Table 16.1B Sphere Gap Sparkover R.M.S. Voltages (One Sphere Grounded) 760 mm, 25°C

	Sphere Diameter (cm)	
	6.25	12.5
Sphere Gap Spacing (cm)	Sixty Cycle (Sinusoidal rms) kV	Sixty Cycle (Sinusoidal rms) kV
0.5	12.0	22.4
1.0	22.1	31.7
1.5	31.5	41.0
2.0	40.3	50.0
2.5	48.6	59.0
3.0	55.7	67.2
3.5	61.2	75.0
4.0	66.2	82.7
4.5	70.6	89.8
5.0	74.6	95.6
5.5	78.0*	101.5
6.0	81.0*	104.3
6.25	83.0*	111.5
7.0		120.5
8.0		128.8
9.0		136.0*
10.0		141.0*
11.0		147.0*
12.0		149.0*
12.5		

	Sphere Diameter (cm)		
	25	50	75
Sphere Gap Spacing (cm)	Sixty Cycle Sinusoidal rms kV	Sixty Cycle Sinusoidal rms kV	Sixty Cycle Sinusoidal rms kV
2.5	51		
5.0	96	96	96
7.5	136	139	141
10.0	170	184	185
12.5	197	224	229
15.0	218	260	269
17.5	239	291	306
20.0	256	319	342
22.5	268	344	373
25.0	278	367	405
30.0		405	462
35.0		435	510
40.0		460	549
45.0		482	585
50.0		500	615
55.0			643
60.0			668
65.0			691
70.0			709
75.0			725

	Sphere Diameter (cm)		
	100	150	200
Sphere Gap Spacing cm	Sixty Cycle Sinusoidal rms kV	Sixty Cycle Sinusoidal rms kV	Sixty Cycle Sinusoidal rms kV
10	185	185	185
20	356	357	358
30	495	520	527
40	610	670	688
50	696	792	829
60	766	887	952
70	822	962	1064
80	873	1031	1156
90	916	1097	1239
100	946	1151	1313
110		1199	
120		1245	
130		1283	
140		1319	
150		1344	

applying precisely the same voltage in magnitude and waveform, it will be found that the breakdown voltage is not univalued, but that the data cluster around a mean. The spread depends on circumstances but is usually a few percent. This must be taken into account when making definitions. For example, the minimum flashover voltage is the value at which flashover occurs 50% of the time.

The statistical effects are most obviously manifest under impulse conditions, where the voltage wave has a steeply rising front. This matter is referred to in Section 12.1 where rod gaps are discussed. The breakdown process depends upon an electron penetrating the high field region and initiating an avalanche, which in turn develops into a spark and finally an arc. There is an initial statistical time lag which is the time that elapses before an electron appears in the gap. There is then a formative time lag during which the spark propagates and the arc is established. The statistical time lag has quite a spread, which is reflected in a spread of breakdown voltage that tends to widen as the rate of rise of voltage increases. This is the same phenomenon that accounts for the upturn in the volt/time curve of protective gaps, referred to in Section 12.1. The sphere gap has considerably less upturn than a rod gap.

Greater consistency in the breakdown voltage level can be obtained by irradiating the gap. The radiation, which can be provided by radioactive material in one of the spheres, or by ultraviolet from a lamp, provides a supply of initiating electrons which shortens the statistical time lag.

Sphere gaps of different sizes are available. A general rule is that such gaps should not be used at spacings greater than the diameter of the spheres. For high voltages a bigger sphere size should be used.

Calibration for sphere gaps are the subject of an American Standards Association publication, "Measurement of Voltage in Dielectric Tests" (15). Data extracted from this publication are given in Table 16.1. This standard also stipulates how corrections shall be made for air density variations and makes recommendations on irradiation of the gap.

REFERENCES

1. H. Tropper, *Electric Circuit Theory*, Longmans, Green and Co., London, New York and Toronto (1949).
2. W. Duddell, "Oscillographs," *Electrician*, Vol. 39 (1897), p. 636.
3. W. Kaiser, "Oscillomink Direct-Writing Jet Oscillograph," *Seimens Review*, Vol. 36, No. 6 (1959), p. 191.
4. H. Sima, "Liquid Jet Oscillographs," *Elektronik*, Vol. 15, No. 6 (1966), p. 179; No. 7, p. 213.
5. J. Millman and S. Seely, *Electronics*, 2nd ed., McGraw-Hill Book Co., New York, Toronto and London (1951).

6. J. H. Park, "Shunts and Inductors for Surge-Current Measurements," *Journ. of Research, Nat. Bureau of Standards*, Vol. 39 (1947), p. 191.
7. B. Hague, *Instrument Transformers, Their Theory, Characteristics and Testing*, Isaac Pitman and Sons, London (1936).
8. F. P. Burch, "On Potential Dividers for Cathode Ray Oscillographs," *Phil. Mag.*, Vol. 13, No. 86 (1932), p. 760.
9. J. D. Craggs and J. M. Meek, *High Voltage Laboratory Technique*, Butterworth Scientific Publications, London (1954).
10. M.I.T. Radio School Staff, *Principles of Radar*, McGraw-Hill Book Co., New York (1946).
11. D. Rudolfsson and S. Skogstedt, "The Recording of Transient Overvoltages in Industrial Networks," *ASEA Journ.*, Vol. 40, No. 6–7 (1967), p. 86.
12. H. Wurz, "Betreibemässige Registrierung von Schaltuberspannugen, insbesondere nach einen neuartigen Messerverfahren," *Electro. Tech. Zeit.—A*, Vol. 81, No. 17 (1960), p. 597.
13. J. F. Peters, "The Klydonograph," *Electrical World* (1924), p. 769.
14. J. G. Anderson and R. V. Giocomoni, "The Teinograph—A New High Voltage Surge Recorder," *Trans. AIEE*, Vol. 78 (1959), p. 1800.
15. American Standard for Measurement of Voltage in Dielectric Tests, *AIEE* (1953).

17 Measuring Techniques and Surge Testing

17.1 Introduction

In Chapter 16 attention was focused on instruments and equipment for measuring electrical transients. The present chapter concerns techniques and measuring systems rather than individual measuring devices. We also consider methods for generating surges of various forms in order to test the ability of power equipments to sustain such surges in practice.

17.2 Minimizing Problems of Interference

The subject of interference, or pickup, was discussed from a rather fundamental point of view in Chapter 8. Here we relate interference specifically to measuring circuits and discuss what steps can be taken to minimize the problem. In principle, the problem is simple enough; stray electric and magnetic fluxes from power circuits couple into the measuring circuits and induce spurious voltages. Difficulty often arises in determining just what constitutes the measuring circuit. Induced transient currents often take some unexpected routes.

The environment in which transient measurements have to be made is frequently quite hostile, in that the transient voltages and currents are high and often rapidly changing; the mutually induced voltages are therefore quite high. The situation is often aggravated by the fact that for safety in high-voltage circuits, it is difficult to keep measuring leads short. The most difficult conditions are those in which two or more measurements must be made simultaneously, especially if the measuring points are some distance apart. These points will be illustrated with a simple example.

Suppose we want to make a simultaneous determination of the inrush current and the transient voltage when a capacitor bank is energized. One possible arrangement is shown in Fig. 17.1. Current is measured by means

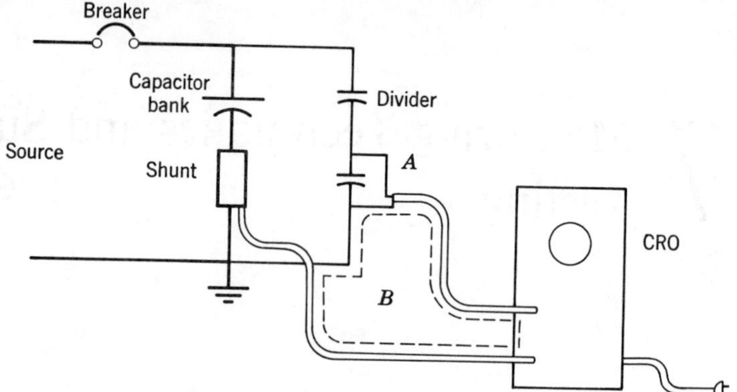

Fig. 17.1. Instrumentation to measure the transients that occur when a capacitor bank is energized.

of a shunt placed, for obvious reasons, on the ground side of the capacitors. To obtain high fidelity a coaxial shunt of the type described in Section 16.5 would be used. On the right is shown a potential divider for measuring the capacitor voltage. As a matter of course one would use coaxial cables to connect the shunt and divider to a dual beam oscilloscope where the phenomena could be recorded. In this way pickup in these connecting leads could be avoided.

But potential problems in this system are manifold. Care has probably been taken to avoid loops in the measuring circuit such as that shown at *A* on the lower end of the potential divider, where stray magnetic flux from the power circuit current could induce unwanted signals. However, all too often much bigger loops, such as indicated by the dotted line at *B*, are overlooked. Here a closed loop, formed by the sheaths of the signal cables and part of the power circuit, is completed through the chassis of the oscilloscope. Electromotive forces induced in this loop will cause circulating currents, which in turn will give rise to a resistive volt drop as these currents flow through the resistance of the sheaths. To whatever extent this occurs, it represents an error in the signal being recorded. It would be most serious in the case of the current measurement, where the signal level is likely to be low. This effect can be reduced by making the loop *B* itself as small as possible. This is accomplished by carrying along the voltage divider cable beside the power circuit and then beside the cable from the shunt, taping the two together if need be. If this is insufficient, a further reduction can be obtained by slipping doughnut-shaped ferrite cores over the individual signal cables, or by taking an individual core and looping the signal cable through it several times (1). These cores introduce considerable impedance into the loop circuit and dramatically reduce the currents that circulate therein.

Being ferrite, they remain effective up to quite a high frequency. They do not, however, attenuate the signals being measured since the signal circuits are wholly within the magnetic core and are not linked by it.

These steps are normally adequate to take care of this particular problem. In extreme cases it may be necessary to enclose both cables, together, in a grounded copper conduit. More discussion of this matter will be found in Section 17.5.

Looking further at the circuit, it is apparent that the oscilloscope receives its power through a line cord from a circuit which is probably grounded. Perhaps the only other ground is the power circuit ground, to which the oscilloscope is connected by way of the cable sheath. The new ground completes a circuit, conceivably of considerable extent, in which transient currents may be induced with the same deleterious effects as before. It is true that the oscilloscope power supply probably contains a transformer with no connection between primary and secondary circuits. However, there is considerable capacitance between the transformer windings, which offers a relatively low impedance to high-frequency currents. When problems of this kind are encountered it is again possible to reduce the magnitude of the induced current by placing ferrites around the line cord. A different solution is to supply the oscilloscope through a low-capacitance isolation transformer (Fig. 17.2). The low interwinding capacitance is achieved by physical spacing between the windings. This means that the leakage inductance is high, which

Fig. 17.2. Isolation transformer with low interwinding capacitance.

increases the regulation, a fact that may be a problem when the supply voltage is low.

For one reason or another it is not always possible to use a shunt; some form of current transformer (Section 16.5) must then be employed. A problem not unlike that just described can arise under these conditions. The potential of the conductor around which the CT is placed may fluctuate violently during transient disturbances. Capacitance between the power conductor and the CT winding can electrostatically couple corresponding voltage disturbances into the measuring circuit where they may distort or even swamp the true signal. To mitigate this particular problem a grounded static shield, perhaps made of metallic foil, should be interposed between the power conductor and the CT. This should not form a closed loop, for it would then have a confining effect on the magnetic field of the conductor current and therefore introduce an error in the CT signal.

In order to synchronize the transient event with the oscilloscope sweep a trip signal must be introduced into the oscilloscope trigger terminal at the appropriate instant. In the example being considered, this may well be derived from an auxiliary switch on the circuit-breaker or closing switch being used to energize the capacitor bank. Such trigger sources are discussed in Section 16.3. The point of interest here is that any such connection represents yet another antenna through which unwanted signals may be fed to the oscilloscope. This particular connection is different from others we have discussed in that it does not carry any detailed intelligence, only a single command, probably in the form of a step change of voltage. This allows a different solution to avoid the pickup problem: we can resort to the use of *fiber optics*. With this technique the signal is transmitted as a light pulse, down a "light pipe," rather than as an electrical impulse over a metallic conductor. This has the advantage that the light pipe, which comprises a bunch of fibers made of glass or some other transparent material, is insensitive to electric and magnetic pickup. The system is quite simple: the electrical impulse generated as a synchronizing signal is converted into a pulse of light at its point of origin by means of a light-emitting solid-state diode. The pulse of light travels down the light pipe to the oscilloscope, where it is converted back into an electrical signal by a light-sensitive transistor. The light is confined within the pipe by the refractive index of the fiber. Light that would attempt to escape is reflected from the pipe wall. The signal, when received, is amplified as needed and delivered to the oscilloscope circuit. The electronics for such an arrangement are relatively simple. Figure 17.3*a* shows circuit diagrams for a transmitter and receiver, while Fig. 17.3*b* is a photograph of an actual equipment. This would be perfectly adequate for triggering an oscilloscope. Where more output is needed, or where because of a long light pipe the signal is weak, amplification stages may have to be added.

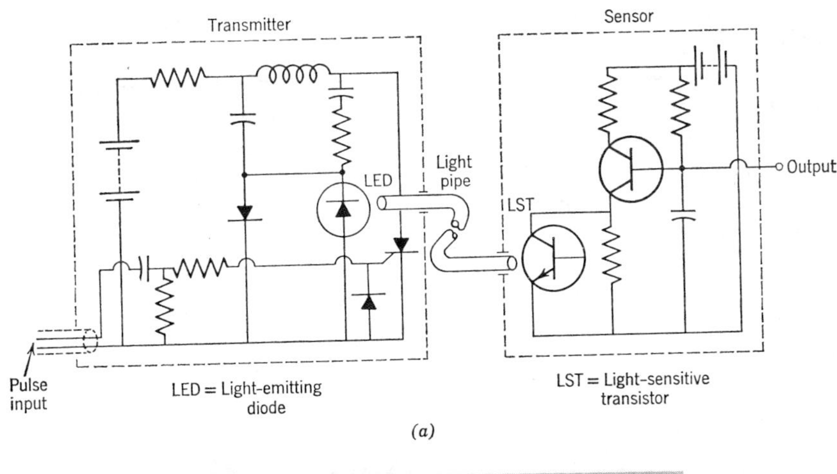

(a)

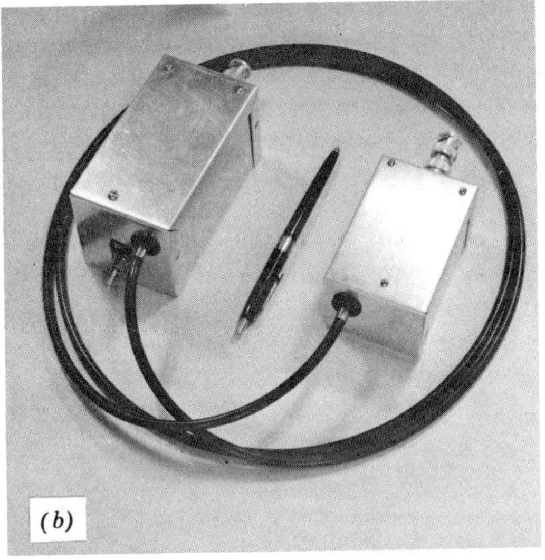

(b)

Fig. 17.3. Fiber optical system for transmitting intelligence. (*a*) Circuit diagram of transmitter and receiver. (*b*) Actual equipment.

Fiber optics are a great boon to the business of transient measurements, for besides the advantage just described, it is a means of providing a signal channel between points separated by a considerable potential difference. There are instances where it is necessary to place measuring equipment at high potential to attain voltage isolation. When using fiber optics, it is a simple matter to turn on an instrument, start a sweep, open a camera shutter, or what have you, from ground potential. Light pipes of ten feet or more are possible, bifurcations can be arranged, and, of course, the pipe can take a circuitous path since the light is entrapped within it by reflections from the walls.

The circuits within the oscilloscope are protected to a considerable extent by the metal case of the instrument, and the cathode ray tube itself usually has a mumetal shield around it in addition, to protect the electron beam from radiated interference. These measures notwithstanding, pickup still occurs sometimes where the environment is severe. On occasions, for example, voltages will be induced in the sweep circuit that may momentarily reverse the direction of the sweep and cause what is known as "back-writing." In extreme cases, it may be necessary to place all of the measuring equipment in a house shielded by a close copper mesh. This point will be touched on again in discussing high-frequency measurements in Section 17.5.

17.3 Differential Measurements

Sometimes it is necessary to measure a transient voltage between two points, both of which may be permanently, or momentarily, above ground potential. The possibility of doing this directly should not be ignored. To solve the problem in this way requires that the measuring equipment be floated above ground and be well insulated. Small battery-operated oscilloscopes are available; these have a very low power consumption and are well-suited for many jobs of this kind. On occasions the author has actually used a battery and inverter to power an oscilloscope. Figure 17.4 is a case in point. This shows the recovery voltages appearing across two interrupters in a circuit-breaker following a switching operation. The measuring points were at a potential to ground in excess of 100 kV.

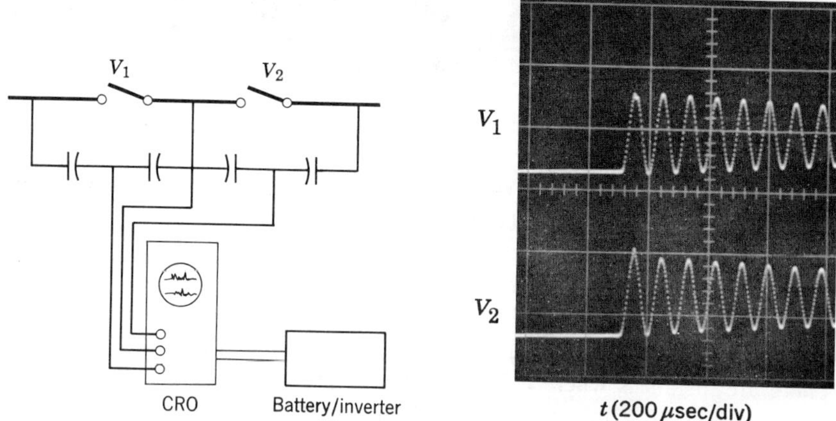

Fig. 17.4. Monitoring the recovery voltage across series interrupters of a circuit breaker by means of electrically isolated instrumentation.

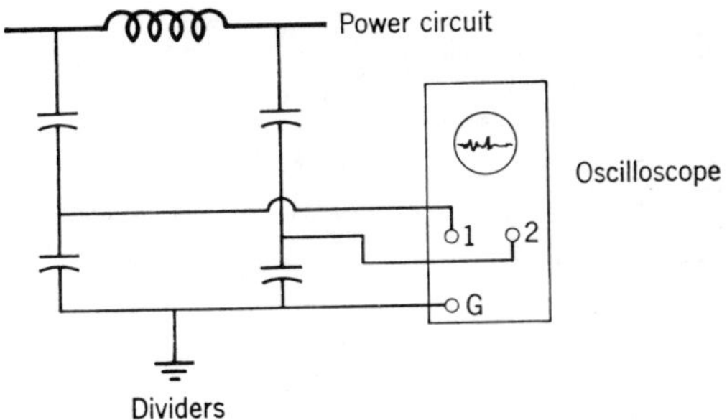

Fig. 17.5. Circuit arrangement for making a differential voltage measurement.

A gasoline engine driving a generator can also be operated in isolation to power measuring equipment.

A more usual approach to this problem is to make a differential measurement, that is, to measure the potential of both points to ground and take the difference between these measurements. The principle is illustrated in Fig. 17.5. The oscilloscope is equipped with a differential amplifier, which applies to the oscilloscope plates the difference in voltage between the two inputs. It is important to balance this carefully before making a measurement to assure that it rejects any *common mode* voltage, that is, any signal that is common to both inputs. This might be a good point to stress the importance of oscilloscope adjustments and calibrations in general. Amplifier and time base calibrations should be checked at frequent intervals to make certain the dial indications are correct. Probes should be checked for their compensation. Many oscilloscopes provide a variable-amplitude 1-kHz square wave for this purpose. This too should be compared from time to time with a laboratory standard.

17.4 Multichannel Sequence Timer

Often in the course of making transient measurements it is necessary to initiate a number of events at predetermined intervals. These range from low-energy functions, such as triggering the sweep of an oscilloscope, to opening or closing a circuit breaker. The basic requirement is to be able to time and synchronize a number of unrelated events in order to obtain accurate records of the transients produced. When transients occur, the oscilloscope spots must have started their sweeps, the camera shutters must be open, and the oscillograph films must be moving. All of the measuring equipment and the circuit proper may not be juxtaposed; different elements may be at different potentials.

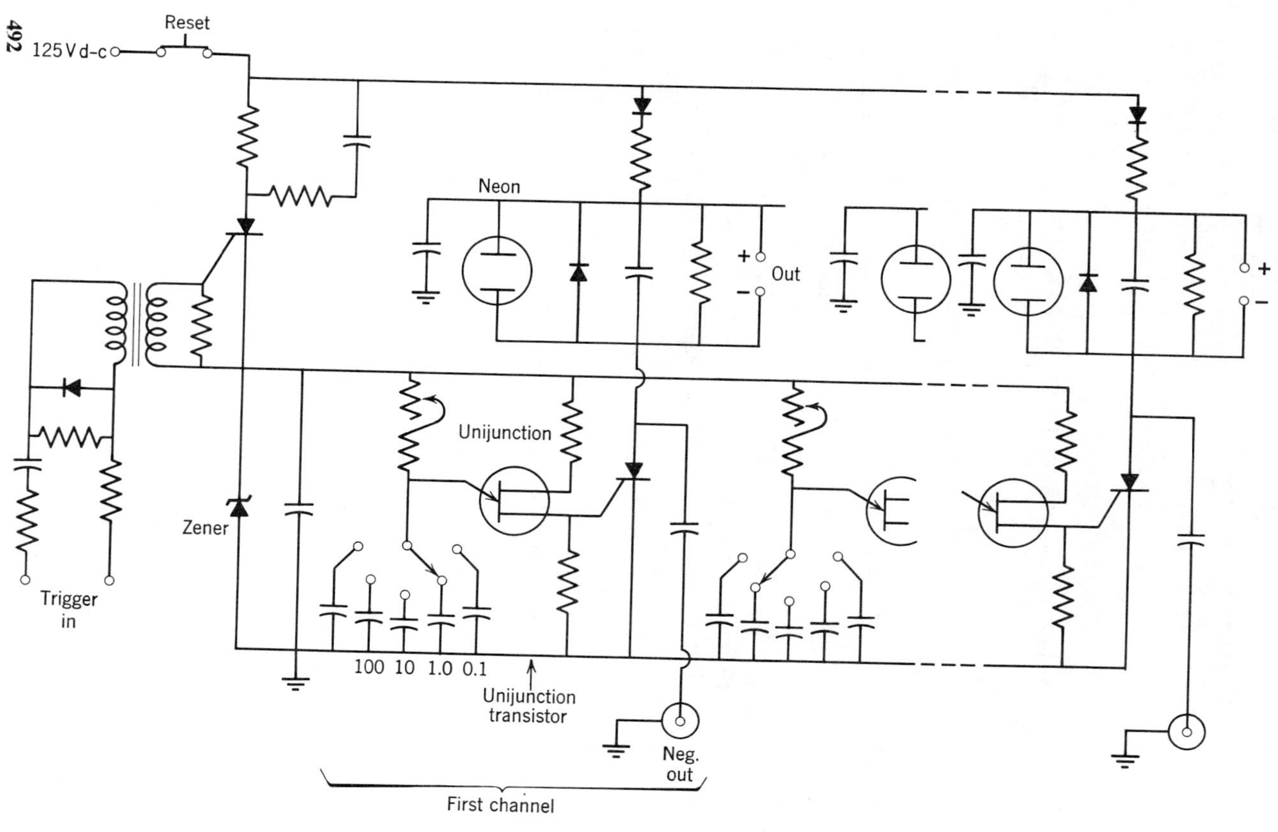

Fig. 17.6. Multichannel sequence timer.

To do this job some master sequence timer is required, which has several output channels. A circuit diagram for such a device is shown in Fig. 17.6. The stages are identical and it will be seen that each is capable of putting out a high- and a low-energy signal. The high-energy outputs can drive solenoids or relays, while the low-energy outputs can start electronic equipments.

Initiation (input) may be provided manually or can come from some independent circuit. Timing of events thereafter depends upon the *RC* time constants of the individual stages. These are adjustable in steps, with a vernier potentiameter for the range between the steps. The capacitors are each charged through their respective resistors; when the threshold level of the unijunction transistor is reached it turns on and gates its thyristor. The stages are decoupled, each having its own energy storage capacitor.

17.5 High-Frequency Transient Measurements

There are occasions in power systems work where, in studying the system itself, individual components, or control circuits, it is necessary to have equipment capable of measuring up to very high frequencies. This may be because such frequencies are excited as an oscillation, or more likely because only by such means can the initial rise time of a particular transient be resolved.

In Chapter 16 high-speed measuring equipment such as traveling-wave oscilloscopes were described, and the frequency response of potential dividers and amplifiers was discussed. However, to make the kind of measurements being presently considered, these devices must be used in combination to form a measuring system. The overall performance of such a system determines the fidelity of the measurement. It is also true that many of the problems we have discussed such as pickup in ground loops and the penetration of radiation into the oscilloscope become particularly severe when very high-frequency disturbances are being studied.

In this section we describe one or two approaches to the subject of high-frequency transient measurements, and try to point out some of the pitfalls. We are concerned with systems capable of measuring rise-times of a fraction of a microsecond. As mentioned above, one problem to be overcome is that of avoiding spurious responses due to signals radiated by the circuit under test. Another is the selection and arrangement of components to minimize the effect of strays. A third is to have adequate recording capability in the oscilloscope and camera.

One approach is the differentiation-integration technique described by Lord (1), which is well suited to situations where the available signal is high and considerable attenuation is desired. In this method the signal is first differentiated and is then reconstituted by integration. To go through such a procedure would not seem justifiable unless it helps to solve some of the problems described. In fact it does, and how it does will be described. But first the principle of the arrangement will be discussed.

If a voltage is applied to the *CR* circuit or the *RC* circuit of Fig. 17.7, the voltage across the *R* will be proportional to the current flowing through it, whereas the voltage across the capacitor will be proportional to the integral of that current. If in Fig. 17.7*a* the capacitor is made the dominant impedance, that is, at the frequency being considered, $1/\omega C_1 \gg R_1$, then the input voltage and the voltage across *C* will be practically in phase. The output voltage across *R* will therefore be practically in quadrature with the input voltage and therefore be proportional to its derivative. This is represented in the appended phasor diagram and what we have described is the simple differentiating circuit so familiar to the electronics engineer.

Similarly, referring to Fig. 17.7*b*, if $R_2 \gg 1/\omega C_2$, the situation is reversed, and the output is practically proportional to the integral of the input. This is the familiar integrating circuit.

If these circuits are put together in the manner shown in Fig. 17.8, where the output of one becomes the input of the other, it is apparent that the output will be proportional to the input but will be considerably attenuated. If the angles designated θ_1 and θ_2 in Fig. 17.7 are equal, the input and output

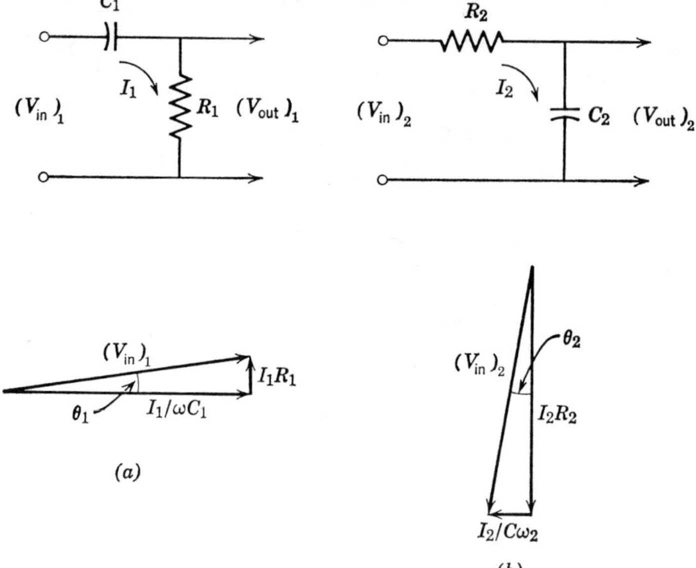

Fig. 17.7. Simple differentiating circuit (*a*) and integrating circuit (*b*).

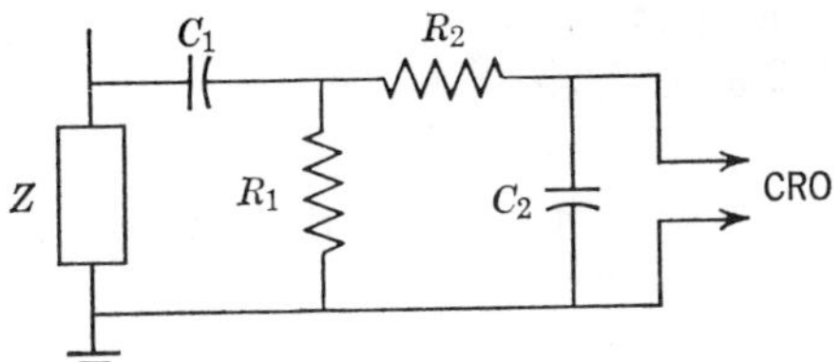

Fig. 17.8. Basic elements of the differentiator/integrator circuit.

voltages will be in phase or, more correctly, directly antiphase. This occurs when

$$R_1\omega C_1 = \frac{1}{R_2\omega C_2}$$

or

$$\omega = (R_1C_1R_2C_2)^{-\frac{1}{2}} \tag{17.5.1}$$

This situation will change slightly for a considerable band of frequencies on either side.

The actual ratio of output to input voltage of the C_1R_1 circuit at a frequency ω is

$$\left[\frac{V_{out}}{V_{in}}\right]_1 = \frac{R_1}{[(1/\omega C_1)^2 + R_1^{\,2}]^{\frac{1}{2}}} \tag{17.5.2}$$

which varies with frequency, increasing as frequency increases. Inasmuch as the first term in the denominator is much bigger than the second, according to the conditions we have presented, Eq. 17.5.2 can be approximated by

$$\left[\frac{V_{out}}{V_{in}}\right]_1 \approx R_1\omega C_1 \tag{17.5.3}$$

For the R_2C_2 circuit the ratio is

$$\left[\frac{V_{out}}{V_{in}}\right]_2 = \frac{1/\omega C_2}{[(1/\omega C_2)^2 + R_2^{\,2}]^{\frac{1}{2}}} \tag{17.5.4}$$

which for the prescribed conditions approximates to

$$\left[\frac{V_{out}}{V_{in}}\right]_2 \approx \frac{1}{R_2\omega C_2} \tag{17.5.5}$$

This decreases with frequency. When the C_1R_1 and R_2C_2 circuits are combined, the two effects compensate, and to a first approximation the overall voltage ratio is given by the product of Eqs. 17.5.3 and 17.5.5:

$$\left[\frac{V_{out}}{V_{in}}\right] \approx \frac{R_1C_1}{R_2C_2} \tag{17.5.6}$$

which is independent of frequency.

A more accurate idea of the overall response can be obtained by re-expressing Eqs. 17.5.2 and 17.5.4:

$$\frac{R_1}{[(1/\omega C_1)^2 + R_1^2]^{1/2}} = R_1\omega C_1[1 + (R_1\omega C_1)^2]^{-1/2}$$

and (17.5.7)

$$\frac{1/\omega C_2}{[R_2^2 + (1/\omega C_2)^2]^{1/2}} = \frac{1}{R_2\omega C_2}\left[1 + \frac{1}{(R_2\omega C_2)^2}\right]^{-1/2}$$

where the second term in the bracket is much smaller than the first in both instances. Taking the product of expressions (17.5.7) and expanding by the binomial,

$$\left[\frac{V_{\text{out}}}{V_{\text{in}}}\right] = \frac{R_1C_1}{R_2C_2}\Bigg\{[1 - \tfrac{1}{2}(R_1\omega C_1)^2 + \frac{3}{8}(R_1\omega C_1)^4 - \cdots]$$

$$\times\left[1 - \frac{1}{2}\left(\frac{1}{R_2\omega C_2}\right)^2 + \frac{3}{8}\left(\frac{1}{R_2\omega C_2}\right)^4 - \cdots\right]\Bigg\} \qquad (17.5.8)$$

Succeeding terms of this series rapidly diminish as long as our assumption that $R_1 \ll 1/\omega C_1$ and $R_2 \gg 1/\omega C_2$ holds. We can obtain an idea of the frequency at which the voltage ratio is a maximum by considering when the most significant terms in the brackets become a minimum. Consider the second term in each case:

$$y = \tfrac{1}{2}(R_1\omega C_1)^2 + \frac{1}{2}\left(\frac{1}{R_2\omega C_2}\right)^2$$

This will be a minimum when $dy/d\omega = 0$:

$$\frac{dy}{d\omega} = R_1^2\omega C_1^2 - \frac{1}{R_2^2\omega^3 C_2^2}$$

if $dy/d\omega = 0$, then

$$R_1^2C_1^2\omega = \frac{1}{R_2^2C_2^2\omega^3}$$

$$\omega^2 = \frac{1}{R_1C_1R_2C_2} \qquad (17.5.9)$$

which corresponds to the case given in Eq. 17.5.1 for no phase angle error.

For frequencies greater or less than this value the differentiator/integrator will produce more attenuation, or, viewed as a potential divider, its ratio will increase, but over a wide range the difference will be small, as will be the phase angle error. It is possible to construct a frequency response curve along the lines of that shown in Fig. 16.1 for a galvanometer, but of course for likely values of the constants the frequency values will be very much higher. It can be shown that the frequencies for the upper and lower "roll-off points," where the response is 3 dB or less, are defined by

$$\left.\begin{aligned} \omega_1 &= \frac{1}{R_1C_1} \\ \omega_2 &= \frac{1}{R_2C_2} \end{aligned}\right\} \qquad (17.5.10)$$

These, together with the attenuation or voltage ratio given in Eq. 17.5.6,

$$\left[\frac{V_{\text{out}}}{V_{\text{in}}}\right] = \frac{R_1C_1}{R_2C_2}$$

describe the performance of the differentiator/integrator system. Thus, with a fixed value of R_1 and C_1, extending the low-frequency response of the system can be done only at the expense of the system gain. The desired system gain is determined by the magnitude of the signals to be recorded and by the input requirements of the oscilloscope.

We now turn to the advantages that accrue from having a potential divider of this form. Later, a numerical example will be given.

A practical embodiment of such a system due to Martzloff (2) is illustrated in Fig. 17.9. The differentiator consists of a high-voltage capacitor C_1 and an arbitrary length of 50-Ω cable terminated into a 50-Ω load. The fact that this cable length can be arbitrary makes possible two desirable features: the oscilloscope can be placed at the end of the long cable, far away from the strong radiation caused by the circuit under test, and a cable filter of ferrite cores, of the type described in Section 17.2, which requires additional cable length, can be inserted to block circulation of ground currents. The effectiveness of such filters is well documented by Lord (1). In contrast, most commercial high-voltage probes are limited to 10–12 ft and have limited built-in

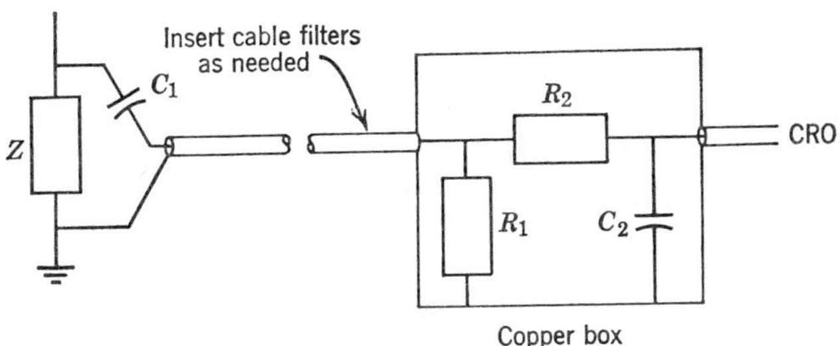

Fig. 17.9. Physical layout of differentiator/integrator.

cable filters. Martzloff (2) has reported on the spurious effects of this limited filtering capability.

To illustrate the potentialities of this approach, consider the following numerical example. Suppose

$R_1 = 50\ \Omega$ (a typical surge impedance for a cable likely to be used)

$C_1 = 5$ pF (for minimum circuit loading)

then

$$R_1C_1 = 2.5 \times 10^{-10}$$

From Eq. 17.5.10,

$$\omega_1 = 4 \times 10^9$$

$$f_1 = \frac{\omega_1}{2\pi} = 640\ \text{MHz}$$

For a desired attenuation of 5000:1, Eq. 17.5.6 indicates that

$$R_2C_2 = 5000R_1C_1$$
$$= 1.25 \times 10^{-6}$$

then, from Eq. 17.5.10,

$$\omega_2 = 8 \times 10^5 \quad \text{or} \quad f_2 = 0.13\ \text{MHz}$$

The point for maximum gain (minimum attenuation) and zero phase error is given by Eq. 17.5.9:

$$\omega = \left[\frac{1}{2.5 \times 10^{-10} \times 1.25 \times 10^{-6}}\right]^{1/2}$$
$$= 5.66 \times 10^7 \quad \text{or} \quad f = 9\text{MHz}$$

In this example, the high upper-frequency limit, f_1, resulted from the desire to have minimum loading of the test circuit, while the relatively high low-frequency limit, f_2, resulted from the desire to produce the mild attenuation. A lower value for f_2 could be obtained by more attenuation or by a larger coupling capacitor C_1, if the resultant circuit loading is acceptable.

Precautions must be taken to keep the input connection inductance low (capacitor C_1 and leads) so that these will not resonate at frequencies within the pass-band. For instance, assuming $C_1 = 5$ pF, resonance occurs for an input inductance L_1 at ω_r such that

$$\omega_r^2 = \frac{1}{L_1C_1} \qquad (17.5.11)$$

For systems with an upper frequency limit of 100 MHz ($\omega_1 = 6.28 \times 10^8$ rad/sec), ω_r should be 2×10^9 or higher, whence

$$4 \times 10^{18} = \frac{1}{L_1 C_1}$$

or

$$L_1 = \frac{1}{4 \times 10^{18} \times 5 \times 10^{-12}} = 0.05\,\mu\text{H}$$

which represents a very small loop inductance and points out the need for very close coupling between the probe and the circuit.

Circuits of this kind can be refined by the addition of further elements to provide flat response over the widest possible band. An example devised by Martzloff (2) is shown in Fig. 17.10. The complete circuit is enclosed in a copper box, starting with the filter comprising 20 turns of coaxial cable around a ferrite core. The major integration is performed by the resistor R_2 and the capacitor C_2. Additional compensation is obtained by the R_3/C_3 integrator, the output being fed through the R_4/50-Ω divider, where the 50-Ω resistance also acts as terminating impedance for the output cable at the oscilloscope input. This terminating resistance is added so that the box can be connected to any oscilloscope through a short cable. As this produces additional attenuation (the $R_3 + R_4/50\ \Omega$ divider) it becomes necessary to increase the gain of the integrator so that R_2C_2 was made smaller. The corresponding increase of the low-frequency limit was then compensated by introducing the R_1C_4 integrator instead of a single resistor R_1. Proper matching of these elements is achieved by actual response curve plotting.

In this regard it must be remembered that the oscilloscope and its amplifier, where one is used, are all part of the complete measuring system. The response of the oscilloscope amplifier must be at least as good as the attenuator equipment. Figure 17.11 shows the complete response of the combination of the equipment shown in Fig. 17.10 and a modified Tektronix type 544 oscilloscope with a 1A1 preamplifier. The amplifier settings are as indicated.

For work of this kind a sharp, high-intensity spot is required on the oscilloscope if the desired writing speed is to be obtained. As explained in Section 16.3, this requires an oscilloscope with a high accelerating voltage. It is possible to convert some commercially available oscilloscopes for operation at higher voltage to improve their writing speed. Figure 17.12 shows the quality that can be achieved by such an arrangement with a differentiater/integrater attenuation system like that in Fig. 17.10. This trace shows that a single transient with a 25 nsec front can be clearly resolved with such a combination.

A different approach from that just described was used by Pollard (3) for making high-speed transient measurements. The objective was to study the

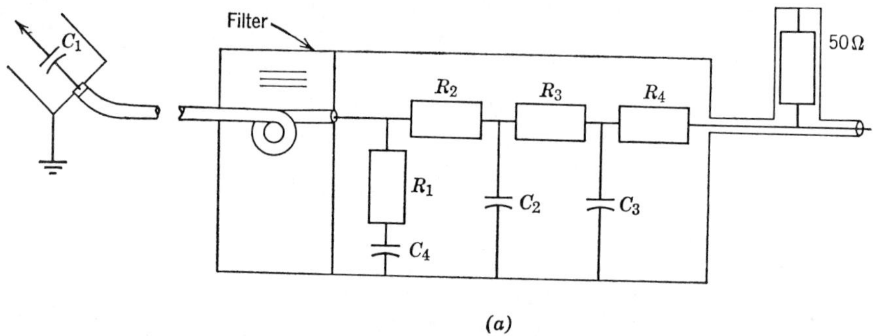

(a)

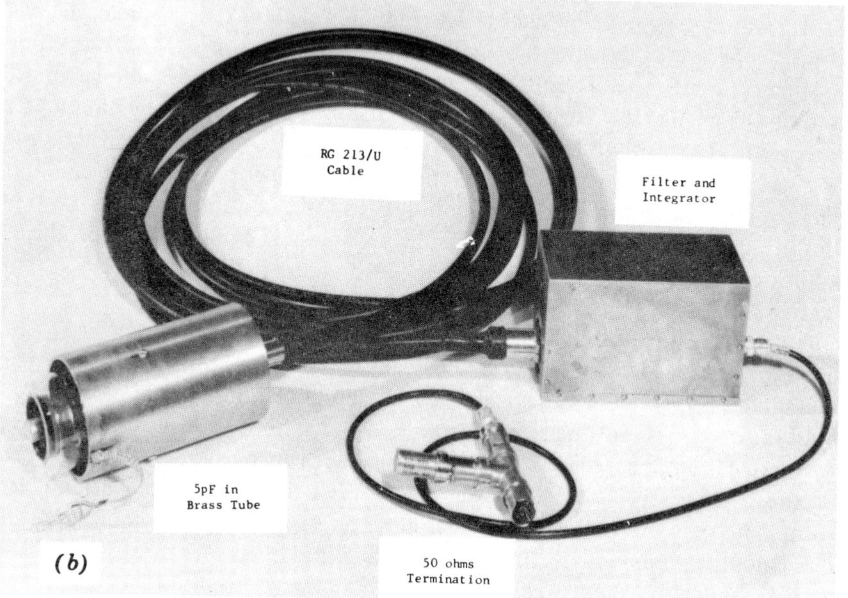

(b)

Fig. 17.10. Wide band attenuator. (*a*) Circuit diagram. (*b*) Physical layout.

response of solid-state power components (diodes and thyristors) to the application of steep wavefronts. The equipment undoubtedly has much wider application. The problems were much the same as before. The design sought compactness to minimize strays and together with shielding, reduce unwanted pickup.

The waveforms were ultimately displayed on a Tektronix type 519 oscilloscope. This has a traveling wave deflection system (see Section 16.3) with a 125-Ω characteristic impedance. Deflection sensitivity is about 9.5 V/cm. The rise time is about 3×10^{-10} sec and the fastest sweep is 2 nsec/cm.

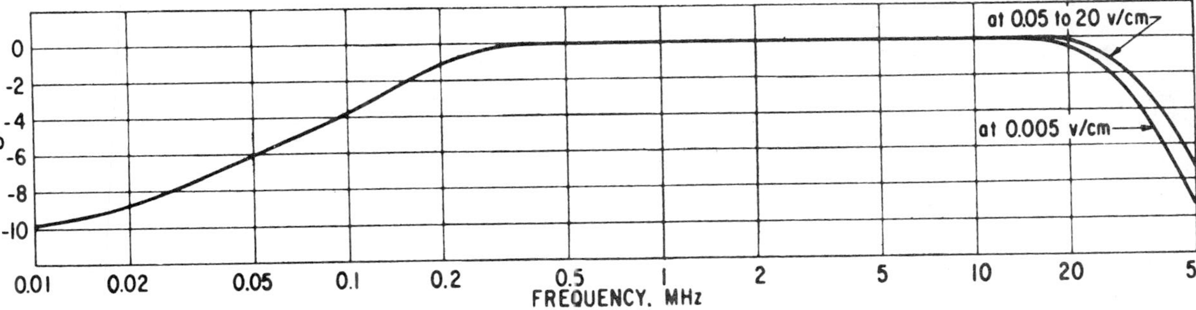

Fig. 17.11. Response of complete system—differentiator, integrator, 1A1 amplifier with settings as shown and Tektronix modified type 544 oscilloscope.

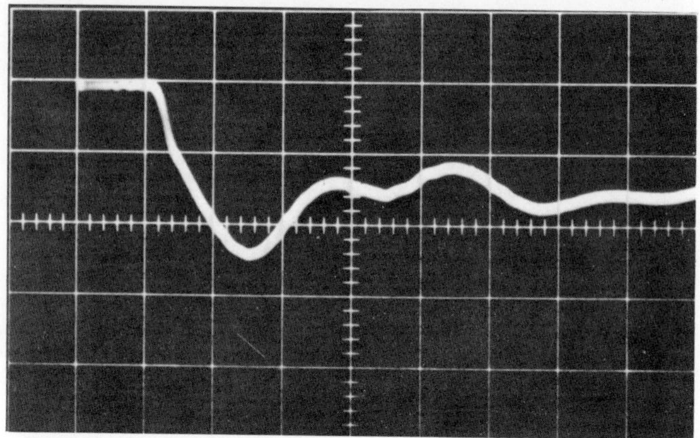

Fig. 17.12. Resolution attainable with the equipment of Fig. 17.10 and a modified type 544 Tektronix oscilloscope; 0.02 μsec per major division. (Courtesy F. D. Martzloff.)

Since the voltages involved in this study were of the order of a few kilovolts, a reduction for measuring purposes of 250:1 was appropriate. This was accomplished by building a 25:1 resistance voltage divider and 10:1 attenuator. These were made so compact that they could be all but contained within the cable fittings. Both units were designed for use either individually or in cascade. Accordingly, the 25:1 probe has an output impedance of 125 Ω, and the 10:1 attenuator is of symmetrical design having a 125-Ω input impedance when its output is loaded with 125 Ω. The probe and attenuator circuit diagrams are shown in Fig. 17.13, while Figs. 17.14 and 17.15 show photographs of the physical layout. Both units were readily compensated by proper shaping of the small copper shields. The efficacy of this is portrayed in the oscillograms of their responses accompanying each illustration.

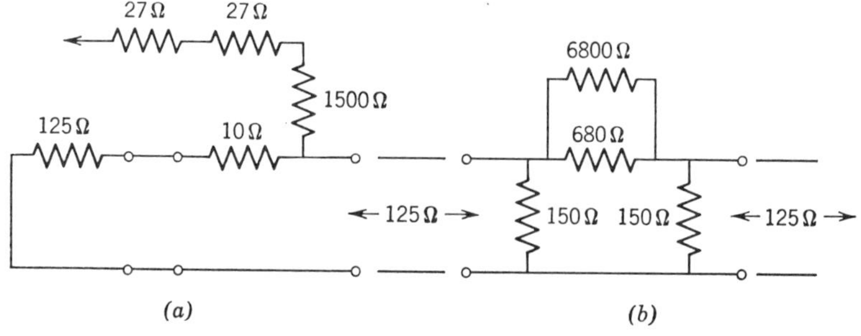

Fig. 17.13. (*a*) 25:1 divider. (*b*) 10.1 attenuator.

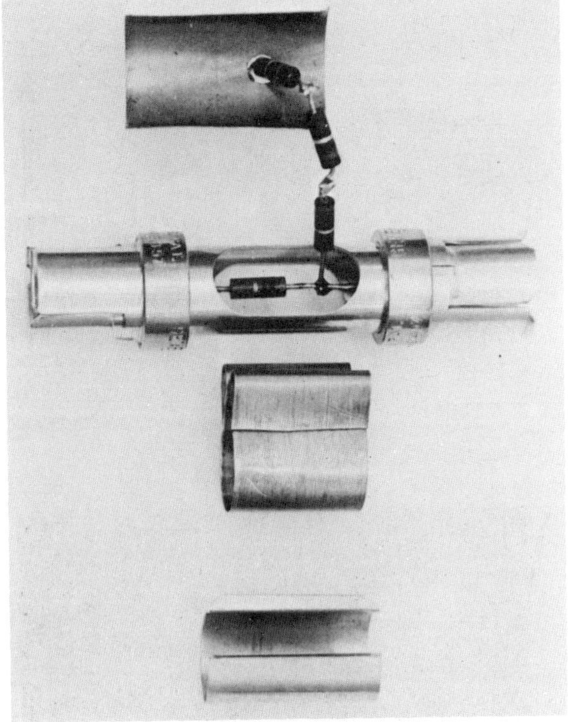

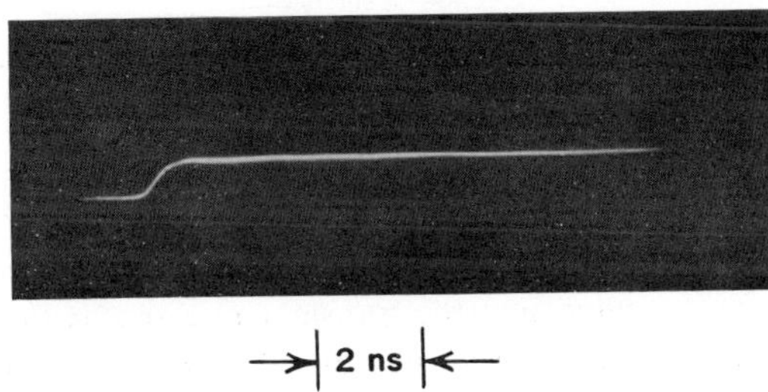

Fig. 17.14. 25:1 voltage divider and its response to a voltage step.

17.6 Measuring the Transient Response of a System

Very early in this book it was pointed out that when an electric circuit is shocked or otherwise disturbed, it responds in its own peculiar way, exhibiting its own particular transient characteristics. Its natural frequencies

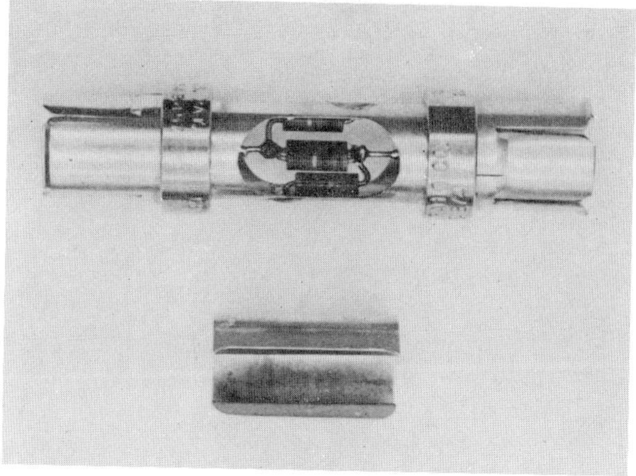

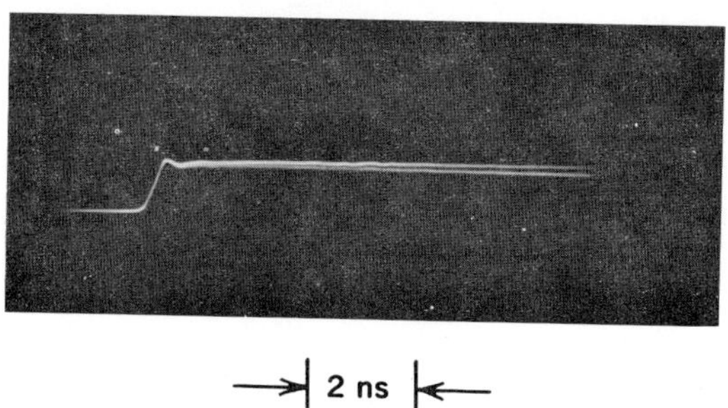

Fig. 17.15. 10:1 attenuator and its response to a voltage step.

are revealed, for example, and the degree to which these are damped will be apparent. The way to determine the transient response is therefore to shock the circuit or system in some way. Techniques for doing this will be the subject of this section and the next.

There are many different ways in which a system can be shocked. We can suddenly apply a voltage to a circuit, or suddenly inject a current into it. This would be the equivalent of closing or opening a switch. This will be discussed further in Section 17.7. If in fact, the ultimate objective is to determine what happens when a particular switch or circuit breaker is closed or opened, it is possible, at least in theory, to close or open it to find out. However, if the equipment involved is part of a functioning power system, such action may not be convenient. If the purpose of the investigation is to study what happens when a short circuit is thrown on a power circuit, direct testing, by

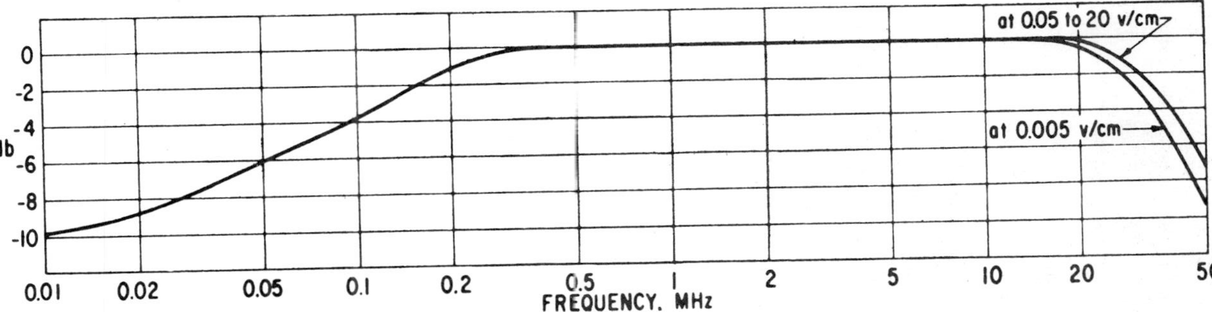

Fig. 17.11. Response of complete system—differentiator, integrator, 1A1 amplifier with settings as shown and Tektronix modified type 544 oscilloscope.

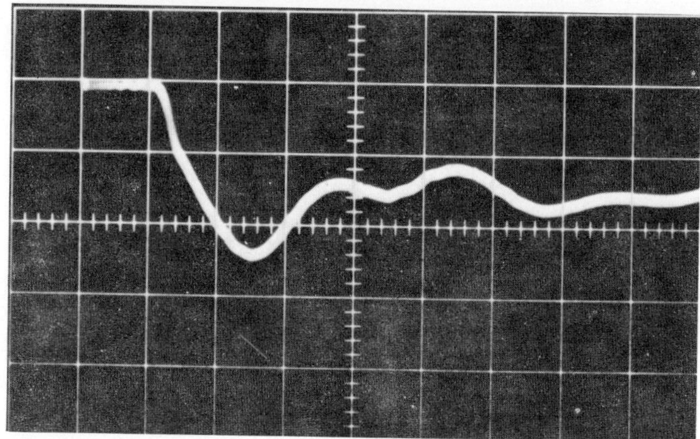

Fig. 17.12. Resolution attainable with the equipment of Fig. 17.10 and a modified type 544 Tektronix oscilloscope; 0.02 μsec per major division. (Courtesy F. D. Martzloff.)

Since the voltages involved in this study were of the order of a few kilovolts, a reduction for measuring purposes of 250:1 was appropriate. This was accomplished by building a 25:1 resistance voltage divider and 10:1 attenuator. These were made so compact that they could be all but contained within the cable fittings. Both units were designed for use either individually or in cascade. Accordingly, the 25:1 probe has an output impedance of 125 Ω, and the 10:1 attenuator is of symmetrical design having a 125-Ω input impedance when its output is loaded with 125 Ω. The probe and attenuator circuit diagrams are shown in Fig. 17.13, while Figs. 17.14 and 17.15 show photographs of the physical layout. Both units were readily compensated by proper shaping of the small copper shields. The efficacy of this is portrayed in the oscillograms of their responses accompanying each illustration.

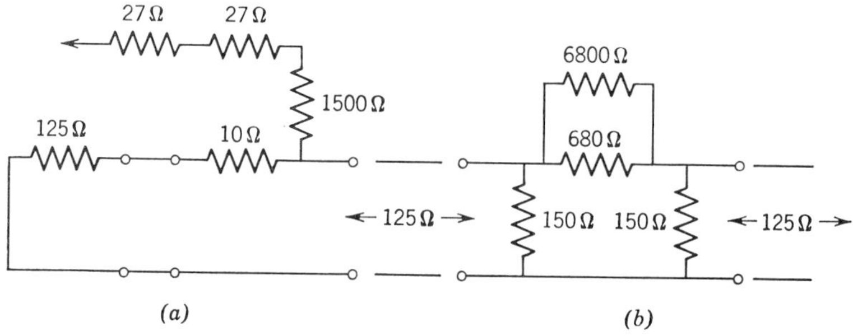

Fig. 17.13. (*a*) 25:1 divider. (*b*) 10.1 attenuator.

For systems with an upper frequency limit of 100 MHz ($\omega_1 = 6.28 \times 10^8$ rad/sec), ω_r should be 2×10^9 or higher, whence

$$4 \times 10^{18} = \frac{1}{L_1 C_1}$$

or

$$L_1 = \frac{1}{4 \times 10^{18} \times 5 \times 10^{-12}} = 0.05\,\mu\text{H}$$

which represents a very small loop inductance and points out the need for very close coupling between the probe and the circuit.

Circuits of this kind can be refined by the addition of further elements to provide flat response over the widest possible band. An example devised by Martzloff (2) is shown in Fig. 17.10. The complete circuit is enclosed in a copper box, starting with the filter comprising 20 turns of coaxial cable around a ferrite core. The major integration is performed by the resistor R_2 and the capacitor C_2. Additional compensation is obtained by the R_3/C_3 integrator, the output being fed through the R_4/50-Ω divider, where the 50-Ω resistance also acts as terminating impedance for the output cable at the oscilloscope input. This terminating resistance is added so that the box can be connected to any oscilloscope through a short cable. As this produces additional attenuation (the $R_3 + R_4/50\ \Omega$ divider) it becomes necessary to increase the gain of the integrator so that R_2C_2 was made smaller. The corresponding increase of the low-frequency limit was then compensated by introducing the R_1C_4 integrator instead of a single resistor R_1. Proper matching of these elements is achieved by actual response curve plotting.

In this regard it must be remembered that the oscilloscope and its amplifier, where one is used, are all part of the complete measuring system. The response of the oscilloscope amplifier must be at least as good as the attenuator equipment. Figure 17.11 shows the complete response of the combination of the equipment shown in Fig. 17.10 and a modified Tektronix type 544 oscilloscope with a 1A1 preamplifier. The amplifier settings are as indicated.

For work of this kind a sharp, high-intensity spot is required on the oscilloscope if the desired writing speed is to be obtained. As explained in Section 16.3, this requires an oscilloscope with a high accelerating voltage. It is possible to convert some commercially available oscilloscopes for operation at higher voltage to improve their writing speed. Figure 17.12 shows the quality that can be achieved by such an arrangement with a differentiater/integrater attenuation system like that in Fig. 17.10. This trace shows that a single transient with a 25 nsec front can be clearly resolved with such a combination.

A different approach from that just described was used by Pollard (3) for making high-speed transient measurements. The objective was to study the

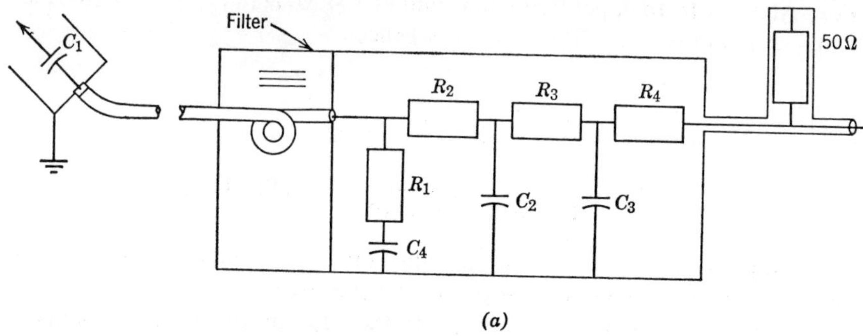

(a)

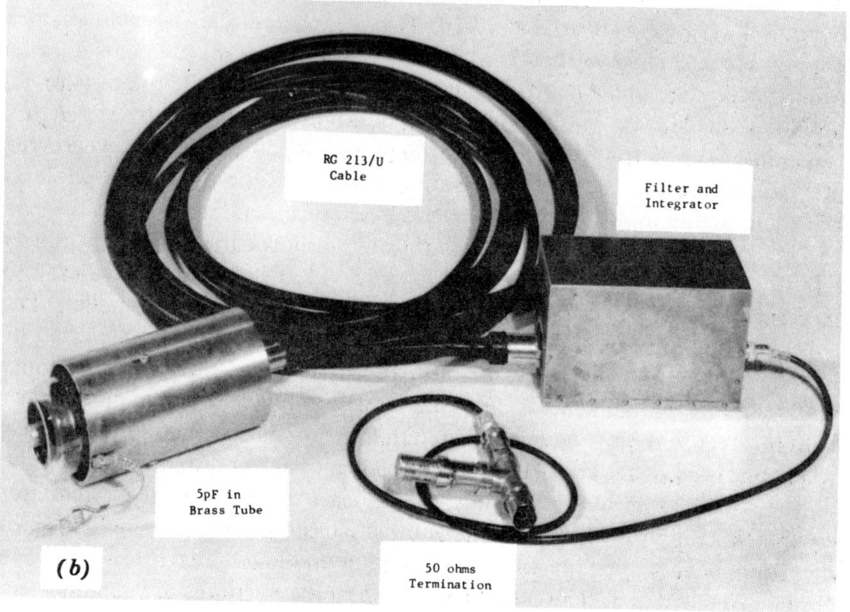

Fig. 17.10. Wide band attenuator. (*a*) Circuit diagram. (*b*) Physical layout.

response of solid-state power components (diodes and thyristors) to the application of steep wavefronts. The equipment undoubtedly has much wider application. The problems were much the same as before. The design sought compactness to minimize strays and together with shielding, reduce unwanted pickup.

The waveforms were ultimately displayed on a Tektronix type 519 oscilloscope. This has a traveling wave deflection system (see Section 16.3) with a 125-Ω characteristic impedance. Deflection sensitivity is about 9.5 V/cm. The rise time is about 3×10^{-10} sec and the fastest sweep is 2 nsec/cm.

phase angle error. It is possible to construct a frequency response curve along the lines of that shown in Fig. 16.1 for a galvanometer, but of course for likely values of the constants the frequency values will be very much higher. It can be shown that the frequencies for the upper and lower "roll-off points," where the response is 3 dB or less, are defined by

$$\left.\begin{aligned} \omega_1 &= \frac{1}{R_1C_1} \\ \omega_2 &= \frac{1}{R_2C_2} \end{aligned}\right\} \tag{17.5.10}$$

These, together with the attenuation or voltage ratio given in Eq. 17.5.6,

$$\left[\frac{V_{\text{out}}}{V_{\text{in}}}\right] = \frac{R_1C_1}{R_2C_2}$$

describe the performance of the differentiator/integrator system. Thus, with a fixed value of R_1 and C_1, extending the low-frequency response of the system can be done only at the expense of the system gain. The desired system gain is determined by the magnitude of the signals to be recorded and by the input requirements of the oscilloscope.

We now turn to the advantages that accrue from having a potential divider of this form. Later, a numerical example will be given.

A practical embodiment of such a system due to Martzloff (2) is illustrated in Fig. 17.9. The differentiator consists of a high-voltage capacitor C_1 and an arbitrary length of 50-Ω cable terminated into a 50-Ω load. The fact that this cable length can be arbitrary makes possible two desirable features: the oscilloscope can be placed at the end of the long cable, far away from the strong radiation caused by the circuit under test, and a cable filter of ferrite cores, of the type described in Section 17.2, which requires additional cable length, can be inserted to block circulation of ground currents. The effectiveness of such filters is well documented by Lord (1). In contrast, most commercial high-voltage probes are limited to 10–12 ft and have limited built-in

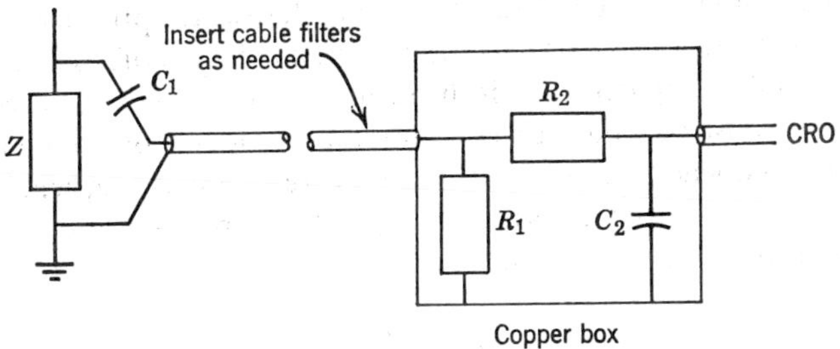

Fig. 17.9. Physical layout of differentiator/integrator.

cable filters. Martzloff (2) has reported on the spurious effects of this limited filtering capability.

To illustrate the potentialities of this approach, consider the following numerical example. Suppose

$R_1 = 50\ \Omega$ (a typical surge impedance for a cable likely to be used)

$C_1 = 5$ pF (for minimum circuit loading)

then

$$R_1C_1 = 2.5 \times 10^{-10}$$

From Eq. 17.5.10,

$$\omega_1 = 4 \times 10^9$$

$$f_1 = \frac{\omega_1}{2\pi} = 640\ \text{MHz}$$

For a desired attenuation of 5000:1, Eq. 17.5.6 indicates that

$$R_2C_2 = 5000R_1C_1$$

$$= 1.25 \times 10^{-6}$$

then, from Eq. 17.5.10,

$$\omega_2 = 8 \times 10^5 \quad \text{or} \quad f_2 = 0.13\ \text{MHz}$$

The point for maximum gain (minimum attenuation) and zero phase error is given by Eq. 17.5.9:

$$\omega = \left[\frac{1}{2.5 \times 10^{-10} \times 1.25 \times 10^{-6}}\right]^{1/2}$$

$$= 5.66 \times 10^7 \quad \text{or} \quad f = 9\text{MHz}$$

In this example, the high upper-frequency limit, f_1, resulted from the desire to have minimum loading of the test circuit, while the relatively high low-frequency limit, f_2, resulted from the desire to produce the mild attenuation. A lower value for f_2 could be obtained by more attenuation or by a larger coupling capacitor C_1, if the resultant circuit loading is acceptable.

Precautions must be taken to keep the input connection inductance low (capacitor C_1 and leads) so that these will not resonate at frequencies within the pass-band. For instance, assuming $C_1 = 5$ pF, resonance occurs for an input inductance L_1 at ω_r such that

$$\omega_r^2 = \frac{1}{L_1C_1} \tag{17.5.11}$$

of Voltage in Dielectric Tests (8). The front of the wave, t_f, is taken as 1.6 × 30% and 90% points on the wavefront. The tail is given by the time from the virtual origin, shown as point 0_1 in Fig. 17.20*b*, and the half peak value point on the back of the wave. If we call this t_t, it is usual to express an impulse wave in terms of $t_f \times t_t$. The accepted standard wave is the 1.2 × 50 wave, which indicates that $t_f = 1.2\ \mu\text{sec}$ and $t_t = 50\ \mu\text{sec}$. Surges of other proportions, such as 5 × 20, are called for from time to time and are used in impulse testing.

Reference has already been made to the chopped wave. Tests with such a wave are made by shunting the test object with a rod gap which flashes over when the impulse is applied. The resulting waveform may look like Fig.

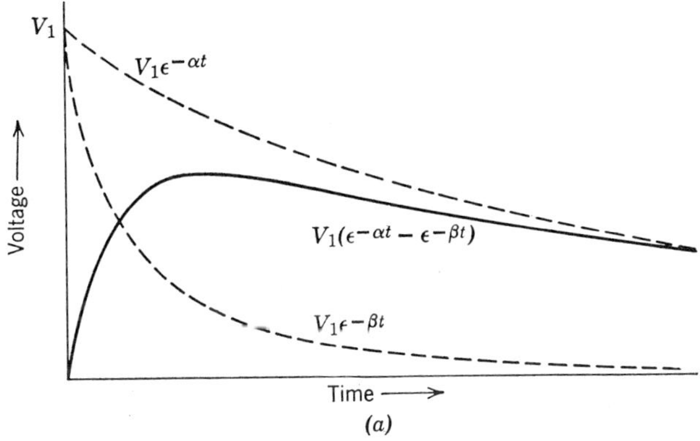

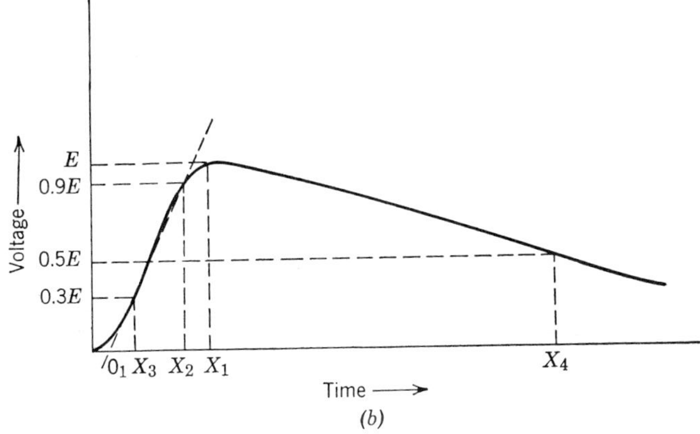

Fig. 17.20. (*a*) Double exponential wave, $V = V(\epsilon^{-\alpha t} - \epsilon^{-\beta t})$. (*b*) Typical surge waveform impulse generator.

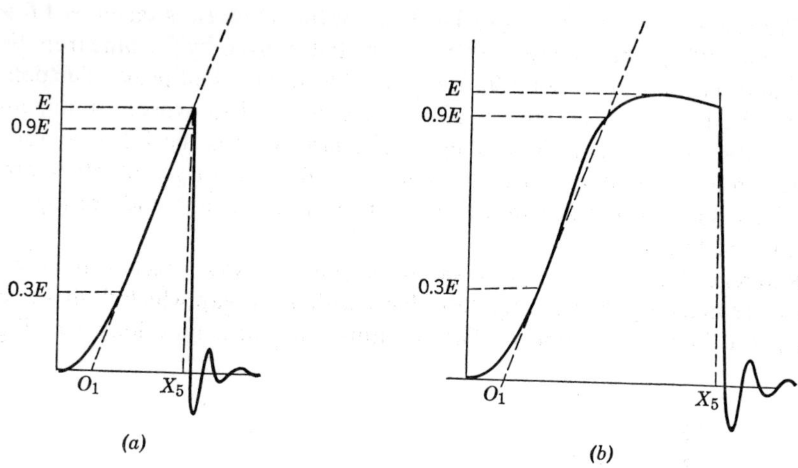

Fig. 17.21. Chopped waves. (*a*) Front of wave flashover. (*b*) Flashover on tail.

17.21*a* or 17.21*b*, depending on whether the chop occurs on the front or the tail. The former is frequently referred to as a "front-of-wave" test. Voltage is applied at a very fast rate (as measured by the slope of the 30–90% line), which may be as high as 1 MV/μsec. The gap is set to limit the voltage to a predetermined value.

Whereas reference (8) describes the general principles of dielectric testing and supplies many details of measurements, the precise tests and methods to be applied to individual types of equipment are spelled out in Standards appropriate to the apparatus concerned. For example, impulse testing of transformers is prescribed in ASA C57 (9).

The double exponential form of the full impulse wave can be generated, in principle, by a relatively simple circuit (Fig. 17.22). Here C_1 is charged from a source (not indicated) to a level which causes the gap G to flash over; C_2 then charges from C_1 through R_1. Simultaneously, but more slowly, both capacitors discharge through R_2. Alternatively, G could be a triggered spark

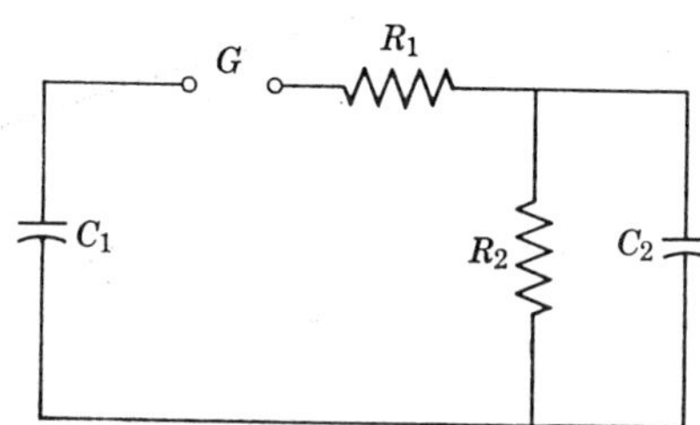

Fig. 17.22. Rudimentary single-impulse generator.

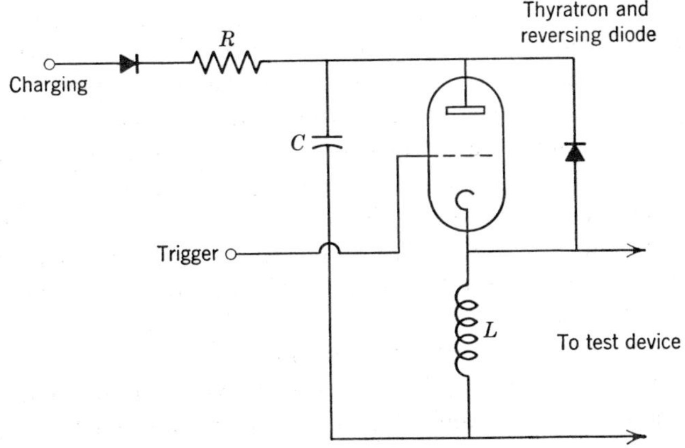

Fig. 17.19. Simple recurrent surge generator.

With each signal applied to the grid of the thyratron a triggering pulse is sent to the CRO to initiate its sweep, synchronizing it with the signal to be measured.

The potential of any point on the winding can be observed, or, as indicated in Fig. 17.18, the interturn potential can be measured, providing taps are available. From data of this kind it is possible to construct curves of the type shown in Fig 11.5. As explained in Chapter 11, the voltage gradient is often very nonuniform. It is important with this kind of measurement that the impedance of the measuring circuit be very high. A relatively small amount of capacitance will upset the natural capacitance of the system, on which the voltage distribution depends, thereby vitiating the results.

The use of cable to form pulses is very satisfactory except that for anything but a short pulse, it takes a lot of cable. A simple, but effective alternative is shown in Fig. 17.19. This surge generator comprises a capacitor, an inductor, and a thyratron or a thyristor. When the thyristor is gated, the capacitor voltage appears across the inductor and therefore across the load. The rate of rise of the surge front depends upon the stray capacitance, but it can be quite steep. The general shape of the voltage surge is a cosine wave. If the period of the LC circuit is long compared with the period of interest in the measurement, the beginning of a cosine wave is a good approximation to a step function.

Where a fair amount of work of this kind is to be done, the surge generator can be incorporated into the oscilloscope to form one piece of relatively portable equipment.

17.8 Impulse Testing

In the course of its operational life a piece of power system equipment will be subjected to overvoltages from a number of different causes. It is important that it be able to sustain these overvoltages without being damaged. In order to establish the adequacy of the equipment in this regard it must be tested in the design stage and perhaps as a routine test before shipping, by applying to it voltages of the appropriate amplitude and waveshape. In this section we consider what forms these voltages should take and how they are generated and applied. A general treatise on surge testing is given by Hawley (7).

The specifications for power equipment almost invariably call for some minimum capability for the insulation system as far as this matter is concerned. It is essential that there be some uniformity in the manner in which this is specified. Standards committees have addressed themselves to this subject and standards do exist for this purpose.

We will start by considering the kind of overvoltages that apparatus such as transformers, reactors, and generators can be subjected to and then consider how they can be reproduced.

The most frequent surges are undoubtedly switching surges but, potentially more dangerous because of their magnitude, are surges due to lightning. The switching surge is usually oscillatory and the period can vary widely. In general the steepness of the front on such a surge is less than that of a typical lightning surge and it will be recalled from Chapter 11 that this is an important consideration in the uniformity or nonuniformity of the surge voltage distribution within a piece of equipment. In Chapter 10 we indicated that more is being continually learned about lightning. The waveforms of lightning surges can vary quite significantly. Finally, it is recognized that waveforms are often modified by protective equipment, as when an arrester operates, or a rod gap flashes over. This latter event cuts off the surge waveform very abruptly as if an extremely steep-fronted surge of opposite polarity were suddenly superimposed on the initial voltage wave, with all the attendant effects of such a rapid change of voltage.

For many years the standard impulse wave has been of double exponential form, that is, it can be represented by an equation of the form

$$V(t) = V_1(\epsilon^{-\alpha t} - \epsilon^{-\beta t}) \tag{17.8.1}$$

The waveshape of this voltage can be understood from Fig. 17.20*a*. It will be seen that the steepness of the front depends largely on the coefficient β, whereas the length of the tail is primarily dictated by the coefficient α. In fact, such waves are defined in terms of the duration of their front and the time to half value in their tail. Because of practical difficulties, notably stray capacitance, it is not possible to produce a voltage waveform of precisely that given in Eq. 17.8.1; actual impulse waves are more like Fig. 17.20*b* and are specified in accordance with the notation shown in that figure. American practice has been revised to be more in line with internationally accepted standards; details can be found in the American Standard for *Measurement*

17.7 The Recurrent Surge Technique

Examples of this technique have been presented at several points in the text, notably in Chapter 13. The principle is simple and very effective and the applications are manifold. It is well-suited, for example, for measuring the transient response of simple or complex circuits or for studying the voltage distribution down a transformer winding under surge conditions.

The basic idea is described in a paper by White and Nethercot (5) and in a discussion thereon by Wilkinson (6). The circuit under study is shocked as proposed in Section 17.6, but in this instance the shock is administered repeatedly. The sweep of the oscilloscope being used to measure the response is initiated by the device generating the shock, so that the response is synchronized with the stimulus at all times. What is observed on the CRO screen is an apparently steady trace, which is in fact the superposition of many identical traces. In summary, the surge is repeatedly applied and the response is repeatedly observed. By synchronizing the two, a very high-speed transient appears as a steady picture on the CRO. Recognizing that if a system is linear, the *form* of its response to a stimulus is independent of the magnitude of the stimulus, it is possible to determine the response by a relatively low amplitude surge.

To simulate the closing of a switch, a square-fronted voltage surge is applied repeatedly at the switch contacts. To simulate the opening of a switch, a ramp of current is repeatedly injected at the switch contacts. A repetition rate of 50–100 pulses/sec is normally adequate. Sufficient time must be allowed between the surges for the transient from one to subside before the transient of the next is provoked. The result of such a recurrent operation is portrayed in Fig. 17.17, which shows the response of a 1000-kVA transformer to the closing of a switch.

One of the advantages of this approach is that circuit changes can be made while the investigation is in progress and the consequences of these changes become immediately apparent. Wilkinson (6) puts this very succinctly:

> If the recurrent surge oscillograph is like a tutor with whom problems can be discussed, the single record oscillograph is to be compared with a system of correspondence by field postcard.

Figure 17.18 shows how the method is used to study the voltage distribution down the winding of a transformer under impulse conditions. In this instance the recurrent impulses are formed by discharging a line into its surge impedance. When the thyratron is triggered, the voltage rises quickly across Z_0 to one-half the value to which the line is charged. It persists at this level until a wave has traveled down the line and back. Thus short, square pulses of voltage are applied to the transformer winding.

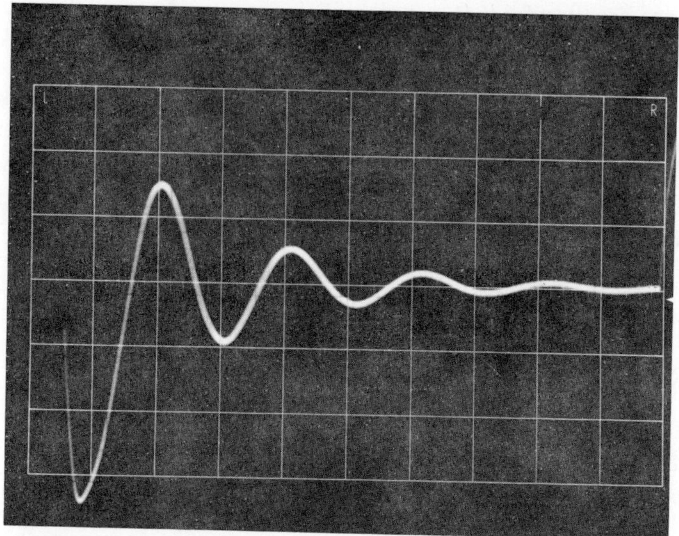

Fig. 17.17. Response of a 1000-kVA, 13.8-kV transformer to the closing of a switch: determined by the recurrent surge technique.

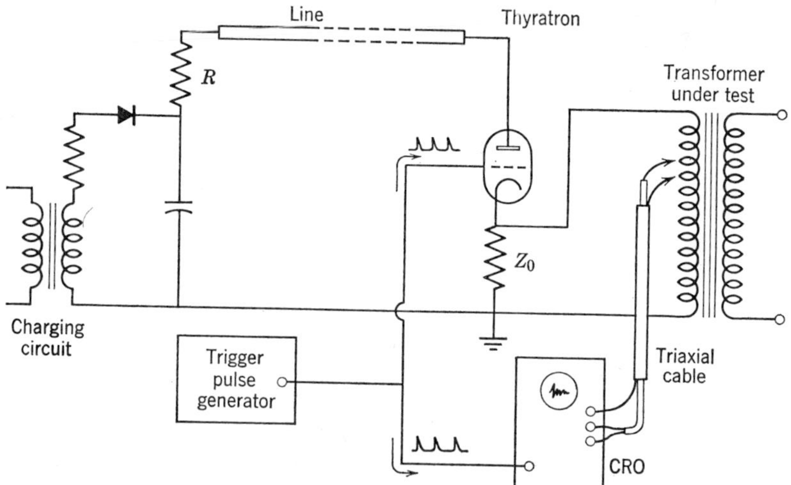

Fig. 17.18. Recurrent surge oscillograph being used to measure the surge voltage distribution down a transformer winding.

actually applying short circuits, would be even less popular with the system operator. However, it is possible to simulate such conditions without actually creating a fault. This is done by suddenly connecting a large discharged capacitor at the point of fault, since such a capacitor behaves like a short circuit, at least momentarily.

As an example of the method, tests will be described whose objective was to determine what happens when a circuit-breaker clears a fault on the 13.8-kV bus in a metropolitan substation (4). The layout of the substation is shown schematically in Fig. 17.16*a*. Power is supplied from three transformers to buses to which feeders are connected through circuit-breakers. These buses are joined by a common bus at the top, which in turn can be connected to similar adjacent sections by tie breakers, the only limitation being that the total available current must not exceed the interrupting capability of the feeder breakers. The feeders in the installation are cables; some of them are quite long so that, in total, they constitute a considerable amount of capacitance on the bus. However, there are capacitor banks which can be connected to the bus for power factor improvement; these possess considerably more capacitance.

For the purpose of the tests one of these capacitor banks was suddenly switched onto the bus by closing the circuit breaker at *A*. The effect of the operation was measured by an oscilloscope connected through capacitance potential dividers to the bus. The dividers had previously been mounted in a breaker framework and had been racked into the bus at one of the breaker locations as indicated by *B* in the figure. An oscillogram typifying the results is shown in Fig. 17.16*b*. Each trace represents the response of one phase; it will be noted that the three poles of the breaker do not close in unison. There are apparently two frequencies involved in the disturbance. Initially, the cables share their charge with the capacitor bank; this is indicated by the higher frequency oscillation. This effectively pulls down the bus voltage. Subsequently the cables and capacitor bank are charged from the source in an oscillatory manner, as indicated by the low frequency transient. Apparently there is not a great deal of damping in the system, because the voltage traces indicate a considerable overshoot. One of the remarkable features of these oscillographic records is that a system, so complex, with multiple feeds, many connected cables, and a vast number of loads, should give such a simple response.

If a fault occurred on this bus or on a feeder adjacent to the bus, with the capacitor bank energized, the transient recovery voltage following the interruption of the fault current would closely resemble the low frequency part of Fig. 17.16*b*. In the event that the capacitor is not connected it is still possible to deduce the response of the system from the data obtained from these tests.

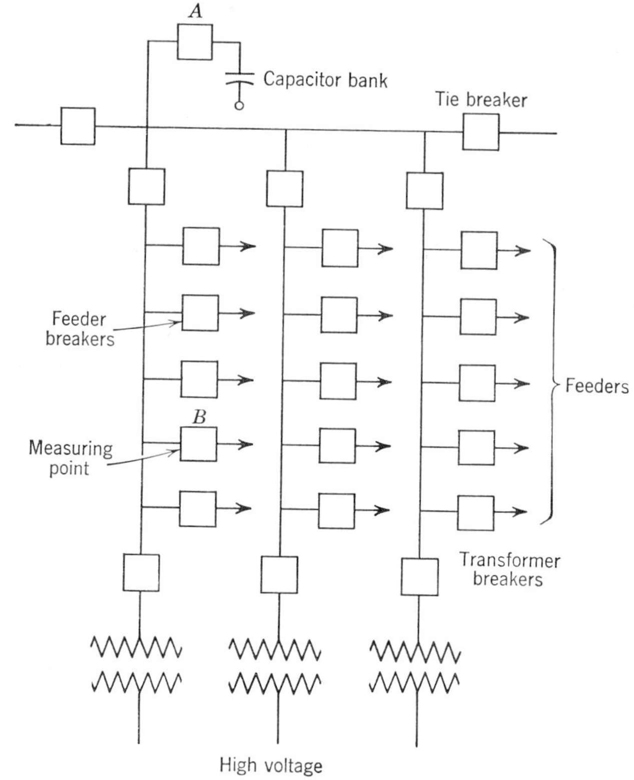

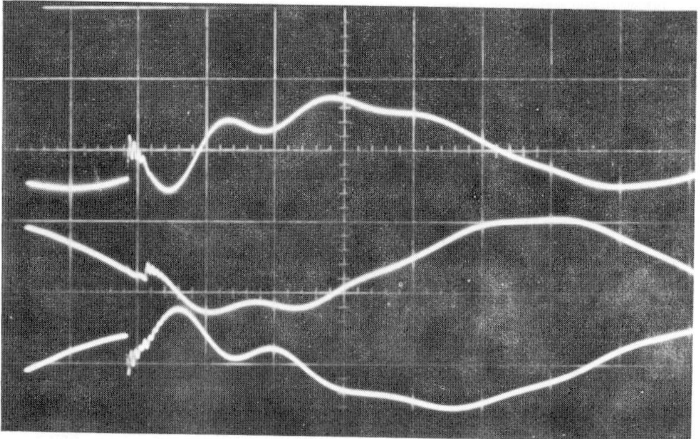

Fig. 17.16. (*a*) Layout of typical metropolitan substation. (*b*) The consequence of shocking the circuit shown in Fig. 17.16*a* by closing the circuit breaker at *A*.

17.7 The Recurrent Surge Technique

Examples of this technique have been presented at several points in the text, notably in Chapter 13. The principle is simple and very effective and the applications are manifold. It is well-suited, for example, for measuring the transient response of simple or complex circuits or for studying the voltage distribution down a transformer winding under surge conditions.

The basic idea is described in a paper by White and Nethercot (5) and in a discussion thereon by Wilkinson (6). The circuit under study is shocked as proposed in Section 17.6, but in this instance the shock is administered repeatedly. The sweep of the oscilloscope being used to measure the response is initiated by the device generating the shock, so that the response is synchronized with the stimulus at all times. What is observed on the CRO screen is an apparently steady trace, which is in fact the superposition of many identical traces. In summary, the surge is repeatedly applied and the response is repeatedly observed. By synchronizing the two, a very high-speed transient appears as a steady picture on the CRO. Recognizing that if a system is linear, the *form* of its response to a stimulus is independent of the magnitude of the stimulus, it is possible to determine the response by a relatively low amplitude surge.

To simulate the closing of a switch, a square-fronted voltage surge is applied repeatedly at the switch contacts. To simulate the opening of a switch, a ramp of current is repeatedly injected at the switch contacts. A repetition rate of 50–100 pulses/sec is normally adequate. Sufficient time must be allowed between the surges for the transient from one to subside before the transient of the next is provoked. The result of such a recurrent operation is portrayed in Fig. 17.17, which shows the response of a 1000-kVA transformer to the closing of a switch.

One of the advantages of this approach is that circuit changes can be made while the investigation is in progress and the consequences of these changes become immediately apparent. Wilkinson (6) puts this very succinctly:

> If the recurrent surge oscillograph is like a tutor with whom problems can be discussed, the single record oscillograph is to be compared with a system of correspondence by field postcard.

Figure 17.18 shows how the method is used to study the voltage distribution down the winding of a transformer under impulse conditions. In this instance the recurrent impulses are formed by discharging a line into its surge impedance. When the thyratron is triggered, the voltage rises quickly across Z_0 to one-half the value to which the line is charged. It persists at this level until a wave has traveled down the line and back. Thus short, square pulses of voltage are applied to the transformer winding.

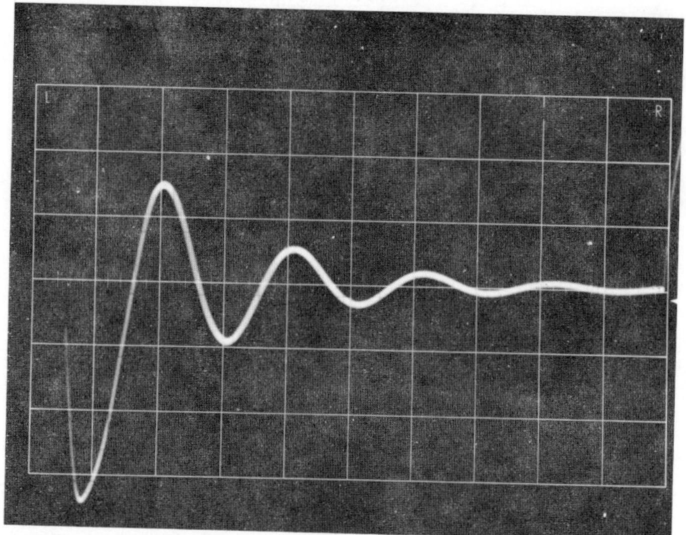

Fig. 17.17. Response of a 1000-kVA, 13.8-kV transformer to the closing of a switch: determined by the recurrent surge technique.

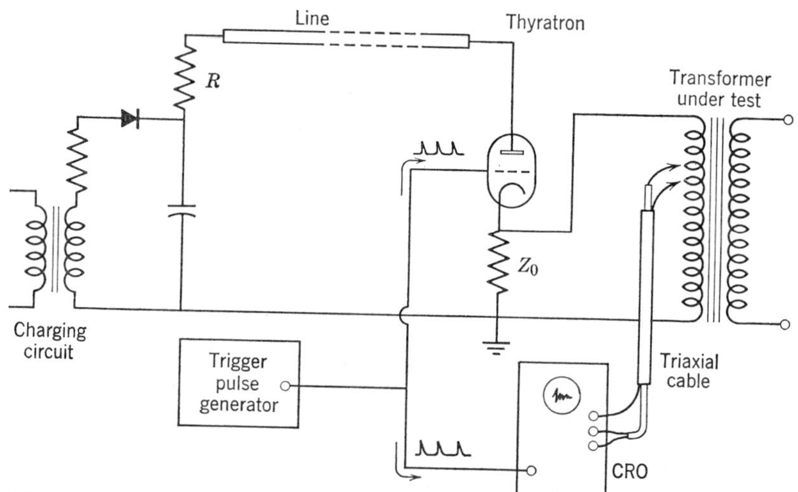

Fig. 17.18. Recurrent surge oscillograph being used to measure the surge voltage distribution down a transformer winding.

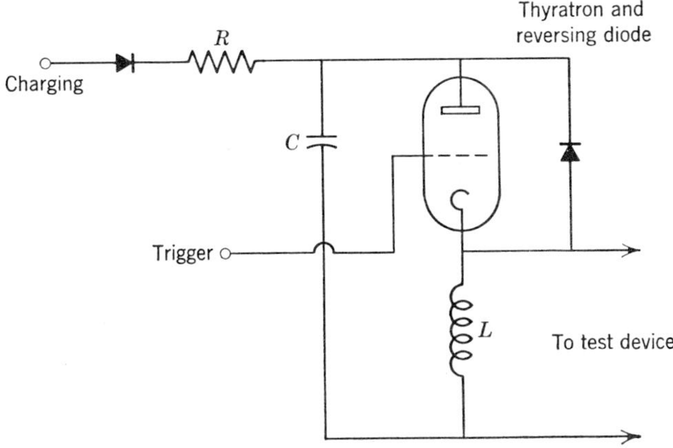

Fig. 17.19. Simple recurrent surge generator.

With each signal applied to the grid of the thyratron a triggering pulse is sent to the CRO to initiate its sweep, synchronizing it with the signal to be measured.

The potential of any point on the winding can be observed, or, as indicated in Fig. 17.18, the interturn potential can be measured, providing taps are available. From data of this kind it is possible to construct curves of the type shown in Fig. 11.5. As explained in Chapter 11, the voltage gradient is often very nonuniform. It is important with this kind of measurement that the impedance of the measuring circuit be very high. A relatively small amount of capacitance will upset the natural capacitance of the system, on which the voltage distribution depends, thereby vitiating the results.

The use of cable to form pulses is very satisfactory except that for anything but a short pulse, it takes a lot of cable. A simple, but effective alternative is shown in Fig. 17.19. This surge generator comprises a capacitor, an inductor, and a thyratron or a thyristor. When the thyristor is gated, the capacitor voltage appears across the inductor and therefore across the load. The rate of rise of the surge front depends upon the stray capacitance, but it can be quite steep. The general shape of the voltage surge is a cosine wave. If the period of the LC circuit is long compared with the period of interest in the measurement, the beginning of a cosine wave is a good approximation to a step function.

Where a fair amount of work of this kind is to be done, the surge generator can be incorporated into the oscilloscope to form one piece of relatively portable equipment.

17.8 Impulse Testing

In the course of its operational life a piece of power system equipment will be subjected to overvoltages from a number of different causes. It is important that it be able to sustain these overvoltages without being damaged. In order to establish the adequacy of the equipment in this regard it must be tested in the design stage and perhaps as a routine test before shipping, by applying to it voltages of the appropriate amplitude and waveshape. In this section we consider what forms these voltages should take and how they are generated and applied. A general treatise on surge testing is given by Hawley (7).

The specifications for power equipment almost invariably call for some minimum capability for the insulation system as far as this matter is concerned. It is essential that there be some uniformity in the manner in which this is specified. Standards committees have addressed themselves to this subject and standards do exist for this purpose.

We will start by considering the kind of overvoltages that apparatus such as transformers, reactors, and generators can be subjected to and then consider how they can be reproduced.

The most frequent surges are undoubtedly switching surges but, potentially more dangerous because of their magnitude, are surges due to lightning. The switching surge is usually oscillatory and the period can vary widely. In general the steepness of the front on such a surge is less than that of a typical lightning surge and it will be recalled from Chapter 11 that this is an important consideration in the uniformity or nonuniformity of the surge voltage distribution within a piece of equipment. In Chapter 10 we indicated that more is being continually learned about lightning. The waveforms of lightning surges can vary quite significantly. Finally, it is recognized that waveforms are often modified by protective equipment, as when an arrester operates, or a rod gap flashes over. This latter event cuts off the surge waveform very abruptly as if an extremely steep-fronted surge of opposite polarity were suddenly superimposed on the initial voltage wave, with all the attendant effects of such a rapid change of voltage.

For many years the standard impulse wave has been of double exponential form, that is, it can be represented by an equation of the form

$$V(t) = V_1(\epsilon^{-\alpha t} - \epsilon^{-\beta t}) \qquad (17.8.1)$$

The waveshape of this voltage can be understood from Fig. 17.20*a*. It will be seen that the steepness of the front depends largely on the coefficient β, whereas the length of the tail is primarily dictated by the coefficient α. In fact, such waves are defined in terms of the duration of their front and the time to half value in their tail. Because of practical difficulties, notably stray capacitance, it is not possible to produce a voltage waveform of precisely that given in Eq. 17.8.1; actual impulse waves are more like Fig. 17.20*b* and are specified in accordance with the notation shown in that figure. American practice has been revised to be more in line with internationally accepted standards; details can be found in the American Standard for *Measurement* *of Voltage in Dielectric Tests* (8). The front of the wave, t_f, is taken as 1.6 × 30% and 90% points on the wavefront. The tail is given by the time from the virtual origin, shown as point 0_1 in Fig. 17.20*b*, and the half peak value point on the back of the wave. If we call this t_t, it is usual to express an impulse wave in terms of $t_f \times t_t$. The accepted standard wave is the 1.2 × 50 wave, which indicates that $t_f = 1.2\ \mu\text{sec}$ and $t_t = 50\ \mu\text{sec}$. Surges of other proportions, such as 5 × 20, are called for from time to time and are used in impulse testing.

Reference has already been made to the chopped wave. Tests with such a wave are made by shunting the test object with a rod gap which flashes over when the impulse is applied. The resulting waveform may look like Fig.

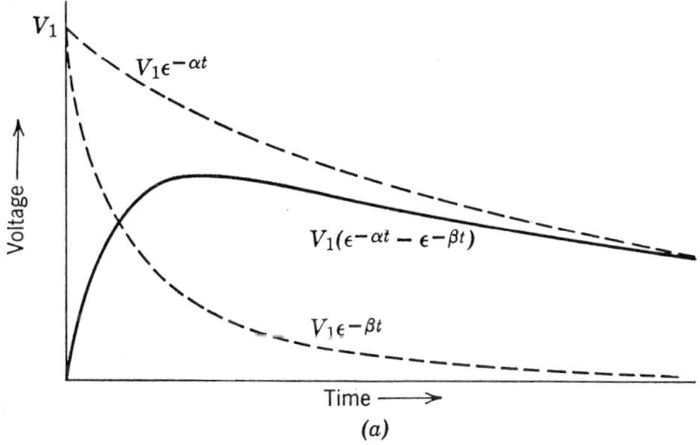

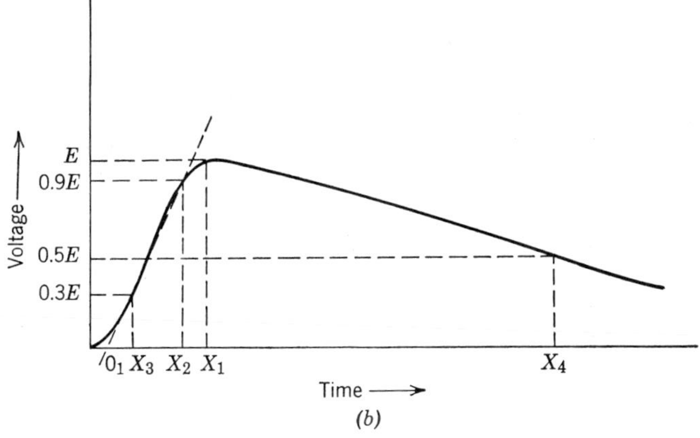

Fig. 17.20. (*a*) Double exponential wave, $V = V(\epsilon^{-\alpha t} - \epsilon^{-\beta t})$. (*b*) Typical surge waveform impulse generator.

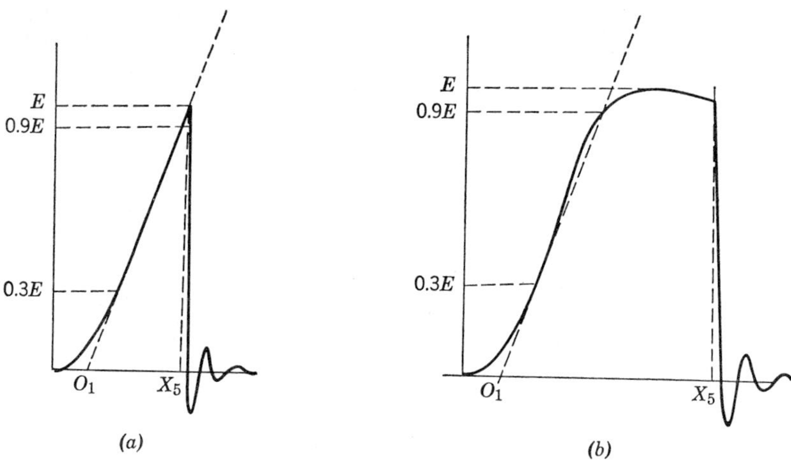

Fig. 17.21. Chopped waves. (*a*) Front of wave flashover. (*b*) Flashover on tail.

17.21*a* or 17.21*b*, depending on whether the chop occurs on the front or the tail. The former is frequently referred to as a "front-of-wave" test. Voltage is applied at a very fast rate (as measured by the slope of the 30–90% line), which may be as high as 1 MV/μsec. The gap is set to limit the voltage to a predetermined value.

Whereas reference (8) describes the general principles of dielectric testing and supplies many details of measurements, the precise tests and methods to be applied to individual types of equipment are spelled out in Standards appropriate to the apparatus concerned. For example, impulse testing of transformers is prescribed in ASA C57 (9).

The double exponential form of the full impulse wave can be generated, in principle, by a relatively simple circuit (Fig. 17.22). Here C_1 is charged from a source (not indicated) to a level which causes the gap G to flash over; C_2 then charges from C_1 through R_1. Simultaneously, but more slowly, both capacitors discharge through R_2. Alternatively, G could be a triggered spark

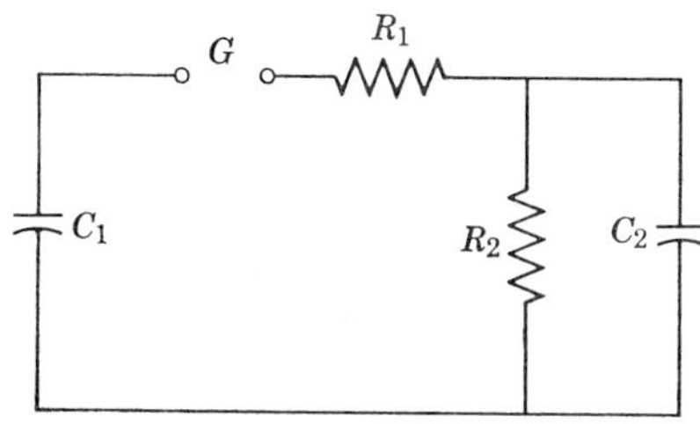

Fig. 17.22. Rudimentary single-impulse generator.

gap, sometimes called a trigatron, that can be switched on at will by a voltage pulse applied to its third electrode. The output voltage is the voltage appearing across C_2.

The solution for this voltage can be found fairly readily by the Laplace transform method developed in the early chapters of this book. But rather than do that, we will obtain a feel for the solution by physical reasoning.

Since R_2 will always be much greater than R_1, when the circuit is completed and C_1 shares its charge with C_2, we can momentarily neglect R_2. The common voltage after such an event would be $[C_1/(C_1 + C_2)]V$. Moreover, C_1 and C_2 are effectively in series, and so present a combined capacitance $C_1C_2/(C_1 + C_2)$. This means that to a close approximation the charging time constant for C_2 will be $R_1C_1C_2/(C_1 + C_2)$. To the same approximation, the output voltage can be expressed as

$$V_1 = \frac{C_1C_2V}{C_1 + C_2}\left[1 - \exp\left(-\frac{C_1 + C_2}{R_1C_1C_2}t\right)\right] \tag{17.8.2}$$

However, simultaneous with the charging of C_2 described by Eq. 17.8.2, both capacitors will be discharging through R_2. In this process C_1 and C_2 will be effectively in parallel and will have a capacitance of $C_1 + C_2$, making the discharging time constant $R_2(C_1 + C_2)$. Assuming that C_2 reaches the voltage $[C_1/(C_1 + C_2)]V$ before the discharge commences, the discharge can be expressed as

$$V_1 = \frac{C_1}{C_1 + C_2}V\exp\left[-\frac{t}{R_2(C_1 + C_2)}\right] \tag{17.8.3}$$

Equations 17.8.2 and 17.8.3 correspond to the parts of Eq. 17.8.1 that are shown in Fig. 17.9*a*:

$$\left.\begin{aligned} \alpha &\approx \frac{1}{R_2(C_1 + C_2)} \\ \beta &\approx \frac{C_1 + C_2}{R_1C_1C_2} \end{aligned}\right\} \tag{17.8.4}$$

This differs slightly from the rigorous solution for the circuit but is a very good approximation when $R_2 \gg R_1$.

The primary effect of circuit inductance, which we have not yet considered, is to delay the front of the wave; this, again, is apparent from physical reasoning. Equation 17.8.1 would suggest that at $t = 0$, the impulse wave has a finite slope. But if the circuit has inductance, the current must start from zero when the gap strikes. Since the current in the capacitor C_2 is given by $C_2(dV/dt)$, it follows that initially $dV/dt = 0$.

The piece of equipment under test, which forms the load of the impulse generator, frequently has an important influence on the output waveform.

The load impedance will shunt R_2 and C_2 and can be factored into the analysis where appropriate.

The circuit shown in Fig. 17.22 might be described as a single-stage impulse generator. Such devices are rarely used in practice. Instead, the multistage impulse generator is employed. The pioneering development work for this device was done by Marx (10), whose name the circuit bears. It has the virtue of producing a high-output voltage from a source of relatively low-charging voltage. This is achieved by charging a number of capacitors in parallel and then reconnecting them in series at the moment of discharge, so that their voltages add. Thus, if the individual capacitors of a stage are charged to 50 kV and there are 20 stages, it is theoretically possible on reconnecting the capacitors in series to attain a peak voltage of 1000 kV. For a detailed treatment of high-voltage impulse generators, the reader is referred to a quite exhaustive treatment by Edwards, Husbands, and Perry (11), which includes a history of their development. We briefly consider a typical installation and review the principle of its operation.

The diagram in Fig. 17.23 is the circuit of a five-stage impulse generator; $C_1, C_2, \ldots . . C_5$ are the storage capacitors and they are charged to any desired potential within their rating from the source on the left-hand side. The switch from parallel to series connection is accomplished automatically by the flashing over of the series of ball gaps shown. These are adjustable and are preset to be close to their breakdown when the capacitors are charged. A simple, although not very good way of operating the generator is to set these gaps with the bottom one somewhat closer than the rest, and raise the charging voltage until this bottom gap goes over. As will be explained shortly, this at once initiates the breakdown of the remaining gaps, and the entire bank discharges in series into the load circuit.

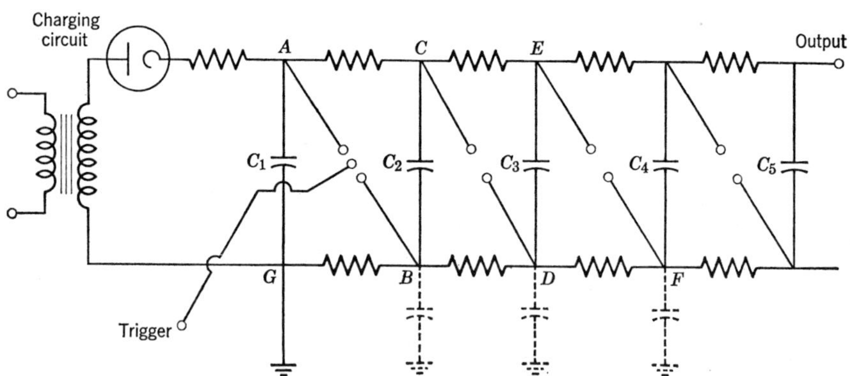

Fig. 17.23. Schematic diagram of a five-stage impulse generator.

A better method is to use some kind of triggered gap for the bottom gap, perhaps a three-ball gap as shown in Fig. 17.23. The generator is then charged to a predetermined value and only discharged when a suitable signal is applied to the third ball of the triggered gap. This signal is usually a short high-voltage pulse, from a thyratron for example. It is sufficient to spill over one-half of the bottom gap; the other half then takes the stress and immediately follows suit.

The part played by stray capacitance in the ensuing flashover of the remaining gaps is not always fully appreciated. When the first gap fires the stray capacitance between B and ground is very rapidly charged from C_1 through the spark. The stray capacitance at D, however, must be charged through the resistance BD, which retards the process. Thus, while B quickly assumes the potential of A, D is slower in responding. Meanwhile, the voltage across C_2 cannot suddenly change so that C rises with B. A consequence of this is that a voltage approaching 2 V appears across the second gap CD and it breaks down. The stray capacitance between D and ground can now be charged directly through two sparks, so D quickly assumes the potential of C. The process then repeats with C_3, the gap EF coming into the act. The whole process is cumulative, so that in short order all gaps fire and the output terminal flies up in potential to a value approaching nV where n is the number of stages and V is the charging voltage. For consistent operation it is found advisable to irradiate the gaps. This can be done with an ultraviolet lamp or by a radioactive pellet in one ball of each gap. It is also customary to put the gaps in line of sight of each other, for then the radiation from the spark in the lower gaps helps to trigger the upper gaps.

To use an impulse generator to impulse test a piece of equipment one further feature is required: the impulse generator must be isolated from the load until the instant of testing. Otherwise, of course, it would not be possible to charge up the generator, at least not if the load were a transformer or similar device possessing a d-c path to ground. Isolation is achieved by a ball gap between the generator and the load, its spacing being such that it will support the stage voltage but break down when the stages are connected in series, so joining the generator to the load.

The arrangements for impulse testing a transformer are shown in Fig. 17.23*a*. The Marx generator is represented by a single capacitor, C_1, which, neglecting strays, will be the series sum of the several stage capacitors. Figure 17.24*b* shows the equivalent circuit, which is like Fig. 17.22 except that the internal inductance of the generator L_1 and the external inductance of the leads to the load have been added. In this circuit C_2 may well comprise only the capacitance of the load itself. This would achieve the maximum voltage from the generator. The transformer is represented by a capacitor

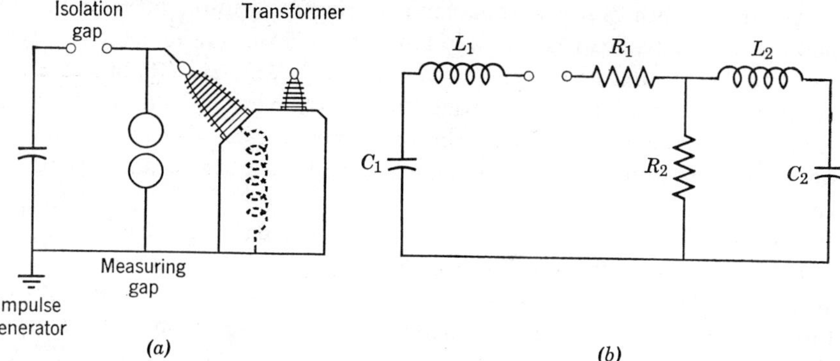

Fig. 17.24. Impulse testing a transformer. (*a*) Physical arrangement. (*b*) Equivalent circuit.

since it is the capacitive attributes of the transformer that predominate under impulse conditions, as was explained in Chapter 12.

The circuits described are capable of generating impulse waves of double exponential form, varying over a wide range in magnitude and shape, depending upon the value of the constants used. It is also possible to generate chopped waves by placing a gap across the output between the generator and the device under test. This can be set to spill over beyond the peak, or, with a high prospective peak, to flash over on the wavefront.

Not all impulse testing requires high-voltage equipment of the type described above; sometimes a few kilovolts, or even a few hundred volts, is all that is required to test a component. On the other hand, the wavefront may be more critical in such tests; in particular a very fast front may be called for. Where this is not the case, a small-scale, single-stage impulse generator can be put together from components readily available in a laboratory. A thyristor or thyratron can be used for the switch in lieu of a spark gap.

When very fast front surges are required they can be obtained by discharging a cable into its surge impedance. A square wave of voltage is developed across this impedance, equal in magnitude to half the voltage to which the cable was precharged (see Fig. 17.25). The pulse will have a duration of one round-trip travel time for a wave along the cable (typically 1 μsec

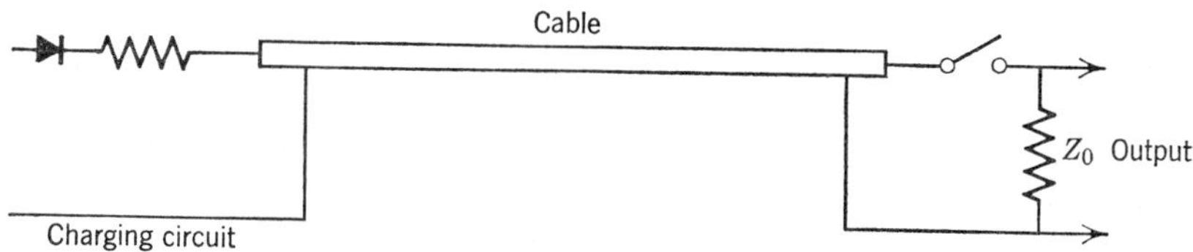

Fig. 17.25. Pulse forming by discharging a cable into its surge impedance.

for a 250-ft length of cable). The cable surge impedance is low (25–125 Ω), making it a relatively stiff source of voltage.

When longer duration square waves are needed, the length of cable becomes unacceptable. In such cases an artificial line, made up of T or π sections of inductors and capacitors, can be used. Such a line is imperfect; to what extent depends upon the number of sections and the quality of the components. Some discussion to this point was made in Section 14.5. The theory of such lines is discussed by Kimbark (12).

The steepness of the wavefront depends upon the characteristics of the switching device. For example, a mechanical switch is useless, except at very low voltages, because it prestrikes and introduces a spark impedance before the contacts close. Hydrogen thyratrons are better, but they still take a significant time to build up the current; furthermore, they are relatively bulky. Pollard (3) describes a satisfactory solution wherein he was attempting to produce pulses of 1–6 kV magnitude and 0.02 to 1 μsec in duration.

The scheme adopted is shown in Fig. 17.26. A shorted cable is used to determine the pulse length, in this instance. Upon completing the circuit through the triggered vacuum gap, a step voltage of magnitude E_c propagates down the line and eventually passes the output terminal, thus starting the output pulse. Beyond, at the shorted end of the line, the wave is reflected with sign inverted. As the reflection travels back along the line it cancels the plus E_c voltage already there, leaving zero voltage in its wake. The magnitude of the voltage wave is equal to the voltage to which the capacitor is charged, instead of half that value, as required by the circuit in Fig. 17.25. However, had the voltage been applied through a resistor, equal to the surge impedance of the cable to absorb the reflections from the other end of the cable, it would have been necessary to charge to twice the voltage as before.

The principal switching device is a triggered vacuum gap. The wavefront generated by using this alone, with conditions optimized to obtain the minimum delay time possible, is shown in Fig. 17.27; it indicates a serious problem. The partial conduction phase seems to start almost immediately, but full conduction is delayed. The 300-ns partial conduction period, and the 50-ns transition time from partial to full conduction, are much too long if

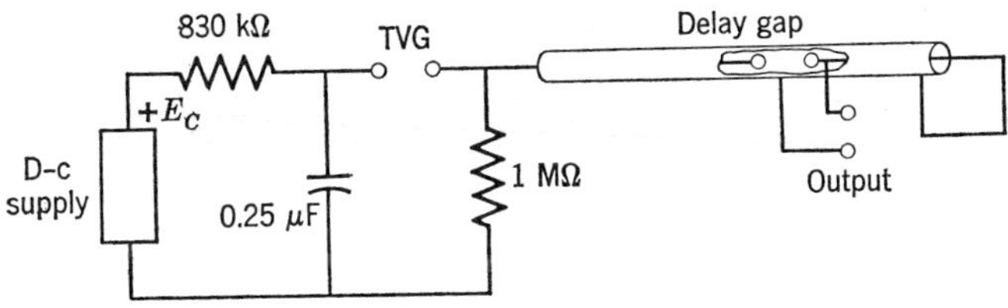

Fig. 17.26. Shorted line pulse-forming circuit with delay gap to sharpen output pulse rise time.

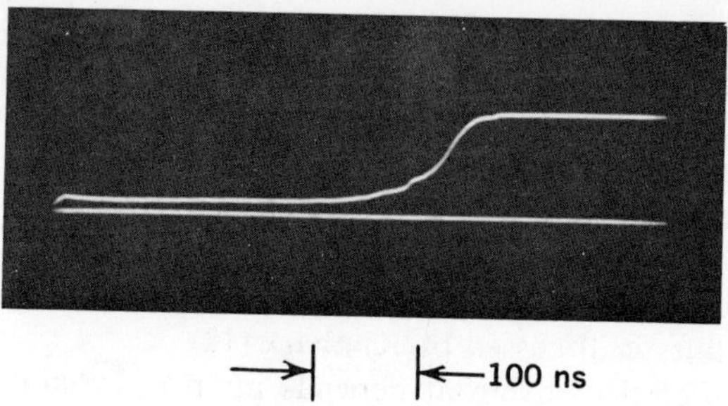

Fig. 17.27. Wavefront achieved with a triggered vacuum gap.

20-ns pulses are to be generated. This intolerable situation is corrected by adding a second gap to the circuit, in series with the vacuum gap, as indicated in Fig. 17.26. The result is shown in Fig. 17.28. The series gap is set for a d-c flashover level of 2 kV when using a charge line voltage of 5 kV. The gap firing delay associated with overvolting a 2 kV gap to 5 kV results in the load not being connected to the line until the vacuum gap is turned on, to all intents and purposes. Turnon action in the series gap is much faster than in the vacuum gap, resulting in an open circuit rise-time of less than 2 ns. It should be pointed out that the actual rise-time across the load itself will depend upon the load capacitance. If this is, say, 100 pF and the surge impedance of the cable is 50 Ω, the time constant for the combination will be

$$RC = 25 \times 10^{-10} = 2.5 \text{ ns.}$$

(we use 25 Ω, since looking back from the load we see two lengths of cable in parallel).

The need for careful physical layout of the circuit is obvious. The vacuum gap and the input end of the cable are mounted directly on the capacitor

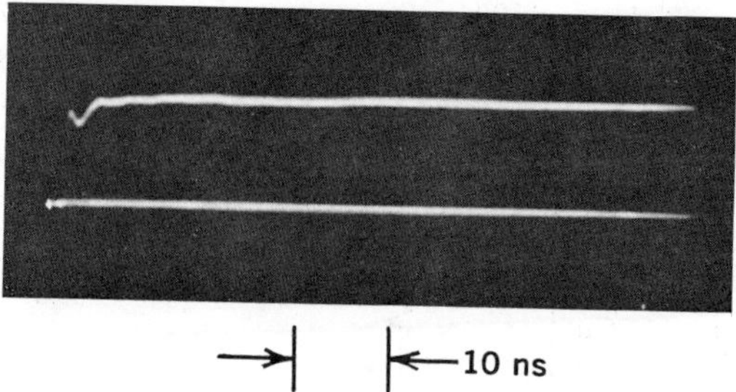

Fig. 17.28. Wavefront obtained with combination of vacuum gap and series air gap.

terminal to minimize inductance. The auxiliary gap is made integral within the cable so as not to disturb the smooth properties of the circuit. While the square pulse waveform can be generated entirely within a coaxial transmission line structure, some means must be provided to apply this voltage to the load. If the load itself is small, it too can be integrated into the structure, perhaps in a coaxial tee, otherwise it must be connected close to the cable in a way that introduces the minimum of capacitance and inductance. Care must be exercised to see to it that the lead inductance does not resonate with the capacitance of the load. An equivalent resistance of 25 Ω, provided by the surge impedance of two 50-Ω cables, is sufficient to damp a series resonant circuit of 12.5-Ω surge impedance. With a load capacitance of 100 pF, this represents a load inductance of approximately 16 nH, which is the inductance of a loop of wire about the size of a nickel!

This example serves to show the care that must be taken in high-frequency testing.

17.9 Testing of Circuit Breakers by Circuit Simulation

Circuit breakers are subjected to all manner of transient voltages, not least of which are those which they generate themselves while interrupting current. This voltage appears across the contact gap at a particularly vulnerable time, when the arc products associated with interruption have just been, or are being, dispersed. Testing is required to establish that breakers can perform their required tasks.

Although the source of power for some circuit-breaker testing stations comes directly from a power system, this is not the way in which most circuit breakers are tested during their design development or to verify their performance. If it is desired to investigate what current a circuit-breaker can interrupt and what transient recovery voltage it can subsequently support, it is more usual to test from a generator or generator/transformer combination specifically designed for short circuit testing. With circuit breakers of higher interrupting rating, the equipment required becomes bigger and correspondingly more expensive, to the point that it is quite uneconomical to test the very biggest circuit breakers directly by this means. An alternative method is to use so-called *synthetic testing* or testing by circuit simulation.

The basic idea in this approach is to have the short-circuit current supplied by one source and the transient recovery voltage impressed across the breaker from another. One of the earliest circuits for this purpose was described by Skeats (13); current and voltage were supplied by transformers at different voltage levels. Subsequently, capacitors have been used for the voltage source as illustrated in Fig. 17.29. Current is supplied from a generator, through an isolation breaker, to the breaker under test. The contacts of the test circuit breaker separate and the current is interrupted at a subsequent current zero.

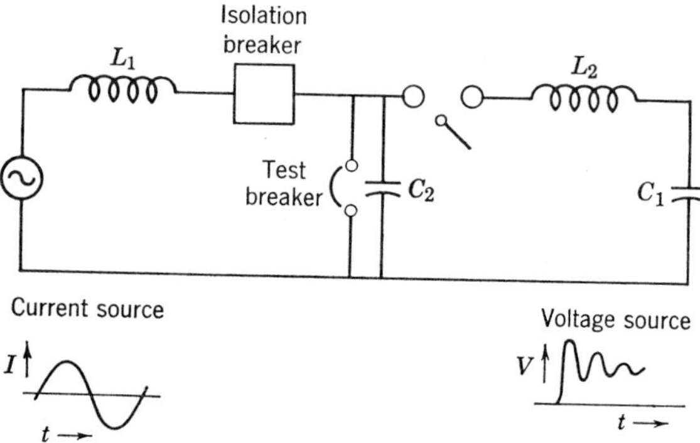

Fig. 17.29. Rudimentary synthetic circuit for testing circuit breakers.

At this instant the trigatron in the voltage circuit is fired, causing a transient recovery voltage, of (1 − cosine) form, to be impressed across the opening breaker. The frequency of this voltage is readily controlled by the parameters L_2 and C_2; it can be damped as desired by the addition of resistance. The magnitude is fixed by the voltage to which the source capacitor C_1 is charged. The purpose of the isolation circuit breaker is to protect the current source from the voltage transient of the voltage source. In performing this function this breaker is subjected to the same stresses as the test breaker, it must therefore have more capability than the breaker under test.

The simple circuit just described indicates how circuit breakers can be tested without the power capacity of their own rating. The generator must simply supply the amperes, its voltage can be quite modest, although it must be several times the arc voltage that the circuit breaker is likely to develop. The voltage for the test is provided by C_1, which is a source of very modest power. The object, of course, is to make the test circuit breaker "believe" that it is in an actual system environment.

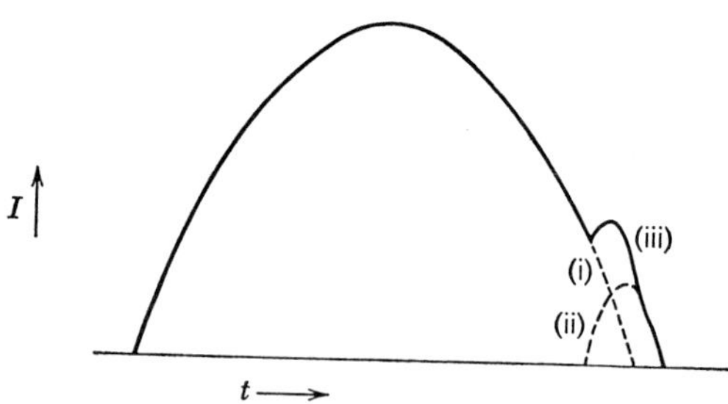

Fig. 17.30. Distortion of the current waveforms in the Weil circuit. (i) Current from current source. (ii) Current from voltage source. (iii) Composite current.

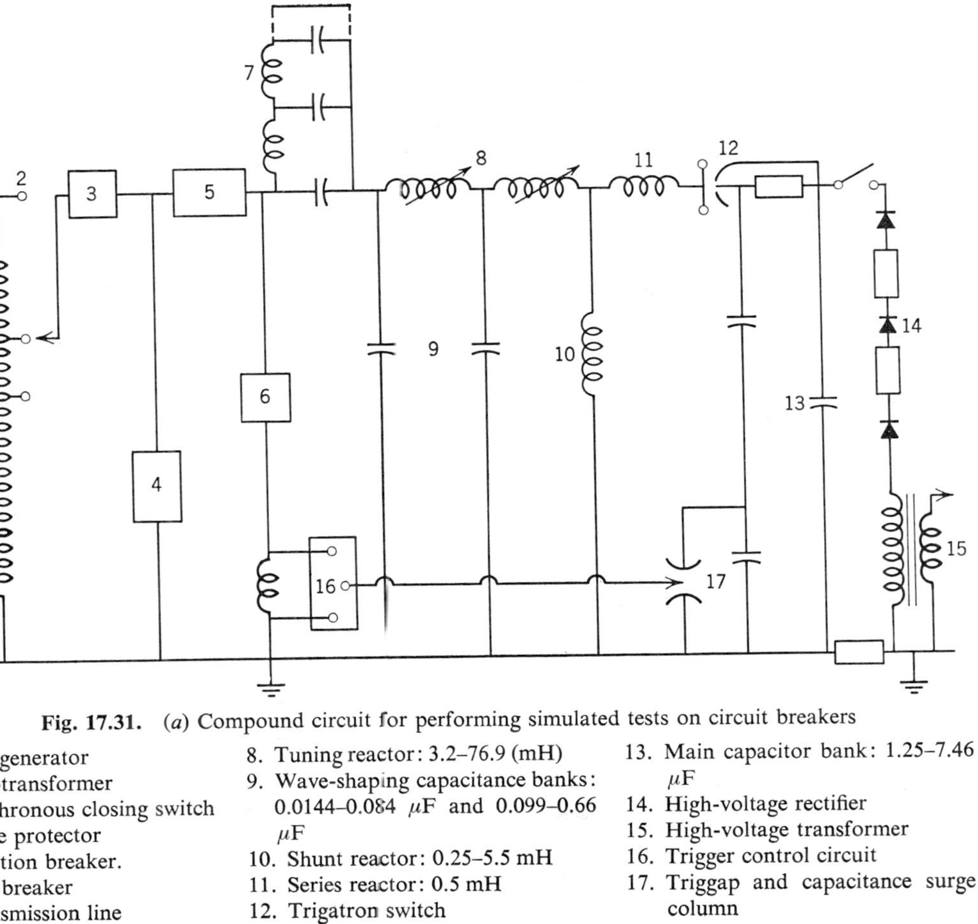

Fig. 17.31. (*a*) Compound circuit for performing simulated tests on circuit breakers

1. Test generator
2. Autotransformer
3. Synchronous closing switch
4. Surge protector
5. Isolation breaker.
6. Test breaker
7. Transmission line
8. Tuning reactor: 3.2–76.9 (mH)
9. Wave-shaping capacitance banks: 0.0144–0.084 μF and 0.099–0.66 μF
10. Shunt reactor: 0.25–5.5 mH
11. Series reactor: 0.5 mH
12. Trigatron switch
13. Main capacitor bank: 1.25–7.46 μF
14. High-voltage rectifier
15. High-voltage transformer
16. Trigger control circuit
17. Triggap and capacitance surge column

Fig. 17.31 (*b*) Compound circuit test facility for simulated tests on circuit breakers. (Courtesy of the General Electric Co.)

The limitation of the above circuit is in the precise synchronization that is required between the current and voltage sources. The triggered gap must be fired at just the right instant for the recovery voltage to be applied immediately upon the current coming to zero. The performance of a circuit breaker, whether it will succeed or fail to interrupt a current, depends very critically upon events in the neighborhood of current zero. This demands that synchronization be obtained to the microsecond.

Many improvements and refinements have been introduced over the years to improve the simulation; the literature abounds with papers on the subject. A good review article on the subject is the paper by Anderson et al. (14), which describes British practice. One approach, the so-called Weil circuit (15), deserves mention. Here the problem of close synchronization is avoided by stiffening the voltage source somewhat and firing the trigatron ahead of the would-be current zero of the current from the generator. The voltage source discharges through the breaker, adding a higher frequency component of current to the generator current. This distorts the waveform somewhat as indicated in Fig. 17.30, but the circuit has the advantage that the voltage source is automatically synchronized with the current source, since the two are one and the same for a brief period before current zero.

It will be seen that the current in the current source comes to zero ahead of the current in the test circuit-breaker, which allows a good opportunity for the protective breaker to clear before having a very high voltage impressed across it. A circuit of this kind, with considerable capacity, has been described by Liao et al. (16). It includes an artificial transmission line, to simulate the effects of short line faults described in Section 13.2. The diagram for this circuit is shown in Fig. 17.31*a*, while Fig. 17.31*b* shows part of the circuit itself.

REFERENCES

1. H. W. Lord, "High-Frequency Transient Voltage Measuring Techniques," *Trans. IEEE* (Communications and Electronics), Vol. 82 (1963), p. 602.
2. F. D. Martzloff, Unpublished report.
3. E. M. Pollard, Unpublished report.
4. A. N. Greenwood, E. T. McCurry and G. Frind, "Identification and Evaluation of a System Condition which can Affect the Performance of Air-Magnetic Circuit Breakers," IEEE Trans. Paper 70 TP 593 PWR (1970).
5. E. L. White and W. Nethercot, "The Recurrent Surge Oscillograph and its Application to the Study of Surge Phenomena in Transformers," *Proc. IEE*, Vol. 96, Part II (1949), p. 269.
6. K. J. R. Wilkinson, Discussion on reference (5), p. 281.
7. W. G. Hawley, *Impulse Voltage Testing*, Chapman & Hall, London (1959).
8. *American Standard for Measurement of Voltage in Dielectric Tests*, ASAC68.1 (1953), published by the AIEE, New York.
9. U.S.A.S. C57.12.90-1968, *Test Code for Distribution, Power and Regulating Transformers and Shunt Reactors.*
10. E. Marx, "Versuche über die Prüfung von Isolataren mit Spannungsstossen," *Electro. Tech. Zeit.*, Vol. 45 (1924), p. 652.
11. F. S. Edwards, A. S. Husbands and F. R. Perry, "The Development and Design of High Voltage Impulse Generators," *Proc. IEE*, Vol. 98 (1951), p. 155.
12. E. W. Kimbark, *Electrical Transmission of Power and Signals*, John Wiley & Sons, New York (1949).
13. W. F. Skeats, "Special Tests on Impulse Circuit Breakers," *Trans. AIEE*, Vol. 55 (1936), p. 710.
14. J. G. P. Anderson, N. S. Ellis, F. O. Mason, R. G. Noble, L. H. Orton, M. P. Reece and J. G. Steel, "Synthetic Testing of AC Circuit Breakers," *Proc. IEE*, Vol. 113 (1966), p. 611.
15. J. Biermanns, "The Weil Circuit for Testing High Voltage Circuit Breakers with Very High Interrupting Capacities," CIGRE Report No. 102 (1954).
16. J. W. Liao, H. N. Schneider, W. F. Skeats and C. H. Titus, "Switching of Extra High Voltage Circuits IV—Compound Circuit Test Facilities for High Voltage Circuit Breakers," *Trans. IEEE*, Vol. 83 (1964), p. 1213.

Appendix 1 Table of Laplace Transform Pairs

	$f(s)$	$F(t)$
1	$\frac{1}{s}$	1
2	$\frac{1}{s^2}$	t
3	$\frac{1}{s^n}\ (n = 1, 2, \ldots)$	$\frac{t^{n-1}}{(n-1)!}$
4	$\frac{1}{(s+\alpha)}$	$\epsilon^{-\alpha t}$
5	$\frac{1}{s(s+\alpha)}$	$\frac{1}{\alpha}(1 - \epsilon^{-\alpha t})$
6	$\frac{1}{(s+\alpha)^2}$	$t\epsilon^{-\alpha t}$
7	$\frac{1}{(s+\alpha)^n}\ (n = 1, 2, \ldots)$	$\frac{t^{n-1}\epsilon^{-\alpha t}}{(n-1)!}$
8	$\frac{1}{s^2(s+\alpha)}$	$\frac{1}{\alpha^2}[\epsilon^{-\alpha t} + \alpha t - 1]$
9	$\frac{1}{s^3(s+\alpha)}$	$\frac{1}{\alpha^2}\left[\frac{1}{\alpha} - t + \frac{\alpha t^2}{2} - \frac{1}{\alpha}\epsilon^{-\alpha t}\right]$
10	$\frac{1}{(s+\alpha)(s+\beta)}$	$\frac{1}{(\beta-\alpha)}[\epsilon^{-\alpha t} - \epsilon^{-\beta t}]$
11	$\frac{s}{(s+\alpha)(s+\beta)}$	$\frac{1}{(\beta-\alpha)}[\beta\epsilon^{-\beta t} - \alpha\epsilon^{-\alpha t}]$

	$f(s)$	$F(t)$
12	$\dfrac{1}{s(s+\alpha)^2}$	$\dfrac{1}{\alpha^2}[1-\epsilon^{-\alpha t}-\alpha t\epsilon^{-\alpha t}]$
13	$\dfrac{1}{s^2(s+\alpha)^2}$	$\dfrac{1}{\alpha^3}[\alpha t-2+\alpha t\epsilon^{-\alpha t}+2\epsilon^{-\alpha t}]$
14	$\dfrac{1}{(s+\alpha)(s+\beta)^2}$	$\dfrac{1}{(\alpha-\beta)^2}\{\epsilon^{-\alpha t}+[(\alpha-\beta)t-1]\epsilon^{-\beta t}\}$
15	$\dfrac{1}{(s+\alpha)(s+\beta)(s+\gamma)}$	$\dfrac{1}{(\beta-\alpha)(\gamma-\alpha)}\epsilon^{-\alpha t}+\dfrac{1}{(\alpha-\beta)(\gamma-\beta)}\epsilon^{-\beta}$ $+\dfrac{1}{(\alpha-\gamma)(\beta-\gamma)}\epsilon^{-\gamma t}$
16	$\dfrac{1}{s^2+\alpha^2}$	$\dfrac{1}{\alpha}\sin\alpha t$
17	$\dfrac{s}{s^2+\alpha^2}$	$\cos\alpha t$
18	$\dfrac{1}{s(s^2+\alpha^2)}$	$\dfrac{1}{\alpha^2}(1-\cos\alpha t)$
19	$\dfrac{1}{s^2(s^2+\alpha^2)}$	$\dfrac{1}{\alpha^2}(\alpha t-\sin\alpha t)$
20	$\dfrac{1}{(s^2+\alpha^2)^2}$	$\dfrac{1}{2\alpha^3}(\sin\alpha t-at\cos\alpha t)$
21	$\dfrac{s^2}{(s^2+\alpha^2)^2}$	$\dfrac{1}{2\alpha}(\sin\alpha t+\alpha t\cos\alpha t)$
22	$\dfrac{s^2-\alpha^2}{(s^2+\alpha^2)^2}$	$t\cos\alpha t$
23	$\dfrac{1}{(s^2-\alpha^2)}$	$\dfrac{1}{\alpha}\sinh\alpha t$
24	$\dfrac{s}{(s^2-\alpha^2)}$	$\cosh\alpha t$

Appendix 2 Natural Cosines

[*Numbers in difference columns to be subtracted, not added.*]

Degrees	0′ 0°.0	6′ 0°.1	12′ 0°.2	18′ 0°.3	24′ 0°.4	30′ 0°.5	36′ 0°.6	42′ 0°.7	48′ 0°.8	54′ 0°.9	Mean Differences 1	2	3	4	5
0	1.000	1.000	1.000	1.000	1.000	1.000	.9999	9999	9999	9999	0	0	0	0	0
1	.9998	9998	9998	9997	9997	9997	9996	9996	9995	9995	0	0	0	0	0
2	.9994	9993	9993	9992	9991	9990	9990	9989	9988	9987	0	0	0	1	1
3	.9986	9985	9984	9983	9982	9981	9980	9979	9978	9977	0	0	1	1	1
4	.9976	9974	9973	9972	9971	9969	9968	9966	9965	9963	0	0	1	1	1
5	.9962	9960	9959	9957	9956	9954	9952	9951	9949	9947	0	1	1	1	1
6	.9945	9943	9942	9940	9938	9936	9934	9932	9930	9928	0	1	1	1	2
7	.9925	9923	9921	9919	9917	9914	9912	9910	9907	9905	0	1	1	2	2
8	.9903	9900	9898	9895	9893	9890	9888	9885	9882	9880	0	1	1	2	2
9	.9877	9874	9871	9869	9866	9863	9860	9857	9854	9851	0	1	1	2	2
10	.9848	9845	9842	9839	9836	9833	9829	9826	9823	9820	1	1	2	2	3
11	.9816	9813	9810	9806	9803	9799	9796	9792	9789	9785	1	1	2	2	3
12	.9781	9778	9774	9770	9767	9763	9759	9755	9751	9748	1	1	2	3	3
13	.9744	9740	9736	9732	9728	9724	9720	9715	9711	9707	1	1	2	3	3
14	.9703	9699	9694	9690	9686	9681	9677	9673	9668	9664	1	1	2	3	4
15	.9659	9655	9650	9646	9641	9636	9632	9627	9622	9617	1	2	2	3	4
16	.9613	9608	9603	9598	9593	9588	9583	9578	9573	9568	1	2	2	3	4
17	.9563	9558	9553	9548	9542	9537	9532	9527	9521	9516	1	2	3	3	4
18	.9511	9505	9500	9494	9489	9483	9478	9472	9466	9461	1	2	3	4	5
19	.9455	9449	9444	9438	9432	9426	9421	9415	9409	9403	1	2	3	4	5
20	.9397	9391	9385	9379	9373	9367	9361	9354	9348	9342	1	2	3	4	5
21	.9336	9330	9323	9317	9311	9304	9298	9291	9285	9278	1	2	3	4	5
22	.9272	9265	9259	9252	9245	9239	9232	9225	9219	9212	1	2	3	4	6
23	.9205	9198	9191	9184	9178	9171	9164	9157	9150	9143	1	2	3	5	6
24	.9135	9128	9121	9114	9107	9100	9092	9085	9078	9070	1	2	4	5	6
25	.9063	9056	9048	9041	9033	9026	9018	9011	9003	8996	1	3	4	5	6
26	.8988	8980	8973	8965	8957	8949	8942	8934	8926	8918	1	3	4	5	6
27	.8910	8902	8894	8886	8878	8870	8862	8854	8846	8838	1	3	4	5	7
28	.8829	8821	8813	8805	8796	8788	8780	8771	8763	8755	1	3	4	6	7
29	.8746	8738	8729	8721	8712	8704	8695	8686	8678	8669	1	3	4	6	7
30	.8660	8652	8643	8634	8625	8616	8607	8599	8590	8581	1	3	4	6	7
31	.8572	8563	8554	8545	8536	8526	8517	8508	8499	8490	2	3	5	6	8
32	.8480	8471	8462	8453	8443	8434	8425	8415	8406	8396	2	3	5	6	8
33	.8387	8377	8368	8358	8348	8339	8329	8320	8310	8300	2	3	5	6	8
34	.8290	8281	8271	8261	8251	8241	8231	8221	8211	8202	2	3	5	7	8
35	.8192	8181	8171	8161	8151	8141	8131	8121	8111	8100	2	3	5	7	8
36	.8090	8080	8070	8059	8049	8039	8028	8018	8007	7997	2	3	5	7	9
37	.7986	7976	7965	7955	7944	7934	7923	7912	7902	7891	2	4	5	7	9
38	.7880	7869	7859	7848	7837	7826	7815	7804	7793	7782	2	4	5	7	9
39	.7771	7760	7749	7738	7727	7716	7705	7694	7683	7672	2	4	6	7	9
40	.7660	7649	7638	7627	7615	7604	7593	7581	7570	7559	2	4	6	8	9
41	.7547	7536	7524	7513	7501	7490	7478	7466	7455	7443	2	4	6	8	10
42	.7431	7420	7408	7396	7385	7373	7361	7349	7337	7325	2	4	6	8	10
43	.7314	7302	7290	7278	7266	7254	7242	7230	7218	7206	2	4	6	8	10
44	.7193	7181	7169	7157	7145	7133	7120	7108	7096	7083	2	4	6		10

[*Numbers in difference columns to be subtracted, not added.*]

Degrees	0′ 0°.0	6′ 0°.1	12′ 0°.2	18′ 0°.3	24′ 0°.4	30′ 0°.5	36′ 0°.6	42′ 0°.7	48′ 0°.8	54′ 0°.9	Mean Differences				
											1	**2**	**3**	**4**	**5**
45	.7071	7059	7046	7034	7022	7009	6997	6984	6972	6959	2	4	6	8	10
46	6947	6934	6921	6909	6896	6884	6871	6858	6845	6833	2	4	6	8	11
47	.6820	6807	6794	6782	6769	6756	6743	6730	6717	6704	2	4	6	9	11
48	.6691	6678	6665	6652	6639	6626	6613	6600	6587	6574	2	4	7	9	11
49	.6561	6547	6534	6521	6508	6494	6481	6468	6455	6441	2	4	7	9	11
50	.6428	6414	6401	6388	6374	6361	6347	6334	6320	6307	2	4	7	9	11
51	.6293	6280	6266	6252	6239	6225	6211	6198	6184	6170	2	5	7	9	11
52	.6157	6143	6129	6115	6101	6088	6074	6060	6046	6032	2	5	7	9	12
53	.6018	6004	5990	5976	5962	5948	5934	5920	5906	5892	2	5	7	9	12
54	.5878	5864	5850	5835	5821	5807	5793	5779	5764	5750	2	5	7	9	12
55	.5736	5721	5707	5693	5678	5664	5650	5635	5621	5606	2	5	7	10	12
56	.5592	5577	5563	5548	5534	5519	5505	5490	5476	5461	2	5	7	10	12
57	.5446	5432	5417	5402	5388	5373	5358	5344	5329	5314	2	5	7	10	12
58	.5299	5284	5270	5255	5240	5225	5210	5195	5180	5165	2	5	7	10	12
59	.5150	5135	5120	5105	5090	5075	5060	5045	5030	5015	3	5	8	10	13
60	.5000	4985	4970	4955	4939	4924	4909	4894	4879	4863	3	5	8	10	13
61	.4848	4833	4818	4802	4787	4772	4756	4741	4726	4710	3	5	8	10	13
62	.4695	4679	4664	4648	4633	4617	4602	4586	4571	4555	3	5	8	10	13
63	.4540	4524	4509	4493	4478	4462	4446	4431	4415	4399	3	5	8	10	13
64	.4384	4368	4352	4337	4321	4305	4289	4274	4258	4242	3	5	8	11	13
65	.4226	4210	4195	4179	4163	4147	4131	4115	4099	4083	3	5	8	11	13
66	.4067	4051	4035	4019	4003	3987	3971	3955	3939	3923	3	5	8	11	14
67	.3907	3891	3875	3859	3843	3827	3811	3795	3778	3762	3	5	8	11	14
68	.3746	3730	3714	3697	3681	3665	3649	3633	3616	3600	3	5	8	11	14
69	.3584	3567	3551	3535	3518	3502	3486	3469	3453	3437	3	5	8	11	14
70	.3420	3404	3387	3371	3355	3338	3322	3305	3289	3272	3	5	8	11	14
71	.3256	3239	3223	3206	3190	3173	3156	3140	3123	3107	3	6	8	11	14
72	.3090	3074	3057	3040	3024	3007	2990	2974	2957	2940	3	6	8	11	14
73	.2924	2907	2890	2874	2857	2840	2823	2807	2790	2773	3	6	8	11	14
74	.2756	2740	2723	2706	2689	2672	2656	2639	2622	2605	3	6	8	11	14
75	.2588	2571	2554	2538	2521	2504	2487	2470	2453	2436	3	6	8	11	14
76	.2419	2402	2385	2368	2351	2334	2317	2300	2284	2267	3	6	8	11	14
77	.2250	2233	2215	2198	2181	2164	2147	2130	2113	2096	3	6	9	11	14
78	.2079	2062	2045	2028	2011	1994	1977	1959	1942	1925	3	6	9	11	14
79	.1908	1891	1874	1857	1840	1822	1805	1788	1771	1754	3	6	9	11	14
80	.1736	1719	1702	1685	1668	1650	1633	1616	1599	1582	3	6	9	12	14
81	.1564	1547	1530	1513	1495	1478	1461	1444	1426	1409	3	6	9	12	14
82	.1392	1374	1357	1340	1323	1305	1288	1271	1253	1236	3	6	9	12	14
83	.1219	1201	1184	1167	1149	1132	1115	1097	1080	1063	3	6	9	12	14
84	.1045	1028	1011	0993	0976	0958	0941	0924	0906	0889	3	6	9	12	14
85	.0872	0854	0837	0819	0802	0785	0767	0750	0732	0715	3	6	9	12	15
86	.0698	0680	0663	0645	0628	0610	0593	0576	0558	0541	3	6	9	12	15
87	.0523	0506	0488	0471	0454	0436	0419	0401	0384	0366	3	6	9	12	15
88	.0349	0332	0314	0297	0279	0262	0244	0227	0209	0192	3	6	9	12	15
89	.0175	0157	0140	0122	0105	0087	0070	0052	0035	0017	3	6	9	12	15

Appendix 3 Natural Sines

Degrees	0′ 0°.0	6′ 0°.1	12′ 0°.2	18′ 0°.3	24′ 0°.4	30′ 0°.5	36′ 0°.6	42′ 0°.7	48′ 0°.8	54′ 0°.9	Mean Differences 1	2	3	4	5
0	.0000	0017	0035	0052	0070	0087	0105	0122	0140	0157	3	6	9	12	15
1	.0175	0192	0209	0227	0244	0262	0279	0297	0314	0332	3	6	9	12	15
2	.0349	0366	0384	0401	0419	0436	0454	0471	0488	0506	3	6	9	12	15
3	.0523	0541	0558	0576	0593	0610	0628	0645	0663	0680	3	6	9	12	15
4	.0698	0715	0732	0750	0767	0785	0802	0819	0837	0854	3	6	9	12	15
5	.0872	0889	0906	0924	0941	0958	0976	0993	1011	1028	3	6	9	12	14
6	.1045	1063	1080	1097	1115	1132	1149	1167	1184	1201	3	6	9	12	14
7	.1219	1236	1253	1271	1288	1305	1323	1340	1357	1374	3	6	9	12	14
8	.1392	1409	1426	1444	1461	1478	1495	1513	1530	1547	3	6	9	12	14
9	.1564	1582	1599	1616	1633	1650	1668	1685	1702	1719	3	6	9	12	14
10	.1736	1754	1771	1788	1805	1822	1840	1857	1874	1891	3	6	9	11	14
11	.1908	1925	1942	1959	1977	1994	2011	2028	2045	2062	3	6	9	11	14
12	.2079	2096	2113	2130	2147	2164	2181	2198	2215	2233	3	6	9	11	14
13	.2250	2267	2284	2300	2317	2334	2351	2368	2385	2402	3	6	8	11	14
14	.2419	2436	2453	2470	2487	2504	2521	2538	2554	2571	3	6	8	11	14
15	.2588	2605	2622	2639	2656	2672	2689	2706	2723	2740	3	6	8	11	14
16	.2756	2773	2790	2807	2823	2840	2857	2874	2890	2907	3	6	8	11	14
17	.2924	2940	2957	2974	2990	3007	3024	3040	3057	3074	3	6	8	11	14
18	.3090	3107	3123	3140	3156	3173	3190	3206	3223	3239	3	6	8	11	14
19	.3256	3272	3289	3305	3322	3338	3355	3371	3387	3404	3	5	8	11	14
20	.3420	3437	3453	3469	3486	3502	3518	3535	3551	3567	3	5	8	11	14
21	.3584	3600	3616	3633	3649	3665	3681	3697	3714	3730	3	5	8	11	14
22	.3746	3762	3778	3795	3811	3827	3843	3859	3875	3891	3	5	8	11	14
23	.3907	3923	3939	3955	3971	3987	4003	4019	4035	4051	3	5	8	11	14
24	.4067	4083	4099	4115	4131	4147	4163	4179	4195	4210	3	5	8	11	13
25	.4226	4242	4258	4274	4289	4305	4321	4337	4352	4368	3	5	8	11	13
26	.4384	4399	4415	4431	4446	4462	4478	4493	4509	4524	3	5	8	10	13
27	.4540	4555	4571	4586	4602	4617	4633	4648	4664	4679	3	5	8	10	13
28	.4695	4710	4726	4741	4756	4772	4787	4802	4818	4833	3	5	8	10	13
29	.4848	4863	4879	4894	4909	4924	4939	4955	4970	4985	3	5	8	10	13
30	.5000	5015	5030	5045	5060	5075	5090	5105	5120	5135	3	5	8	10	13
31	.5150	5165	5180	5195	5210	5225	5240	5255	5270	5284	2	5	7	10	12
32	.5299	5314	5329	5344	5358	5373	5388	5402	5417	5432	2	5	7	10	12
33	.5446	5461	5476	5490	5505	5519	5534	5548	5563	5577	2	5	7	10	12
34	.5592	5606	5621	5635	5650	5664	5678	5693	5707	5721	2	5	7	10	12
35	.5736	5750	5764	5779	5793	5807	5821	5835	5850	5864	2	5	7	9	12
36	.5878	5892	5906	5920	5934	5948	5962	5976	5990	6004	2	5	7	9	12
37	.6018	6032	6046	6060	6074	6088	6101	6115	6129	6143	2	5	7	9	12
38	.6157	6170	6184	6198	6211	6225	6239	6252	6266	6280	2	5	7	9	11
39	.6293	6307	6320	6334	6347	6361	6374	6388	6401	6414	2	4	7	9	11
40	.6428	6441	6455	6468	6481	6494	6508	6521	6534	6547	2	4	7	9	11
41	.6561	6574	6587	6600	6613	6626	6639	6652	6665	6678	2	4	7	9	11
42	.6691	6704	6717	6730	6743	6756	6769	6782	6794	6807	2	4	6	9	11
43	.6820	6833	6845	6858	6871	6884	6896	6909	6921	6934	2	4	6	8	11
44	.6947	6959	6972	6984	6997	7009	7022	7034	7046	7059	2	4	6	8	10

Degrees	0′ 0°.0	6′ 0°.1	12′ 0°.2	18′ 0°.3	24′ 0°.4	30′ 0°.5	36′ 0°.6	42′ 0°.7	48′ 0°.8	54′ 0°.9	Mean Differences 1	2	3	4	5
45	.7071	7083	7096	7108	7120	7133	7145	7157	7169	7181	2	4	6	8	10
46	.7193	7206	7218	7230	7242	7254	7266	7278	7290	7302	2	4	6	8	10
47	.7314	7325	7337	7349	7361	7373	7385	7396	7408	7420	2	4	6	8	10
48	.7431	7443	7455	7466	7478	7490	7501	7513	7524	7536	2	4	6	8	10
49	.7547	7559	7570	7581	7593	7604	7615	7627	7638	7649	2	4	6	8	9
50	.7660	7672	7683	7694	7705	7716	7727	7738	7749	7760	2	4	6	7	9
51	.7771	7782	7793	7804	7815	7826	7837	7848	7859	7869	2	4	5	7	9
52	.7880	7891	7902	7912	7923	7934	7944	7955	7965	7976	2	4	5	7	9
53	.7986	7997	8007	8018	8028	8039	8049	8059	8070	8080	2	3	5	7	9
54	.8090	8100	8111	8121	8131	8141	8151	8161	8171	8181	2	3	5	7	8
55	.8192	8202	8211	8221	8231	8241	8251	8261	8271	8281	2	3	5	7	8
56	.8290	8300	8310	8320	8329	8339	8348	8358	8368	8377	2	3	5	6	8
57	.8387	8396	8406	8415	8425	8434	8443	8453	8462	8471	2	3	5	6	8
58	.8480	8490	8499	8508	8517	8526	8536	8545	8554	8563	2	3	5	6	8
59	.8572	8581	8590	8599	8607	8616	8625	8634	8643	8652	1	3	4	6	7
60	.8660	8669	8678	8686	8695	8704	8712	8721	8729	8738	1	3	4	6	7
61	.8746	8755	8763	8771	8780	8788	8796	8805	8813	8821	1	3	4	6	7
62	.8829	8838	8846	8854	8862	8870	8878	8886	8894	8902	1	3	4	5	7
63	.8910	8918	8926	8934	8942	8949	8957	8965	8973	8980	1	3	4	5	6
64	.8988	8996	9003	9011	9018	9026	9033	9041	9048	9056	1	3	4	5	6
65	.9063	9070	9078	9085	9092	9100	9107	9114	9121	9128	1	2	4	5	6
66	.9135	9143	9150	9157	9164	9171	9178	9184	9191	9198	1	2	3	5	6
67	.9205	9212	9219	9225	9232	9239	9245	9252	9259	9265	1	2	3	4	6
68	.9272	9278	9285	9291	9298	9304	9311	9317	9323	9330	1	2	3	4	5
69	.9336	9342	9348	9354	9361	9367	9373	9379	9385	9391	1	2	3	4	5
70	.9397	9403	9409	9415	9421	9426	9432	9438	9444	9449	1	2	3	4	5
71	.9455	9461	9466	9472	9478	9483	9489	9494	9500	9505	1	2	3	4	5
72	.9511	9516	9521	9527	9532	9537	9542	9548	9553	9558	1	2	3	3	4
73	.9563	9568	9573	9578	9583	9588	9593	9598	9603	9608	1	2	2	3	4
74	.9613	9617	9622	9627	9632	9636	9641	9646	9650	9655	1	2	2	3	4
75	.9659	9664	9668	9673	9677	9681	9686	9690	9694	9699	1	1	2	3	4
76	.9703	9707	9711	9715	9720	9724	9728	9732	9736	9740	1	1	2	3	3
77	.9744	9748	9751	9755	9759	9763	9767	9770	9774	9778	1	1	2	3	3
78	.9781	9785	9789	9792	9796	9799	9803	9806	9810	9813	1	1	2	2	3
79	.9816	9820	9823	9826	9829	9833	9836	9839	9842	9845	1	1	2	2	3
80	.9848	9851	9854	9857	9860	9863	9866	9869	9871	9874	0	1	1	2	2
81	.9877	9880	9882	9885	9888	9890	9893	9895	9898	9900	0	1	1	2	2
82	.9903	9905	9907	9910	9912	9914	9917	9919	9921	9923	0	1	1	2	2
83	.9925	9928	9930	9932	9934	9936	9938	9940	9942	9943	0	1	1	1	2
84	.9945	9947	9949	9951	9952	9954	9956	9957	9959	9960	0	1	1	1	1
85	.9962	9963	9965	9966	9968	9969	9971	9972	9973	9974	0	0	1	1	1
86	.9976	9977	9978	9979	9980	9981	9982	9983	9984	9985	0	0	1	1	1
87	.9986	9987	9988	9989	9990	9990	9991	9992	9993	9993	0	0	0	1	1
88	.9994	9995	9995	9996	9996	9997	9997	9997	9998	9998	0	0	0	0	0
89	.9998	9999	9999	9999	9999	1.000	1.000	1.000	1.000	1.000	0	0	0	0	0

Appendix 4 Exponential and Hyperbolic Functions

x	e^x	e^{-x}	sinh x	cosh	x	e^x	e^{-x}	sinh x	cosh x
.00	1.0000	1.0000	.0000	1.0000	**1.0**	2.7183	.3679	1.1752	1.5431
.02	1.0202	.9802	.0200	1.0002	**1.1**	3.0042	.3329	1.3356	1.6685
.04	1.0408	.9608	.0400	1.0008	**1.2**	3.3201	.3012	1.5095	1.8107
.06	1.0618	.9418	.0600	1.0018	**1.3**	3.6693	.2725	1.6984	1.9709
.08	1.0833	.9231	.0801	1.0032	**1.4**	4.0552	.2466	1.9043	2.1509
.10	1.1052	.9048	.1002	1.0050	**1.5**	4.4817	.2231	2.1293	2.3524
.11	1.1163	.8958	.1102	1.0061	**1.6**	4.9530	.2019	2.3756	2.5775
.12	1.1275	.8869	.1203	1.0072	**1.7**	5.4739	.1827	2.6456	2.8283
.13	1.1388	.8781	.1304	1.0085	**1.8**	6.0496	.1653	2.9422	3.1075
.14	1.1503	.8694	.1405	1.0098	**1.9**	6.6859	.1496	3.2682	3.4177
.15	1.1618	.8607	.1506	1.0113	**2.0**	7.3891	.1353	3.6269	3.7622
.16	1.1735	.8521	.1607	1.0128	**2.1**	8.1662	.1225	4.0219	4.1443
.17	1.1853	.8437	.1708	1.0145	**2.2**	9.0250	.1108	4.4571	4.5679
.18	1.1972	.8353	.1810	1.0162	**2.3**	9.9742	.1003	4.9370	5.0372
.19	1.2092	.8270	.1911	1.0181	**2.4**	11.023	.0907	5.4662	5.5569
.20	1.2214	.8187	.2013	1.0201	**2.5**	12.182	.0821	6.0502	6.1323
.21	1.2337	.8106	.2115	1.0221	**2.6**	13.464	.0743	6.6947	6.7690
.22	1.2461	.8025	.2218	1.0243	**2.7**	14.880	.0672	7.4063	7.4735
.23	1.2586	.7945	.2320	1.0266	**2.8**	16.445	.0608	8.1919	8.2527
.24	1.2712	.7866	.2423	1.0289	**2.9**	18.174	.0550	9.0596	9.1146
.25	1.2840	.7788	.2526	1.0314	**3.0**	20.086	.0498	10.018	10.068
.26	1.2969	.7711	.2629	1.0340	**3.1**	22.198	.0450	11.076	11.122
.27	1.3100	.7634	.2733	1.0367	**3.2**	24.533	.0408	12.246	12.287
.28	1.3231	.7558	.2837	1.0395	**3.3**	27.113	.0369	13.538	13.575
.29	1.3364	.7483	.2941	1.0423	**3.4**	29.964	.0334	14.965	14.999
.30	1.3499	.7408	.3045	1.0453	**3.5**	33.115	.0302	16.543	16.573
.31	1.3634	.7334	.3150	1.0484	**3.6**	36.598	.0273	18.285	18.313
.32	1.3771	.7261	.3255	1.0516	**3.7**	40.447	.0247	20.211	20.236
.33	1.3910	.7189	.3360	1.0549	**3.8**	44.701	.0224	22.339	22.362
.34	1.4049	.7118	.3466	1.0584	**3.9**	49.402	.0202	24.691	24.711
.35	1.4191	.7047	.3572	1.0619	**4.0**	54.598	.0183	27.290	27.308
.36	1.4333	.6977	.3678	1.0655	**4.1**	60.340	.0166	30.162	30.178
.37	1.4477	.6907	.3785	1.0692	**4.2**	66.686	.0150	33.336	33.351
.38	1.4623	.6839	.3892	1.0731	**4.3**	73.700	.0136	36.843	36.857
.39	1.4770	.6771	.4000	1.0770	**4.4**	81.451	.0123	40.719	40.732
.40	1.4918	.6703	.4108	1.0811	**4.5**	90.017	.0111	45.003	45.014
.41	1.5068	.6637	.4216	1.0852	**4.6**	99.484	.0101	49.737	49.747
.42	1.5220	.6570	.4325	1.0895	**4.7**	109.95	.00910	54.969	54.978
.43	1.5373	.6505	.4434	1.0939	**4.8**	121.51	.00823	60.751	60.759
.44	1.5527	.6440	.4543	1.0984	**4.9**	134.29	.00745	67.141	67.149
.45	1.5683	.6376	.4653	1.1030	**5.0**	148.41	.00674	74.203	74.210
.46	1.5841	.6313	.4764	1.1077	**5.1**	164.02	.00610	82.008	82.014
.47	1.6000	.6250	.4875	1.1125	**5.2**	181.27	.00552	90.633	90.639
.48	1.6161	.6188	.4986	1.1174	**5.3**	200.34	.00499	100.17	100.17
.49	1.6323	.6126	.5098	1.1225	**5.4**	221.41	.00452	110.70	110.71
.50	1.6487	.6065	.5211	1.1276	**5.5**	244.69	.00409	122.34	122.35
.60	1.8221	.5488	.6367	1.1855	**5.6**	270.43	.00370	135.21	135.22
.70	2.0138	.4966	.7586	1.2552	**5.7**	298.87	.00335	149.43	149.44
.80	2.2255	.4493	.8881	1.3374	**5.8**	330.30	.00303	165.15	165.15
.90	2.4596	.4066	1.0265	1.4331	**5.9**	365.04	.00274	182.52	182.52
1.00	2.7183	.3679	1.1752	1.5431	**6.0**	403.43	.00248	201.71	2017.2

$\cosh x = \frac{1}{2}(e^x + e^{-x})$, $\sinh x = \frac{1}{2}(e^x - e^{-x})$.

Author Index

Subject Index